D1629027

Bioinorganic Medicinal Chemistry

Edited by
Enzo Alessio

Related Titles

Dunn, Peter / Wells, Andrew / Williams, Michael T. (eds.)

Green Chemistry in the Pharmaceutical Industry

2010
ISBN: 978-3-527-32418-7

Mohr, Fabian (ed.)

Gold Chemistry
Applications and Future Directions in the Life Sciences

2009
ISBN: 978-3-527-32086-8

Thompson, K.

Medicinal Inorganic Chemistry

ISBN: 978-0-470-72544-3

Abraham, D. J. (ed.)

Burger's Medicinal Chemistry and Drug Discovery, Academic Version

ISBN: 978-0-471-37027-7

Rehder, D.

Bioinorganic Vanadium Chemistry

2008
ISBN: 978-0-470-06516-7

John Wiley & Sons, Inc.

Wiley Handbook of Current and Emerging Drug Therapies

Volumes 1-4
ISBN: 978-0-470-04098-0

Bioinorganic Medicinal Chemistry

Edited by
Enzo Alessio

WILEY-
VCH

WILEY-VCH Verlag GmbH & Co. KGaA

The Editor

Prof. Dr. Enzo Alessio
Università di Trieste
Dipt. di Scienze Chimiche
Via L.Giorgieri 1
34127 Trieste
Italy

All books published by **Wiley-VCH** are carefully produced. Nevertheless, authors, editors, and publisher do not warrant the information contained in these books, including this book, to be free of errors. Readers are advised to keep in mind that statements, data, illustrations, procedural details or other items may inadvertently be inaccurate.

Library of Congress Card No.: applied for

British Library Cataloguing-in-Publication Data
A catalogue record for this book is available from the British Library.

Bibliographic information published by the Deutsche Nationalbibliothek
The Deutsche Nationalbibliothek lists this publication in the Deutsche Nationalbibliografie; detailed bibliographic data are available on the Internet at http://dnb.d-nb.de

© 2011 Wiley-VCH Verlag & Co. KGaA, Boschstr. 12, 69469 Weinheim, Germany

All rights reserved (including those of translation into other languages). No part of this book may be reproduced in any form – by photoprinting, microfilm, or any other means – nor transmitted or translated into a machine language without written permission from the publishers. Registered names, trademarks, etc. used in this book, even when not specifically marked as such, are not to be considered unprotected by law.

Composition   MPS Limited, Chennai
Printed and bound by Fabulous Printers Pte Ltd

Cover Design   Schulz Grafik-Design, Fußgönheim

Printed in Singapore
Printed on acid-free paper

ISBN: 978-3-527-32631-0

Contents

Bioinorganic Medicinal Chemistry. Edited by Enzo Alessio
Copyright © 2011 WILEY-VCH Verlag GmbH & Co. KGaA, Weinheim
ISBN: 978-3-527-32631-0

List of Contributors

Silvio Aime
University of Torino
Department of Chemistry IFM &
Molecular Imaging Center
Via.P. Giuria 7
10125 Torino
Italy

Roger Alberto
University of Zürich
Institute of Inorganic Chemistry
Winterthurerstr. 190
8057 Zürich
Switzerland

Enzo Alessio
University of Trieste
Department of Chemical Sciences
Via Giorgieri 1
34127 Trieste
Italy

Susan J. Berners-Price
The University of Western Australia
School of Biomedical, Biomolecular
and Chemical Sciences
35 Stirling Highway, Crawley
Perth WA 6010
Australia

Angela Boccarelli
University of Bari
Department of Biomedical
Sciences and Human Oncology
Piazza Giulio Cesare 11
70124 Bari
Italy

Ioannis Bratsos
University of Trieste
Department of Chemical Sciences
Via Giorgieri 1
34127 Trieste
Italy

Daniela Delli Castelli
University of Torino
Department of Chemistry IFM
& Molecular Imaging Center
Via.P. Giuria 7
10125 Torino
Italy

H.Y. Vincent Ching
The University of Sydney
School of Chemistry
Sydney NSW 2006
Australia

Bioinorganic Medicinal Chemistry. Edited by Enzo Alessio
Copyright © 2011 WILEY-VCH Verlag GmbH & Co. KGaA, Weinheim
ISBN: 978-3-527-32631-0

Mauro Coluccia
University of Bari
Department of Biomedical Sciences
and Human Oncology
Piazza Giulio Cesare 11
70124 Bari
Italy

Ellen L. Crossley
The University of Sydney
School of Chemistry
Sydney NSW 2006
Australia

Shanta Dhar
Massachusetts Institute of Technology
Department of Chemistry
Room 18-498
Cambridge, MA 02139
USA

Nicola J. Farrer
University of Warwick
Department of Chemistry
Coventry CV4 7AL
UK

Teresa Gianferrara
University of Trieste
Department of Pharmaceutical
Sciences
Piazzale Europa 1
34127 Trieste
Italy

Eliana Gianolio
University of Torino
Department of Chemistry IFM &
Molecular Imaging Center
Via.P. Giuria 7
10125 Torino
Italy

Gilles Gasser
University of Zürich
Institute of Inorganic Chemistry
Winterthurerstrasse 190
8057 Zürich
Switzerland

Zijian Guo
Nanjing University
School of Chemistry and
Chemical Engineering, State Key
Laboratory of Coordination
Chemistry
Nanjing 210093
P.R. China

Trevor W. Hambley
University of Sydney
School of Chemistry
Sydney NSW 2006
Australia

Christian G. Hartinger
University of Vienna
Institute of Inorganic Chemistry
Waehringer Str. 42
1090 Vienna
Austria

Joseph A. Ioppolo
The University of Sydney
School of Chemistry
Sydney NSW 2006
Australia

Michael A. Jakupec
University of Vienna
Institute of Inorganic Chemistry
Waehringer Str. 42
1090 Vienna
Austria

Bernhard K. Keppler
University of Vienna
Institute of Inorganic Chemistry
Waehringer Str. 42
1090 Vienna
Austria

Stephen J. Lippard
Massachusetts Institute of Technology
Department of Chemistry
Room 18-498
Cambridge, MA 02139
USA

Yasmin Mawani
University of British Columbia
Medicinal Inorganic Chemistry
Group, Department of Chemistry
2036 Main Mall
Vancouver V6T 1Z1
BC
Canada

Nils Metzler-Nolte
Ruhr-University Bochum
Faculty of Chemistry and
Biochemistry
Universitätsstrasse 150
44801 Bochum
Germany

Julia F. Norman
University of Sydney
School of Chemistry
Sydney NSW 2006
Australia

Chris Orvig
University of British Columbia
Medicinal Inorganic Chemistry
Group, Department of Chemistry
2036 Main Mall
Vancouver V6T 1Z1
BC
Canada

Alessandra Pannunzio
University of Bari
Department of Biomedical Sciences
and Human Oncology
Piazza Giulio Cesare 11
70124 Bari
Italy

David N. Reinhoudt
University of Twente
Faculty of Science and Technology
(TNW) and MESA + Institute for
Nanotechnology
7500 AE Enschede
The Netherlands

Louis M. Rendina
The University of Sydney
School of Chemistry
Sydney NSW 2006
Australia

Albert Ruggi
University of Twente
Faculty of Science and Technology
(TNW) and MESA + Institute for
Nanotechnology
7500 AE Enschede
The Netherlands

Peter J. Sadler
University of Warwick
Department of Chemistry
Coventry CV4 7AL
UK

Aldrik H. Velders
University of Twente
Faculty of Science and Technology
(TNW) and MESA + Institute for
Nanotechnology
7500 AE Enschede
The Netherlands

Xiaoyong Wang
Nanjing University
School of Life Sciences, State Key
Laboratory of Pharmaceutical
Biotechnology
Nanjing 210093
P.R. China

1
Medicinal Inorganic Chemistry: State of the Art, New Trends, and a Vision of the Future

Nicola J. Farrer and Peter J. Sadler

1.1
Introduction

Inorganic chemistry is an essential part of life. It is not just the chemistry of dead or inanimate things. It was probably even inorganic chemistry that started it all off. For example, iron sulfides may have been the energy sources for early forms of life [1]. There is currently emerging interest in the medicinal chemistry of the elements of the periodic table (Table 1.1).

Currently 24 elements are thought to be essential for mammalian biochemistry (H, C, N, O, F, Na, Mg, Si, P, S, Cl, K, Ca, V, Mn, Fe, Co, Ni, Cu, Zn, Se, Mo, Sn, and I). However the biochemistry of some elements, particularly F, Si, V, Ni, and Sn is poorly understood. It has even been suggested that the biological requirement for Si is merely to protect against Al toxicity [2]. Interestingly, aluminum compounds are widely used as adjuvants in human and veterinary vaccines (helping and enhancing the pharmacological effect) although the chemical basis of the mechanism for this effect is not understood [3].

This situation with aluminum serves to illustrate the problem we face with inorganic medicines. They are often used on a mass scale, but with little rational basis and limited understanding of their molecular mechanism of action. Often we find this is because the methods and techniques available for the study of inorganic agents are either inadequate or are not fully exploited. In particular, determining the speciation of inorganic compounds under biological conditions remains a major challenge. Inorganic and metal compounds are often prodrugs that may not only be transformed on the way to target sites but also when attempts are made to extract the biologically active form from biological media.

This list of essential elements is probably not complete; for example, Cr and B may prove to be essential but the current evidence is unclear. Importantly, essentiality is not just about the element itself, but particular compounds of that element. For example, we need cobalt, but probably only in the form of the vitamin B_{12}. Similarly, toxicity (often used as an argument against using metal compounds

Bioinorganic Medicinal Chemistry. Edited by Enzo Alessio
Copyright © 2011 WILEY-VCH Verlag GmbH & Co. KGaA, Weinheim
ISBN: 978-3-527-32631-0

Table 1.1 Some areas of medical interest in the elements of the periodic table. Entries are restricted to a few comments about each element and no attempt is made to be comprehensive. Elements thought to be essential for man are in italics.

Z	Symbol	Element	Some medically-relevant uses
1	H	*Hydrogen*	pH tightly controlled but variable: blood $\sim$7.4, lysosomes 4–5, tumor tissue 6–7, endosomes (transferrin) 5.5; duodenum 6–6.5, large intestine 5.5–7, stomach 1–3; ^{2}H for kinetic control of organic drugs
2	He	Helium	He-O_2; for treatment of chronic obstructive pulmonary disease; hyperpolarized ^{3}He for MRI
3	Li	Lithium	Li_2CO_3: drug for treatment of bipolar disorders
4	Be	Beryllium	Compounds can provoke severe immune response (chronic beryllium disease)
5	B	Boron	Boromycin: bacteriocidal polyether-macrolide antibiotic. Is B essential? B associated with Ca and steroid metabolism? Neutron capture therapy
6	C	*Carbon*	Carbon nanotubes for drug delivery; body produces 3–6 ml CO per day – suppresses organ rejection, "neurotransmitter"
7	N	*Nitrogen*	NO is a muscle relaxant, vasodilator, hypotensive, "neurotransmitter"
8	O	*Oxygen*	Reactive oxygen species (ROS) (e.g., $^1O_2$, O_2^-, H_2O_2, O_3, ONO_2^-)
9	F	*Fluorine*	Toughens tooth enamel as component of (fluoro)apatite
10	Ne	Neon	^{20}Ne ion therapy
11	Na	*Sodium*	About 0.14 M in blood; excessive NaCl intake increases arterial hypertension
12	Mg	*Magnesium*	Mg^{II} is a laxative (sulfate, Epsom salts); $Mg(OH)_2$ (milk of magnesia) antacid; Mg^{II}(aspartate)$_2$ dietary supplement
13	Al	Aluminum	Added to some vaccines as an adjuvant; $Al(OH)_3$ antacid
14	Si	*Silicon*	Role in connective tissue? Silicates essential to prevent Al toxicity? Contained in some types asbestos (polysilicate mineral fibers containing $Na^+/Mg^{2+}/Fe^{2+/3+}$) that are hazardous to health; intake from plant-based foods (e.g., rice) about 10 mg d^{-1}? Silatranes stimulate hair growth?
15	P	*Phosphorus*	Polyphosphate abundant in all cells; oral sodium phosphate for bowel cleansing; phytate (inositol hexaphosphate) in plants can control metal uptake (e.g., Zn^{2+})
16	S	*Sulfur*	H_2S as signaling molecule (vasodilator and regulator of blood pressure); SO_2 signaling? Sulfite as food preservative
17	Cl	*Chlorine*	Defect in membrane Cl^- transport in cystic fibrosis (CFTR gene); $HOCl/OCl^-$, generated by myeloperoxidase in neutrophils, disinfectant, also ClO_2 for water treatment
18	Ar	Argon	Argon plasma coagulation to control bleeding from lesions in gastrointestinal tract
19	K	*Potassium*	110–140 g in body; ^{40}K 0.012% β emitter, $t_{1/2}$ 1.3 $\times$ 10^9 y; 5000 ^{40}K atoms disintegrate in body every second; KCl supplements

20	Ca	*Calcium*	$CaCO_3$ antacid; Ca oxalate/phosphate/carbonate kidney stones
21	Sc	Scandium	[47]Sc therapeutic radionuclide
22	Ti	Titanium	Budotitane and Cp_2TiCl_2 clinical anticancer trials abandoned; new Cp derivatives in development
23	V	*Vanadium*	Insulin-enhancing drugs, V^{IV} bis(2-ethyl-3-hydroxy-4-pyronato) complexes in clinical trials for type-2 diabetes; vanadate as phosphate mimic
24	Cr	Chromium	Cr^{III} tris(picolinate) sold as nutritional supplement; essentiality of Cr unclear
25	Mn	*Manganese*	SOD mimetics (e.g., Mn^{III} salen chloride), Mn^{II} macrocycles for treatment of pain, neuroprotection; Mn^{II} dipyridoxyl diphosphate (MnDPDP) clinical MRI contrast agent
26	Fe	*Iron*	Fe^{II} compounds for Fe deficiency (succinate, fumarate), also Fe^{III} with $E^{\circ} < -324$ mV at pH 7 (dextran, dextrin); ferroquine antimalarial; superparamagnetic iron oxide MRI contrast agent; $Na_2[Fe^{II}(CN)_5NO]$ hypotensive
27	Co	*Cobalt*	Coenzyme vitamin B_{12} essential (2–3 µg d^{-1}); treatment of pernicious anemia; CTC-96 Co^{III} bis(2-methylimidazole) acacen derivative (Doxovir) shows antiviral activity
28	Ni	*Nickel*	Potential allergen (ear rings); used in trace mineral supplements; role in body poorly understood
29	Cu	*Copper*	Cu^{II} bis(histidine) for Menke's disease; ^{64}Cu PET
30	Zn	*Zinc*	Zn^{II}(gluconate)$_2$ dietary supplement; ZnO skin ointment; Zn^{II} citrate anti-plaque (toothpastes)
31	Ga	Gallium	[67]Ga γ-ray radio-imaging; [66/68]Ga PET; $[Ga^{III}(malolate)_3]$ and [Ga(hydroxyquinolinate)$_3$] in anticancer clinical trials
32	Ge	Germanium	Ge nanoparticles as radiosensitizers; GeO_2 as dietary supplement (despite no evidence of medical benefit) – absorbed orally and through skin (mitochondria-mediated apoptosis?)
33	As	Arsenic	As_2O_3 approved drug for treatment of leukemia; arsenobetaine in marine organisms; Roxarsone (3-nitro-4-hydroxyphenyl arsonic acid) growth promoter in poultry
34	Se	*Selenium*	Human selenoproteome consists of 25 selenoproteins; selenocysteine tRNA; Se^{IV} sulfide active ingredient in some anti-dandruff shampoos (antifungal)
35	Br	Bromine	Daily dietary intake of bromide about 2–8 mg (fish, grains, and nuts). Concentration in blood 10–100 µM; substrate for eosinophil peroxidase $Br/H_2O_2 \rightarrow HOBr$
36	Kr	Krypton	Hyperpolarized [83]Kr for MR imaging of airways
37	Rb	Rubidium	[82]Rb PET
38	Sr	Strontium	$SrCl_2$ in toothpastes (for sensitive teeth); ^{89}Sr therapeutic radionuclide
39	Y	Yttrium	[90]Y therapeutic radionuclide; [86]Y PET
40	Zr	Zirconium	"Aluminum zirconium tetrachlorohydrex gly" (Zr^{4+}/Al^{3+} OH/Cl/glycine complexes) used in antiperspirants; zirconia (ZrO_2) ceramics for orthopedic surgery; functionalized Cp Zr^{IV} compounds potential anticancer
41	Nb	Niobium	Heteropolyniobates $[SiNb_{12}O_{40}]^{16-}$ can immobilize viruses

(Continued)

Table 1.1 *(Continued)*

Z	Symbol	Element	Some medically-relevant uses
42	Mo	*Molybdenum*	MoO_4^{2-} transport pathways; tetrathiomolybdate, $[MoS_4]^{2-}$: copper chelator for overload (e.g., Wilson's) diseases
43	Tc	Technetium	^{99m}Tc γ-ray radio-imaging
44	Ru	Ruthenium	Two tetrachlorido bis(N-heterocycle) Ru^{III} complexes in clinical trials as anticancer and antimetastatic agents
45	Rh	Rhodium	^{105}Rh therapeutic radionuclide; dinuclear Rh^{II} anticancer; photochemotherapeutic complexes
46	Pd	Palladium	^{103}Pd radiotherapy
47	Ag	Silver	Antimicrobial; treatment of burn wounds; sulfadiazine, carbene complexes, nanoparticles
48	Cd	Cadmium	Induces metallothionein synthesis (detoxification); cadmium carbonic anhydrase active in marine diatoms
49	In	Indium	^{111}In γ-ray radio-imaging
50	Sn	*Tin*	Little known about biochemistry as essential element; Sn^{IV} ethyl etiopurpurin (Purlytin) photosensitizer for photodynamic therapy of psoriasis and restenosis (Phase II)
51	Sb	Antimony	Antileishmanial Sb^V drugs: meglumine antimoniate (Glucantime) and sodium stibogluconate (Pentostam)
52	Te	Tellurium	Ammonium trichloro(dioxoethylene-O,O') tellurate is an immunomodulator
53	I	*Iodine*	Thyroid hormones; iodo-organics as X-ray contrast agents; ^{123}I, ^{125}I radio-imaging; ^{131}I radiotherapy
54	Xe	Xenon	"Ideal" anesthetic; ^{133}Xe scintigraphy (SPECT; imaging heart, lungs, brain); hyperpolarized ^{129}Xe as MRI contrast agent
55	Cs	Cesium	^{131}Cs prostate brachytherapy
56	Ba	Barium	$BaSO_4$ (barium sulfate meal) for radiographs of esophagus, stomach, and duodenum
57	La	Lanthanum	La_2CO_3 approved drug Oct 2004 (Fosrenol) for hyperphosphatemia
58	Ce	Cerium	Flammacerium [cerium(III) nitrate-silver sulfadiazine] for treatment of burn wounds; Ce^{IV} sulfate antiseptic; ^{141}Ce, cerebral blood flow
59	Pr	Praseodymium	Is Pr^{III} anti-inflammatory? (and other Ln^{III}?)
60	Nd	Neodymium	
61	Pm	Promethium	^{149}Pm therapeutic radionuclide
62	Sm	Samarium	^{153}Sm therapeutic radionuclide
63	Eu	Europium	PARACEST MRI contrast agent
64	Gd	Gadolinium	Chelated Gd^{III} complexes as contrast agents for MRI (e.g., DTPA, DOTA); ^{157}Gd for neutron capture therapy; Gd^{III} texaphyrin clinical trials for brain tumors. $GdCl_3$ for treatment of liver fibrosis?
65	Tb	Terbium	
66	Dy	Dysprosium	
67	Ho	Holmium	^{166}Ho therapeutic radiopharmaceutical
68	Er	Erbium	
69	Tm	Thulium	^{167}Tm for bone scanning (density)

70	Yb	Ytterbium	^{90}Y therapeutic radionuclide
71	Lu	Lutetium	^{177}Lu therapeutic radionuclide; Lu^{III} texaphyrin photosensitizer Phase II trials for breast cancer, malignant melanomas atherosclerotic plaque in coronary heart disease
72	Hf	Hafnium	
73	Ta	Tantalum	^{178}Ta short lived radionuclide (half-life = 9.3 min) for blood pool, imaging
74	W	Tungsten	Tungstate, $[WO_4]^{2-}$: antidiabetic; antiviral polyoxotungstates
75	Re	Rhenium	β-Emitters ^{186}Re, ^{188}Re for radiotherapy
76	Os	Osmium	OsO$_4$ injection into knee joints for synovectomy (Scandinavia); Os^{II} arene anticancer complexes
77	Ir	Iridium	^{192}Ir γ emitter used clinically for vascular brachytherapy
78	Pt	Platinum	Clinically established anticancer drugs: cisplatin, carboplatin, oxaliplatin, etc.
79	Au	Gold	Aurothiomalate (injectable) and auranofin (oral) anti-rheumatoid arthritic drugs; $[Au(CN)_2]^-$ has anti-HIV activity
80	Hg	Mercury	Declining use in diuretics; antimicrobial (thiomersal); vaccine preservative
81	Tl	Thallium	Tl^{I} often toxic, can substitute for K^{I}; ^{201}Tl used for radiodiagnostic imaging (SPECT)
82	Pb	Lead	Can inhibit heme synthesis and cause anemia, neurotoxicity
83	Bi	Bismuth	Bi^{III} subsalicylate, subgallate, subcitrate used for gastrointestinal disorders; Ranitidine bismuth citrate for antibacterial applications; ^{212}Bi for radiotherapy, α and β$^-$ emitter (generated *in vivo* from ^{212}Pb by β$^-$ decay, $t_{1/2}$ = 10.6 h)
84	Po	Polonium	
85	At	Astatine	^{211}At ($t_{1/2}$ = 7.2 h, α-particle emitter) used in labeled agents for targeted radiotherapy
86	Rn	Radon	About 150 atoms in each ml of air; may accumulate in basements of dwellings; second major cause of lung cancer after smoking; most stable isotope ^{222}Rn ($t_{1/2}$ 3.8 d, α-particle emitter) used in radiotherapy
87	Fr	Francium	
88	Ra	Radium	^{223}Ra ($t_{1/2}$ = 11.4 d, α-particle emitter), is a bone-seeking radionuclide used for the treatment of skeletal metastases
89	Ac	Actinium	^{225}Ac ($t_{1/2}$ = 10.0 d, α-particle emitter) used for tagging antibodies in radioimmunotherapy

as drugs) is typically related not just to the metal itself but also to the ligands and to the type of complex.

Whilst a given metal does exhibit recurring features peculiar to itself (e.g., a preference for particular oxidation states and ligand geometries) the ligand environment can have a marked effect on the overall reactivity of the complex. Furthermore, the behavior of a metal complex is dependent on both its composition and the environment in which it finds itself. Predicting and controlling that behavior is one of the challenges for advancing the rational design of inorganic pharmaceuticals.

In this chapter we discuss transformations of metallodrugs by ligand exchange and/or redox processes, drawing on a wide range of examples. We attempt to relate these transformations to mechanisms of action with the aim of introducing rational design concepts to as many areas of medicinal inorganic chemistry as possible.

1.1.1
Metals in the Body: Essential Elements and Diseases of Metabolism

The homeostasis and control of metal ions in the body is an area of research in itself [4, 5]. Evolution has incorporated many metals into essential biological functions, using the variable oxidation states exhibited by the metal center (e.g., Fe^{II}/Fe^{III} in heme) to control reversible binding of small molecules (e.g., O_2) and implement structural changes. The ligand binding in a typical coordination M–L bond (50–150 kJ mol^{-1}) is much weaker than covalent bonding (the energy of a single C-C bond is 300–400 kJ mol^{-1}) [6]; for hemes and vitamin B_{12}, for example, this allows much more flexibility in small molecule binding and dissociation (signaling) under biological conditions; the energies involved are much smaller. Other, even weaker, interactions such as hydrogen bonding (20–60 kJ mol^{-1}) and van der Waal's interactions (<50 kJ mol^{-1}) are crucial for correct structure and functioning of biological systems. For metallodrugs such non-covalent interactions can play vital roles in target recognition.

A knowledge of the transport of metals and their complexes *in vivo* (particularly cell uptake and efflux) is important for understanding the metabolism of inorganic (and organic) drugs, and also for understanding the nature of diseases caused by erroneous metal transport.

1.1.2
Metals as Therapeutic Agents

Medicinal inorganic chemistry is a relatively young, interdisciplinary research area that has grown primarily due to the success of cisplatin, a Pt-based anticancer drug developed in the late 1960s. In addition to metal-centered therapies, metals may also be used to enhance the efficacy of organic drugs (such as the cyclams, e.g., AMD3100) and as small-molecule delivery vehicles (e.g., for NO, CO).

Organic compounds used in medicine may be activated by metal ions or metalloenzymes, and others can have a direct or indirect effect on metal ion metabolism. Since many organic drugs follow conventional design rules (e.g., Lipinski's rule of 5), they typically incorporate groups with the ability to act as electron donors (H-bond acceptors), endowing them with potential metal-binding sites. This bioinorganic reactivity needs to be considered when new organic drugs are designed. The rational design of metal-based drugs is a relatively new concept, bolstered by improvements in characterization and imaging techniques. In general a metal complex that is administered is likely to be a "prodrug" that undergoes a transformation *in vivo* before reaching its target site. Such

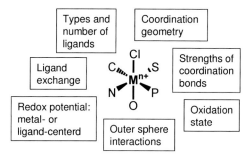

Figure 1.1 Some features of metal coordination complexes that can play a role in biological activity. Control of these characteristics is important in rational design.

transformations can include reduction or oxidation of the metal ion, ligand substitution, or reactions of the ligands at sites remote from the metal. Elucidating the precise mechanisms of action of these new drugs is perhaps the most challenging and complicated aspect of the research; it requires drawing together knowledge concerning the reactivity of the metal complex (outlined in Figure 1.1 and Table 1.2)

Table 1.2 Features of metals and metal complexes that can be used in the design of therapeutic and diagnostic agents.

Feature	Comments (examples)
Coordination number	Full range 2–10; transition metals typically 4–6, can be more variable for main group metals (e.g., Bi) and larger for Ln (e.g., 9)
Geometry	Examples; linear (Au^I), square-planar (Pt^{II}), tetrahedral (e.g., Ti^{IV} in $TiCp_2Cl_2$, distorted), trigonal bipyramidal, octahedral (Ti^{IV}, Ru^{III}, Pt^{IV}), possible metal-centered chirality (Co^{III}, Rh^{III})
Oxidation state	Wide range (typically 0–7 in biological media), with different oxidation states favoring different coordination numbers and rates of exchange (e.g., Pt^{IV} vs. Pt^{II})
Ligand type	Wide range of donors, e.g., C, N, O, halides, P, S, Se. Chelating ligands; denticity, e.g., (κ^2) 1,2-diaminoethane, (κ^6), EDTA; hapticity, e.g., η^6 and η^4 binding for benzene
Thermodynamic stability	Wide range of M–L bond strengths (typically 50–150 kJ mol^{-1})
Kinetic stability	Lifetimes of M–L bonds cover wide range (ns–years). Highly dependent on metal oxidation state and other ligands, can be stereospecific, e.g., *trans* effect in Pt^{II}
Properties of ligands	Outer sphere interactions, e.g., H-bonding, hydrophobic interactions (<50 kJ mol^{-1}) for receptor recognition (including use of chirality); may undergo transformation *in vivo*, e.g., by redox, hydrolysis, enzymatic reactions (e.g., by P450 in the liver).
Nuclear stability	Radioactive nuclides can be used to track metabolism of drugs, e.g., ^{195m}Pt ($t_{1/2} = 4$ d) and ^{99m}Tc ($t_{1/2} = 6$ h). Appropriate nuclide depends on decay pathway (α, β, γ) and half-life

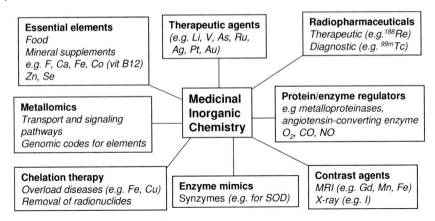

Figure 1.2 Some key areas of medicinal inorganic chemistry.

to control features such as biochemical stability, coupled with an appreciation of the biochemical pathways governing cell uptake, metabolism, and excretion. By appropriate choice of the ligands and metal oxidation state, it is possible to control the thermodynamic and kinetic properties of metal complexes and to attempt to control their biological activity [7].

The use of metals in medicine is varied [8–12] (Figure 1.2). Fundamental wide-reaching medical problems such as bacterial, viral (particularly HIV), and parasitic infections such as malaria are being addressed by research on metal-based medicines. There are also promising developments for tackling the main diseases affecting an affluent, aging western population: cardiovascular, age-related inflammatory diseases, neurological diseases (e.g., Parkinson's, Alzheimer's), cancer (Chapters 3–5) [13–15], diabetes, and arthritis (Chapter 7). Bipolar and gastrointestinal disorders are addressed by metal-based medicines and metal-based diagnostic agents for MRI and X-ray are in routine clinical use; for example, gadolinium MRI contrast agents have now been administered to ~50 million patients worldwide (Chapter 8).

In the following sections, using selected examples, we provide an overview of the uses of metal-based drugs in these aforementioned diverse medical fields and, where appropriate, we guide the reader to chapters that provide a more in-depth analysis.

1.2
Antimicrobial Agents

Many metal compounds show appreciable antimicrobial activity, some established examples are based on silver, bismuth, mercury, and antimony.

Under an inert atmosphere, *silver* has no effect on microorganisms; however, in the presence of oxygen it exhibits a broad spectrum of antimicrobial activities. Several different mechanisms are thought to be responsible: (i) Ag^+ interacts with thiol groups of L-cysteine residues of proteins, inactivating their enzymatic functions; (ii) Ag^+ causes potassium release [16]; (iii) Ag^+ binds to nucleic acids [17]; and (iv) Ag^+ generates superoxide ($O_2^{\bullet-}$) intracellularly [18]. This interaction with bacterial proteins and nucleic acids causes structural changes in membranes (blocking respiration), and nucleic acids (blocking transcription). Morphologically, treatment of *Escherichia coli* and *Staphylococcus aureus* (model strains for Gram-negative and Gram-positive bacteria, respectively) with $AgNO_3$ has been shown to cause detachment of the bacterial membrane from the cell wall, condensation of the nuclear DNA, and deposition of S- and Ag-rich granules both around the cell wall and within the cytoplasm [19]. The same effects are not seen in mammalian systems.

Simple compounds of silver (e.g., $AgNO_3$) have long been used as antibacterial agents, with particular efficacy in the treatment of burn wounds. Although the use of silver diminished in the 1940s following the development of penicillin, there has been a recent resurgence of interest due to the emergence of strains of bacteria resistant to the current organic-based drugs [20]. The combination of silver with a sulfonamide antibiotic in 1968 produced silver sulfadiazine (SSD) cream (an insoluble polymeric Ag^+ compound, see Figure 1.3a), a broad spectrum silver-based antibacterial that also exhibits antiviral and antifungal properties. Wound dressings and materials for medical devices such as catheters are often manufactured with the incorporation of silver to improve sterility.

Since silver acts biologically as "Ag^+," the main role of the ligands is to tailor the solubility and pharmacokinetic profile (i.e., release and distribution of Ag^+). Recent developments in silver-based antimicrobials have focused on the use of sophisticated ligands (e.g., imidazolium N-heterocyclic carbenes, NHC) and also on the potential offered by silver nanoparticles.

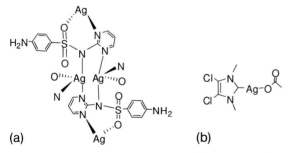

(a) (b)

Figure 1.3 Structures of polymeric silver sulfadiazine (a) and (1,3-dimethyl-4,5-dichloroimidazole-2-ylidene)silver(I) acetate (b). The latter shows bacteriostatic and bacteriocidal properties against a panel of pathogens associated with cystic fibrosis and chronic lung infections.

Silver complexes of NHC with electron-withdrawing groups in the 4- and 5-positions [such as (1,3-dimethyl-4,5-dichloroimidazole-2-ylidene)silver(I) acetate, see Figure 1.3b] have demonstrated activity against bacterial strains associated with cystic fibrosis and chronic lung infections [21]. Typically, NHC-Ag complexes decompose rapidly in aqueous solution, but incorporation of Cl in the 4,5 positions of the imidazole ring withdraws electron density from the carbene carbon, making it less susceptible to attack and slowing the rate of hydrolysis. Silver nanoparticles appear to possess greater antimicrobial activity (nM vs. µM) than conventional forms of silver [22, 23]. Although the mammalian response to silver nanoparticles is not well-characterized, they have been shown to be toxic towards a mouse spermatogonial stem cell line [24].

Development of bacterial resistance to silver is thought to be due to increased use of efflux pumps [25]. Silver toxicity in humans is seen in the form of argyria (a permanent blue-tinting of the skin) following administration of particularly high doses.

Bismuth is regarded as a borderline metal and typically exists as Bi^{III} *in vivo*. Bismuth exhibits variable coordination numbers (3–10) with irregular geometries, the structure of the aqua complex $[Bi(H_2O)_9]^{3+}$ is similar to those of Ln^{III} complexes. Bismuth compounds show low toxicity towards mammalian cells (probably due to the protection afforded by the thiol-rich proteins, metallothioneins) and have been in use for over 200 years as antimicrobial agents, for treating syphilis and other infections, including colitis, gastritis, and diarrhea [26–30]. Various Bi compounds – colloidal bismuth sub-citrate, ranitidine bismuth citrate, bismuth sub-salicyclate (Pepto-Bismol), and ammonium potassium Bi^{III} citrate (De-Nol™) – are used, often in combination therapy with other antibiotics, for the treatment of gastrointestinal disorders caused by *Helicobacter pylori*. This bacterium causes inflammation of the stomach lining and is linked to the development of duodenal and gastric ulcers [31]. Bismuth salts also show activity against several other gastrointestinal tract pathogens, including *Escherichia coli*, *Vibrio cholerae*, *Campylobacter jejuni*, and those of the *Yersinia*, *Salmonella*, and *Shigella* genera [32]. The major biological targets for Bi^{III} are proteins; Bi^{III} is known to bind to both Fe^{III}- and Zn^{II}-coordination sites of proteins [33] and, in particular, is thought to inhibit the nickel-binding protein urease and bind to the histidine-rich and cysteine-rich metal-binding domains of heat-shock protein A, which are both crucial to the survival of *H. pylori* in the gut [34]. In addition to the antibacterial action shown by Bi^{III}, the derivatives formed in the acid environment of the stomach bind strongly to the proteins in ulcerated tissue to form a protective layer, allowing it to heal. They also cause an increase in local prostaglandin levels, which stimulates the production of bicarbonate and mucin thereby protecting the stomach [32].

Bismuth complexes have shown inhibition of HIV-1 virus production and activity against SARS coronavirus [35]. In other applications, the radioactive nuclide ^{212}Bi (α-emitter, $t_{1/2} \approx 1$ h) is used in targeted radiotherapy through complexation to monoclonal antibodies. A long unsolved mystery in bismuth pharmacology is the nature of so-called bismuth inclusion bodies found in the nuclei of kidney tubular proximal cells, hepatocytes, and pneumocytes, usually assumed to be associated with toxic side-effects of Bi therapy [36].

Compounds of *mercury* show significant variation in bioavailability, bioaccumulation, and metabolism in humans and as such can be divided into three groups: elemental mercury (Hg), organometallic complexes (i.e., containing at least one Hg—C bond), and non-organometallic (often called "inorganic" complexes, e.g., sulfides). Here we focus on the latter two groups, discussing their use in medicine and briefly exploring the suggested mechanisms of toxicity towards both humans and microorganisms.

The antibacterial and antifungal properties of organometallic mercurials have resulted in their application as topical disinfectants (thiomersal and merbromin), preservatives in vaccines (thiomersal), and grain products (methyl and ethyl mercurials). Despite several high-profile fatalities (Iraq and China in 1970s, Minamata Bay, Japan in 1950s and 1960s) organometallic mercurial compounds are still in widespread use; thiomersal (sodium ethylmercurithiosalicylate) (Scheme 1.1) is contained in GlaxoSmithKline's recently released influenza pandemic (swine flu) vaccines Pandemrix and Arepanrix. In Pandemrix, thiomersal is present at 5 µg per 0.5 ml dose [37], falling well within the World Health Organization (WHO) recommended maximum limit for the similar but more toxic organometallic mercurial compound, methyl mercury (1.6 µg per kg body weight per week) [38]. After injection, thiomersal rapidly dissociates to produce ethyl mercury, which binds to the available thiol ligands present in tissue proteins – a mechanism that is corroborated by the fact that the nature of the ligand attached to the ethyl mercury group (thiosalicyclate in the case of thiomersal) makes little difference to the ultimate bodily distribution of mercury. Merbromin (Mercurochrome) (Figure 1.4) is thought to react in a similar fashion. The toxicology of methyl mercury is better understood than that of ethyl mercury; methyl mercury is generated in nature by microorganisms in aquatic environments from mercury salts, and its bioaccumulation up the food chain has long been known; the mean level of Hg in tinned white tuna in one US study was found to be 0.407 µg per g of

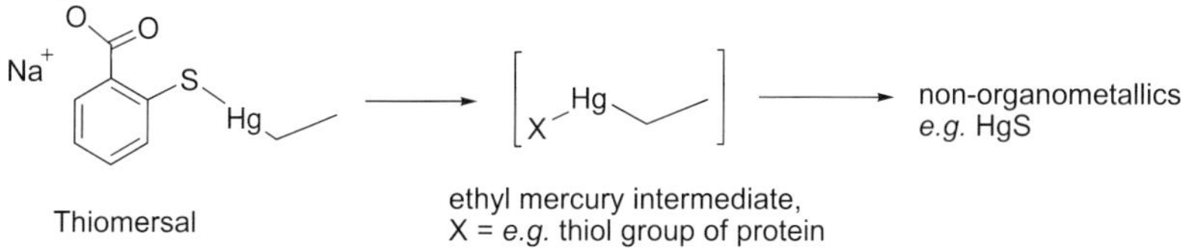

Thiomersal

ethyl mercury intermediate,
X = *e.g.* thiol group of protein

non-organometallics
e.g. HgS

Scheme 1.1 Thiomersal is degraded rapidly *in vivo* to ethyl mercury, which in turn is slowly metabolized to non-organometallic mercurial compounds and excreted in the feces.

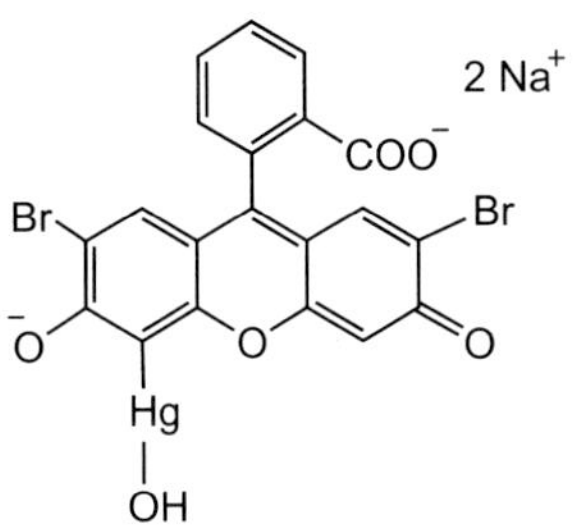

Figure 1.4 Structure of merbromin.

tuna [39]. In the body, methyl mercury is typically found attached to the sulfur of thiolate ligands, entering the endothelial cells of the blood–brain barrier complexed to L-cysteine. It is removed from mammalian cells as a complex of reduced glutathione, but is then recycled *in vivo* to the L-cysteine complex. If any is converted into inorganic complexes of mercury (which are typically insoluble in biological fluids) it can be accumulated in the CNS in the inert form of mercury selenide, or excreted in the feces. The precise mechanism of damage to the brain is still unclear but it appears to involve inhibition of protein synthesis with specific damage to the granule cells in the cerebellum, which have a critical absence of protective mechanisms exhibited by neighboring cell types. Thiomersal has been demonstrated to cause oxidative stress in leukemia cells, leading to the activation of execution caspases and apoptotic cell death [40]. Organometallic mercurials are slowly metabolized to inorganic complexes of mercury mainly by microflora in the intestines, although some demethylation also occurs in phagocytic cells.

Several thiol-containing complexing agents, including the orally-available *N*-acetyl-L-cysteine (Figure 1.5), have shown promise in enhancing methyl mercury excretion [41]. There is evidence that administration of selenium compounds can delay the onset of the toxic effects of methyl mercury in animals.

Mercurous chloride (calomel, Cl-Hg^I-Hg^I-Cl) has been used for centuries as a diuretic, laxative, antiseptic, skin ointment, and to treat vitiligo, although its use has largely been superseded by modern medicines [42]. The traditional Chinese medicine Cinnabar, which is used to achieve sedative and hypnotic effects, contains mercury sulfide (HgS). Since purified mercury sulfide shows poor bioavailability and low absorption from the gastrointestinal tract, it is postulated that the major medicinal benefit of cinnabar might be due to its interactions with other

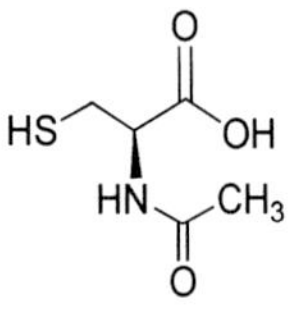

Figure 1.5 Structure of *N*-acetyl-L-cysteine.

components of traditional Chinese medicines. Once absorbed into the blood, mercury disposition from cinnabar follows the pattern for other inorganic mercury complexes and it is preferentially distributed to the kidneys, with a small portion to the brain [43].

Aquaporins (first reported by Benga and co-workers in 1985) are attractive targets for the development of novel drug therapies for disorders that involve aberrant water movement, such as edema and kidney disease. Compounds such as mercurous chloride are known inhibitors of aquaporins, sterically blocking the water transport pore by binding to a cysteine thiolate sulfur (Cys189 in CHIP28/AQP1) [44–46].

As with silver, the antibacterial action of mercury is attributed largely, but not exclusively, to the strong affinity of the metal for thiol groups of proteins. Evidence suggests that mercurials cause structural and functional changes of bacterial cell walls and inhibit membrane bound proteins, interfering with respiration, ATP synthesis, and transport processes [47]. Bacterial resistance to mercury anti-microbial agents involves induction of enzymes (such as mucuric reductase) that are capable of converting Hg^{II} in the cytoplasm into the less toxic and more volatile Hg [48].

Antimony (Sb^V) compounds such as sodium stibogluconate (Pentostam) and meglumine antimonite (Glucantime) are used to treat leishmaniasis, a disease caused by a parasite that is transmitted by the bite of a certain species of sand fly. The parasites replicate within mammalian macrophage phagolysosomes, initially causing skin sores, although some forms of the disease exhibit graver effects such as anemia and damage to the spleen and liver, which can be fatal. The precise mechanism of action of these Sb^V drugs, which have been in use for 60 years, is still unclear. Several effects have been noted: inhibition of glycolysis and fatty acid oxidation, fragmentation of parasitic DNA, and externalization of phosphati-dylserine on the outer surface of membranes via a caspase-independent pathway [49]. In the parasite, reduced trypanothione $T(SH)_2$ rapidly reduces Sb^V to Sb^{III}, which is thought to be the active form. New insights into the molecular basis for the antiparasitic activity of Sb^{III} come from the recent X-ray crystal structures of reduced trypanothione reductase (TR) from *Leishmania infantum* with NADPH and Sb^{III}. TR is essential for parasite survival and virulence and is absent from mammalian cells. Sb^{III} is coordinated by the two redox-active catalytic cysteine residues, a threonine residue, and a histidine, and has been shown to strongly inhibit TR activity, blocking trypanothione reduction [50]. Toxicity of the drugs, as well as increasing resistance, are leading to more research into their mechanism of action [51]. In one study, resistance towards sodium stibogluconate has been attributed to the upregulation of multidrug resistance-associated protein 1 (MRP1) and permeability glycoprotein (P-gp) in host cells, resulting in a non-accumulation of intracellular Sb and favoring parasite replication. Co-administration of the drug lovastatin (Figure 1.6) has been shown to inhibit this resistance *in vivo*, restoring the curative properties of the Sb agent [52]. The exact formulation of the antimony drugs is not well defined; sodium stibogluconate is a mixture of Sb^V and carbo-hydrate. Improvement of efficacy of Sb^V-based antiparasitic agents is anticipated to focus on controlling the activation-by-reduction step of Sb^V to Sb^{III}.

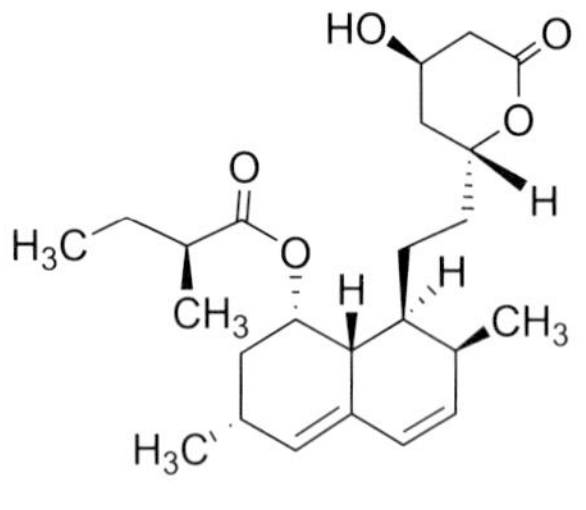

Figure 1.6 Structure of lovastatin.

The *arsenic*-based antimicrobial agent Salvarsan ({RAs}$_n$, R = 3-amino-4-hydroxyphenyl, Figure 1.7) was used historically to treat syphilis and trypanosomiasis, although in recent times it has been superseded by penicillin. It is suggested that oxidation *in vivo* generates the active form of the drug, such that Salvarsan serves as a slow-release source of RAs(OH)$_2$ [53].

Malaria is caused by a protozoan parasite of the genus *Plasmodium* and is responsible for about 2 million deaths per year [54]. The most commonly used antimalarial drug is chloroquine (CQ), which accumulates within the parasite and interferes with the function of its digestive vacuole (where the host hemoglobin is digested). Cases of malaria are on the increase, and it is as yet unclear whether climatic changes may [55] or may not [56] be responsible. These increases may also be attributed to increased resistance towards established antimalarials [54].

Metal complexation of established antimalarials shows promise in overcoming resistance. The ferrocene derivative Ferroquine (FQ, see Figure 1.8b), a 4-aminoquinoline antimalarial that contains a quinoline nucleus similar to

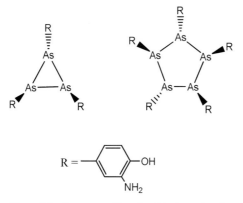

Figure 1.7 Structure of 3-amino-4-hydroxyphenylarsenic(III) (Salvarsan). This has been the subject of much debate, but in solution it is now thought to consist of cyclic species (AsR)$_n$ (R = 3-amino-4-hydroxyphenyl) where $n = 3$ and $n = 5$ are the most abundant species.

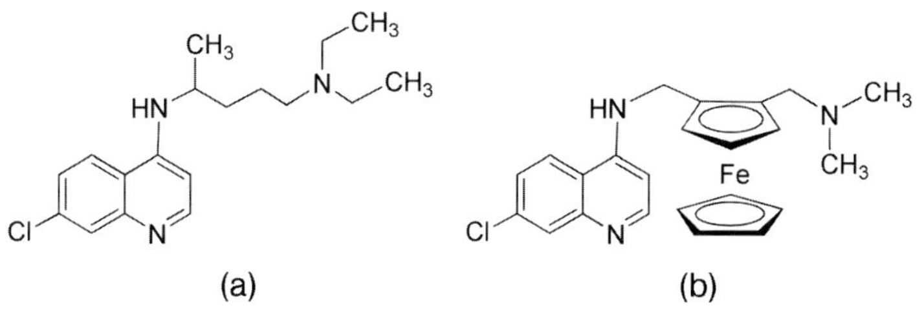

Figure 1.8 (a) Structure of the antimalarials chloroquine (CQ), a component of [RuCl$_2$(CQ)]$_2$, and (b) ferroquine.

chloroquine (for structure of CQ see Figure 1.8a), is being developed by Sanofi-Aventis and has recently entered phase II clinical trials (2007), showing excellent activity against CQ-resistant *Plasmodium falciparum*, both *in vitro* and *in vivo*. Its mechanism of action is probably similar to that of chloroquine itself, involving hematin as the target and inhibition of hemozoin formation [57]. The metal fragment in the Ru^{II} chloroquine complex [RuCl$_2$(CQ)]$_2$ has been shown to alter the structure, the basicity, and most importantly the lipophilicity of CQ, to make it less recognizable to the parasite's defense mechanism [58]. Metal complexes of the form [M(madd)]$^+$ (M = Al, Ga, Fe^{III}; madd = 1,12-bis(2-hydroxy-3-methoxybenzyl)-1,5,8,12-tetraazadodecane) show high activity against CQ-resistant *P. falciparum* (the strain that accounts for most instances of morbidity and mortality) and inhibit heme polymerization in the digestive vacuole of the parasite [59].

1.3
Antiviral Agents

Polynuclear, early-transition metal oxyanions (polyoxometalates or POMs [60] have demonstrated antiviral, anticancer, and antibacterial activity [61]. The anti-HIV activity stems from the binding of the anionic POM to the viral envelope glycoprotein (gp120); the negative charge on the POM shields positively-charged sites on the glycoprotein that are necessary for viral attachment to a cell surface glycopolysaccharide, heparan sulfate. Because POMs prevent viral entry to cells, it is suggested that their main role in the management of HIV infections may reside in the prevention of sexual transmission of HIV infection, blocking HIV infection through both virus-to-cell and cell-to-cell contact [62]. Furthermore, Nb-containing POMs α_1-K$_7$[P$_2$W$_{17}$(NbO$_2$)O$_{61}$], α_2-K$_7$[P$_2$W$_{17}$(NbO$_2$)O$_{61}$], α_1-K$_7$[P$_2$W$_{17}$NbO$_{62}$], and α_2-K$_7$[P$_2$W$_{17}$NbO$_{62}$] selectively inhibit HIV-1 protease (HIV-1P) with IC$_{50}$ values of 1–2 µM and display high activity in cell culture against HIV-1 (EC$_{50}$ = 0.17–0.83 µM) [63]. The heteropolyoxotungstate K$_7$[PTi$_2$W$_{10}$O$_{40}$].6H$_2$O (PM-19) (Figure 1.9) inhibits the replication of herpes simplex virus (HSV) strain 169 both *in vitro* and *in vivo* [64].

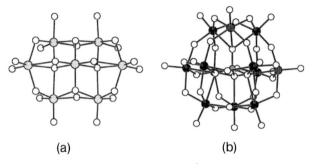

(a)   (b)

Figure 1.9 POMs: (a) structure of $[NH_3Pr^i]_6[Mo_7O_{24}] \cdot 3H_2O$
(PM-8), which shows potent antitumor activity, and (b)
$K_7[PTi_2W_{10}O_{40}] \cdot 6H_2O$ (PM-19), which shows both antiviral
and antibacterial activity.
Figure reproduced with permission from Reference [61].

Polyoxotungstates such as PM-19 also have been shown to enhance the effect of β-lactam antibiotics on methicillin-resistant *S. aureus* (MRSA) by depressing the formation of a protein (PBP2′) that is selectively expressed in the resistant strain, and which is essential for cell wall construction. They also suppress the production of β-lactamase which hydrolyzes the β-lactam antibiotics.

Cyclams are 14-membered tetraamine macrocycles that bind strongly to a wide range of metal ions [65]. Medical interest has centered on clinical trials of a bicyclam for the treatment of AIDS and for stem cell mobilization, and on adducts with Tc and Cu radionuclides for diagnosis and therapy. Other potential applications, particularly for Cr, Mn, Zn, and Ru cyclams, are also emerging. Macrocyclic bicyclams ligands such as AMD3100 (Figure 1.10) are amongst the most potent inhibitors of HIV ever described. AMD3100 targets the early stages of the retrovirus replicative cycle and blocks HIV-1 cell entry by interacting with the seven-helix transmembrane, G-protein coupled, coreceptor protein CXCR4 [66].

Although clinical trials of AMD3100 for the treatment of AIDS were halted due to adverse side-effects, it has been clinically approved (as the drug Plerixafor, Mozobil) for mobilization of stem cells and stem cell transplantation in cancer patients [67]. The same coreceptor CXCR4 that mediates cell entry of HIV-1 also

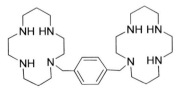

Figure 1.10 Structure of AMD3100 (previously also known as JM3100).

anchors stem cells in the bone marrow. It is possible that a metalated form of the drug is the active species *in vivo*, constraining the macrocycle configuration, which is important in receptor recognition. Configurationally-restricted metallomacrocycles can be even more potent anti-HIV agents than AMD3100 [68, 69]. Both zinc and nickel cyclams can also interact with carboxylates to give unusual folded *cis* configurations [70].

The CXCR4 transmembrane receptor protein is widely expressed on cells of the immune, central nervous, and gastrointestinal systems, and also on many different types of cancer cells, besides playing a central role on the anchorage of $CD34^+$ stem cells in bone marrow. Thus, cyclam derivatives may find use in the treatment of various pathogenic disorders, including HIV infections, rheumatoid, allergic and malignant diseases, and in other diseases that can benefit from stem-cell mobilization.

The Co^{III} bis(2-methylimidazole) acacen complex [acacen = bis(acetylacetone) ethylenediimine] CTC-96 (Doxovir, Figure 1.11) has been shown to be a potent topical microbicide and has successfully completed phase II clinical trials for treatment of herpes simplex virus (HSV) [71]. It has also shown activity against epithelial herpetic keratitis, adenovirus keratoconjunctivitis, and HIV [71, 72]. A range of CTC-96 analogs has been shown to inhibit human α-thrombin by substitution of an axial ligand at the Co^{III} centre with an active site histidine residue in the protein [73]. Interestingly, $[Co^{III}(NH_3)_6]^{3+}$ also possesses antiviral activity [74].

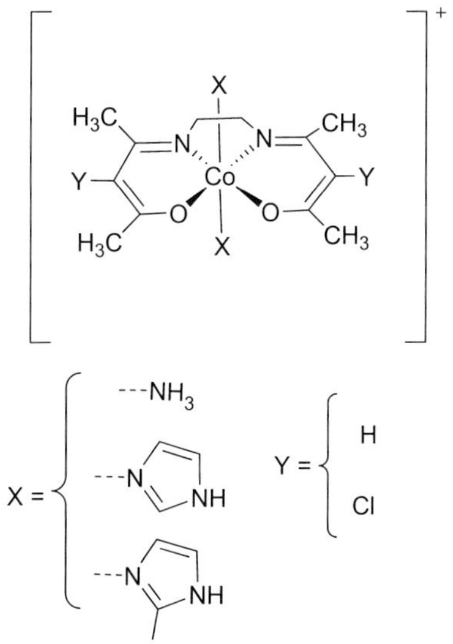

Figure 1.11 Cobalt(III) Schiff base complexes that have been shown to inhibit human α-thrombin by substitution of an axial ligand for an active site histidine residue. Doxovir (CTC-96) has groups X = 2-methylimidazole, Y = H.

1.4
Systemic and Metabolic Diseases Including Inflammation

In this section we discuss treatments for diabetes, obesity, Alzheimer's, Parkinson's, arthritis, and bipolar disorder. Passivation and removal of aberrant metal ions in disease, including Parkinson's, Friedreich's Ataxia, iron overload, and Wilson's disease, have recently been reviewed [75]. Metabolic disorders related to essential metals are also discussed later in this book (Chapter 11).

1.4.1
Diabetes and Obesity

Orally-available vanadium complexes can be used to mitigate insufficient insulin response in diabetes. Although they cannot entirely substitute for lack of insulin (a feature of type 1 diabetes), they can reduce reliance on injected insulin. They can also substitute for oral hypoglycaemic agents that are currently used to treat type 2 diabetes, a condition that is characterized by resistance to insulin. Early investigations of purely inorganic vanadyl salts such as vanadyl sulfate ($VOSO_4$) were hampered by gastrointestinal distress and low absorptive efficiency [76]. The five-coordinate complex bis (2-ethyl-3-hydroxy-4-pyronato)oxovanadium(IV) (BEOV, see Figure 1.12) has been developed specifically to overcome these problems; inclusion of ionizable ligands allowed formation of a neutral complex, with the use of oxygen-rich ligands contributing to enhanced water solubility. BEOV is currently in Phase IIa clinical trials as an insulin-enhancing agent for treatment of diabetes. It has an intermediate stability *in vivo*, dissociating within hours after ingestion, but is stable enough to avoid potential gastrointestinal intolerance and demonstrates a bioavailability two to three times higher than vanadyl sulfate. It is suggested that the oxidation state of the administered drug is crucial; V^{IV} compounds, as a group, have been found to be more potent than complexes of V^{III} and V^V. Eventual ligand dissociation is necessary for the activity of the complex, and so non-toxic ligands are essential. BEOV has been suggested to be a multifactorial insulin sensitizer, inhibiting phosphotyrosine phosphatase 1B (PTP1B) within the insulin signaling cascade, and affecting insulin signaling to regulate glucose homeostasis favorably [77]. Other V^{IV} complexes such as bis(allixinato)oxovanadium(IV) are also being investigated. [78].

Sodium tungstate ($Na_2WO_4 \cdot 2H_2O$) is a novel agent in the treatment of obesity. Oral administration of tungstate significantly decreases body weight gain and

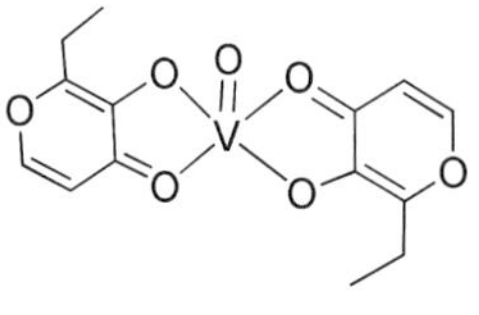

Figure 1.12 Structure of bis(2-ethyl-3-hydroxy-4-pyronato) oxovanadium(IV) (BEOV) which is in Phase II trials for treating diabetes.

adiposity without modifying caloric intake, intestinal fat absorption, or growth rate in obese rats [79]. Sodium tungstate reduces glycemia and reverts the diabetic phenotype in several induced and genetic animal models of diabetes. Importantly, tungstate has also shown promise as an effective treatment for diabetes. Tungstate appears to mimic most of the metabolic effects of insulin and exerts insulin-like actions in primary cultured rat hepatocytes by increasing glycogen deposition [80]. The *in vivo* transport of tungstate is thought to be mediated by plasma proteins such as human serum albumin [81].

1.4.2
Metal Homeostasis and Related Diseases

Several neurological disorders may result from errors in the distribution of metal ions in the brain, including Zn, Cu, Fe, and Mn (see Chapter 11 for more detail). Zinc is a cofactor for many proteins involved in cellular processes such as differentiation, proliferation, and apoptosis. Zinc homeostasis is tightly regulated; one mechanism of control includes storage in, and retrieval from, vesicles, so-called zincosomes [82]. Disturbance of zinc homeostasis due to genetic defects or dietary availability influences the development of diseases such as diabetes (insulin is stored as a zinc complex), cirrhosis of the liver, cancer, asthma, bowel diseases, and rheumatoid arthritis. Because zinc is important in cell division, zinc deprivation during the development of T-lymphocytes impairs their polarization into effector cells, and their function, leading to a reduction in T-cell numbers. As a consequence, zinc deficiency can also affect the immune system, increasing susceptibility to bacterial infections [83]. Zinc homeostasis in the brain is integral for normal brain function. Alterations in zinc levels can have devastating results with zinc becoming a pathogenic agent that mediates neuronal death in conditions such as Alzheimer's disease, amyotrophic lateral sclerosis, and ischemia (a shortage of blood supply). Although zinc-chelating agents are being investigated to treat these neurological disorders, zinc homeostasis needs to be better understood to aid rational intervention [84].

Copper, Zn and Fe are known to promote the aggregation of amyloid beta (A^β) deposits that cause Alzheimer's disease. [85]. Furthermore, copper binds to both the prion protein, the amyloid precursor protein and alpha-synuclein [86]. In addition, vanadium compounds have been shown to have interesting properties for treatment of Alzheimer's [87].

1.5
Metal Chelating Agents

Chelating agents can be used either to deliver metal ions for medical applications or for the removal of unwanted metals from the body. In diagnostic applications, the metal ions can be radioactive (γ-emitter), paramagnetic, or luminescent, whereas therapeutic applications generally require metals that either emit ionizing radiation (α- or β-emitter) or are chemically active. The stability of the chelate *in vivo* is crucial for the metal ion to be delivered safely and effectively to its target. (See later section on diagnostic agents and Chapters 8 and 9.) The use of macrobicyclic caged metal complexes (clathrochelates) is one avenue that is currently being explored [88].

It is useful to be able to reduce the levels of unwanted forms of metals in the human body; for example, long-term exposure to manganese particulates through inhalation can be neurotoxic, with deficits in neuromotor and cognitive domains [89]. Chelating agents vary widely according to metal-binding properties: dithiocarbamate derivatives have shown promise in polonium chelation *in vivo* [90], whereas catechol-3,6-bis(methyliminodiacetic acid) (CBMIDA) has been used to chelate and increase excretion of uranium [91]. Iron chelators such as Desferal (Desferrioxamine mesylate) are neuroprotective against iron-overload. However, since Desferal is not able to cross the blood–brain barrier (BBB), there is a need for further development if chelating agents such as these are to be used to treat cases of abnormal iron accumulation in the brain, such as Parkinson's disease [92].

A potential therapy for Alzheimer's involves targeting Cu and Zn in the brain with chelators. The antifungal drug and antiprotozoal drug clioquinol (5-chloro-7-iodo-quinolin-8-ol, Figure 1.13) has been on clinical trial for treatment of Alzheimer's disease but is likely to be replaced soon by a related 8-hydroxy quinoline, PBT2, which lacks the iodine (structure not explicitly reported) [93], a once per day, orally bioavailable, second-generation derivative of clioquinol [94]. Phase IIa studies of PBT2 [95] have established a favorable safety profile for the drug and suggest a central effect on Aβ metabolism. PBT2 may not act as a metal chelator, but rather as an ionophore that makes copper and zinc more available for normal neuronal function [96].

Tetrathiomolybdate ($[MoS_4]^{2-}$) has anticancer and anti-angiogenic activities, and acts through copper chelation and NF-kappa-B inhibition. Phase I and II clinical trials in solid tumors have demonstrated efficacy with a favorable toxicity profile [97]. Although initially used for treatment of neurologically-presenting Wilson's disease [98], $[MoS_4]^{2-}$ has efficacy in animal models of fibrotic [99], metastatic colorectal cancer [100], inflammatory, autoimmune, and neoplastic diseases, as well as Alzheimer's disease. A Cu^I ion can bind strongly to a pair of MoS_4^{2-} sulfurs. Additional coordination of Cu^I to sulfur of glutathione may facilitate transport from the liver to the kidney [101].

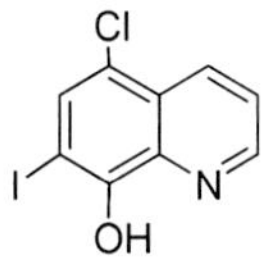

Figure 1.13 Structure of clioquinol.

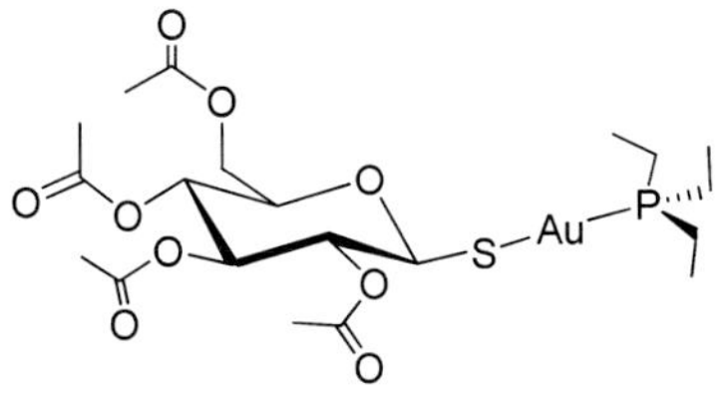

Figure 1.14 Structure of triethylphosphine gold-2,3,4,6-tetra-
o-acetyl-L-thio-D-glucopyranoside (auranofin).

1.6
Antiarthritic Drugs and Inflammation

Gold-based drugs, including the injectable thiolate polymer aurothiomalate and oral triethylphosphine complex auranofin (Figure 1.14), are widely used to treat rheumatoid arthritis (see Chapter 7 for more detail). In the blood, Au^I is transported by albumin on Cys34, and inside cells binds to the thiol of glutathione. It is thought that the pharmacologic effect may be achieved through inhibition of the selenoenzyme thioredoxin reductase [102]. The use of thiol-reactive probes and photocrosslinking methods for detecting gold binding sites on proteins has recently been proposed [103]. The finding that $[Au(CN)_2]^-$ is excreted not only by patients receiving gold therapy who are smokers (about 35 µg HCN/cigarette) but also by non-smokers, emphasizes that strong small ligands such as cyanide can play a role in the metabolism of metallodrugs. $[Au(CN)_2]^-$ is readily taken up by cells (and also has anti-HIV activity). Whitehouse *et al.* have recently reported that colloidal (purple) gold exhibits potent antiarthritic activity in rats, approximately ten-times that of the clinically-used drug sodium aurothiomalate (Myocrisin) [104].

Metaphore Pharmaceuticals Inc. has investigated M-40403 (Figure 1.15) [105], a Mn-based superoxide dismutase (SOD) mimetic, for the potential treatment of post-operative ileus, oral mucositis [106], pain, dermatological disease, inflammation [107], and arthritis [108]. M-40403 is said to improve the effectiveness

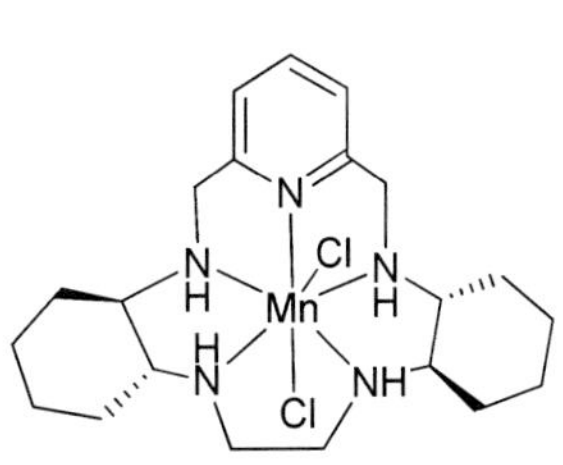

Figure 1.15 Structure of the pentagonal bipyramidal complex M-40403, which shows potential as a superoxide dismutase (SOD) mimetic.

and predictability of morphine. A potential target indication for M-40403 is the treatment of cancer pain and post-operative pain in patients in acute care settings who are on, or are candidates for, opioid therapy. In Metaphore's earlier Phase II trial (completed May 2003) the company found that 20 mg of M-40403 as a single agent was significantly superior to placebo for up to one hour ($p < 0.05$) following molar tooth extraction. No serious adverse events were reported. Results from the trial provided clinical proof of concept that SOD mimetics are well-tolerated and have therapeutic potential [109]. It is suggested that M-40403 removes superoxide anions without interfering with other reactive species known to be involved in inflammatory responses, for example, nitric oxide (NO) and peroxynitrite ($ONOO^-$). M-40403 treatment prevents activation of poly (ADP-ribose) polymerase – having an anti-inflammatory effect – in the gingivomucosal tissue during ligature-induced periodontitis [110]. Other Mn complexes such as Mn^{III} salen chloride [salen = N,N' bis(salicylidene)ethylenediamine] also show SOD and catalase mimetic functionality [111].

1.7
Bipolar Disorder

Lithium salts such as Li_2CO_3 have been widely used for treating bipolar disorder since the early 1950s, and are administered in near-gram quantities on a daily basis. Millimolar levels of Li^+ build up in the blood significantly reducing intracellular sodium concentration in electrically activated cells, a marker that is typically elevated (two to five times higher) in bipolar patients as compared to control subjects [112]. Lithium may also modify monoamine neurotransmitter concentrations in the CNS, inducing changes in neurotransmitter signaling and attenuating fluctuations of cAMP levels (by raising the lowest basal levels and decreasing the highest stimulated increases), stabilizing the activity of this signaling system [113]. Response to lithium appears to cluster in families and can be used to predict the recurrence of bipolar disorder symptoms [114]. Li^+ is similar in size to Mg^{2+} and can therefore inhibit Mg-activated enzymes such as *myo*-inositol monophosphatase. As a consequence it can also interfere with calcium signaling, since Ca^{2+} is mobilized by inositol phosphates [115].

1.8
Anticancer Agents

The discovery of the anticancer activity of cisplatin (*cis*-$[Pt^{II}(NH_3)_2Cl_2]$) by Rosenberg and coworkers some 40 years ago and its subsequent approval for clinical use ten years later has done more for the advancement of inorganic medicinal chemistry than almost any other development in the field. It is worth considering what has happened since and where platinum drugs are heading in the future (see also Chapters 3 and 4).

1.8.1
Pt^{II} Cytotoxic Agents

Both the thermodynamics (ligand arrangement, *trans* influences) and kinetics (rates of ligand substitution, *trans* effects) are crucial for the rational design of platinum anticancer drugs. Clearly, both the metal and the ligands determine the biological activity, not just the metal (Pt). Platinum occupies a unique place in the recent development of inorganic medicinal chemistry due to the clinical success of cisplatin, subsequently carboplatin, and more recently oxaliplatin (Figure 1.16a). The latter is currently a billion-dollar ("blockbuster") drug, even though it is used mainly only for colorectal cancer in combination with organic drugs [116]. These successes have provided enormous incentive for exploration of the chemistry of platinum, which has uncovered new and interesting findings, both thermodynamic and kinetic. The ligands on Pt^{II} play important roles not only in determining the activation of the administered complex (e.g., by hydrolysis) but also in the recognition of the drugs and their DNA adducts by proteins and repair enzymes, and hence the ligands play a pivotal role in the development of resistance mechanisms.

The precise mechanism of cell uptake of platinum complexes remains a topic of great curiosity. The copper transporter Ctr1 has been implicated in the transport of cisplatin, (carboplatin and oxaliplatin) across the cell membrane [117] and Ctr2 has been shown to regulate the cellular accumulation and cytotoxicity of cisplatin and carboplatin [118]. In general, copper transporters appear to regulate the cellular pharmacology and sensitivity to Pt drugs [119]. However, reactions of

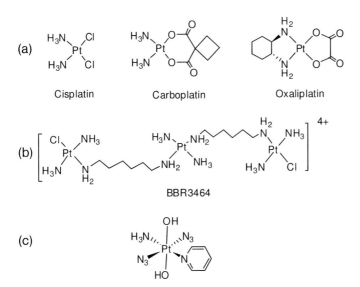

Figure 1.16 Structures of (a) cisplatin, carboplatin, and oxaliplatin; (b) BBR3464; and (c) *trans,trans,trans*-[Pt (N$_3$)$_2$(OH)$_2$(py)(NH$_3$)].

cisplatin with the extracellular methionine-rich N-terminal domain of Ctr1 can lead to the release of the ammine ligands, thus deactivating the drug [120]. This has precedent, stemming from the high trans effect of sulfur (L-methionine) [121] and such reactions have also been detected in cancer cell extracts [122]. The hypothesis put forward by Arnesano and Natile [123] – that interaction of cisplatin with Ctr1 leads to pinocytosis and delivery of cisplatin into the cell in vesicles, helping to shield it from reactions with nucleophiles such as glutathione and metallothionein, which lead to drug inactivation – is therefore an attractive one.

Not only does protein recognition appear to be important in cell entry of platinum drugs, but also in the processing of platinated DNA lesions and excision-repair resistance mechanisms. For example, platinated *intra*-strand crosslinks are recognized by high mobility group (HMG) proteins, which shield them from repair, whereas *inter*strand crosslinks are recognized by the mismatch repair (MMR) family [124].

An early design rule for platinum drugs was that they should be neutral in charge to allow uptake by cells. Now there are several exceptions to this rule, notable examples being the 1+ complex cis-[Pt(NH$_3$)$_2$(pyridine)Cl]$^+$ [125] and the 4+ trinuclear complex [{$trans$-PtCl(NH$_3$)$_2$}$_2$(μ-$trans$-Pt(NH$_3$)$_2$\{NH$_2$(CH$_2$)$_6$NH$_2$\}$_2$)] (NO$_3$)$_4$ (BBR3464) (Figure 1.16b) [126], both of which contain only monofunctional Pt centers. Both examples also break one of the early rules that complexes should be bifunctional or potentially bifunctional after loss of two *cis* monodentate ligands or a weakly chelated (e.g., oxygen donor) bidentate ligand. The belief that the am(m)ines in square-planar Pt^{II} complexes need to be of *cis* geometry for activity is also no longer valid [127, 128]. It is apparent that the inactivity of transplatin is not a feature of *trans* diamine Pt^{II} complexes in general, and several *trans*-Pt^{II} complexes are being developed that rival, or improve upon, the cytotoxicity of cisplatin [127].

Recent advances in formulation of platinum drugs and means for delivery and for targeting have included attaching Pt^{IV} prodrugs to single-walled carbon nanotubes [129] and platinum–polymer conjugates, dendrimers, micelles, and microparticulates [130], including Pt^{IV}-encapsulated prostate-specific membrane antigen (PSMA) targeted nanoparticles [131].

1.8.2
Pt^{IV} Prodrugs

Several Pt^{IV} prodrugs have been on clinical trials but none has been successful enough to date to gain widespread approval. Clinical trials of tetraplatin ([Pt (DACH)Cl$_4$]) and iproplatin ([$cis,trans,cis$-[Pt(i-PrNH$_2$)(OH)$_2$Cl$_2$]) were abandoned some years ago although satraplatin (JM216, $cis,trans,cis$-[Pt(cyclohexylamine) (NH$_3$)(acetate)$_2$Cl$_2$]), which is orally active, is in Phase III trials for hormone-refractory prostate cancer. One of the difficulties perhaps is the requirement for *in vivo* activation by reduction at the tumor; the levels of reductants such as ascorbate and thiols may be too variable for the activation to occur in a controlled and predictable way in patients.

1.8.3
Photoactivatable Pt^{IV} Complexes

Adverse side effects and development of acquired resistance to platinum diam(m)ine (Pt^{II}) complexes pose a serious problem; improved targeting is crucial for the clinical success of new anticancer agents. Octahedral Pt^{IV} complexes are typically more inert to reaction than square-planar Pt^{II} complexes; Pt^{IV} complexes that can be specifically activated only at a tumor site provide a route to a better targeted treatment. Sadler *et al.* have developed Pt^{IV} azido complexes that are non-toxic in the dark but which become highly toxic to cells following irradiation [132]. For example, *trans,trans,trans*-[Pt(N$_3$)$_2$(OH)$_2$(py)(NH$_3$)] (Figure 1.16c) has little or no dark toxicity and is an order of magnitude more potent towards human ovarian cancer cells (A2780) when photoactivated than cisplatin under similar conditions [133]. Moreover, recent work has shown that the *trans* bis-pyridine adduct can be activated in cells by visible light [133b]. The mechanism of action of these complexes is postulated to involve platination of nuclear DNA, but such excited state drugs can undergo unusual reactions [134] and other targets (proteins) may be involved. The use of light in this way provides a degree of spatial and temporal control over drug activation [135].

1.8.4
Ruthenium

Soon after the discovery of the anticancer properties of platinum complexes, ruthenium compounds were investigated. They often have similarly slow kinetics of substitution reactions as platinum. NAMI-A (Figure 1.17a) is a tetrachlorido imidazole/dmso Ru^{III} compound that inhibits matrix metalloproteinases and prevents tumor invasion of nearby tissues [136]. The related bis-indazole complex Ru^{III} (KP1019) (Figure 1.17a) showed no dose-limiting toxicity in Phase I studies [137]. It has been demonstrated to be largely protein bound in blood (to albumin and transferrin) and is found on DNA in peripheral leukocytes. The results of Phase II colorectal trials are anticipated with interest. The kinetically-inert organometallic Ru^{II} complex, DW1/2 (Figure 1.17b) (Table 1.3), which mimics staurosporine, targets a signal transduction pathway; its action involves inhibition of the beta form of glycogen synthase kinase-3 (binding to its ATP site), activation of p53, and apoptosis via the mitochondrial pathway [138].

1.8.4.1
Interaction with Plasma Proteins

Binding to plasma proteins causes a drastic decrease of NAMI-A bioavailability and a subsequent reduction of its biological activity, implying that association with plasma proteins essentially represents a mechanism of drug inactivation [139]. Contrastingly, an important step in the mode of action of KP1019 is thought to be binding to the serum protein transferrin and transport into the cell via the transferrin pathway [163]. In the blood, transferrin is only one-third saturated with its natural metal ion Fe^{III} and cancer cells have a higher density of transferrin

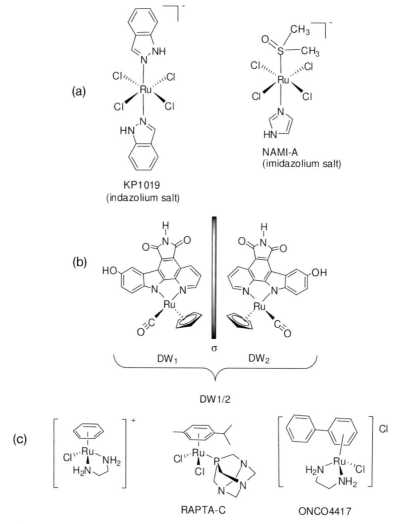

Figure 1.17 (a) KP1019, NAMI-A; (b) DW$_{1/2}$; and (c) half-sandwich organometallic Ru^{II} complexes of the type [(arene) Ru(X)(Y)(Z)]; inclusion of a chelating ligand (1,2-diaminoethane, en) can improve aqueous stability.

receptors than normal cells; so hijacking this route into the cell with therapeutic agents such as KP1019 provides a potentially selective uptake mechanism.

1.8.4.2
Ruthenium Arenes
Half-sandwich organometallic Ru^{II} complexes of the type [(arene)Ru(XY)Z] provide a versatile platform for anticancer drug design (Figure 1.17c, for example). Some structure–activity relationships have been described [164–166]. If X, Y, and Z

Table 1.3 Some examples of the dependence of the biological activity of ruthenium, platinum, and gold complexes on oxidation state and coordinated ligands.

Complex	Activity	Reference
Ruthenium		
DW1/2 [CpRu^{II}(pyridocarbazole)CO)]	Inert ATP-competitive protein kinase inhibitors	Figure 1.17b (Meggers) [140]
NAMI-A *trans*-[Ru^{III}Cl$_4$(DMSO)(Im)] (ImH)	Antimetastatic	Figure 1.17a (Alessio, Sava) [141]
KP1019 *trans*-[Ru^{III}Cl$_4$(Ind)$_2$](IndH)	Cancer cell cytotoxic	Figure 1.17a (Keppler) [137]
[(η^6-biphenyl)Ru(en)Cl]$^+$	Organometallic anticancer complex	Figure 1.17c (Sadler) [142]
[Ru^{II}(CO)$_3$Cl(glycinate)]	CO delivery; cytoprotectant	(Mann) [143]
[Ru^{II}(bpy)$_2$(dipyridophenazine)]$^{2+}$	Luminescent DNA intercalator, oxidative damage	(Barton) [144]
[Ru^{III}(HEDTA)Cl]$^-$	NO scavenger (alleviation of NO-mediated disease states)	(Review by Fricker) [145]
Ruthenium red [Ru$_3$Cl$_8$(OH)$_3$(NH$_3$)$_{12}$(OH$_2$)$_3$]	Blocker of mitochondrial calcium uptake/efflux: potential application for prevention of ischemia reperfusion injury in liver transplantation.	Crystal structure [146] Medical application [147]
Platinum		
[PtCl$_4$]$^{2-}$, [PtCl$_6$]$^{2-}$	Potent immunogens (allergens, skin sensitizers), but not [Pt(NH$_3$)$_4$]$^{2+}$	(General review) [148]
Cisplatin *cis*-[PtCl$_2$(NH$_3$)$_2$]	Cytotoxic anticancer drug; bifunctional, crosslinks, and kinks DNA	Figure 1.16a (Review) [123], [149]
trans-[PtCl$_2$(NH$_3$)$_2$]	Inactive as anticancer agent	
cis-[Pt(NH$_3$)$_2$(py)Cl]$^+$	Monofunctional anticancer agent; substrate for organic cation transporters SLC22A1 and SLC22A2	(Lippard) [125]
trans-[PtCl$_2$(iminoether)$_2$] where iminoether is HN=C(OR)R′	Active *trans* anticancer complex	(Natile) [128]
cis-[Pt^{II}(H$_2$N(CH$_2$)$_2$PPh$_2$)$_2$]Cl$_2$	Cytotoxic, reversible chelate ring-opening, and binding to G	(Sadler) [150]
[Pt^{II}(en)Cl(*S*-thiourea-acridine)]	DNA intercalator; inactive *in vivo*	(Bierbach)[151]
[Pt^{II}(en)Cl(*N*-amidine-acridine)]	DNA intercalator; active *in vivo* (lung cancer)	
Pt^{IV} prodrugs	On reduction can release active agents, e.g., estrogen or enzyme inhibitors	(Lippard) [152] (Dyson) [153]

(Continued)

Table 1.3   (*Continued*)

Complex	Activity	Reference
t,t,t-[Pt^{IV}(N$_3$)$_2$(OH)$_2$(NH$_3$)(py)]	Photoactivatable anticancer complex; only cytotoxic following irradiation	See Figure 1.16c (Sadler) [133]
TriplatinNC [{$trans$-Pt^{II}-(NH$_3$)$_2$(NH$_2$(CH$_2$)$_6$(NH$_3$$^+$)}$_2$-μ-{$trans$-Pt^{II}(NH$_3$)$_2$(NH$_2$(CH$_2$)$_6NH_2$)$_2$}]	Trinuclear; cytotoxic; trinuclear phosphate clamp, electrostatic binding to DNA backbone	(Farrell) [154]
BBR3464 [{$trans$-Pt^{II}Cl(NH$_3$)$_2$}$_2$-μ-{$trans$-Pt^{II}(NH$_3$)$_2$(NH$_2$(CH$_2$)$_6$NH$_2$)$_2$}] (NO$_3$)$_4$	Trinuclear; cytotoxic; monofunctional centers, long-range DNA crosslinking	Figure 1.16b (Farrell) [155]
[Pt^{II}(S,S-dach)(phen)]$^{2+}$	DNA intercalator; leukemia, non-toxic *in vivo*, and reduces tumor growth	(Aldrich-Wright) [156]
[(cis-{Pt^{II}(NH$_3$)$_2$})$_2$(μ-OH)(μ-pyrazolate)]$^{2+}$	Potent cytotoxic; DNA GG crosslinks but little distortion	(Reedijk) [157]
Gold		
Et$_3$PAu^{I}SR (Auranofin) and [Au^I-S-CH(CO$_2$Na)(CH$_2$CO$_2$Na)]$_n$, (sodium aurothiomalate, Myochrysine)	Oral antiarthritic and injectable antiarthritic	Figure 1.14 (Review) [158]
[Au^I(CN)$_2$]$^-$	Stable metabolite of gold antiarthritic drugs	(Graham) [159]
[Au^I(diphosphine)$_2$]$^+$; [Au^I(N-heterocyclic carbene)$_2$]$^+$	Anticancer; antimitochondrial; thioredoxin and thioredoxin reductase inhibitor	Figure 1.20b (Berners-Price) [160, 161]
[Au^{III}(dithiocarbamate-R)X$_2$]	Anticancer; proteasome inhibitor	Figure 1.20c (Fregona) [162]

are monodentate ligands, the complexes demonstrate low cytotoxicity on account of rapid hydrolysis and weak binding to DNA [167], although some 1,3,5-triaza-7-phosphaadamantane adducts have interesting antimetastatic activity (RAPTA-C, Figure 1.17c) [168].

Activity tends to increase with the size of the arene, and extended arenes can intercalate between DNA bases [169, 170]. For XY = ethane-1,2-diamine (en, an N-chelating ligand), and Z = halide, activation occurs via aquation and the aqua adduct selectively binds to guanine (G) residues in DNA, at the N$_7$ ring position. This is accompanied by C$_6$O (G) hydrogen bonding to an NH group of the en ligand. Chelated ligands with X and/or Y = O (H-bond acceptor) can also bind to adenine (H-bond donor).

The half-sandwich Ru complex [(η^6-biphenyl)Ru(en)Cl]Cl (ONCO4417, Figure 1.17c) exerts antiproliferative effects in H460 lung cancer cells by inducing apoptosis, with levels of DNA damage comparable to those produced by cisplatin. Cell death appears to occur prior to entry into the G2/M phase [171]. The complex is currently in preclinical development. Such complexes are non-cross-resistant with cisplatin in line with the different lesions produced on DNA. The loss of cytotoxicity in complexes of this type when XY = phenanthroline or bipyridine, both strong π-acceptor ligands, is intriguing; in the case of bipyridine activity can be restored by 3,3'-hydroxylation of the rings [172]. Use of another strong π-acceptor chelating ligand hydroxy- or N,N-dimethyl-phenylazopyridine, Z = iodide, leads to hydrolytically inert complexes that appear to kill cancer cells by a different mechanism: catalysis of glutathione oxidation and production of reactive oxygen species (ROS) in cells [173].

Ligand oxidation appears to provide a route to activation of Z = thiolato complexes, a route that may be important when the intracellular thiol glutathione binds to these ruthenium arene complexes. Mono- and bis-oxygenation appear to be facile but surprisingly do not weaken the Ru–S bond [174] even though this provides a route to nucleobase binding [175]. Protonation of the sulfenate oxygen on the other hand does labilize this bond [176].

1.8.5
Osmium

Organometallic half-sandwich Os^{II} complexes that are isostructural with their Ru^{II} counterparts have been synthesized [166, 177]. Although structurally similar, their properties differ in some important ways, influencing their biological activity. For example, Z = Cl complexes hydrolyze about 100 times more slowly, and the resultant bound water is about 1.5 pK_a units more acidic. The latter feature favors formation of hydroxo adducts that readily associate to give very stable hydroxo-bridged dimers, even in cell culture media, for example, for the Os^{II} complex where XY = acetylacetonate (Figure 1.18). Picolinate (XY = N,O) ligands give rise to complexes with promising cancer cell cytotoxicity [178] and cause significant DNA unwinding but little DNA bending, in contrast to Ru [179]. It is possible to switch off the cytotoxic activity of these complexes, by use of substituents on the pyridine ring. Intriguingly, the activity of picolinamido complexes is switched off by conversion from N,O into N,N coordination. This switch is accompanied by more rapid hydrolysis and a greater extent of nucleobase (G) binding [180].

In general, osmium arene chemistry has been little studied in comparison with the related ruthenium arene chemistry, especially in aqueous solution, but in the anticancer field is now attracting increasing attention [181, 182].

1.8.6
Titanium

Two Ti^{IV} complexes entered clinical trials as anticancer drugs in the 1990s, a tris-acetylacetonate derivative complex (Budotitane) and titanocene dichloride [Cp$_2$TiCl$_2$] (Figure 1.19). Both contain two labile cis coordination positions, as does cisplatin. However, hydrolysis is not only faster for these Ti complexes but also bound water is more acidic, leading to ready formation of hydroxo-bridged species

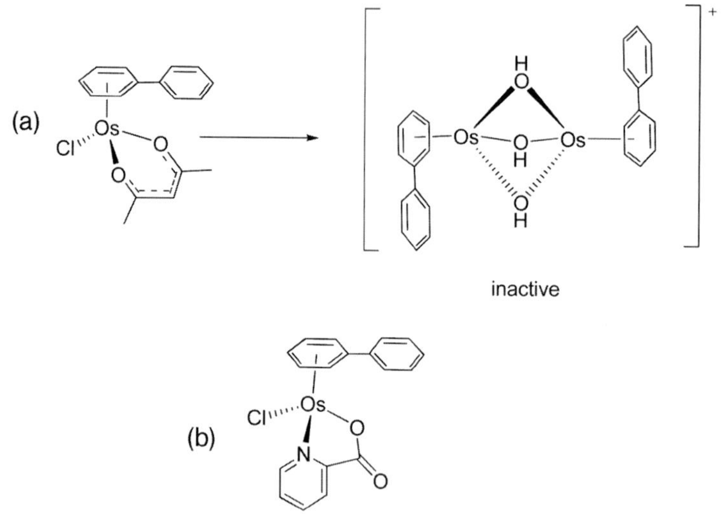

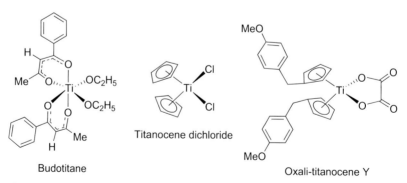

(a)

inactive

(b)

Figure 1.18 (a) Complexes such as [Os^{II}(acac)(biphenyl)Cl]
(acac = acetylacetonate) give rise to oxo-bridged dimers,
which are inactive in cancer cell lines, while (b) those
including picolinate derivatives show significant activity
(IC$_{50}$ ≈ 5 µM against A2780 cancer cells).

and eventually TiO$_2$. It is perhaps not surprising therefore that Budotitane did not
progress beyond Phase I due to formulation problems and titanocene dichloride
did not progress beyond Phase II.

Curiously, Jaouen *et al.* have discovered that [Cp$_2$TiCl$_2$] has strong estrogenic activity and a proliferative effect on hormone-dependent cell lines [183]. It seems

Budotitane

Titanocene dichloride

Oxali-titanocene Y

Figure 1.19 Structures of anticancer agents based
on titanium: budotitane, titanocene dichloride, and
oxali-titanocene Y.

unlikely that DNA bases are the target for Ti^{IV} anticancer complexes. It is possible that Ti^{IV} is delivered to cancer cells by the serum Fe^{III} transport protein transferrin [184]. Tinoco *et al.* have shown that Ti_2^{IV}-transferrin binds to the transferrin receptor very strongly [185].

Attempts are now being made to design new titanocene dichloride analogs that are more stable to hydrolytic reactions [186]. For example, ring-substituted cationic titanocene dichloride derivatives are more active than the parent $[Cp_2TiCl_2]$ [187]. Bis[(*p*-methoxybenzyl)cyclopentadienyl] titanium dichloride (Titanocene Y) exhibits significant activity against renal cell cancer both in vitro and also in xenografted tumors in mice [188]. Bis[(*p*-methoxybenzyl)cyclopentadienyl]titanium(IV) oxalate (oxali-titanocene Y; Figure 1.19) is twice as potent as cisplatin towards pig kidney epithelial (LLC-PK) cells [189]. In view of these results, exploration of the activity of other derivatized metallocenes, for example, of Zr and Hf might now be worthwhile.

1.8.7
Gold

The anticancer activity of tetrahedral Au^I tetraphosphine complexes such as [Au (dppe)$_2$]$^+$ (dppe = diphenylphosphinoethane) (Figure 1.20a) [190] is intriguing not only because the stereochemistry is unusual for gold(I) complexes (usually linear) but also because of its proposed mechanism of action; positively charged lipophilic complexes such as this can destroy the membrane potentials of mitochondria. The attraction of Au^I for anticancer therapy is that it is unlikely to cause DNA mutations (unlike cisplatin) since it binds only weakly to DNA bases. Berners-Price *et al.* have successfully optimized the lipophilicity (partition coefficients) of this class of complexes to achieve selective cell uptake [191] (see also Chapter 7).

Lipophilic, cationic Au^I complexes of N-heterocyclic carbenes (Figure 1.20b) can also act as mitochondria-targeted antitumor agents causing cell death through a mitochondrial apoptotic pathway and inhibiting the activity of selenoprotein thioredoxin reductase (TrxR), but not Se-free enzyme glutathione reductase [192]. The isoelectronic and isostructural nature of Au^{III} complexes with those of Pt^{II} presents a challenge for anticancer drug design since, in general, Au^{III} reacts (e.g., towards hydrolysis) more quickly and is readily reduced. Notable success has been obtained recently with Au^{III} dithiocarbamates (Figure 1.20c) by Fregona *et al.* that target the proteasome and are candidates for clinical trials [193]. Porphyrin ligands can also stabilize Au^{III}. *In vivo*, intraperitoneal injections of Au^{III} meso-tetraarylporphyrin (gold TPP, Figure 1.20d) can significantly inhibit tumor cell proliferation, induce apoptosis, and suppress colon cancer tumor growth [194]. Gold TPP is stable towards demetallation under physiological conditions and has been shown to bind to calf thymus DNA with a binding constant of $K_b = 2.79 \pm 0.34 \times 10^6\,dm^3\,mol^{-1}$ at 293 K; fragmentation of genomic DNA was observed after treating HeLa cells with gold TPP for 15 h [195].

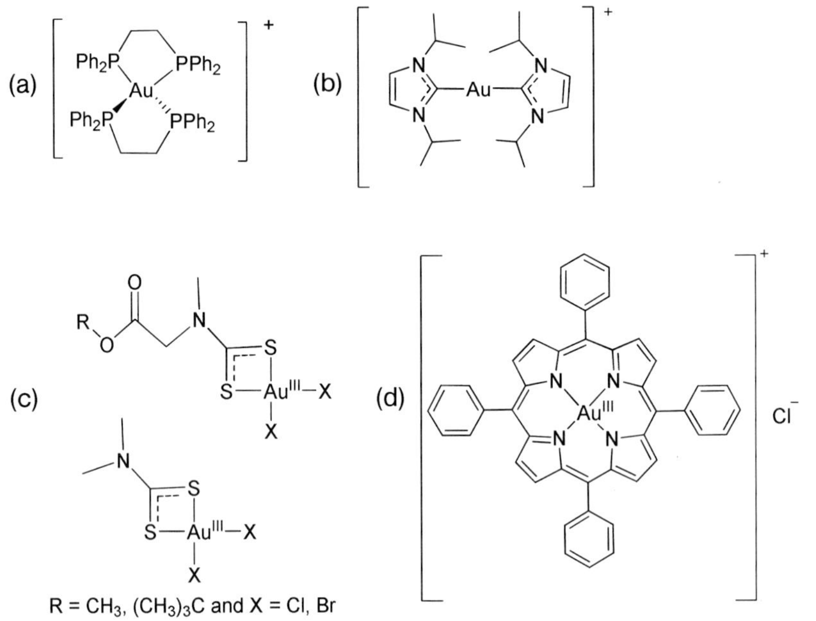

Figure 1.20 Structures of (a) [Au(dppe)$_2$]$^+$; (b) cationic Au^I complexes of N-heterocyclic carbenes; (c) dithiocarbamates; and (d) gold TPP a Au^{III} meso-tetraarylporphyrin complex.

1.8.8
Tin

A wide range of organometallic Sn^{IV} complexes have long been known to be cytotoxic to cancer cells but few have activity *in vivo*. Interesting are recent findings for the tributyl complex tri-*n*-butyltin(IV)lupinylsulfide hydrogen fumarate (IST-FS 35, Figure 1.21) which inhibits both P388 myelomonocytic leukemia and the B16-F10 melanoma, implanted subcutaneously in BDF1 mice, by 96% in tumor volume at day 11 following a single intravenous injection. The mechanism is not understood [196]. Triorganotin carboxylates may exist in monomeric or polymeric forms, while diorganotin derivatives may exist as true dicarboxylates or as distannooxane salts [(R$_2$SnO-COR')$_2$O] and may further aggregate in several ways that influence both solubility and bioavailability. Inhibition of macromolecular synthesis, mitochondrial energy metabolism, and reduction of DNA synthesis, as well as direct interaction with the cell membrane (increase in cytosolic Ca^{2+} concentration), have been implicated in organotin-induced cytotoxicity. Promotion of DNA damage *in vivo* has been detected. Oxidative damage and increased concentration of intracellular calcium ions seem to be the major factors contributing to triorganotin-induced apoptosis in

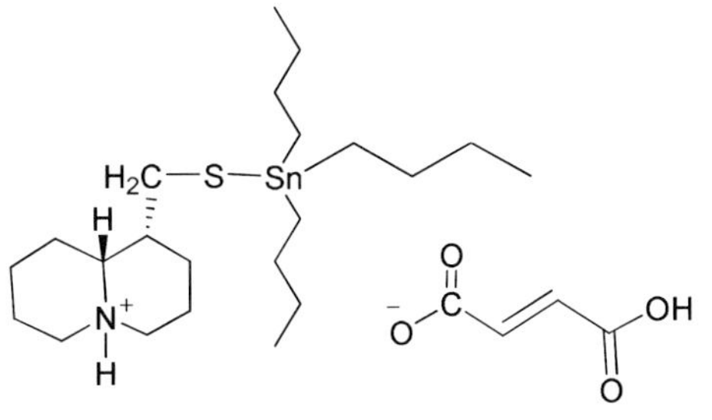

Figure 1.21 Structure of the tributyl Sn^{IV} complex tri-*n*-butyltin(IV)lupinylsulfide hydrogen fumarate (IST-FS 35).

many cell lines [197]. More work on the molecular basis for the activity of tin complexes is needed.

1.8.9
Gallium

Ganite® (which contains Ga^{III} nitrate and citrate in a 1 : 1 mol ratio at pH 6–7) is used to treat cancer-related hypercalcemia, so far a unique FDA-approved gallium drug [198]. However, it has poor bioavailability [199]. The orally bioavailable gallium complex KP46 [tris(8-quinolinolato)gallium(III)] has been in Phase I clinical trials for treatment of solid tumors. In melanoma cells the complex causes S-phase cell cycle arrest and apoptosis [200]. Gallium maltolate, tris(3-hydroxy-2-methyl-4*H*-pyran-4-onato)gallium (GaM), is also an orally-active gallium compound. Gallium complexes can act as potent proteasome inhibitors in prostate cancer cells [201]. Gallium(III) has similar chemistry to Fe^{III} although it is not redox-active and is transported to cells bound to the serum protein transferrin. Cancer cells have a high density of transferrin receptors. In cells it may interfere with iron metabolism, for example, by binding to ribonucleotide reductase [202].

1.8.10
Arsenic

The drug Trisenox as a 1 mg ml^{-1} solution of arsenic trioxide (As$_2$O$_3$) is given by infusion to adult patients with relapsed/refractory acute promyelocytic leukemia (APL) at a dose of 0.15 mg kg^{-1} per day for several weeks [203]. There is also activity in multiple myeloma (MM). In aqueous solution it exists as neutral As(OH)$_3$ (often mistakenly referred to as arsenite; pK_a values 9.3, 13.5, and 14.0) [204, 205], and is transported into cells by the membrane protein aquaglyceroporin [206], which normally transports water and glycerol. Oxidative methylation of As^{III} to As^V can occur *in vivo* to give [(CH$_3$)As^V(O)(OH)$_2$] and [(CH$_3$)$_2$As^V(O)(OH)], primarily in the liver. These methylation products are then excreted, with half-lives of 32–70 h.

Attempts to improve the targeting and delivery of arsenic anticancer compounds include those of Boise *et al.* with darinaparsin (ZIO-101, *S*-dimethylarsino-glutathione) [207]. Like As_2O_3 this induces up-regulation of BH3-only proteins (members of the Bcl-2 protein family containing *only* one of the Bcl-2 homology regions, essential initiators of programmed cell death, required for apoptosis). It is in Phase II trials in patients with primary liver cancer, advanced myeloma and lymphomas, and a Phase I oral trial is in progress. O'Halloran *et al.* [208] have shown that liposomal encapsulated (100 nm-scale) nanoparticles of, for example, Co^{II} or Ni^{II} arsenite allow pH-controlled release of active drug. Folate-targeted liposomes potentiate As_2O_3 efficacy in relatively insensitive solid tumor-derived cells [209].

1.8.11
Copper

The potential for use of copper complexes as anticancer agents has yet to be fully exploited [210–212]. Mixed chelate copper(II) complexes (known as Casiopeínas) with general formulae $[Cu(N-N)(\alpha\text{-}L\text{-amino acidato})](NO_3)$ and $[Cu(N-N)(O-O)](NO_3)$, where the N-N donor is an aromatic substituted diimine [1,10-phenanthroline (phen) or 2,2′-bipyridine (bpy)] and the O-O donor is acetylacetonate (acac) or salicylalde-hydate (salal), are in preclinical development as antineoplastic agents. The central fused aromatic ring in the phen-containing complexes is necessary for the anti-proliferative activity [213].

There are many possible targets for copper complexes in cells, complicated by the natural pathways involved in copper chaperoning. Copper 2,4-diiodo-6-[(pyr-idine-2-ylmethylamino)methyl]phenol complexes are proteasome inhibitors in prostate cancer cells [214].

1.8.12
Zinc

Zinc complexes of 1-hydroxypyridine-2-thione derivatives inhibit the growth of A549 lung and PC3 prostate cancer cells in xenograft models. Zinc bis-pyrithione itself is a widely-used antidandruff compound. Gene expression profiles of A549 cultures treated with one of these complexes revealed activation of stress response pathways under the control of metal-responsive transcription factor 1 (MTF-1), hypoxia-inducible transcription factor 1 (HIF-1), and heat shock transcription factors [215].

1.8.13
Bismuth

Anticancer activity is demonstrated by thioguanine, 6-mercaptopurine complexes of Bi and, more recently, dithiocarbamate complexes of general formula $Bi(S_2CNR_2)_3$ have shown the ability to reduce the rate of tumor growth in animal models [216].

1.8.14
Molybdenum

Polyoxomolybdates such as $[NH_3Pr^i]_6[Mo_7O_{24}] \cdot 3H_2O$ (PM-8) (Figure 1.9a) exhibit potent antitumor activity against a range of cancer cell lines. The mechanism appears to involve preferential uptake of PM-8 into tumor cells, and conversion into tenfold more toxic species through biological reduction (in the mitochondria), resulting in inhibition of ATP generation and apoptosis [61]. Polyoxomolybdates have also shown promising anticancer activity *in vivo* [217].

1.8.15
Photosensitizers: Porphyrins

Photodynamic therapy (PDT) involves the delivery of a photosensitizing agent to tumor cells, followed by activation with light, which generates cytotoxic singlet oxygen. Porphyrins, used in PDT treatment of cancer, are strong metal chelators, and there is evidence to suggest that the first-generation PDT photosensitizer haematoporphyrin derivative (HpD) (commercialized as Photofrin, a mixture of monomeric and oligomeric porphyrins) may form Zn^{II} complexes *in vivo* [218]. Such complexation is thought to lead to enhanced solubility in aqueous media, and, through interfering with π–π stacking of the porphyrins, to de-aggregation and improved PDT activity [219]. Intentional metal chelation can be used to influence the tumor-localizing properties of the photosensitizer and to tune the photophysical properties. The presence of a metal can increase the efficiency of singlet oxygen generation owing to the enhancement of spin–orbit coupling and intersystem crossing by the metal ion [220]. Expanded porphyrins (texaphyrins) form stable 1 : 1 complexes with a range of larger metal ions, such as Ln^{III} [221]. These complexes exhibit strong absorptions at longer wavelengths than conventional porphyrins (600–900 nm) with high quantum yields, making them attractive for PDT applications. Lutetium texaphyrin (Motexafin Lutetium, Lutrin, Lu-Tex) is in Phase IIb clinical trials for treatment of recurrent breast cancer and shows promise for treating cervical, prostate, and brain tumors. Other metal-containing photosensitizers for cancer treatment are beginning to enter clinical trials [222]. Iron(III) complexes incorporating ligands such as dipyridoquinoxaline and 2,2-bis(3,5-di-*tert*-butyl-2-hydroxybenzyl)amino-acetate show DNA photocleavage when irradiated with light of wavelength ≥ 630 nm but this work has not yet been extended to cytotoxicity studies [223].

1.9
Small Molecule Delivery and Control

1.9.1
Nitric Oxide (NO)

Peroxynitrite ($ONOO^-$), the product of the interaction between O_2^- and NO, can act as a signaling mediator. Peroxynitrite decomposition catalysts may have therapeutic potential as adjuncts to opiates in relieving suffering from chronic pain [224, 225].

NO is recognized to be an important signaling molecule with a wide range of functions in the cardiovascular, nervous, and immune systems [226]. The interactions of NO with metal complexes *in vivo* – heme and non-heme Fe in particular – are of prime importance to its physiological role. Metal complexes show promise for both controlled release and scavenging of NO [227].

Sodium nitroprusside, $Na_2[Fe(CN)_5NO] \cdot 2H_2O$, is used clinically to treat cardiovascular disorders and lower blood pressure through release of NO. The hypotensive effect is rapid and controls blood pressure within minutes; however, toxicity involving cyanide accumulation limits its application. Photoactive Fe complexes incorporating large photon-capturing "antenna" ligands attached to an iron nitroso center are being investigated as photoactivatable NO-releasing agents [228], and Ru agents have been developed that can release [229] or scavenge [230] NO.

1.9.2
Carbon Monoxide (CO)

Metal carbonyls have been extensively investigated as potential CO-donating pharmaceuticals [231, 232]. *In vivo*, CO appears to have a role as a messenger. It also demonstrates anti-inflammatory properties and has been investigated for its ability to suppress organ graft rejection. The design of metal complexes that can release CO at a predicable rate is therefore valuable as a relatively non-toxic source of CO. $[Ru(CO)_3Cl(glycinate)]$ has been investigated in some detail and, interestingly, $[Fe(bpy)(SPh)_2(CO)_2]$ appears to liberate CO only intracellularly. Cell viability studies of HT29 colon cancer cells treated with the CO-releasing compound $[Mn(CO)_3(tpm)]PF_6$, where tpm = tris(pyrazolyl)methane, revealed a significant photo-induced cytotoxicity, comparable to that of the established anticancer agent 5-fluorouracil, whilst controls kept in the dark were unaffected at up to 100 μM [233].

1.10
Diagnostic Agents

Both diagnostic (γ-emitting, e.g., ^{99m}Tc, ^{67}Ga, ^{111}In), and therapeutic (α-emitting, e.g., ^{211}At, ^{212}Bi and β-emitting, e.g., ^{90}Y, ^{188}Re) radiopharmaceuticals are clinically important (Chapter 9). The synthesis of Tc complexes from $[TcO_4]^-$ requires careful control of the Tc oxidation state by the ligands, which also play a major role in determining the distribution of the complex in the body (e.g., extra- vs. intracellular, heart, brain, kidneys, etc.). These compounds have the advantage of a rapid passage from the laboratory into the clinic, since very small doses are usually administered, posing a negligible toxicity hazard.

Recent advances include the targeting of Tc complexes to specific cellular receptors. The stable Tc^I fragment $\{Tc(CO)_3\}^+$ used by Alberto *et al.* [234] is very useful for labeling purposes.

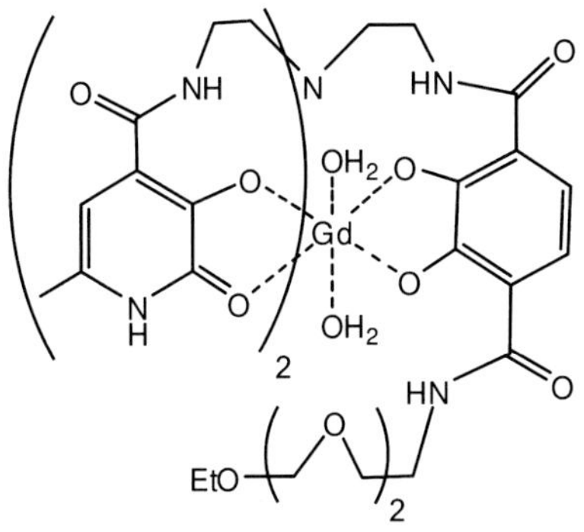

Figure 1.22 An example of a promising Gd-hydroxypyridinone (HOPO) complex: Gd-TREN-bis-6-CH$_3$-3,2-HOPO-TAM-TRI.

For imaging purposes, the relatively insoluble BaSO$_4$ is widely used as an X-ray contrast agent. For contrast enhancement in magnetic resonance imaging (MRI) there are four Gd^{III} complexes currently in clinical use, and manganese dipyridoxyl-diphosphate (MnDPDP) [235] and superparamagnetic iron oxide nanoparticles [236] show promise in this regard (Chapter 8). The nature of the ligands is very important in the Gd^{III} complexes for controlling the access of water to the paramagnetic center, the rotational correlation time of the complex, and its biodistribution and toxicity. Recent toxic side-effects have arisen from some newer Gd^{III} complexes which have less-tightly bound chelating ligands. In this respect the highly stable hydroxypyridinone complexes of Raymond *et al.* (Figure 1.22) appear to be especially promising [237]. Future advances will be concerned especially with combined imaging and therapeutic agents (dual modality).

1.11
Veterinary Medicinal Inorganic Chemistry

An important area of study is that of therapeutic inorganic agents for farm and other animals. For example, the carboxylate-bridged dimer copper(II) indomethacin is used (as Cu-Algesic) as a non-steroidal anti-inflammatory and analgesic for treatment of acute and sub-acute musculoskeletal/locomotor inflammatory conditions in horses [238]. Arsenic-based compounds such as Roxarsone (2-nitrophenol-4-arsonic acid, Figure 1.23) [239] are widely used as growth factors for poultry, which has implications for both the environment and public health [240].

1.12
Conclusions and Vision

The application of inorganic chemistry to medicine and in particular the design of inorganic therapeutic and diagnostic agents is in its infancy, despite the use of a

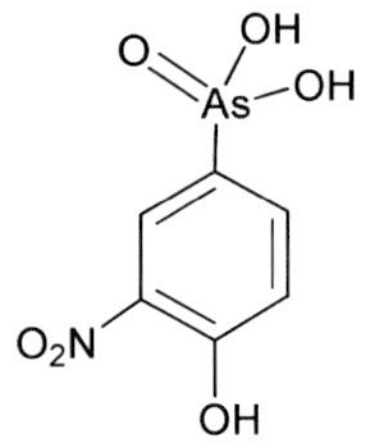

Figure 1.23 Structure of 3-nitro-4-hydroxyphenylarsonic acid (Roxarsone), which is used in veterinary medicine.

few such compounds for several centuries. These older uses were usually based on empirical observations (a valid drug discovery route) and there is often little understanding of their mechanism of action, especially at the molecular level. Such problems still exist today and need to be tackled in the future so that we enter an age where inorganic drugs can be said to be rationally designed.

Our discussions have largely been on an element-by-element (mostly metal-by-metal) basis with the aim of highlighting the enormous differences between metals and the need for advances in technology that will allow both thermodynamic and kinetic speciation of metals to be carried out under biologically-relevant conditions, and preferably in intact cells and tissues. Metal complexes may be transformed by ligand substitution or redox reactions before they reach their target sites (Figure 1.24). This includes *in vitro* tests and transformations in cell culture media.

Most importantly, the biological properties of metal complexes depend not only on the metal and its oxidation state but also on the ligands. Ligands can render either the whole complex inert to ligand substitution reactions or can activate other coordination positions stereospecifically. Besides the metal itself, the ligands can also be the centers of redox reactions, or hydrolytic and other reactions.

The rational design of metal complexes as drugs requires a greater understanding of their effect on cell signaling pathways. An interaction of the metal complex with a constituent (e.g., protein) in one pathway may have consequences for a seemingly unrelated target for which the metal has little affinity. Drugs based on the essential elements need to take into account the pathways programmed by genomes for the uptake, transport, and excretion of those elements (homeostasis). We need to make use of metallomics in therapy: corrective procedures are needed when there are metals in the wrong place at the wrong time, or indeed are not present in the right place at the right time. Clearly, the biochemistry of essential metals is carefully controlled both in terms of thermodynamics (distribution amongst available binding sites) and kinetics (rates of transfer).

We can expect to see new developments relating to the use of radionuclides for diagnosis and therapy, including dual modality, as procedures for their production and handling improve. In addition, radiolabeling is particularly effective for studies of biodistribution.

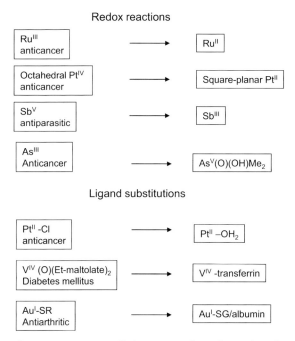

Redox reactions

Ru^{III} anticancer	$\longrightarrow$	Ru^{II}
Octahedral Pt^{IV} anticancer	$\longrightarrow$	Square-planar Pt^{II}
Sb^V antiparasitic	$\longrightarrow$	Sb^{III}
As^{III} Anticancer	$\longrightarrow$	$As^V(O)(OH)Me_2$

Ligand substitutions

Pt^{II} -Cl anticancer	$\longrightarrow$	Pt^{II} –OH_2
$V^{IV}(O)(Et\text{-}maltolate)_2$ Diabetes mellitus	$\longrightarrow$	V^{IV} -transferrin
Au^I-SR Antiarthritic	$\longrightarrow$	Au^I-SG/albumin

Figure 1.24 Most metallodrugs are prodrugs that undergo ligand exchange and redox reactions before they reach the target site. The challenge is to control these processes so as to achieve the required uptake and distribution of the drug followed by activation in the right place at the right time. Some examples of these processes are shown in the figure and are discussed in the text.

As methods for preparing, derivatizing, and characterizing nanoparticles become more advanced we can expect to see them used more extensively as vehicles for drug delivery or as agents themselves. In general the key to the wider use of metal complexes in medicine will depend on the development of new transport, delivery, and targeting methods and on gaining deeper insights into their molecular mechanisms of action. Inorganic chemistry and metal coordination chemistry in particular offers the exciting prospect of novel drugs with novel mechanisms of action.

Acknowledgments

We thank the bodies that support our research, including the EC, MRC, EPSRC, BBSRC, ERC, and Science City (AWM/ERDF), and many colleagues and co-workers for stimulating discussions.

References

1 Pósfai, M., Buseck, P.R., Bazylinski, D.A., and Frankel, R.B. (1998) *Science*, **280**, 880–883.

2 (a) Birchall, J.D., Exley, C., Chappell, J.S., and Phillips, M.J. (1989) *Nature*, **338**, 146–148;
(b) Birchall, J.D. (1991) The toxicity of aluminium and the effect of silicon on its bioavailability, in *Aluminium in Chemistry, Biology and Medicine – A Series of Advances*, Vol. **1** (eds M. Nicolini, P.F. Zatta, and B. Corain), Raven Press, New York, pp. 53–69.

3 Gupta, R.K. (1998) *Adv. Drug Del. Rev.*, **32**, 155–172.

4 Waldron, K.J., Rutherford, J.C., Ford, D., and Robinson, N.J. (2009) *Nature*, **460**, 823–830.

5 Finney, L.A. and O'Halloran, T.V. (2003) *Science*, **300**, 931–936.

6 Reedijk, J. (2008) *Plat. Met. Rev.*, **52**, 2–11.

7 Hambley, T.W. (2007) *Dalton Trans.*, 4929–4937.

8 Ronconi, L. and Sadler, P.J. (2007) *Coord. Chem. Rev.*, **251**, 1633–1648.

9 Guo, Z. and Sadler, P.J. (2000) *Adv. Inorg. Chem.*, **49**, 183–306.

10 Orvig, C. and Abrams, M.J. (1999) *Chem. Rev.*, **99**, 2201–2203.

11 Guo, Z. and Sadler, P.J. (1999) *Angew. Chem. Int. Ed.*, **38**, 1512–1531.

12 Mukherjee, A. and Sadler, P.J. (2009) *Metals in medicine: therapeutic agents*, in *Wiley Encyclopedia of Chemical Biology*, Wiley-Blackwell, Vol. **3**, pp. 80–126.

13 Bruijnincx, P.C.A. and Sadler, P.J. (2008) *Curr. Opin. Chem. Biol.*, **12**, 197–206.

14 Desoize, B. (2004) *Anticancer Res.*, **24**, 1529–1544.

15 Pizarro, A.M. and Sadler, P.J. (2009) *Biochimie*, **91**, 1198–1211.

16 Russell, A.D. and Hugo, W.B. (1994) *Prog. Med. Chem.*, **31**, 351–371.

17 Arakawa, H., Neault, J.F., and Tajmir-Riahi, H.A. (2001) *Biophys. J.*, **81**, 1580–1587.

18 Park, H.-L., Kim, J.Y., Kim, J., Lee, J. H., Hahn, J.-S., Gu, M.B., and Yoon, J. (2009) *Water Res.*, **43**, 1027–1032.

19 Feng, Q.L., Wu, J., Chen, G.Q., Cui, F. Z., Kim, T.N., and Kim, J.O. (2000) *J. Biomed. Mater. Res.*, **52**, 662–668.

20 Chopra, I. (2007) *J. Antimicrob. Chemother.*, **59**, 587–590.

21 Hindi, K.M., Siciliano, T.J., Durmus, S., Panzner, M.J., Medvetz, D.A., Reddy, D.V., Hogue, L.A., Hovis, C.E., Hilliard, J.K., Mallet, R.J., Tessier, C.A., Cannon, C.L., and Youngs, W.J. (2008) *J. Med. Chem.*, **51**, 1577–1583.

22 Sharma, V.K., Yngard, R.A., and Lin, Y. (2009) *Adv. Colloid. Interface Sci.*, **145**, 83–96.

23 Rai, M., Yadav, A., and Gade, A. (2009) *Biotechnol. Adv.*, **27**, 76–83.

24 Braydich-Stolle, L., Hussain, S., Schlager, J.J., and Hofmann, M.-C. (2005) *Toxicol. Sci.*, **88**, 412–419.

25 Gupta, A., Matsui, K., Lo, J.-F., and Silver, S. (1999) *Nat. Med.*, **5**, 183–188.

26 Yang, N. and Sun, H. (2007) *Coord. Chem. Rev.*, **251**, 2354–2366.

27 Sadler, P.J., Li, H., and Sun, H. (1999) *Coord. Chem. Rev.*, **185–186**, 689–709.

28 Ge, R. and Sun, H. (2007) *Acc. Chem. Res.*, **40**, 267–274.

29 Moody, L. and Holder, A.A. (2008) *Annu. Rep. Prog. Chem. Sect. A Inorg. Chem.*, **104**, 477–497.

30 Briand, G.G. and Burford, N. (1999) *Chem. Rev.*, **99**, 2601–2658.

31 Miehlke, S. and Graham, D.Y. (1997) *Int. J. Antimicrob. Agents*, **8**, 171–178.

32 Lambert, J.R. and Midolo, P. (1997) *Aliment. Pharmacol. Ther.*, **11**, 27–33.

33 Jin, L., Szeto, K.Y., Zhang, L., Du, W., and Sun, H. (2004) *J. Inorg. Biochem.*, **98**, 1331–1337.

34 Cun, S., Li, H., Ge, R., Lin, M.C.M., and Sun, H. (2008) *J. Biol. Chem.*, **283**, 15142–15151.

35 Yang, N., Tanner, J.A., Wang, Z., Huang, J.-D., Zheng, B.-J., Zhu, N., and Sun, H. (2007) *Chem. Commun.*, 4413–4415.

36 Boulahdour, H. and Berry, J.-P. (1996) *Cell. Mol. Biol.*, **42**, 421–429.

37 (a) Jones, T. (2009) *Curr. Opin. Mol. Ther.*, **11**, 337–345.

(b) http://hcp.gsk.co.uk/therapy-areas/
pandemic/pandemrix/ (accessed 15
Sept 2010)

38 Food and Agriculture Organization of
the United Nations and World Health
Organization (2003) Summary and
conclusions. Presented at the Sixty-
First Meeting of the Joint FAO/WHO
Expert Committee on Food Additives
Rome, Italy, 10–19 June 2003.
Available at http://www.who.int/ipcs/
food/jecfa/summaries/en/summary_61.
pdf.

39 Burger, J. and Gochfeld, M. (2004)
Environ. Res., **96**, 239–249.

40 Woo, K.J., Lee, T.-J., Bae, J.H., Jang,
B.-C., Song, D.-K., Cho, J.-W., Suh, S.-I.,
Park, J.-W., and Kwon, T.K. (2006) *Mol.
Carcinog.*, **45**, 657–666.

41 Ornaghi, F., Ferrini, S., Prati, M., and
Giavini, E. (1993) *Fund. Appl. Toxicol.*,
20, 437–445.

42 Clarkson, T.W. (2002) *Environ. Health
Perspect.*, **110**, 11–23.

43 Liu, J., Shi, J.-Z., Yu, L.-M., Goyer, R.A.,
and Waalkes, M.P. (2008) *Exp. Biol.
Med.*, **233**, 810–817.

44 Castle, N.A. (2005) Aquaporins as
targets for drug discovery. *Drug Discov.
Today*, **10**, 485–493.

45 Preston, G.M., Jung, J.S., Guggino, W.B.,
and Agre, P. (1993) *J. Biol. Chem.*, **268**,
17–20.

46 Savage, D.F. and Stroud, R.M. (2007)
J. Mol. Biol., **368**, 607–617.

47 Denyer, S.P. (1995) *Int. Biodeterior.
Biodegrad.*, **36**, 227–245.

48 Brown, N.L., Shih, Y.-C., Leang, C.,
Glendinning, K.J., Hobman, J.L., and
Wilson, J.R. (2002) *Biochem. Soc.
Trans.*, **30**, 715–718.

49 Ashutosh, S.S. and Goyal, N. (2007)
J. Med. Microbiol., **56**, 143–153.

50 Baiocco, P., Colotti, G., Franceschini,
S., and Ilari, A. (2009) *J. Med. Chem.*,
52, 2603–2612.

51 Leprohon, P., Legare, D., Raymond, F.,
Madore, E., Hardiman, G., Corbeil, J.,
and Ouellette, M. (2009) *Nucleic Acids
Res.*, **37**, 1387–1399.

52 Mookerjee, B.J., Mookerjee, A.,
Banerjee R., Saha, M., Singh, S.,
Naskar, K., Tripathy, G., Sinha, P.K.,
Pandey, K., Sundar, S., Bimal, S., Das
Pradip, K., Choudhuri, S.K., and Roy,
S. (2008) *Antimicrob. Agents Chemother.*,
52, 1080–1093.

53 Lloyd, N.C., Morgan, H.W., Nicholson,
B.K., and Ronimus, R.S. (2005) *Angew.
Chem. Int. Ed.*, **44**, 941–944.

54 White, N.J. (2004) *J. Clin. Invest.*, **113**,
1084–1092.

55 Reeves, W.C., Hardy, J.L., Reisen, W.K.,
Milby, M.M. (1994) *J. Med. Entomol.*
31, 323–332.

56 Rogers, D.J. and Randolph, S.E. (2000)
Science, **289**, 1760–1766.

57 Dive, D. and Biot, C. (2008) *Chem.
Med. Chem.*, **3**, 383–391.

58 (a) Martinez, A., Rajapakse, C.S.K.,
Naoulou, B., Kopkalli, Y., Davenport,
L., and Sanchez-Delgado, R.A. (2008) *J.
Biol. Inorg. Chem.*, **13**, 703–712;
(b) Rajapakse, C.S.K., Martinez, A.,
Naoulou, B., Jarzecki, A.A., Suarez, L.,
Deregnaucourt, C., Sinou, V.,
Schrevel, J., Musi, E., Ambrosini, G.,
Schwartz, G.K., and Sanchez-Delgado,
R.A. (2009) *Inorg. Chem.*, **48**, 1122–
1131.

59 Sharma, V., Beatty, A., Goldberg, D.E.,
and Piwnica-Worms, D. (1997) *Chem.
Commun.*, 2223–2224.

60 Hill, C.L. (2004) *Polyoxometalates:
reactivity*, in *Comprehensive Coordination
Chemistry II*, Vol. **4** (eds J.A.
McCleverty and T.J. Meyer), pp. 679–
759.

61 Yamase, T. (2005) *J. Mater. Chem.*, **15**,
4773–4782.

62 De Clercq, E. (2002) *Med. Res. Rev.*, **22**,
531–565.

63 Judd, D.A., Nettles, J.H., Nevins, N.,
Snyder, J.P., Liotta, D.C., Tang, J.,
Ermolieff, J., Schinazi, R.F., and Hill,
C.L. (2001) *J. Am. Chem. Soc.*, **123**,
886–897.

64 Dan, K. and Yamase, T. (2006) *Biomed.
Pharmacother.*, **60**, 169–173.

65 Liang, X. and Sadler, P.J. (2004) *Chem.
Soc. Rev.*, **33**, 246–266.

66 Bridger, G.J., Skerlj, R.T.,
Padmanabhan, S., Martellucci, S.A.,
Henson, G.W., Struyf, S., Witvrouw,
M., Schols, D., and De Clercq, E.
(1999) *J. Med. Chem.*, **42**, 3971–3981.

67 De Clercq, E. (2009) *Biochem.
Pharmacol.*, **77**, 1655–1664.

68 McRobbie, G., Valks, G.C., Empson, C.J.,
Khan, A., Silversides, J.D.,

Pannecouque, C., De Clercq, E., Fiddy, S.G., Bridgeman, A.J., Young, N.A., and Archibald, S.J. (2007) *Dalton Trans.*, 5008–5018.

69 Khan, A., Nicholson, G., Greenman, J., Madden, L., McRobbie, G., Pannecouque, C., De Clercq, E., Ullom, R., Maples, D.L., Maples, R.D., Silversides, J.D., Hubin, T.J., and Archibald, S.J. (2009) *J. Am. Chem. Soc.*, **131**, 3416–3417.

70 Hunter, T.M., McNae, I.W., Simpson, D.P., Smith A.M., Moggach S., White, F., Walkinshaw, M.D., Parsons S., and Sadler, P.J. (2007) *Chemistry*, **13**, 40–50.

71 Schwartz, J.A., Lium, E.K., and Silverstein, S.J. (2001) *J. Virol.*, **75**, 4117–4128.

72 Hall, M.D., Failes, T.W., Yamamoto, N., and Hambley, T.W. (2007) *Dalton Trans.*, 3983–3990.

73 Takeuchi, T., Böttcher, A., Qeuzada, C. M., Meade, T.J., and Gray, H.B. (1999) *Bioorg. Med. Chem.*, **7**, 815–819.

74 Delehanty, J.B., Bongard, J.E., Thach, D.C., Knight, D.A., Hickey, T.E., and Chang, E.L. (2008) *Bioorg. Med. Chem.*, **16**, 830–837.

75 Scott, L.E. and Orvig, C. (2009) *Chem. Rev.*, **109**, 4885–4910.

76 Thompson, K.H., McNeill, J.H., and Orvig, C. (1999) *Chem. Rev.*, **99**, 2561–2571.

77 Thompson, K.H., Lichter, J., LeBel, C., Scaife, M.C., McNeill, J.H., and Orvig, C. (2009) *J. Inorg. Biochem.*, **103**, 554–558.

78 Hiromura, M., Nakayama, A., Adachi, Y., Doi, M., and Sakurai, H. (2007) *J. Biol. Inorg. Chem.*, **12**, 1275–1287.

79 Claret, M., Corominola, H., Canals, I., Saura, J., Barcelo-Batllori, S., Guinovart, J.J., and Gomis, R. (2005) *Endocrinology*, **146**, 4362–4369.

80 Miró-Queralt, M., Guinovart, J.J., and Planas, J.M. (2008) *Am. J. Physiol. Gastrointest. Liver Physiol.*, **295**, G479–G484.

81 Rodríguez-Fariñas, N., Gomez-Gomez, M.M., and Camara-Rica, C. (2008) *Anal. Bioanal. Chem.*, **390**, 29–35.

82 Wellenreuther, G., Cianci, M., Tucoulou, R., Meyer-Klaucke, W., and Haase, H. (2009) *Biochem. Biophys. Res. Commun.*, **380**, 198–203.

83 Jansen, J., Karges, W., and Rink, L. (2009) *J. Nutr. Biochem.*, **20**, 399–417.

84 Bitanihirwe, B.K.Y. and Cunningham, M.G. (2009) *Synapse*, **63**, 1029–1049.

85 Barnham, K.J. and Bush, A.I. (2008) *Curr. Opin. Chem. Biol.*, **12**, 222–228.

86 Brown, D.R. (2009) *Dalton Trans.*, 4069–4076.

87 Han, F., Shioda, N., Moriguchi, S., Qin, Z.-H., and Fukunaga, K. (2008) *Neuroscience*, **151**, 671–679.

88 Voloshin, Ya.Z., Varzatskii, O.A., and Bubnov, Yu. N. (2007) *Russ. Chem. Bull.*, **56**, 577–605.

89 Bouchard, M., Mergler, D., Baldwin, M.E., and Panisset, M. (2008) *Neurotoxicology*, **29**, 577–583.

90 Rencova, J., Volf, V., Jones, M.M., and Singh, P.K. (1994) *Radiat. Prot. Dosim.*, **53**, 311–313.

91 Fukuda, S., Ikeda, M., Nakamura, M., Yan, X., and Xie, Y. (2009) *Radiat. Prot. Dosim.*, **133**, 12–19.

92 Jiang, H., Luan, Z., Wang, J., and Xie, J. (2006) *Neurochem. Int.*, **49**, 605–609.

93 Barnham, K.J., Gautier, E.C.L., Kok, G.B., and Krippner, G. (2004) Preparation of 8-hydroxyquinolines for treatment of neurological conditions. PCT Int. Appl. WO 2004007461.

94 Bush, A.I. and Tanzi, R.E. (2008) *Neurotherapeutics*, **5**, 421–432.

95 Lannfelt, L., Blennow, K., Zetterberg, H., Batsman, S., Ames, D., Harrison, J., Maters, C.L., Targum, S., Bush, A.I., Murdoch, R., Wilson, J., and Ritchie, C.W. (2008) *Lancet Neurol.*, **7**, 779–786.

96 Relkin, N.R. (2008) *Lancet Neurol.*, **7**, 762–763.

97 Khan, G. and Merajver, S. (2009) *Expert Opin. Investig. Drugs*, **18**, 541–548.

98 Brewer, G.J., Askari, F., Dick, R.B., Sitterly, J., Fink, J.K., Carlson, M., Kluin, K.J., and Lorincz, M.T. (2009) *Transl. Res.*, **154**, 70–77.

99 Hou, G., Dick, R., and Brewer, G.J. (2009) *Exp. Biol. Med.*, **234**, 662–665.

100 Gartner, E.M., Griffith, K.A., Pan, Q., Brewer, G.J., Henja, G.F., Merajver, S.D., and Zalupski, M.M. (2009) *Invest. New Drugs*, **27**, 159–165.

101 Zhang, L., Lichtmannegger, J., Summer, K.H., Webb, S., Pickering, I.J., and George, G.N. (2009) *Biochemistry*, **48**, 891–897.

102 Gromer, S., Arscott, L.D., Williams, C.H. Jr., Schirmer R.H., and Becker, K. (1998) *J. Biol. Chem.*, **273**, 20096–20101.

103 Karver, M.R. and Barrios, A.M. (2008) *Anal. Biochem.*, **382**, 63–65.

104 Brown, C.L., Whitehouse, M.W., Tiekink, E.R.T., and Bushell, G.R. (2008) *Inflammopharmacology*, **16**, 133–137.

105 Salvemini, D., Wang, Z.-Q., Zweier, J.L., Samouilov, A., MacArthur, H., Misko, T.P., Currie, M.G., Cuzzocrea, S., Sikorski, J.A., and Riley, D.P. (1999) *Science*, **286**, 304–306.

106 Murphy, C.K., Fey, E.G., Watkins, B.A., Wong, V., Rothstein, D., and Sonis, S.T. (2008) *Clin. Cancer Res.*, **14**, 4292–4297.

107 Di Napoli, M. and Papa, F. (2005) *IDrugs*, **8**, 67–76.

108 Salvemini, D., Mazzon, E., Dugo, L., Serraino, I., De Sarro, A., Caputi, A.P., and Cuzzocrea, S. (2001) *Arthritis Rheum.*, **44**, 2909–2921.

109 PR Newswire, "Metaphore Pharmaceuticals, Inc. Announces Positive Results of Phase II Clinical Trial; SOD Mimetic M40403 Improved Effectiveness and Predictability of Morphine" 27th April, (2004); http://www.highbeam.com/doc/1G1-115898882.html (accessed 17th September 2010).

110 Di Paola, R., Mazzon, E., Rotondo, F., Dattola, F., Britti, D., De Majo, M., Genovese, T., and Cuzzocrea, S. (2005) *Eur. J. Pharmacol.*, **516**, 151–157.

111 Sharpe, M.A., Ollosson, R., Stewart, V.C., and Clark, J.B. (2002) *Biochem. J.*, **366**, 97–107.

112 Birch, N.J. (1999) *Chem. Rev.*, **99**, 2659–2682.

113 Marmol, F. (2008) *Prog. Neuropsychopharmacol. Biol. Psychiatry*, **32**, 1761–1771.

114 Cruceanu, C., Alda, M., Turecki, G. (2009) *Genome Med.*, **1**, 79.

115 Gill, R., Mohammed, F., Badyal, R., Coates, L., Erskine, P., Thompson, D., Cooper, J., Gore, M., and Wood, S. (2005) *Acta Crystallogr. D. Biol. Crystallogr.*, **61**, 545–555.

116 Simpson, D., Dunn, C., Curran., M., and Goa, K.L. (2003) *Drugs*, **63**, 2127–2156.

117 Ishida, S., Lee, J., Thiele, D.J., and Herskowitz, I. (2002) *Proc. Natl. Acad. Sci. USA*, **99**, 14298–14302.

118 Blair, B.G., Larson, C.A., Safaei, R., and Howell, S.B. (2009) *Clin. Cancer Res.*, **15**, 4312–4321.

119 Safaei, R. and Howell, S.B. (2005) *Crit. Rev. Oncol. Hematol.*, **53**, 13–23.

120 Arnesano, F., Scintilla, S., and Natile, G. (2007) *Angew. Chem. Int. Ed. Engl.*, **46**, 9062–9064.

121 Barnham, K.J., Djuran, M.I., del Socorro Murdoch, P., Ranford, J.D., and Sadler, P.J. (1995) *J. Chem. Soc., Dalton Trans.*, 3721–3726.

122 Kasherman, Y., Sturup, S., and Gibson, D. (2009) *J. Biol. Inorg. Chem.*, **14**, 387–399.

123 Arnesano, F. and Natile, G. (2009) *Coord. Chem. Rev.*, **253**, 2070–2081.

124 Zhu, G. and Lippard, S.J. (2009) *Biochemistry*, **48**, 4916–4925.

125 Lovejoy, K.S., Todd, R.C., Zhang, S., McCormick, M.S., D'Aquino, J.A., Reardon, J.T., Sancar, A., Giacomini, K.M., and Lippard, S.J. (2008) *Proc. Natl. Acad. Sci. USA*, **105**, 8902–8907.

126 Kabolizadeh, P., Ryan, J., and Farrell, N. (2007) *Biochem. Pharmacol.*, **73**, 1270–1279.

127 Aris, S.M. and Farrell, N.P. (2009) *Eur. J. Inorg. Chem.*, 1293–1302.

128 Natile, G. and Coluccia, M. (2001) *Coord. Chem. Rev.*, **216–217**, 383–410.

129 Feazell, R.P., Nakayama-Ratchford, N., Dai, H., and Lippard, S.J. (2007) *J. Am. Chem. Soc.*, **129**, 8438–8439.

130 Haxton, K.J. and Burt, H.M. (2009) *J. Pharm. Sci.*, **98**, 2299–2316.

131 Dhar, S., Gu, F.X., Langer, R., Farokhzad, O.C., and Lippard, S.J. (2008) *Proc. Natl. Acad. Sci. USA*, **105**, 17356–17361.

132 Bednarski, P.J., Mackay, F.S., and Sadler, P.J. (2007) *Anti-Cancer Agents Med. Chem.*, **7**, 75–93.

133 (a) Mackay, F.S., Woods, J.A., Heringová, P., Kašparková, J., Pizzaro, A.M., Moggach, S.A., Parsons, S., Brabec V., and Sadler, P.J. (2007) *Proc. Natl. Acad. Sci. USA*, **104**, 20743–20748; (b) N.J. Farrer, N.J., Woods, J.A., Salassa, L., Zhao, Y., Robinson, K.S., Clarkson, G., Mackay, F.S. and Sadler P.J. (2010) *Angew. Chem. Int. Ed.* doi 10.1002/anie.201003399.

134 Ronconi, L. and Sadler, P.J. (2008) *Chem. Commun.*, 235–237.

135 Farrer, N.J. and Sadler, P.J. (2008) *Aust. J. Chem.*, **61**, 669–674.

136 Bergamo, A., Gagliardi, R., Scarcia, V., Furlani, A., Alessio, E., Mestroni, G., and Sava, G. (1999) *J. Pharmacol. Exp. Ther.*, **289**, 559–564.

137 Lentz, F., Drescher, A., Lindauer, A., Henke, M., Hilger, R.A., Hartinger, C. G., Scheulen, M.E., Dittrich, C., Keppler, B.K., and Jaehde, U. (2009) *Anticancer Drugs*, **20**, 97–103.

138 Smalley, K.S.M., Contractor, R., Haass, N.K., Kulp, A.N., Atilla-Gokcumen, G. E., Williams, D.S., Bregman, H., Flaherty, K.T., Soengas, M.S., Meggers, E., and Herlyn, M. (2007) *Cancer Res.*, **67**, 209–217.

139 Bergamo, A., Messori, L., Piccioli, F., Cocchietto, M., and Sava, G. (2003) *Invest. New Drugs.*, **21**, 401–411.

140 Maksimoska, J., Feng, L., Harms, K., Yi, C., Kissil, J., Marmorstein, R., and Meggers, E. (2008) *J. Am. Chem. Soc.*, **130**, 15764–15765.

141 Alessio, E., Mestroni, G., Bergamo, A., and Sava, G. (2004) *Curr. Top. Med. Chem.*, **4**, 1525–1535.

142 Morris, R.E., Aird, R.E., Murdoch, P. del-S., Chen, H., Cummings, J., Hughes, N.D., Parsons, S., Parkin, A., Boyd, G., Jodrell, D.I., and Sadler, P.J. (2001) *J. Med. Chem.*, **44**, 3616–3621.

143 Johnson, T.R., Mann, B.E., Teasdale, I. P., Adams, H., Foresti, R., Green, C.J., and Motterlini, R. (2007) *Dalton Trans.*, 1500–1508.

144 Delaney, S., Pascaly, M., Bhattacharya, P.K., Han, K., and Barton, J.K. (2002) *Inorg. Chem.*, **41**, 1966–1974.

145 Marmion, C.J., Cameron, B., Mulcahy, C., and Fricker, S.P. (2004) *Curr. Top. Med. Chem.*, **4**, 1585–1603.

146 Sterling, C. (1970) *Am. J. Bot.*, **57**, 172–175.

147 Belous, A., Knox, C., Nicoud, I.B., Pierce, J., Anderson, C., Pinson, C.W., and Chari, R.S. (2003) *J. Surg. Res.*, **111**, 284–289.

148 Linnett, P.J. and Hughes, E.G. (1999) *Occup. Environ. Med.*, **56**, 191–196.

149 Gibson, D. (2009) *Dalton Trans*, 10681–10689.

150 Habtemariam, A. and Sadler, P.J. (1996) *Chem. Commun.*, 1785–1786.

151 Ma, Z., Choudhury, J.R., Wright, M. W., Day, C.S., Saluta, G., Kucera, G.L., and Bierbach, U. (2008) *J. Med. Chem.*, **51**, 7574–7580.

152 Barnes, K.R., Kutikov, A., and Lippard, S.J. (2004) *Chem. Biol.*, **11**, 557–564.

153 Ang, W.H., Khalaila, I., Allardyce, C.S., Juillerat-Jeanneret, L., and Dyson, P.J. (2005) *J. Am. Chem. Soc.*, **127**, 1382–1383.

154 Komeda, S., Moulaei, T., Woods, K.K., Chikuma, M., Farrell, N.P., and Williams, L.D. (2006) *J. Am. Chem. Soc.*, **128**, 16092–16103.

155 (a) Farrell, N.P. (2004) *Semin Oncol.*, **31**, 1–9; (b) Billecke, C., Finniss, S., Tahash, L., Miller, C., Mikkelsen, T., Farrell, N.P., and Bögler, O. (2006) *Neuro Oncol.*, **8**, 215–226.

156 Fisher, D.M., Fenton, R.R., and Aldrich-Wright, J.R. (2008) *Chem. Commun.*, 5613–5615.

157 Teletchea, S., Komeda, S., Teuben, J.-M., Elizondo-Riojas, M.-A., Reedijk, J., and Kozelka, J. (2006) *Chem. Eur. J.*, **12**, 3741–3753.

158 Best, S.L. and Sadler, P.J. (1996) *Gold Bull.*, **29**, 87–93.

159 Graham, G.G., Bales, J.R., Grootveld, M.C., and Sadler, P.J. (1985) *J. Inorg. Biochem.*, **25**, 163–173.

160 Rackham, O., Nichols, S.J., Leedman, P.J., Berners-Price, S.J., and Filipovska, A. (2007) *Biochem. Pharm.*, **74**, 992–1002.

161 Baker, M.V., Barnard, P.J., Berners-Price, S.J., Brayshaw, S.K., Hickey, J.L., Skelton, B.W., and White, A.H. (2006). *Dalton Trans.*, 3708–3715.

162 Milacic, V., Chen, D., Ronconi, L., Landis-Piwowar, K.R., Fregona, D., and Dou, Q.P. (2006) *Cancer Res.*, **66**, 10478–10486.

163 (a) Hartinger, C.G., Zorbas-Seifried, S., Jakupec, M.A., Kynast, B., Zorbas, H., and Keppler, B.K. (2006) *J. Inorg. Biochem.*, **100**, 891–904; (b) Frasca, D., Ciampa, J., Emerson, J., Umans, R.S., and Clarke, M.J. (1996) *Met. Based Drugs*, **3**, 197–209.

164 Habtemariam, A., Melchart, M., Fernández, R., Parsons, S., Oswald, I. D.H., Parkin, A., Fabbiani, F.P.A., Davidson, J.E., Dawson, A., Aird, R.E., Jodrell, D.I., and Sadler, P.J. (2006) *J. Med. Chem.*, **49**, 6858–6868.

165 Dougan, S.J. and Sadler, P.J. (2007) *CHIMIA Int. J. Chem.*, **61**, 704–715.

166 Bruijnincx, P.C.A. and Sadler, P.J. (2009) *Adv. Inorg. Chem.*, **61**, 1–62.

167 Melchart, M., Habtemariam, A., Novakova, O., Moggach, S.A., Fabbiani, F.P.A., Parsons, S., Brabec, V., and Sadler, P.J. (2007) *Inorg. Chem.*, **46**, 8950–8962.

168 Scolaro, C., Bergamo, A., Brescacin, L., Delfino, R., Cocchietto, M., Laurenczy, G., Geldbach, T.J., Sava, G., and Dyson, P.J. (2005) *J. Med. Chem.*, **48**, 4161–4171.

169 Liu, H.-K., Berners-Price, S.J., Wang, F., Parkinson, J.A., Xu, J., Bella, J., and Sadler, P.J. (2006) *Angew. Chem. Int. Ed.*, **45**, 8153–8156.

170 Bugarcic, T., Nováková, O., Halámiková, A., Zerzánková, L., Vrána, O., Kašpárková, J., Habtemariam, A., Parsons, S., Sadler, P.J., and Brabec, V. (2008) *J. Med. Chem.*, **51**, 5310–5319.

171 (a) Foster, R.E., Cole, D.A., Mead, S., Sadler, P.J., and Grimshaw, K.M., (2009) *Proc. Am. Assoc. Cancer Res.*, **18–22**, 889. (b) Aird, R.E., Cummings, J., Ritchie, A.A., Muir, M., Morris, R.E., Chen, H., Sadler, P.J., and Jodrell, D.I. (2002) *Br. J. Cancer*, **86**, 1652–1657.

172 Bugarcic, T., Habtemariam, A., Stepankova, J., Heringová, P., Kašpárková, J., Deeth, R.J., Johnstone, R.D.L., Prescimone, A., Parkin, A., Parsons, S., Brabec, V., and Sadler, P.J. (2008) *Inorg. Chem.*, **47**, 11470–11486.

173 Dougan, S.J., Habtemariam, A., McHale, S.E., Parsons, S., and Sadler,

P.J. (2008) *Proc. Natl. Acad. Sci. USA*, **105**, 11628–11633.

174 Sriskandakumar, T., Petzold, H., Bruijnincx, P.C.A., Habtemariam, A., Sadler, P.J., and Kennepohl, P. (2009) *J. Am. Chem. Soc.*, **131**, 13355–13361.

175 Wang, F., Xu, J., Habtemariam, A., Bella, J., and Sadler, P.J. (2005) *J. Am. Chem. Soc.*, **127**, 17734–17743.

176 Petzold, H., Xu, J., and Sadler, P.J. (2008) *Angew. Chem. Int. Ed.*, **47**, 3008–3011.

177 Peacock, A.F.A. and Sadler, P.J. (2008) *Chem. Asian J.*, **3**, 1890–1899.

178 van Rijt, S.H., Peacock, A.F.A., Johnstone, R.D.L., Parsons, S., and Sadler, P.J. (2009) *Inorg. Chem.*, **48**, 1753–1762.

179 Kostrhunova, H., Florian, J., Novakova, O., Peacock, A.F.P., Sadler, P.J., and Brabec, V. (2008) *J. Med. Chem.*, **51**, 3635–3643.

180 van Rijt, S.H., Hebden, A.J., Amarasekera, T., Deeth, R.J., Clarkson, G.J., Parsons, S., McGowan, P.C., and Sadler, P.J. (2009) *J. Med. Chem.*, **52**, 7753–7764.

181 Schmid, W.F., John, R.O., Arion, V.B., Jakupec, M.A., and Keppler, B.K. (2007) *Organometallics*, **26**, 6643–6652.

182 (a) Dorcier, A., Ang, W.H., Bolano, S., Gonsalvi, L., Juillerat-Jeanneret, L., Laurenczy, G., Peruzzini, M., Phillips, A.D., Zanobini, F., and Dyson, P.J. (2006) *Organometallics*, **25**, 4090–4096; (b) Dorcier, A., Dyson, P.J., Gossens, C., Rothlisberger, U., Scopelliti, R., and Tavernelli, I. (2005) *Organometallics*, **24**, 2114–2123.

183 Top, S., Kaloun, E.B., Vessieres, A., Laios, I., Leclercq, G., and Jaouen, G. (2002) *J. Organomet. Chem.*, **643–644**, 350–356.

184 Guo, M., Sun, H., McArdle, H.J., Gambling, L., and Sadler, P.J. (2000) *Biochemistry*, **39**, 10023–10033.

185 Tinoco, A.D., Eames, E.V., and Valentine, A.M. (2008) *J. Am. Chem. Soc.*, **130**, 2262–2270.

186 Tshuva, E.Y. and Ashenhurst, J.A. (2009) *Eur. J. Inorg. Chem.*, **15**, 2203–2218.

187 Allen, O.R., Croll, L., Gott, A.L., Knox, R.J., and McGowan, P.C. (2004) *Organometallics*, **23**, 288–292.

188 Fichtner, I., Pampillón, C., Sweeney, N.J., Strohfeldt, K., and Tacke, M. (2006) *Anticancer Drugs*, **17**, 333–336.

189 Claffey, J., Hogan, M., Müller-Bunz, H., Pampillón, C., and Tacke, M. (2008) *Chem. Med. Chem.*, **3**, 729–731.

190 Berners-Price, S.J., Mirabelli, C.K., Johnson, R.K., Mattern, M.R., McCabe, F.L., Faucette, L.F., Sung, C.M., Mong, S.M., Sadler, P.J., and Crooke, S.T. (1986) *Cancer Res.*, **46**, 5486–5493.

191 Liu, J.J., Galettis, P., Farr, A., Maharaj, L., Samarasinha, H., McGechan, A.C., Baguley, B.C., Bowen, R.J., Berners-Price, S.J., and McKeage, M.J. (2008) *J. Inorg. Biochem.*, **102**, 303–310.

192 Hickey, J.L., Ruhayel, R.A., Barnard, P.J., Baker, M.V., Berners-Price, S.J., and Filipovska, A. (2008) *J. Am. Chem. Soc.*, **130**, 12570–12571.

193 Aldinucci, D., Ronconi, L., and Fregona, D. (2009) *Drug Discov. Today*, **14**, 1075–1076.

194 Tu, S., Sun, R.W.-Y., Lin, M.C.M., Cui, T.J., Zou, B., Gu, Q., Kung H.-F., Che, C.-M., and Wong, B.C.Y. (2009) *Cancer*, **115**, 4459–4469.

195 Che, C.-M., Sun, R.W.-Y., Yu, W.-Y., Ko, C.-B., Zhu, N., and Sun, H. (2003) *Chem. Commun.*, 1718–1719.

196 (a) Alama, A.,Viale, M., Cilli, M., Bruzzo, C., Novelli, F., Tasso, B., and Sparatore, F. (2009) *Invest. New Drugs*, **27**, 124–130
(b) Alama A., Tasso B., Novelli F., and Sparatore, F. (2009) *Drug Discovery Today*, **14**, 500–508.

197 Hadjikakou, S.K. and Hadjiliadis, N. (2009) *Coord. Chem. Rev.*, **253**, 235–249.

198 Bandoli, G., Dolmella, A., Tisato, F., Porchia, M., and Refosco, F. (2009) *Coord. Chem. Rev.*, **253**, 56–77.

199 Jakupec, M.A. and Keppler, B.K. (2004) *Met. Ions Biol. Syst.*, **42**, 425–462.

200 Valiahdi, S.M., Heffeter, P., Jakupec, M.A., Marculescu, R., Berger, W., Rappersberger, K., and Keppler, B.K. (2009) *Melanoma Res.*, **19**, 283–293.

201 Chen, D., Frezza, M., Shakya, R., Cui, Q.C., Milacic, V., Verani, C.N., and Dou, Q.P. (2007) *Cancer Res.*, **67**, 9258–9265.

202 DeLeon, K., Balldin, F., Watters, C., Hamood, A., Griswold, J., Sreedharan, S., and Rumbaugh, K.P. (2009) *Antimicrobl. Agents Chemother.*, **53**, 1331–1337.

203 Yedjou, C., Tchounwou, P., Jenkins, J., McMurray, R. (2010) *J. Hematol. Oncol.* 3, doi:10.1186/1756-8722-3-28

204 Ramírez-Solís, A., Mukopadhyay, R., Rosen, B.P., and Stemmler, T.L. (2004) *Inorg. Chem.*, **43**, 2954–2959.

205 Ni Dhubhghaill, O.M. and Sadler, P.J. (1991) *Struct. Bonding*, **78**, 129–190.

206 Bhattacharjee, H., Rosen, B.P., and Mukhopadhyay, R. (2009), in *Aquaporins* (ed. E. Beitz) Handbook of Experimental Pharmacology, Vol. **190**, pp. 309–325.

207 Matulis, S.M., Morales, A.A., Yehiayan, L., Croutch, C., Gutman, D., Cai, Y., Lee, K.P., and Boise, L.H. (2009) *Mol. Cancer Ther.*, **8**, 1197–1206.

208 Chen, H., MacDonald, R.C., Li, S., Krett, N.L., Rosen, S.T., and O'Halloran, T.V. (2006) *J. Am. Chem. Soc.*, **128**, 13348–13349.

209 Chen, H., Ahn, R., Van den Bossche, J., Thompson, D.H., O'Halloran, T.V. (2009) *Mol. Cancer Ther.*, **8**, 1955–1963.

210 Marzano, C., Pellei, M., Tisato, F., and Santini, C. (2009) *Anticancer Agents Med. Chem.*, **9**, 185–211.

211 Tardito, S., and Marchiò, L. (2009) *Curr. Med. Chem.*, **16**, 1325–1348.

212 Tisato, F., Marzano, C., Porchia, M., Pellei, M., and Santini, C. (2010) *Med. Res. Rev.*, **30**, 708–749.

213 Bravo-Gómez, M.E., García-Ramos, J.C., Gracia-Mora, I., and Ruiz-Azuara, L. (2009) *J. Inorg. Biochem.*, **103**, 299–309.

214 Hindo, S.S., Frezza, M., Tomco, D., Heeg, M.J., Hryhorczuk, L., McGarvey, B.R., Dou, Q.P., and Verani, C.N. (2009) *Eur. J. Med. Chem.*, **44**, 4353–4361.

215 Magda, D., Lecane, P., Wang, Z., Hu, W., Thiemann, P., Ma, X., Dranchak, P.K., Wang, X., Lynch, V., Wei, W., Csokai, V., Hacia, J.G., and Sessler, J.L. (2008) *Cancer Res.*, **68**, 5318–5325.

216 Li, H., Lai, C.S., Wu, J., Ho, P.C., de Vos, D., and Tiekink, E.R.T. (2007) *J. Inorg. Biochem.*, **101**, 809–816.

217 Ogata, A., Yanagie, H., Ishikawa, E., Morishita, Y., Mitsui, S., Yamashita, A., Hasumi, K., Takamoto, S., Yamase, T.,

and Eriguchi, M. (2008) *Br. J. Cancer*, **98**, 399–409.

218 Sommer, S., Rimington, C., and Moan, J. (1984) *FEBS Lett.*, **172**, 267–271.

219 Bonnett, R. (2004) Metal complexes for photodynamic therapy, in *Comprehensive Coordination Chemistry II* (ed. M. McCleverty), Elsevier Pergamon, Oxford, p. 945.

220 Szaciowski, K., Macyk, W., Drewiecka-Matuszek, A., Brindell, M., and Stochel, G. (2005) *Chem. Rev.*, **105**, 2647–2694.

221 (a) Mody, T.D. and Sessler, J.L. (2001) *J. Porphyrins Phthalocyanines*, **5**, 134–142; (b) Detty, M.R., Gibson, S.L., and Wagner, S.J. (2004) *J. Med. Chem.*, **47**, 3897–3915; (c) Kostenich, G., Babushkina, T., Lavi, A., Langzam, Y., Malik, Z., Orenstein, A., and Ehrenberg, B. (1998) *J. Porphyrins Phthalocyanines*, **2**, 383–390.

222 Sharman, W.M., Allen, C.M., and van Lier, J.E. (1999) *Drug Discov. Today*, **4**, 507–517.

223 Roy, M., Saha, S., Patra, A.K., Nethaji, M., and Chakravarty, A.R. (2007) *Inorg. Chem.*, **46**, 4368–4370.

224 Pacher, P., Beckman, J.S., and Liaudet, L. (2007) *Physiol. Rev.*, **87**, 315–424.

225 Muscoli, C., Cuzzocrea, S., Ndengele, M.M., Mollace, V., Porreca, F., Fabrizi, F., Esposito, E., Masini, E., Matuschak, G.M., and Salvemini, D. (2007) *J. Clin. Invest.*, **117**, 3530–3539.

226 Bian, K. and Murad, F. (2003) *Front. Biosci.*, **8**, D264–D278.

227 Fricker, S.P. (2004) *Met. Ions Biol. Syst.*, **41**, 421–480.

228 Ford, P.C. (2008) *Acc. Chem. Res.*, **41**, 190–200.

229 Holanda, A.K.M., da Silva, F.O.N., Sousa, J.R., Diogenes, I.C.N., Carvalho, I.M.M., Moreira, I.S., Clarke, M.J., and Lopes, L.G.F. (2008) *Inorg. Chim. Acta*, **361**, 2929–2933.

230 Cameron, B.R., Darkes, M.C., Yee, H., Olsen, M., Fricker, S.P., Skerlj, R.T., Bridger, G.J., Davies, N.A., Wilson, M.T., Rose, D.J., and Zubieta, J. (2003) *Inorg. Chem.*, **42**, 1868–1876.

231 Johnson, T.R., Mann, B.E., Clark, J.E., Foresti, R., Green, C.J., and Motterlini, R. (2003) *Angew. Chem. Int. Ed.*, **42**, 3722–3729.

232 Motterlini, R., Mann, B.E., and Foresti, R. (2005) *Expert Opin. Investig. Drugs.*, **14**, 1305–1318.

233 Niesel, J., Pinto, A., Peindy N'Dongo, H.W., Merz, K., Ott, I., Gust, R., and Schatzschneider, U. (2008) *Chem. Commun.*, 1798–1800.

234 Alberto, R., Pak, J.K., van Staveren D., Mundwiler, S., and Benny, P. (2004) *Biopolymers*, **76**, 324–333.

235 Skjold, A., Amundsen, B.H., Wiseth, R., Støylen, A., Haraldseth, O., Larsson, H.B.W., and Jynge, P. (2007) *J. Magn. Reson. Imaging.*, **26**, 720–727.

236 Tysiak, E., Asbach, P., Aktas, O., Waiczies, H., Smyth, M., Schnorr, J., Taupitz, M., and Wuerfel, J. (2009) *J. Neuroinflammation*, **6**, 20.

237 Datta, A. and Raymond, K.N. (2009) *Acc. Chem. Res.*, **42**, 938–947.

238 Weder, J.E., Hambley, T.W., Kennedy, B.J., Lay, P.A., MacLachlan, D., Bramley, R., Delfs, C.D., Murray, K.S., Moubaraki, B., Warwick, B., Biffin, J.R., and Regtop, H.L. (1999) *Inorg. Chem.*, **38**, 1736–1744.

239 Lloyd, N.C., Morgan, H.W., Nicholson, B.K., and Ronimus, R.S. (2008) *J. Organomet. Chem.*, **693**, 2443–2450.

240 Silbergeld, E.K. and Nachman, K. (2008) *Ann. N. Y. Acad. Sci.*, **1140**, 346–357.

2
Targeting Strategies for Metal-Based Therapeutics

Julia F. Norman and Trevor W. Hambley

2.1
Introduction

With the notable exception of platinum anticancer drugs, metal-based therapeutics occupy a relatively minor place in the organic dominated history of drug development. Despite platinum chemotherapeutics being used in more treatment regimes than any other class of anticancer drugs [1], there still exists a stigma surrounding the toxicity of metal-based drugs. To this end, medicinal inorganic chemistry has shifted focus away from the cytotoxic "shotgun" therapeutics and towards the targeted "magic bullet" paradigm. In theory, targeting physiological and molecular signatures of a disease should result in higher selectivity and fewer toxic side effects than conventional broad spectrum and cytotoxic drugs. However, one drawback of designing therapeutics with a single target and mechanism of action is the increased likelihood of acquired resistance to the drug developing as influx mechanisms are switched off or exhausted and efflux mechanisms are promoted. Combinative therapy with other agents is therefore likely to emerge as a more prominent therapeutic strategy in the future.

This chapter will address the various targeting strategies employed in the design of metal-based drugs, ranging from established paradigms based on the physical features of diseases to newer concepts that have emerged as knowledge about the molecular signatures of diseases are elucidated.

2.2
Physiological Targeting

Exploiting the physical differences between the morphology of diseased and healthy cells, tissues, and structures is a well established therapeutic strategy. There are essentially three levels of physical targeting considered in the design of therapeutic agents: the first involves the localization of the drug to the desired site within the body, followed by targeting the tissues and cells of interest, then transport across the cell membrane and direction to agent's site of action within

Bioinorganic Medicinal Chemistry. Edited by Enzo Alessio
Copyright © 2011 WILEY-VCH Verlag GmbH & Co. KGaA, Weinheim
ISBN: 978-3-527-32631-0

the cell. Metal complexes, with their ability to coordinate a number of ligands with complementary structures and functions, offer the potential to change the physicochemical properties of the drug to optimize all three stages of targeting. Chemotherapeutic agents used in the treatment of solid tumors are perhaps the most compelling and most thoroughly investigated examples of physiologically targeted agents and for this reason they are the focus of this section.

2.2.1
Antitumor Drugs Targeting Tumor Hypoxia

Tumor hypoxia refers to the state of diminished oxygen concentration (typically <3 μM) found within solid tumors. The dominating factors contributing to chronic tumor hypoxia are the poor and irregular development of vascular tissue and the increased interstitial pressure within the tumor. Transient hypoxia can also result from temporary shutdown of blood vessels [2]. The irregular vascular structure observed in tumors results in a significant proportion of cells being located >100 μm from any blood vessels [3, 4], causing heterogeneous delivery of oxygen and other nutrients within the tumor and localized regions of anoxia and necrosis [5]. Increased interstitial pressure within tumors can be attributed to vascular leakiness, interstitial fibrosis, and increased contractility of the interstitial space mediated by stromal fibroblasts [6] and contributes to limited diffusion of oxygen and other nutrients to some regions of the tumor. Figure 2.1 shows these features of solid tumors.

Hypoxic cells in tumors tend to be inherently resistant to chemotherapy due to the upregulation of genes implicated in drug resistance, including the

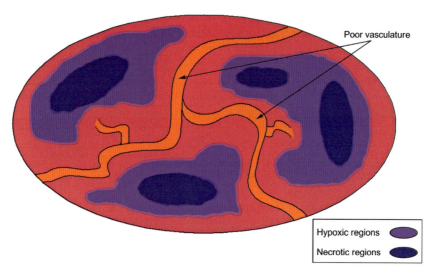

Figure 2.1 Physiology of hypoxia in a solid tumor. Figure adapted from Reference [7].

transcription factor hypoxia inducible factor (HIF)-1 [8]. HIFs in general are involved in the regulation of proliferation, angiogenesis, and metastasis and specifically HIF-1 regulates the multidrug resistance transporter p-glycoprotein [9]. HIF-1 also regulates apoptosis, reducing the effectiveness of therapeutic agents that act by inducing apoptosis [8]. In addition, the decreasing ratio of proliferating to non-proliferating cells in increasingly hypoxic areas of a tumor limits the effectiveness of antitumor drugs that are mainly effective against rapidly proliferating cells [10]. Drugs that generate DNA-damaging free radicals only when in the presence of oxygen are also less effective in hypoxic regions where the cells survive. Oxygen-deprived cells also exhibit resistance to radiotherapy, which is similarly reliant on the generation of oxygen radicals. In normoxic environments, DNA free radicals generated by ionizing radiation are fixed by reaction with oxygen resulting in double strand breaks and cell death. However, in hypoxic cells, the lack of O_2 allows the DNA free radicals to return to their original form via reduction by thiol containing compounds [11].

Targeting the normoxic-hypoxic dichotomy in cancer therapy has the potential to select for tumor tissue over healthy tissue and as such minimize problems regarding systemic toxicity and acquired resistance. Metal-based drugs have the potential to select for hypoxic tissue through administration as an inert, oxidized complex that is activated upon reduction in the oxygen deprived environment. Figure 2.2 shows the possible mechanisms by which this occurs. In normal cells, reduction by endogenous reductases or another reducing agent would be reversible due to reoxidation by transfer of the free electron to available O_2. However, in a hypoxic environment reoxidation is less likely and the labilized form of the drug accumulates. In the case of metal complexes, this manifests as the reduction of the metal center from a higher oxidation state to a lower one, which can involve the release of at least one attached ligand. There are two groups of

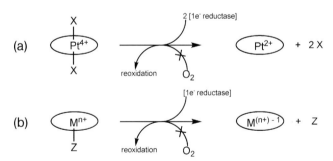

Figure 2.2 Example of reduction *in vivo* by endogenous single electron reductases. (a) The Pt(IV) complex is reduced to Pt(II) and loses the two axial ligands (X). The Pt(II) complex is typically cytotoxic through its reaction with DNA. (b) The metal center (oxidation state $n+$) is reduced to release a single cytotoxic ligand (Z). Here (where M = Co, Ru, etc.) the metal complex acts as a chaperone by deactivating the cytotoxic ligand until it is released.

hypoxia activated metal complexes as shown above: those that lose ligands to yield an active complex or metal ion (i) and those that act as chaperones for delivery of active ligand upon reduction, known as tumor activated prodrugs (TAPs) (ii).

There are various ways in which the leaving ligands can modify the physico-chemical properties of the metal complex. The first is that they can modify the lipophilicity of the complex to control cell membrane permeability and uptake via passive diffusion. Following influx of the compound it is reduced in the hypoxic intracellular environment and the active drug is released. This mechanism of drug uptake has the potential to circumvent drug resistance as it removes reliance on energy dependent drug uptake which is depleted in drug resistant cell lines. Despite being sound in theory, this hypothesis has not been supported by *in vitro* studies, with poor correlation between lipophilicity and cellular accumulation observed across platinum [12–14] and gold [15] complexes. This is with the exception of Kelland and coworkers, who achieved high *in vitro* cytotoxicity of the butyrate analogs of the Pt(IV) prodrug JM221, which was attributed to the highly lipophilic nature of the compounds [16]. Figure 2.3a shows the structure of JM221.

The leaving ligand(s) can also be exploited to alter the redox potential and the rate of reduction of the metal complex. There has been extensive research into the correlation between reduction potential and rate of reduction for platinum(IV) prodrugs, and to a lesser degree Co(III) [17] and Ru(III) complexes. Hambley and coworkers have demonstrated that, in the case of Pt(IV), complexes with halide ions in the axial positions are most readily reduced, followed by those with carboxylato and hydroxido ligands respectively [18, 19]. They have also demonstrated the stabilizing effect of polyfluoroaryl equatorial ligands [20] and concluded that complexes that are very difficult to reduce can achieve toxicities similar to that of the parent Pt(II) compound. It has been established that a correlation between E_p and reduction rate is dependent on the electron-withdrawing power of the co-ligand(s) but there is only minimal evidence of a relationship between E_p and cytotoxicity, as shown in several studies [20–22].

In the case of TAPs, the leaving ligand(s) exert their own cytotoxic effects on tumor cells following reduction and employ the metal center merely as a carrier and deactivator [23]. These metal chaperones tether biologically active molecules and as a result can alter the pharmacokinetics, biodistribution, and biotransformation

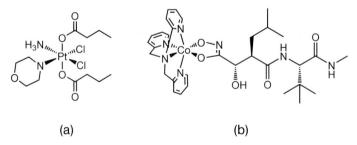

(a)   (b)

Figure 2.3 Hypoxia selective metal complexes: (a) JM22; (b) [Co(mmst)(tpa)].

[23] of the original compounds. Some successful examples of this strategy include the synthesis of Co(III) complexes of nitrogen mustards by Ware and coworkers [24]. Coordinating the mustard ligand to the Co(III) center through the electron lone pair on the nitrogen atom mutes the toxicity of the mustard as the nucleophilic lone pair is unavailable, and gives rise to hypoxia selective toxicity. As a result, the Co(III) mustard complexes were up to 20 times more toxic in hypoxic cells than healthy normoxic cells.

The localized ligand release and cellular uptake of Co(III) complexes has been monitored via confocal microscopy of complexes labeled with fluorescent compounds [25]. [Co(c343)$_2$-(cyclam)]ClO$_4$ and [Co(c343ha)(tpa)]ClO$_4$ – where c343 = coumarin-343, c343ha = deprotonated coumarin-343 hydroxamic acid, cyclam = 1,4,8,11-tetraazacyclotetradecane, and tpa = tris(methylpyridyl)amine – complexes were administered to cells that were also treated with the nuclear acid stain SYTO21. The results are shown in Figure 2.4. [Co(c343)$_2$(cyclam)]ClO$_4$ was found to be much more cytotoxic than both [Co(c343ha)(tpa)]ClO$_4$ and free c343haH$_2$. This corresponded with the degree and localization of observed fluorescence. The ligand release of c343 was believed to have occurred not by a redox cycling mechanism, but rather ligand displacement by reducing species such as cysteine and ascorbate.

Selective delivery of the matrix metalloproteinase (MMP) inhibitor marimastat (mmst), through chelation to chaperone Co(III)-tpa and Fe(III)-salen complexes (where salen = *N,N*-bis(salicylidene)-ethane-1,2-diimine), has been investigated by Failes *et al.* [26–28]. The mechanistic principle behind these complexes is that the hydroxamate moiety on mmst is protected by chelation with the metal center, and liberation of mmst upon hypoxic reduction yields the active MMP inhibitor. The Co(mmst)(tpa) system, depicted in Figure 2.3b, displayed a higher degree of tumor growth inhibition than lone mmst, whereas the Fe(mmst)(salen) system is yet to undergo biological testing.

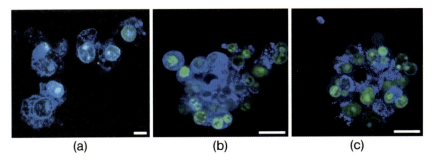

(a)   (b)   (c)

Figure 2.4 Images of A2780 ovarian cancer cells following incubation for 4–5 h. Blue = fluorescence from complex or ligand (405 nm excitation), green = fluorescence from SYTO21 (excitation 488 nm). Co-localization of the complex/ligand with the nucleus appears aqua in color. Scale bar represents 20 μm. (a) Cells treated with [Co(c343)$_2$(cyclam)]ClO$_4$: co-localization is apparent; (b) cells treated with c343haH$_2$; (c) cells treated with [Co(c343ha)(tpa)]ClO$_4$.

Reproduced with kind permission from Springer Science + Business Media.

2.2.2
Antitumor Drugs Targeting Vascular Structure

There are marked differences in the vascular structure and physiology in tumor tissue compared to healthy tissue. The processes of angiogenesis and vasculogenesis in healthy tissue give rise to hierarchically structured vessels [11] capable of delivering oxygen and micronutrients to all cells. Contrastingly, the vascular network in solid tumors is arranged erratically and has arteriovenous shunts, blind ends, a lack of smooth muscle or enervation, and incomplete endothelial linings and basement membranes [2]. This manifests as languid and variable blood flow and localized leakiness of the blood vessels through widening of the fenestrae in the range of 100 to 600 μm [29]. These features may be exploited for targeting solid tumors by taking advantage of size-selective accumulation. Molecules such as serum albumin [30] and liposomes [31–33] have been exploited as carriers for antitumor agents as they are dimensionally similar to macromolecules. Depending on size, extravasation of drug molecules can occur via convective transport through the defective endothelial cells, resulting in selective delivery of the drug to the tumor interstitium. The poor, and in some cases non-existent, lymphatic system in tumors also contributes in part to the extracellular accumulation of the drug [32], resulting in what is known as the enhanced permeability and retention (EPR) effect.

Encapsulation of cisplatin in liposomal carriers has had some success, with the 100 nm diameter SPI-077 formulation by Newman and coworkers exhibiting high drug-tumor efficacy, reduced systemic toxicity, and prolonged circulation in the bloodstream [34, 35], but showed no antitumor activity in humans [36–38]. More recently, a liposomal encapsulation of cisplatin with a diameter of 17 μm showed increased therapeutic efficacy *in vivo* compared to cisplatin alone [39]. These results, amongst others not detailed here, have supported the viability of exploiting the EPR effect for targeted tumor therapy using nanocarriers to transport metal-based cytotoxics.

2.2.3
Antitumor Drugs Targeting Tumor pH

The extracellular pH of tumor tissue is significantly lower than that of surrounding healthy tissue, with a pH range of 5.4–7.4 [40–46]. The morphological characteristics of pH levels in mice brains have been recently imaged using ^{13}C magnetic resonance imaging [47], revealing a marked difference between the interstitial pHs of healthy and solid tumor tissue, with pHs of 7.3 and 6.0 respectively. The acidic environment was originally attributed to the accumulation of lactic acid as a product of glycolytic respiration in hypoxic regions. However, more recent work has demonstrated that the hydrolysis of ATP in oxidative respiration has a significant effect on interstitial pH level due to the build up of CO_2 and carbonic acid by-products [40, 48, 49], which are not removed efficiently due to the poor vascular and lymphatic networks and increased interstitial pressure. The intracellular pH of tumor cells is known to be slightly above neutral, which is consistent with normal healthy cells [42]. The interstitial pH gradient existing between healthy and tumor tissue can be exploited by metal complexes that are inert at an interstitial pH of 7.4 but become cytotoxic, by various mechanisms, following protonation in acidic surroundings. These complexes can potentially improve the specificity of treatment by becoming more labile at the tumor site.

Investigations into pH dependent Pt(II) *N*-hydroxyalkyl prodrugs [50–54] have revealed limited success in terms of cytotoxicity. The activity of these complexes is reliant on their protonation, subsequent ring opening and exposure of the platinum center to react with external ligands, in this case those present on DNA. This will occur readily in the acidic extracellular pH but not in the slightly alkaline intracellular environment, meaning that they may have been delivered to the nucleus in their inactive form, justifying why minimal cytotoxicity was observed.

2.2.4
Light Activated Prodrugs

The development of light dependent metal therapeutics is one of the more contemporary therapeutic targeting strategies being explored. This mode of treatment aims at achieving localized activation of a drug or diagnostic agent by point irradiation with light of optimal wavelength and intensity.

Metal complexes, and in particular those of d-block transition metals, are well suited to photochemically activated therapeutics as their electronic structure allows for the generation of excited states under relatively moderate conditions [55]. Electrons bonded in coordination complexes have multiple unoccupied orbitals available for electronic transitions during photochemical excitation and decay. The excited complex may then decay via the following pathways: physical, radiative (fluorescence, phosphorescence), non-radiative (internal conversion, intersystem crossing), and chemical reaction. The resultant chemical reactions can involve ligand liberation yielding either an active ligand or metal center, reactions with endogenous molecules, or the transfer of energy or electrons to species such as ground state O_2 in photodynamic therapy. As a result, the reaction product is characteristic of irradiation with light of a discrete wavelength.

Nitric oxide (NO) donors are one of the most extensively investigated examples of this class of metal therapeutic. Clinical administration of exogenous NO may be beneficial in a diverse range of physiological and pathological conditions, including cardiovascular related diseases and as a mediator in tumor-induced angiogenesis [56].

Ruthenium complexes have demonstrated potential for photodynamic therapy [57–61] due to ruthenium's high affinity for the linear NO^+ ligand. Amongst the first investigated were the ruthenium nitrosyl chlorides $[Ru(NO)(Cl_5)]^{2-}$ and [Ru(NO)Cl$_3$]. Despite promising results following localized irradiation in animal models, these complexes exhibited toxic side effects due to substitutive aquation at the chloride ion sites and subsequent affinity for DNA, in particular guanine [62]. Future studies in ruthenium nitrosyls will likely focus on the complexation of amine and polydentate ligands to minimize further speciation of the drug with the aim of reducing toxic side effects. Recent work by Eroy-Reveles and coworkers [63] into non-porphyrin polydentate ligands in manganese nitrosyl complexes demonstrated a high NO donating capacity for the complex $[Mn(PaPy_2Q)(NO)]ClO_4$, where $PaPy_2Q = N,N$-bis(2-pyridylmethyl)amine-N-ethyl-2-quinoline-2-carboxamide) (Figure 2.5). This was following irradiation with low intensity near infrared light (800–900 nm), which has the greatest tissue penetration [64].

Photoactivated metal drugs have another compelling application in the treatment of solid tumors but are fundamentally limited by accessibility of the tumor for irradiation. Porphyrin–metal complexes have been investigated for their photosensitizing potential; that is, their ability to act as catalysts for further reactions *in vivo*, the most common of those being the excitation of triplet oxygen (O_2) to its cytotoxic singlet state. Although porphyrin complexes are seemingly ideal therapeutics due to a high tendency to accumulate in cancer cells, hypoxia and limited access to the poorly vascularized regions can result in inefficient photodynamic excitation of triplet oxygen, aggregation of porphyrin bases, and eventual hepatotoxicity as a result of porphyrin conversion into bilirubin [65].

Photoactivated tumor therapy is also capable of targeting DNA through binding or cleavage to induce cell apoptosis. Administration of non-labile, lipophilic metal prodrugs facilitates accumulation within tumor cells and point irradiation of the tumor allows for site-specific activation. Platinum(IV) prodrugs are of great interest due to their kinetic inertness and G-G cross-linking potential in DNA strands upon photocatalyzed reductive elimination of two axial ligands. Studies by Bednarski *et al.* [66–70] have demonstrated the photolability of Pt(IV) azide complexes and their similar nucleotide platination profiles to cisplatin, whereas photolabile Pt(IV) iodo analogs are susceptible to reduction by endogenous reducing agents such as glutathione, which would limit tumor selectivity *in vivo*.

Further work by Sadler and coworkers [71] found that both *trans,trans,trans*-$[Pt(N_3)_2(OH)_2(NH_3)_2]$ (a) and *cis,trans,cis*-$[Pt(N_3)_2(OH)_2(NH_3)_2]$ (b) complexes (Figure 2.6) displayed photoactivity and were nontoxic in the dark. When irradiated with UVA light, their cytotoxicities were greater than transplatin and less than cisplatin. The improved cytotoxicity observed for (a) over transplatin was attributed to its ability to form a bis(5′-guanosine monophosphate) (5′-GMP) adduct, which transplatin cannot.

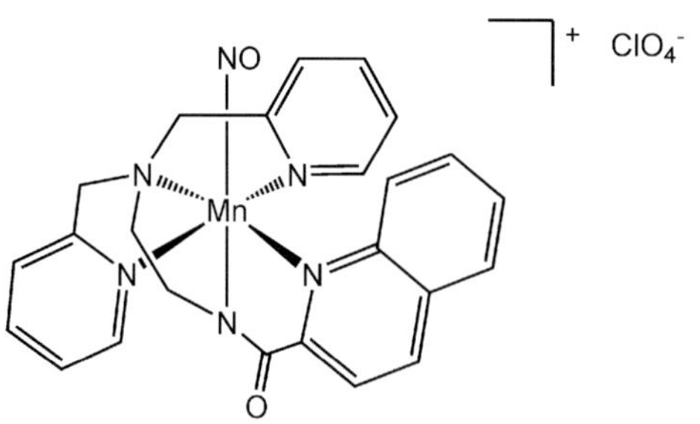

Figure 2.5 $[Mn(PaPy_2Q)(NO)]ClO_4$ complex synthesized by Eroy-Reveles *et al.* [63].

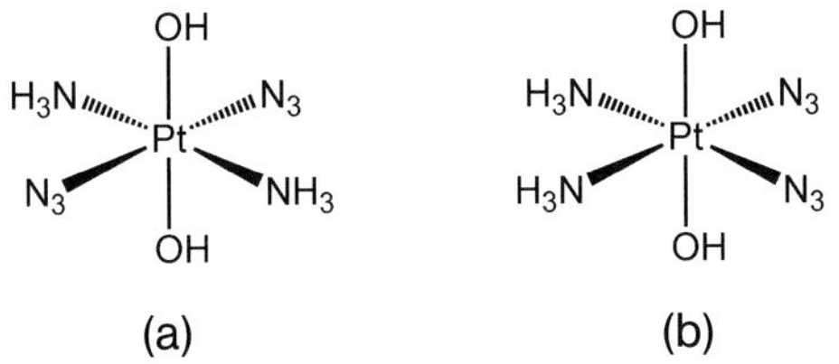

Figure 2.6 Photoactivated Pt(IV) complexes:
(a) *trans,trans,trans*-[Pt(N$_3$)$_2$(OH)$_2$(NH$_3$)$_2$];
(b) *cis,trans,cis*-[Pt(N$_3$)$_2$(OH)$_2$(NH$_3$)$_2$].

To maximize the selectivity of photochemotherapeutics in the future, it is likely that design and synthetic strategies will attempt to incorporate more than one targeting tactic, such as those mentioned elsewhere in this section. Although rational design dictates that this should correlate with increased toxicity towards the targeted tissue and reduced systemic effects, evidence thus far suggests that these assumptions cannot be relied upon, and rigorous testing of all new complexes will become even more crucial.

2.3
Molecular Targeting

Molecularly targeted drug development strategies focus on structure-based design, where the metal center serves as an inert scaffold for various organic and bioorganic ligands. The ability of metals to exhibit multiple oxidation states and coordination geometries not only increases the structural diversity of metal complexes, but allows for conformational modifications that can maximize ligand–target site interactions. This feature of metal complexes is particularly relevant in the targeting of biomolecules where site-binding is expected to be stereospecific.

The potential of these drugs is dependent on knowledge of ligand–target interactions and as such will be augmented by advances in the fields of genomics, proteomics, and metallomics. In addition, the ability of metal centers to coordinate multiple ligands of varying identities means that there is potential for several targeting issues to be addressed by a single compound. For example, future therapeutics may emerge where one ligand is responsible for optimizing redox characteristics, another for physically directing the compound to the desired tissue or intracellular location, and the remaining ligands may exert a therapeutic effect upon release from the metal center.

2.3.1
Protein and Peptide Targeting

2.3.1.1 Protein Receptor Binding
Metallotherapeutics offer immense potential for improving selectivity, specificity, and efficacy of drug action based on selective protein receptor binding. The biology

of various disease states involves discrete cellular process, including the up- and down-regulation of specific genes; the over- and under-expression of peptides, proteins and enzymes, and the stimulation or inhibition of metabolic processes.

The degree of exclusivity provided by targeting one or more of these processes means that there is great potential for synthesizing highly specific therapeutic agents that can minimize systemic toxicity and limit drug resistance. However, by paring down the number of targets with which a particular drug can interact, resistance can develop more rapidly and more easily than for a drug with a greater number of molecular targets.

Target-oriented drug design of these complexes typically follows one or a combination of the following three strategies: optimizing amino acid coordination, exploiting structural selectivity of receptor binding sites, and tethering known protein binders to metal centers.

Amino Acid Coordination The premise of this targeting strategy is that the metal center is not just employed as a structural stabilizer, but can itself undergo coordination to particular amino acid residues as well as coordination through tethered ligands. Rational design of complexes intended to behave according to this principle must take into consideration the intended receptor target and the amino acid expression on the surface or within the active site. Cysteine, histidine, and tyrosine residues contain the major metal binding thiol, imidazole, and hydroxyl groups, respectively, and as such are apt targets for metal complexes with vacant coordination sites [72]. The metal center, its oxidation state, and possible geometries as well as the coordination sphere can be altered for fine tuning of the kinetic properties of the complex and variations in the ancillary ligands can lead to increased binding site recognition by the complex.

Fricker and coworkers have synthesized "3 + 1" mixed-ligand oxoRe(V) complexes that act as cathepsin B inhibitors and may have clinical application in the treatment of some tumors, rheumatoid arthritis, and muscular dystrophy [73]. In cathepsin B, the presence of a unique structural occluding loop containing two histidine residues (His110 and His111) is capable of directing the C-terminal carboxylates within endogenous enzymes towards the active site cysteine. Of the mixed-ligand oxoRe(V) complexes studied, those containing smaller chelating ligands demonstrated greater potency [74, 75] than those containing bulkier groups, most likely due to easier access to the active site. The mechanism of binding to the active site was determined to be nucelophilic attack of the metal center by the cysteine thiolate, resulting in the displacement of the monodentate ligand. A >90-fold increase *in vitro* activity was observed upon changing the monodentate leaving ligand (L) of analogs of [ReO(SSS-2,2)(L)] (where SSS $=$ $SCH_2CH_2SCH_2CH_2S$) from L $=$ (SPhOMe-*p*) to Cl$^-$ as shown in Figure 2.7. This binding was also found to be reversible, which is critical given the tendency of other thiol-containing biomolecules such as glutathione and free cysteine to also bind to the rhenium center, leading to toxic side effects.

Metal complexation of the highly specific CXCR4 receptor (the fourth receptor for natural chemokine proteins containing a conserved Cys-X-Cys disulfide bond

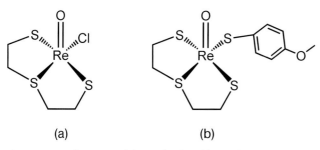

Figure 2.7 Cathepsin B inhibitors developed by Fricker and coworkers:
(a) [ReO(SSS-2,2)(Cl)] and (b) [ReO(SSS-2,2)(SPhOMe-*p*)].

[76]) antagonist AMD3100 has shown considerable potential as an anti-HIV
therapeutic agent. The CXCR4 receptor facilitates entry of HIV into the cell via a
complex multistep process. This is dependent on the binding of the gp120 subunit
of HIV-1 to modified sulfated tyrosines as well as acidic amino acid residues in the
N-terminus of both CCR5 and CXCR4 [77], resulting in the insertion of viral DNA
into the cell via a fusion peptide [78]. AMD3100 (Figure 2.8) is a bicyclam compound composed of two cyclam moieties connected by a structurally constraining
p-phenylenebis(methylene) linker. Studies have shown that complexation to metals enhances the affinity of AMD3100 for the CXCR4 receptor by factors of 50, 36,
and 7 for Ni^{2+}, Zn^{2+}, and Cu^{2+}, respectively [79] as well as having a more
complex mode of interaction with the receptor compared to free AMD3100.
Modeling studies by Sadler *et al.* [76, 80] have shown that amino acid residue
interactions are largely dependent on the chiral conformation adopted by the two
respective cyclam groups around the bound nitrogen atoms. Further work by this
group has also shown that at the physiological pH of 7.4 it is likely that the binding
of free cyclam anti-HIV drugs to Zn^{2+} is both thermodynamically and kinetically
favorable [81]. This may have extensive implications in the development of novel
cyclam-based therapeutics as the coordination to endogenous Zn(II) may enhance
receptor binding strength.

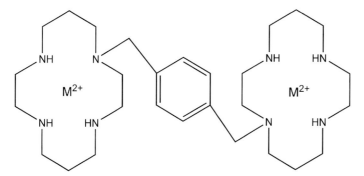

Figure 2.8 CXCR4 receptor antagonist AMD3100, where $M^{2+} = Ni^{2+}$, Zn^{2+}, and Cu^{2+}.

Exploiting Structural Selectivity Inorganic metal complexes have an inherent constitutional advantage over small organic molecules due to their ability to exist in various oxidation states and various structural conformations. Whereas organic molecules are restricted by the three potential binding geometries of carbon – linear (sp-hybridization), trigonal planar (sp^2-hybridization), and tetrahedral (sp^3-hybridization) – metal centers can additionally exist as trigonal bipyramidal, square pyramidal, and octahedral geometries with five, five, and six coordination sites, respectively [82].

The stereochemistry of these complexes is also important, with a metal center coordinated octahedrally to six different ligands capable of at least 30 stereo-isomers, compared to a tetrahedral carbon's equivalent of two. The metal center can provide a degree of structural stability or rigidity, depending on the mode of ligand coordination as well as the non-covalent interactions between moieties on the ligands. By making the complex more rigid, the number of conformations that can be adopted by the overall structure is minimized. Interaction with receptors other than the intended target will be limited by the inflexibility of the molecule and, as a result, the complex will be highly selective (Figure 2.9). This property of metal complexes makes them suitable for high-throughput screening of combi-natorial libraries of compounds following the identification of a lead compound.

The first experimental evidence indicative of the role played by structure in metal complex–protein interactions came from Dwyer and coworkers. Investiga-tions into the biological activity of tris(1,10-phenanthroline) Ru(II) perchlorate demonstrated a high level of cholinesterase inhibition that correlated with a high chemical stability in both acid and alkali solutions [83]. As such it was concluded that the level of dissociation of the metal complex was negligible, and that the observed biological activity was attributable to the intact complex cation and not to the metal or ligand independently [83–85]. To this end, recent work into this area has focused on complexes with high kinetic stability contributed to by either an inert metal center or polydentate ligands.

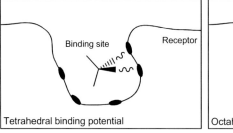
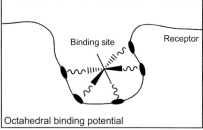

● Binding regions

Figure 2.9 Structural advantages of metal-based therapeutics. The additional coordination geometries allow for maximum ligand–receptor interactions, which improves drug–receptor affinity.

One example of maximizing the kinetic stability of complexes is the development of Ru(III) organometallics as protein kinase inhibitors by Meggers *et al.* [86–93], who envisioned the metal center as a sort of "hypervalent carbon." Protein kinases possess highly selective binding sides for ATP, and the preliminary compounds were based on indolocarbazole alkaloids known to be competitive with ATP. The robust geometry of the compounds was attributable to the inert metal center such as in the Ru(II) octahedral inhibitor Λ-FL172 (Figure 2.10). This stability contributed to a high degree of active site–coordination sphere matching and subsequent max- imizing of the number of hydrogen bonds between enzyme and inhibitor [94] as shown in cocrystal studies [95]. These structural mimics have already shown high levels of efficacy in tumor cell lines and spheroid models [96] with the ability to induce apoptosis through activation of the p53 tumor suppressor gene.

It can also be advantageous to use compounds that mimic the structure of endogenous biomolecules in terms of both dimensionality and chemical func- tionality. Katzenellenbogen *et al.* have had some success using this approach, exploring metals conjugated with the intact androgen [97–99] as well as complexes that imitate the androgenic structure through inclusion of a metal chelate system within the compound [100, 101]. [188]Re complexes of 5α-dihydrotestosterone mimics demonstrated extraordinary structural and stereochemical similarity to the endogenous androgen but were limited *in vitro* by poor lipophilicity and stability due to competitive ligand exchange with water [100]. Later studies into proges- terone mimic complexes resulted in increased lipophilicity to improve cellular uptake. However, the low binding affinity towards estrogen receptors lead to the au- thors concluding that metal complexes must exhibit an electronic complementarity to the receptor ligands in addition to size and shape complementarity to achieve high receptor–ligand affinity [101].

Known Protein Binders The metallocomplexation of established protein and peptide targeting therapeutic molecules offers various benefits, including the ability to tailor complexes to target particular biological attributes of the disease state with greater specificity and reduced systemic toxicity compared to the original

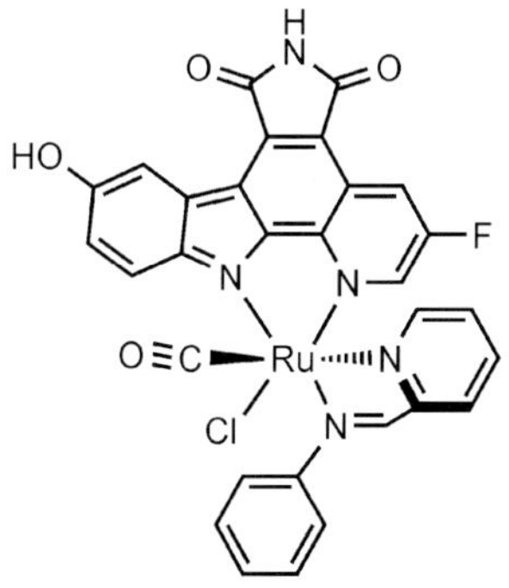

Figure 2.10 Λ-FL172.

organic compound. One of the advantages of modifying drugs with an already established mode of action is that a greater proportion of resources can be spent on improving the drug selectivity, rather than developing therapeutically active compounds from scratch.

Ferrocynal derivatives (ferrocifens) of the anticancer compound tamoxifen (TAM) have been developed over the past decade with the intention of broadening the spectrum of action of the original drug. These compounds are shown in Figure 2.11. Whereas TAM is widely known as a selective estrogen receptor modulator (SERM) capable of inducing apoptosis in hormone-dependent breast cancer cells

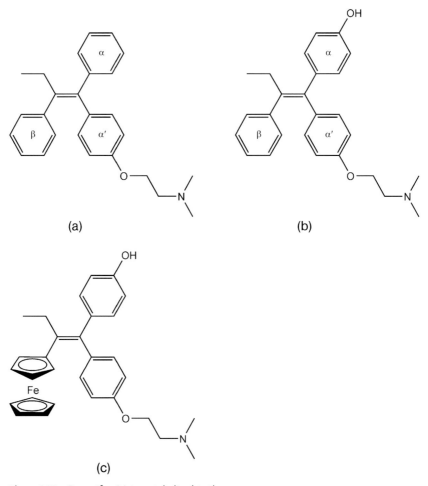

(a)

(b)

(c)

Figure 2.11 Tamoxifen (a) is metabolized to the more hydrophilic hydroxyl-tamoxifen (b), which has a higher ER binding affinity [102]. Ferrocynal derivatives mimic this metabolite by inclusion of a hydroxyl moiety on the α phenyl group, as in ferrocifen (c).

that express the estrogen receptor-α (ER-α(+)), its action is precluded in breast cancer cells that have acquired resistance to TAM or are hormone-independent and are ER-α/β(−). Ferrocifens are globular shape mimics of the original organic drug, where the β phenyl ring has been replaced by a metallocene moiety containing iron [103–106] and, more recently, ruthenium [107].

It is widely understood that the activity of ferrocifen mimics that of TAM and metabolites, binding competitively to ER-α via stabilizing interactions between the basic amino side chain of the TAM molecules and the aspartate 351 residue in the ligand binding domain of the receptor. The ER/bioligand complex can then undergo dimerization to interact directly with DNA at the "estrogen response element" (ERE) or participate in the activated protein pathway (AP1), where it interacts with two proteins (Jun and Fos) to form a DNA binding complex [108]. These respective pathways essentially lead to the inhibition of normal estradiol-mediated DNA transcription. As TAM can function as an estrogen-like ligand, a significant side effect of prolonged TAM use is the development of cancer in uterine tissue [109] and, as such, the development of various structural analogs is of interest. It has been suggested that there are varying mechanisms of action for ferrocifen, depending on ER expression, leading to its activity across hormone dependent and independent cancer cells: TAM-like antiproliferation in ER-α(+) cells, and redox-based cytotoxicity in ER-α(−) and ER-β(+) cells due to the *in situ* oxidation of the ferrocene moiety. Jaouen and coworkers proposed that the single electron facile oxidation of the ferrocene species affords, via Fenton-type chemistry, a quinone methide species in the presence of basic species such as DNA and nucleobases [103]. It is this species that is thought to be responsible for causing cell death in hormone independent breast cancer. Ruthenocene derivatives of TAM showed a higher binding affinity for the ER in ER-α(+) cells over hydroxy-TAM (>53% and 38.5% respectively) but displayed no antiproliferative effect in ER-α(−) cells, lending support to the redox-dependent cytotoxicity of ferrocifen.

2.3.2
Peptide Tethering

The attachment of peptides and peptide motifs to metallochaperones is a rapidly increasing area of interest in medicinal inorganic chemistry. Optimized drug delivery to specific targets is necessary for two main reasons: to minimize undesirable side effects in non-diseased tissue and to maximize drug delivery to the site of action to minimize dosage requirements. The bioconjugation of metals and metal complexes to peptides and peptide motifs has the potential to address both these issues as, from a probabilistic perspective, the more specific the tethered peptide is the less side-reactions should occur *en route* to the target location. This is the case, providing that the rest of the metal complex is inert and non reactive. This concept applies to various extracellular locations within the body but more importantly to the intracellular localization of therapeutic agents. This is because a large proportion of drugs exert their action on specific organelles within the cell – for example, the antitumor drug cisplatin exerts its action on DNA in the cell nucleus.

2.3.2.1 Directing Effects

The most extensively investigated application of peptide tethered metal complexes is in tumor therapy and more specifically the delivery of DNA modifying agents to the cell nucleus. The efficient delivery of these agents to the nucleus of cancer cells requires three stages of localization: targeting to the cancerous tissue, accumulation in the cancer cells, and intracellular localization in the nucleus [110, 111].

Various physiological targeting strategies, as mentioned elsewhere in this chapter, have been employed to cover the first two stages of delivery, while more recent work into the attachment of nuclear localization signal (NLS) peptide sequences has attempted to achieve the third stage of delivery. NLS peptides are short positively charged basic peptides that actively transport large proteins across the nuclear membrane from the cytosol to the cell nucleus [111, 112].

Kirin and coworkers achieved bioconjugation of the chelating bis(picolyl)amine (bpa) ligand with a heptapeptide derivative of the SV 40 virus antigen NLS [113, 114] with Cu^{2+}, Zn^{2+}, and more recently Co^{2+} complexes. In the latest case, the bpa ligand was also conjugated to a peptide nucleic acid (PNA) oligomer sequence prior to the NLS and showed increased cellular accumulation as a result, which was solely attributable to the inclusion of PNA. Cells treated with Co-bpa-NLS and Co-bpa-PNA-NLS (Figure 2.12a) favored nuclear localization of the respective complexes compared to unconjugated Co-bpa [114].

In similar work, Noor *et al.* [115] have synthesized cobaltocenium–peptide bioconjugates containing the same SV 40 antigen NLS. A Co metallocene was chosen over ferrocene due to its higher kinetic stability and redox potential. Nuclear localization was visualized by attaching fluorescent labels to the various

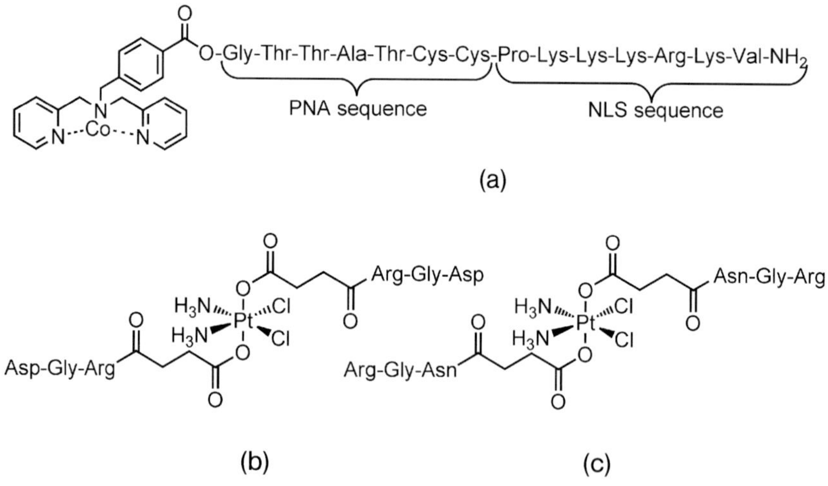

(a)

(b)   (c)

Figure 2.12 Examples of metal complexes containing peptide sequences for improved directing effects: (a) Co-bpa-PNA-NLS complex synthesized by Kirin *et al.* [113, 114]; (b) Pt(IV) complex containing RGD peptide motifs; and (c) NGR peptide motifs synthesized by Mukhopadhyay *et al.* [116].

cobaltocenium compounds and free NLS. It was observed that free NLS was unable to traverse the cell membrane, whereas the cobaltocenium-NLS species showed intracellular accumulation and co-localization with the Hoechst 3342 nuclear stain in HepG2 cells. This study indicated the importance of the organometallic moiety for endocytosis into the cell, and as such the potential for metal-based therapeutics for improving on current ideas and strategies.

NLSs have also been exploited for improving the nuclear localization of established chemotherapeutic compounds, such as carboplatin and its analogs [111]. This study also involved the attachment of a poly(ethylene glycol) (PEG) moiety to modify the lipophilicity and size of the conjugate. Increased levels of nuclear localization and cellular accumulation were observed for the Pt-PEG-NLS conjugates, but fewer intracellular DNA-adducts were found compared to the Pt-PEG compounds. It was therefore recommended that this method of conjugation with NLS alone was not suitable for carboplatin, oxaliplatin, and analogs as it was not able to deliver the cytotoxic agent in a bioavailable form.

Mukhopadhyay *et al.* have tethered peptide motifs to the axially coordinated succinato ligands of a Pt(IV) cisplatin analog via amide linkages. The peptide motifs RGD (Arg-Gly-Asp) and NGR (Asn-Gly-Arg) were employed as "tumor homing devices" intended to recognize the upregulated $\alpha_v\beta_3$ and $\alpha_v\beta_5$ integrins on the membranes of metastatic tumor cells. It was found that both RGD and NGR motif containing complexes had a potent inhibitory effect of cell proliferation, and upregulated integrins were likely mediators in the internalization of the Pt(IV)-peptide complex into the tumor cells [116]. These complexes are shown in Figure 2.12b, c.

2.3.3
Selective Activation

The localized activation of prodrugs can be achieved by targeting pathophysiological molecules indigenous to specific sites or disease phenotypes. The creation of organic prodrugs that are selectively activated at therapeutically significant locations has been widely considered in drug development in the past. However, the application of this more sophisticated design archetype has scarcely been considered in inorganic therapeutic design. Site-specific activation of metal-based prodrugs is a concept that will likely gather interest and momentum in the future as the molecular pathologies of disease states is revealed through work in the fields of proteomics, genomics, and metallomics. Cleavage at a recognized substrate by endogenous enzymes may liberate an active drug complex, ligand, or metal center depending on the position of the substrate, thereby enabling a tailored approach to designing the prodrug.

Tauro and coworkers have used this approach in the design of antitumor therapeutics delivered in hydrogel matrixes. Figure 2.13 depicts the mechanism by which cisplatin is believed to be released in close proximity to the tumor tissue [117–120]. Progressive fine tuning resulted in a hexapeptide substrate for tumor specific MMP-2 and MMP-9, which resulted in improved selectivity for tumor tissue

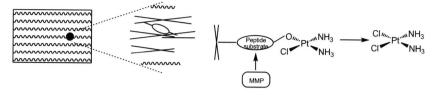

Figure 2.13 Cisplatin molecules are tethered to the peptide (through the oxygen atoms of aspartic acid), which is attached to the hydrogel via poly(ethylene glycol) acrylate spacers. Upon cleavage by the extracellular MMPs, which are overexpressed in tumor tissue, cisplatin and cisplatin–peptide conjugates are released and can form DNA adducts. Figure adapted from Reference [117].

and decreased systemic release. Prior to this study, literature on the selective activation of metal-based drugs is scarce.

Kageyama *et al.* have investigated the activation of cisplatin-like complexes following stereoselective ester hydrolysis by intracellular hydrolase enzymes as a means of selective targeting. Alkyl chains were attached via ester, ether, and amide linkages to various cisplatin derivatives to improve the lipophilicity of the complexes for cell entry via passive diffusion. The greatest cytotoxicity was observed for the complexes with the longest hydrophobic alkyl chains tethered by an ester linkage. Comparison of the growth inhibition observed in rat liver cancer cells (Anr4) and rat hepatoma cells (H4-II-E), which exhibit high and low stereoselectivity for acyl group hydrolysis respectively, suggested that enzyme specificity contributed to stereoselective activation of the Pt(II) complexes [121].

2.3.4
DNA Targeting

The number of therapeutic agents that target nucleic acids is rapidly increasing, and can be divided into two classes – those that exert their action on RNA (ribose nucleic acid) and those that exert their action on DNA (deoxyribose nucleic acid). The latter, and more specifically only those that undergo non-coordinative interactions with DNA, will be discussed here. DNA targeting allows for the potential disruption of the replication, and transcription processes of homeostatic cell functioning. Metal complexes traditionally interact with DNA in one of three ways: the most commonly encountered metallo-intercalators, and groove binders; and the more novel metallo-inserters (Table 2.1).

2.3.4.1 Duplex DNA Sequence Selectivity
The ability to target DNA at specific sequences is relevant to the formulation of new diagnostics and chemotherapeutics, particularly as the molecular basis for diseases is elucidated. Transition metal complexes can achieve explicit DNA site specificity through manipulation of the complex as a modular system – the metal center and the ancillary ligand scaffold. The inherent kinetic inertness of some metal complexes and the expanded possibility for chiral discrimination are

Table 2.1 Properties of DNA targeting therapeutic molecules.

	Site of action	Mode of action
Metallo-intercalators	Major groove of DNA double helix	The intercalator unwinds the DNA double helix and acts as a new base pair. It undergoes π stacking between adjacent base pairs. The intercalator increases the height of the major groove by opening of the phosphate angles
Groove binders	Minor groove of DNA double helix	The groove binding agent is hydrophobically transferred from solution to the minor groove. The subsequent interactions are dependent on the functionality of binder
Metallo-inserters	Minor groove of DNA double helix	The metallo-inserters is a planar ligand that extends and inserts into the base-stack. The ligand ejects bases of single base pairs and replaces them in the π stacking arrangement

contributors to the immense potential for increasing the complexity of the non-covalent DNA binding mechanisms and thereby selectivity for specific DNA sites. This selectivity may be based on the shape of the complex or on the functionality of the ligands themselves.

DNA targeting metallo-compounds of the general formula $[M(chel)_3]^{2+}$ (where M = the metal center and chel = bidentate chelating ligand) are structurally rigid – a result of their strict coordination geometries – and as such the steric bulk and stereochemistry of the attached ligands contributes chiefly to the overall shape of the complex and, in principle, to DNA sequence discrimination. Intercalating ligands with larger aromatic surface spans have a greater π stacking potential than smaller ligands and therefore a potentially greater degree of enantioselectivity. The strength and nature of the non-covalent interactions holding the non-intercalating ligands in position against the adjacent DNA base pairs essentially determines the binding affinity of the entire complex. By varying the appended ligand moieties and their orientation it has proven possible to vary the site selectivity.

Studies into the enantioselectivity exhibited between 9,10-phenanthrenequinone diimine (phi) complexes of Rh(III) and duplex DNA [122, 123] have shown that in the case of B-DNA the Δ enantiomer of $[Rh(phi)(DPB)_2]^{3+}$ (where DPB = 4,4'-diphenylbipyridine) readily cleaves the 5'-CTCTAGAG-3' base pair sequence, but the Λ enantiomer does not. This is because the Δ enantiomer matches the helical symmetry of B-DNA and also undergoes dimerization and simultaneous intercalation at the central 5'-CT-3' site. This promotes non-covalent interactions between the phenyl and bipyridyl rings of adjacent complexes and steric constraints increase the binding affinity of the second intercalator complex [124].

DNA sequence selectivity can also be based on the functionality of the ligands tethered to the metal center. Rational design of complexes of this type seeks to optimize non-covalent interactions such as hydrogen bonding and van der Waals interactions between the ancillary ligands and DNA bases to achieve high site specificity and affinity. Investigations into a group of Rh(III) polyamine complexes [125–127] have demonstrated the success of this strategy. These are shown in Figure 2.14. Hydrogen bonding between the axial amines of the complexes and the O6 of guanine resulted in recognition and ultimately cleavage at the 5'GC-3' site. This was confirmed by alternative binding sites being observed for the control complex [Rh[12] aneS$_4$(phi)]$^{3+}$ (where [12]aneS4 = 1,4,7,10-tetrathiacyclododecane), which cannot undergo hydrogen bonding, as well as the replacement of guanine with O6-methyl guanine, which disrupted site binding [126].

This strategy was expanded on with Δ-α-[Rh(Me$_2$trien)(phi)]$^{3+}$ (where Me$_2$trien = 2,9-diamino-4,7-diazadecane), which was designed to recognize the 5'-TGCA-3' sequence through hydrogen bonding between the axial amines and O6 of guanine. This was also intended to be in combination with van der Waals interactions between the pendant methyl groups on the complex and the methyl groups on the bordering thymine bases [128].

2.3.4.2 Telomeric Targeting

Telomeres are the regions of highly conserved 5'-TTAGGG repeats located at the ends of eukaryotic chromosomes [129]. They are vital protectors of chromosomes during mitotic cell division, preventing homologous recombination and non-homologous end joining, and become progressively shortened over rounds of cell replication with their eventual depletion leading to apoptosis. As such, telomeres have been investigated as indicators of cell senescence [130]. The enzyme telomerase, which counteracts this shortening by adding base pair repeats to the 3' end of DNA strands, is largely absent in healthy cells, but is expressed in 85–90% of tumor cells [131]. This facilitates rapid and infinite proliferation of the tumor cells. Telomeric DNA exists in a G-quadruplex structure, which is the result of hydrogen bonding between tetrads of guanine bases which consequently undergo stacking and eliminate the 3' DNA overhang. Further stabilization of this tetrad

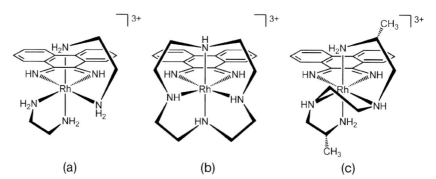

(a) (b) (c)

Figure 2.14 Rhodium(III) polyamine complexes designed for DNA sequence selectivity: (a) Δ-[Rh(en)$_2$(phi)]$^{3+}$; (b) [Rh[12] aneN$_4$(phi)]$^{3+}$; and (c) Δ-α-[Rh(Me$_2$trien)(phi)]$^{3+}$ (where [12] aneN$_4$ = 1,4,7,10-tetraaminecyclododecane).

structure occurs due to electrostatic interactions between the guanine carbonyl groups and a central metal cation.

Chemotherapeutics that selectively recognize and stabilize the G-quadruplex structure of telomeric DNA and inhibit telomerase are of great interest for selectively targeting malignant cells. Until recently, stabilization of these structures was mainly attempted with small organic molecules. However, planar ligands with π-delocalized electrons, such as in chromophores, tethered to a positively charged metal center have appeared to offer some improvements.

Vilar *et al.* have investigated the ability of Ni(II), Cu(II), Pt(II), and V(IV) complexes to stabilize G-quadruplex structures based on geometry and functionality [132–137]. Their general structure is shown in Figure 2.15. By measuring the increase in melting temperature induced by binding with the metal complexes, it was concluded that square-planar metal complexes possess the optimal geometry to maximize π–π stacking between the complex and G-tetrad compared to complexes with square-based pyramidal geometries [133]. The ability of metal complexes to have these geometries is one of the inherent advantages held over small organic molecules and is why there has been such success in targeting DNA with metal complexes.

One of the difficulties in targeting telomeric DNA is the extremely high level of selectivity required by the administered therapeutic given the length of telomeres compared to the remaining chromosome – 6–8 kb pairs compared to >50 million base pairs, respectively. It is likely that, unless the therapeutic agent is entirely specific for the G-quadruplex structure, binding will occur at multiple locations along the chromosome and cause undesirable side effects. As such, *in vitro* testing of telomere binding agents will have to be rigorous and extensive.

2.4
Immunological Targeting

The last decade of the nineteenth century saw the pioneering work of Paul Ehrlich develop the "magic bullet" paradigm of immunological targeting, which is still

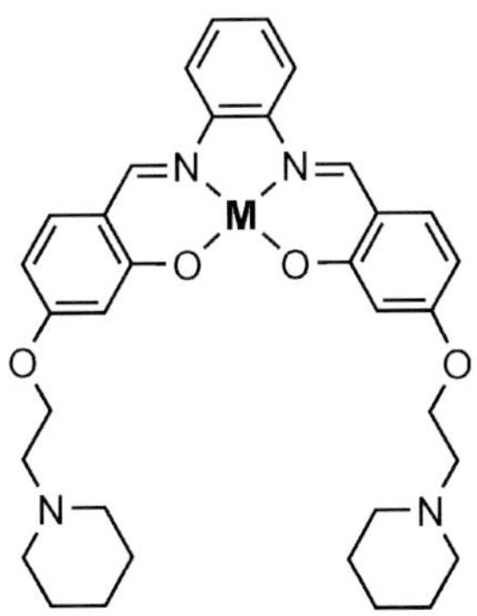

Figure 2.15 General structure of the compounds prepared by Vilar *et al.*, where M $=$ Ni^{2+}, Cu^{2+}, Pt^{2+}, and V^{4+}.

pursued today. Ehrlich's postulates led to the idea that by targeting receptors that are present on the causative agent but are absent in host cells, a more efficacious therapy could be achieved. In the context of this chapter, the term antigen refers to a molecule capable of binding an antibody and invoking an immune response. Although antigen targeting and antibody (Ab) tethering are mutually inclusive facets of immunological targeting, they have been discussed here as separate therapeutic strategies.

2.4.1
Antigen Targeting

Disease states often result in the selective expression or overexpression of particular antigens on the affected cell's surface, establishing a discrepancy between healthy and diseased cells that can be exploited for selective targeting. Employing a molecule that binds the antigen with high selectivity and affinity and can activate a therapeutic agent requires a two-step approach, which in the case of cancer therapy is known as antibody-directed enzyme prodrug therapy (ADEPT) (Figure 2.16) [138]. This form of therapy could also fall fittingly under Section 2.3.

This method of achieving targeted therapy depends upon: the localized expression of disease-specific antigens, minimal binding to healthy cells, the antibody–epitope binding affinity, the catalytic properties of the enzyme compared to endogenous enzymes, and the subsequent drug uptake into the cell [139]. So far, ADEPT has primarily focused on the release of organic cytotoxic agents and only a handful of studies concerning metal-based therapeutics have been published.

Work by Hanessian *et al.* demonstrated that the cleavage of two cephalosporin-carboplatinum based prodrugs by β-lactamase generated the cytotoxic Pt(II) complexes [140], but further *in vitro* studies with an Ab-β-lactamase conjugate to

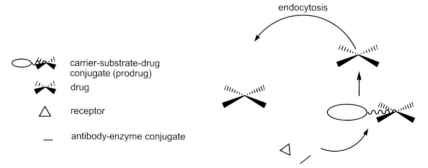

Figure 2.16 ADEPT: The antibody–enzyme conjugate is administered and selectively binds to antigens expressed on the surface of tumor cells. A prodrug is administered and the conjugated enzyme cleaves the prodrug to generate the active drug in the vicinity of the tumor for cellular uptake by endocytosis or other transport mechanisms.

confirm ADEPT success were not conducted. Similar work by Tromp *et al.* explored the cleavage of a glucuronyl-[Pt(NO$_3$)$_2$ (chxn)] (where chxn = cyclohexane-1,2-diamine) conjugate by β-glucuronidase [141]. NMR studies showed that cleavage of the compound into glucuronide and [Pt(NO$_3$)$_2$(chxn)] was virtually instantaneous. *In vitro* studies of an appropriate immunoconjugate were also not performed in this instance. β-Glucuronidase is, however, an endogenous enzyme that is commonly present in regions of necrosis and as such this specific therapeutic mechanism is not strictly ADEPT.

The success of these two attempts is indicative of the potential for metal therapeutics to be employed in ADEPT and administration in prodrug form allows for opportunities to address some of the problems associated with currently used drugs such as poor solubility, acquired resistance, and systemic toxicity.

Not all antigen targeting therapies follow the ADEPT principles, and of particular interest is the attachment of directing aptamers to "drug containers" [142] capable of delivering a large volume of cytotoxic agent from a single container. Dhar and coworkers have synthesized poly(D,L-lactic-*co*-glycolic acid) (PLGA)–poly (ethylene glycol) (PEG) nanoparticles encapsulating *c,t,c*-[Pt(caproate)$_2$Cl$_2$(NH$_3$)$_2$] and functionalized them with prostate-specific membrane antigen (PSMA) A10 aptamers [143] for targeted delivery to prostate cancer cells. This strategy attempts to deliver the drug into the cell in a "parcel" form. PMSA has a high level of expression in both metastatic and hormone-refractory forms of prostate cancer [144, 145] and as such it was expected that the Pt(IV) prodrug would remain encapsulated until localized binding to the antigen occurred at the tumor site. Subsequent endocytosis of the capsules was observed via fluorescence microscopy. The Pt(IV) complex was then released and reduced in the intracellular environment to generate cisplatin and this was confirmed by immunofluorescence studies of the 1,2-d(GpG) intrastrand crosslink adducts.

2.4.2
Antibody Tethering

The use of antibodies to selectively target cells and tissues has been established for some time. Radio-immunoconjugates are the most prominent examples of antibody bound metals and can be used both therapeutically (α and β emitters) or for imaging (γ emitters). These compounds can either involve a bifunctional chelator molecule, which is conjugated to the antibody and coordinated to the metal, or direct attachment of the metal to the antibody [146]. These applications are discussed in more depth in Chapter 9. Antibody–drug conjugates, where the drug is a small organic molecule, have been widely explored as a means of targeting particular cell and tissue types. However, the immunoconjugation of metal-based drugs remains relatively uncharted territory. As with radioimmunoconjugates, tethering of highly specific monoclonal antibodies (mAb) to drug compounds requires that the antibody immunoreactivity is preserved such that the antibody–antigen binding affinity is retained. It is also important that the potency of the drug is not reduced by conjugation; however, inactivation upon conjugation can be desirable, providing that the drug can be reactivated upon entering cells as it limits toxicity to non-targeted cells. Furthermore, the ratio of drug to antibody in the immunoconjugate should ideally be as high as possible, to maximize potency by delivering a greater number of drug molecules per immunoconjugate [147].

Based on these principles, Gao and coworkers have synthesized Herceptin-Pt (II) binding complexes for targeted breast cancer therapy [148] as shown in Figure 2.17.

Herceptin is a humanized monoclonal antibody that binds to the Her2/neu protein, and is overexpressed in 15–30% of breast cancer cases according to the American Cancer Society [149]. *In vitro* studies showed that the Herceptin-Pt(II) compounds displayed a high degree of selectivity for Her2/neu overexpressing cancer cells over normal cells. The apoptotic potential of the antibody–drug compounds was found to be limited by the level of receptor expression, as doubling the dose of Herceptin-10[PtCl$_2$(L)], where L = 2,2'-(1R,2R)-cyclohexane-1,2-diylbis (azanediyl)bis(methylene)bis(4-methylphenol), resulted in only a minor increase in the proportion of apoptotic cells from 31.1% to 33.04% in the Her2/neu overexpressing SK-BR-3 cell line [148]. This study is indicative of the potential of antibody–metal complex therapeutic strategies, as the inherent selectivity of the antibody is coupled to metal complexes with various mechanisms of action.

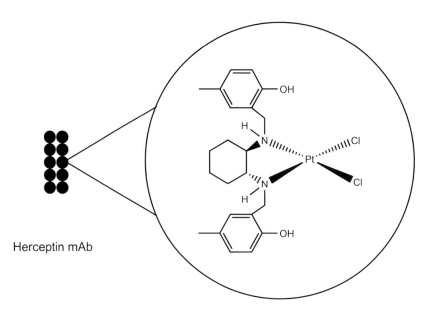

Herceptin mAb

Figure 2.17 Herceptin-10[PtCl$_2$(L)]. The Herceptin antibody (left) is loaded with ten Pt(II) complex units via non-covalent interactions. In principle, this allows for the delivery of ten cytotoxic agents per single antibody in an attempt to improve therapeutic potency, particularly in cases where receptor expression is low.

This type of targeting is likely to become increasingly pursued as the human proteome and molecular hallmarks of diseases are elucidated.

2.5
Concluding Remarks

Progress in the treatment of cancer and other diseases requires targeting of the therapy to the intended site of action by one or more methods. The work described in this chapter shows that metals can play a wide variety of roles in achieving this goal: as part of the targeting system or as the therapeutic agent. Exploitation of the flexibility and control offered by metal-based systems can make a serious contribution to the development of new therapeutics in the coming years.

References

1 Hambley, T.W. (2007) *Science*, **318** (5855), 1392–1393.

2 Brown, J.M. and Giaccia, A.J. (1998) *Cancer Res.*, **58** (7), 1408–1416.

3 Harris, A.L. (2002) *Nat. Rev. Cancer*, **2** (1), 38–47.

4 Tredan, O., Galmarini, C.M., Patel, K., and Tannock, I.F. (2007) *J. Natl. Cancer Inst.*, **99** (19), 1441–1454.

5 Coleman, C.N. (1988) *J. Natl. Cancer Inst.*, **80** (5), 310–317.

6 Heldin, C.-H., Rubin, K., Pietras, K., and Ostman, A. (2004) *Nat. Rev. Cancer*, **4** (10), 806–813.

7 Hambley, T.W. (2008) *Aust. J. Chem.*, **61** (9), 647–653.

8 Rankin, E.B. and Giaccia, A.J. (2008) *Cell Death Differ.*, **15** (4), 678–685.

9 Wartenberg, M., Gronczynska, S., Bekhite, M.M., Saric, T., Niedermeier, W., Hescheler, J., and Sauer, H. (2005) *Int. J. Cancer*, **113** (2), 229–240.

10 Durand, R.E. (1994) *In Vivo*, **8** (5), 691–702.

11 Brown, J.M. and William, W.R. (2004) *Nat. Rev. Cancer*, **4** (6), 437–447.

12 Hall, M.D., Amjadi, S., Zhang, M., Beale, P.J., and Hambley, T.W. (2004) *J. Inorg. Biochem.*, **98** (10), 1614–1624.

13 Platts, J.A., Hibbs, D.E., Hambley, T.W., and Hall, M.D. (2001) *J. Med. Chem.*, **44** (3), 472–474.

14 Ang, W.H., Pilet, S., Scopelliti, R., Bussy, F., Juillerat-Jeanneret, L., and Dyson, P.J. (2005) *J. Med. Chem.*, **48** (25), 8060–8069.

15 McKeage, M.J., Berners-Price, S.J., Galettis, P., Bowen, R.J., Brouwer, W., Ding, L., Zhuang, L., and Baguley, B. C. (2000) *Cancer Chemother. Pharmacol.*, **46** (5), 343–350.

16 Kelland, L.R., Murrer, B.A., Abel, G., Giandomenico, C.M., Mistry, P., and Harrap, K.R. (1992) *Cancer Res.*, **52** (4), 822–828.

17 Failes, T.W. and Hambley, T.W. (2006) *Dalton Trans.*, (15), 1895–1901.

18 Ellis, L.T., Er, H.M., and Hambley, T. W. (1995) *Aust. J. Chem.*, **48** (4), 793–806.

19 Battle, A.R., Deacon, G.B., Dolman, R. C., and Hambley, T.W. (2002) *Aust. J. Chem.*, **55** (11), 699–704.

20 Hambley, T.W., Battle, A.R., Deacon, G.B., Lawrenz, E.T., Fallon, G.D., Gatehouse, B.M., Webster, L.K., and Rainone, S. (1999) *J. Inorg. Biochem.*, **77** (1–2), 3–12.

21 Choi, S., Filotto, C., Bisanzo, M., Delaney, S., Lagasee, D., Whitworth, J. L., Jusko, A., Li, C., Wood, N.A., Willingham, J., Schwenker, A., and Spaulding, K. (1998) *Inorg. Chem.*, **37** (10), 2500–2504.

22 Mellor, H.R., Snelling, S., Hall, M.D., Modok, S., Jaffar, M., Hambley, T.W., and Callaghan, R. (2005) *Biochem. Pharmacol.*, **70** (8), 1137–1146.

23 Hambley, T.W. (2007) *Dalton Trans.*, (43), 4929–4937.

24 Ware, D.C., Palmer, B.D., Wilson, W. R., and Denny, W.A. (1993) *J. Med. Chem.*, **36** (13), 1839–1846.

25 Yamamoto, N., Danos, S., Bonnitcha, P.D., Failes, T.W., New, E.J., and Hambley, T.W. (2008) *J. Biol. Inorg. Chem.*, **13** (6), 861–871.

26 Failes, T.W., Cullinane, C., Diakos, C. I., Yamamoto, N., Lyons, J.G., and Hambley, T.W. (2007) *Chem. Eur. J.*, **13** (10), 2974–2982.

27 Failes, T.W. and Hambley, T.W. (2007) *J. Inorg. Biochem.*, **101** (3), 396–403.

28 Failes, T.W., Diakos, C.I., Underwood, C.K., Hambley, T.W., Cullinane, C.M., and Lyons, J.G. (2003) *J. Inorg. Biochem.*, **96** (1), 128.

29 Yuan, F., Dellian, M., Fukumura, D., Leunig, M., Berk, D.A., Torchilin, V.P., and Jain, R.K. (1995) *Cancer Res.*, **55** (17), 3752–3756.

30 Schilling, U., Friedrich, E.A., Sinn, H., Schrenk, H.H., Clorius, J.H., and Maier-Borst, W. (1992) *Int. J. Rad. Appl. Instrum. B*, **19** (6), 685–995.

31 Jehn, C.F., Boulikas, T., Kourvetaris, A., Possinger, K., and Lueftner, D. (2007) *Anticancer Res.*, **27** (1A), 471–476.

32 Suzuki, R., Takizawa, T., Kuwata, Y., Mutoh, M., Ishiguro, N., Utoguchi, N., Shinohara, A., Eriguchi, M., Yanagie, H., and Maruyama, K. (2008) *Int. J. Pharm.*, **346** (1–2), 143–150.

33 Hamelers, I.H.L. and de Kroon, A.I.P. M. (2007) *J. Liposome Res.*, **17** (3–4), 183–189.

34 Newman, M.S., Colbern, G.T., Working, P.K., Engbers, C., and Amantea, M.A. (1999) *Cancer Chemother. Pharm.*, **43** (1), 1–7.

35 Peleg-Shulman, T., Gibson, D., Cohen, R., Abra, R., and Barenholz, Y. (2001) *Biochim. Biophys. Acta-Biomembr.*, **1510** (1–2), 278–291.

36 Harrington, K.J., Lewanski, C.R., Northcote, A.D., Whittaker, J., Wellbank, H., Vile, R.G., Peters, A.M., and Stewart, J.S.W. (2001) *Ann. Oncol.*, **12** (4), 493–496.

37 Terwogt, J.M.M., Groenewegen, G., Pluim, D., Maliepaard, M., Tibben, M. M., Huisman, A., Huinink, W.W.T., Schot, M., Welbank, H., Voest, E.E., Beijnen, J.H., and Schellens, J.H.M. (2002) *Cancer Chemother. Pharm.*, **49** (3), 201–210.

38 Vail, D.M., Kurzman, I.D., Glawe, P.C., O'Brien, M.G., Chun, R., Garrett, L.D., Obradovich, J.E., Fred, R.M., Khanna, C., Colbern, G.T., and Working, P.K. (2002) *Cancer Chemother. Pharm.*, **50** (2), 131–136.

39 Xiao, C.J., Qi, X.R., Maitani, Y., and Nagai, T. (2004) *J. Pharm. Sci.*, **93** (7), 1718–1724.

40 Helmlinger, G., Schell, A., Dellian, M., Forbes, N.S., and Jain, R.K. (2002) *Clin. Cancer Res.*, **8** (4), 1284–1291.

41 Tannock, I.F. and Rotin, D. (1989) *Cancer Res.*, **49** (16), 4373–4384.

42 Wike-Hooley, J.L., Haveman, J., and Reinhold, H.S. (1984) *Radiother. Oncol.*, **2** (4), 343–366.

43 Warburg, O. (1931) *The Metabolism of Tumors*, R. R. Smith Inc., New York.

44 Thistlethwaite, A.J., Leeper, D.B., Moylan, D.J., and Nerlinger, R.E. (1985) *Int. J. Radiat. Oncol. Biol. Phys.*, **11** (9), 1647–1652.

45 Vaupel, P., Kallinowski, F., and Okunieff, P. (1989) *Cancer Res.*, **49** (23), 6449–6465.

46 Lindner, D. and Raghavan, D. (2009) *Br. J. Cancer*, **100** (8), 1287–1291.

47 Gallagher, F.A., Kettunen, M.I., Day, S. E., Hu, D.E., Ardenkjaer-Larsen, J.H., in't Zandt, R., Jensen, P.R., Karlsson, M., Golman, K., Lerche, M.H., and Brindle, K.M. (2008) *Nature*, **453** (7197), 940–943.

48 Newell, K., Franchi, A., Pouyssegur, J., and Tannock, I. (1993) *Proc. Natl. Acad. Sci. USA*, **90** (3), 1127–1131.

49 Yamagata, M., Hasuda, K., Stamato, T., and Tannock, I.F. (1998) *Br. J. Cancer*, **77** (11), 1726–1731.

50 Galanski, M., Baumgartner, C., Meelich, K., Arion, V.B., Fremuth, M., Jakupec, M.A., Schluga, P., Hartinger, C.G., Von Keyserlingk, N.G., and Keppler, B.K. (2004) *Inorg. Chim. Acta*, **357** (11), 3237–3244.

51 Habtemariam, A., Parkinson, J.A., Margiotta, N., Hambley, T.W., Parsons, S., and Sadler, P.J. (2001) *J. Chem. Soc., Dalton Trans*, (4), 362–372.

52 Habtemariam, A. and Sadler, P.J. (1996) *Chem. Commun.*, (15), 1785–1786.

53 Habtemariam, A., Watchman, B., Potter, B.S., Palmer, R., Parsons, S., Parkin, A., and Sadler, P.J. (2001) *J. Chem. Soc., Dalton Trans*, (8), 1306–1318.

54 Zorbas-Seifried, S., Hartinger, C.G., Meelich, K., Galanski, M., Keppler, B. K., and Zorbas, H. (2006) *Biochemistry*, **45** (49), 14817–14825.

55 Szacilowski, K., Macyk, W., Drzewiecka-Matuszek, A., Brindell, M., and Stochel, G. (2005) *Chem. Rev.*, **105** (6), 2647–2694.

56 Ziche, M., Morbidelli, L., Choudhuri, R., Zhang, H.T., Donnini, S., Granger, H.J., and Bicknell, R. (1997) *J. Clin. Invest.*, **99** (11), 2625–2634.

57 Tfouni, E., Krieger, M., McGarvey, B. R., and Franco, D.W. (2003) *Coord. Chem. Rev.*, **236** (1–2), 57–69.

58 Serli, B., Zangrando, E., Gianferrara, T., Yellowlees, L., and Alessio, E. (2003) *Coord. Chem. Rev.*, **245** (1–2), 73–83.

59 Marcondes, F.G., Ferro, A.A., Souza-Torsoni, A., Sumitani, M., Clarke, M.J., Franco, D.W., Tfouni, E., and Krieger, M.H. (2002) *Life Sci.*, **70** (23), 2735–2752.

60 Cox, A.B. and Wallace, R.M. (1971) *Inorg. Nucl. Chem. Lett.*, **7** (12), 1191–1194.

61 Works, C.F. and Ford, P.C. (2000) *J. Am. Chem. Soc.*, **122** (31), 7592–7593.

62 Rose, M.J. and Mascharak, P.K. (2008) *Coord. Chem. Rev.*, **252** (18–20), 2093–2114.

63 Eroy-Reveles, A.A., Leung, Y., Beavers, C.M., Olmstead, M.M., and Mascharak, P.K. (2008) *J. Am. Chem. Soc.*, **130** (13), 4447–4458.

64 Brancaleon, L. and Moseley, H. (2002) *Lasers Med. Sci.*, **17** (3), 173–186.

65 Bonnett, R. (2004) *Metal complexes for photodynamic therapy*, in *Comprehensive Coordination Chemistry II* (eds J.A. McCleverty and T.J. Meyer), Elsevier Pergamon, Oxford, p. 945.

66 Kratochwil, N.A., Parkinson, J.A., Bednarski, P.J., and Sadler, P.J. (1999) *Angew. Chem. Int. Ed.*, **38** (10), 1460–1463.

67 Bednarski, P.J., Grunert, R., Zielzki, M., Wellner, A., Mackay, F.S., and Sadler, P.J. (2006) *Chem. Biol.*, **13** (1), 61–67.

68 Kasparkova, J., Mackay, F.S., Brabec, V., and Sadler, P.J. (2003) *J. Biol. Inorg. Chem.*, **8** (7), 741–745.

69 Kratochwil, N.A., Zabel, M., Range, K.-J., and Bednarski, P.J. (1996) *J. Med. Chem.*, **39** (13), 2499–2507.

70 Kratochwil, N.A., Bednarski, P.J., Mrozek, H., Vogler, A., and Nagle, J.K. (1996) *Anti-Cancer Drug Des.*, **11** (2), 155–171.

71 Mackay, F.S., Woods, J.A., Moseley, H., Ferguson, J., Dawson, A., Parsons, S., and Sadler, P.J. (2006) *Chem. Eur. J.*, **12** (11), 3155–3161.

72 Meggers, E. (2009) *Chem. Commun.*, (9), 1001–1010.

73 Fricker, S.P. (2007) *Dalton Trans.*, (43), 4903–4917.

74 Mosi, R., Baird, I.R., Cox, J., Anastassov, V., Cameron, B., Skerlj, R. T., and Fricker, S.P. (2006) *J. Med. Chem.*, **49** (17), 5262–5272.

75 Baird, I.R., Mosi, R., Olsen, M., Cameron, B.R., Fricker, S.P., and Skerlj, R.T. (2006) *Inorg. Chim. Acta*, **359** (9), 2736–2750.

76 Liang, X.Y., Parkinson, J.A., Weishaupl, M., Gould, R.O., Paisey, S. J., Park, H.S., Hunter, T.M., Blindauer, C.A., Parsons, S., and Sadler, P.J. (2002) *J. Am. Chem. Soc.*, **124** (31), 9105–9112.

77 Farzan, M., Mirzabekov, T., Kolchinsky, P., Wyatt, R., Cayabyab, M., Gerard, N. P., Gerard, C., Sodroski, J., and Choe, H. (1999) *Cell*, **96** (5), 667–676.

78 Murdoch, C. (2000) *Immunol. Rev.*, **177**, 175–184.

79 Gerlach, L.O., Jakobsen, J.S., Jensen, K.P., Rosenkilde, M.R., Skerlj, R.T., Ryde, U., Bridger, G.J., and Schwartz, T.W. (2003) *Biochemistry*, **42** (3), 710–717.

80 Liang, X.Y., Weishaupl, M., Parkinson, J.A., Parsons, S., McGregor, P.A., and Sadler, P.J. (2003) *Chem. Eur. J.*, **9** (19), 4709–4717.

81 Paisey, S.J. and Sadler, P.J. (2004) *Chem. Commun.*, (3), 306–307.

82 Meggers, E. (2007) *Curr. Opin. Chem. Biol.*, **11** (3), 287–292.

83 Brandt, W.W., Dwyer, F.P., and Gyarfas, E.C. (1954) *Chem. Rev.*, **54** (6), 959–1017.

84 Dwyer, F.P., Gyarfas, E.C., Rogers, W. P., and Koch, J.H. (1952) *Nature*, **170** (4318), 190–191.

85 Dwyer, F.P., Reid, I.K., Shulman, A., Laycock, G.M., and Dixson, S. (1969) *Aust. J. Exp. Biol. Med.*, **47**, 203–218.

86 Atilla-Gokcumen, G.E., Pagano, N., Streu, C., Maksimoska, J., Filippakopoulos, P., Knapp, S., and Meggers, E. (2008) *ChemBioChem.*, **9** (18), 2933–2936.

87 Xie, P., Williams, D.S., Atilla-Gokcumen, G.E., Milk, L., Xiao, M., Smalley, K.S.M., Herlyn, M., Meggers, E., and Marmorstein, R. (2008) *ACS Chem. Biol.*, **3** (5), 305–316.

88 Pagano, N., Maksimoska, J., Bregman, H., Williams, D.S., Webster, R.D., Xue, F., and Meggers, E. (2007) *Org. Biomol. Chem.*, **5** (8), 1218–1227.

89 Bregman, H., Carroll, P.J., and Meggers, E. (2006) *J. Am. Chem. Soc.*, **128** (3), 877–884.

90 Bregman, H. and Meggers, E. (2006) *Org. Lett.*, **8** (24), 5465–5468.

91 Bregman, H., Williams, D.S., and Meggers, E. (2005) *Synthesis*, (9), 1521–1527.

92 Bregman, H., Williams, D.S., Atilla, G. E., Carroll, P.J., and Meggers, E. (2004) *J. Am. Chem. Soc.*, **126** (42), 13594–13595.

93 Zhang, L., Carroll, P., and Meggers, E. (2004) *Org. Lett.*, **6** (4), 521–523.

94 Maksimoska, J., Feng, L., Harms, K., Yi, C., Kissil, J., Marmorstein, R., and Meggers, E. (2008) *J. Am. Chem. Soc.*, **130** (47), 15764–15765.

95 Debreczeni, J.E., Bullock, A.N., Atilla, G.E., Williams, D.S., Bregman, H., Knapp, S., and Meggers, E. (2006) *Angew. Chem. Int. Ed.*, **45** (10), 1580–1585.

96 Smalley, K.S.M., Contractor, R., Haass, N.K., Kulp, A.N., Atilla-Gokcumen, G. E., Williams, D.S., Bregman, H., Flaherty, K.T., Soengas, M.S., Meggers, E., and Herlyn, M. (2006) *Cancer Res.*, **67** (1), 209–217.

97 Chi, D.Y. and Katzenellenbogen, J.A. (1993) *J. Am. Chem. Soc.*, **115** (15), 7045–7046.

98 Chi, D.Y., Oneil, J.P., Anderson, C.J., Welch, M.J., and Katzenellenbogen, J. A. (1994) *J. Med. Chem.*, **37** (7), 928–937.

99 Katzenellenbogen, J.A. (1999) *Abstr. Pap. Am. Chem. Soc.*, **217**, 182-NUCL.

100 Hom, R.K., Chi, D.Y., and Katzenellenbogen, J.A. (1996) *J. Org. Chem.*, **61** (8), 2624–2631.

101 Hom, R.K. and Katzenellenbogen, J.A. (1997) *J. Org. Chem.*, **62** (18), 6290–6297.

102 Lien, E.A., Solheim, E., and Ueland, P. M. (1991) *Cancer Res.*, **51** (18), 4837–4844.

103 Hillard, E., Vessieres, A., Thouin, L., Jaouen, G., and Amatore, C. (2006) *Angew. Chem. Int. Ed.*, **45** (2), 285–290.

104 Nguyen, A., Top, S., Vessieres, A., Pigeon, P., Huche, M., Hillard, E.A., and Jaouen, G. (2007) *J. Organomet. Chem.*, **692** (6), 1219–1225.

105 Top, S., Vessieres, A., Cabestaing, C., Laios, I., Leclercq, G., Provot, C., and Jaouen, G. (2001) *J. Organomet. Chem.*, **637**, 500–506.

106 Jaouen, G., Top, S., Vessieres, A., Leclercq, G., and McGlinchey, M.J. (2004) *Curr. Med. Chem.*, **11** (18), 2505–2517.

107 Pigeon, P., Top, S., Vessieres, A., Huche, M., Hillard, E.A., Salomon, E., and Jaouen, G. (2005) *J. Med. Chem.*, **48** (8), 2814–2821.

108 Paech, K., Webb, P., Kuiper, G., Nilsson, S., Gustafsson, J.A., Kushner, P.J., and Scanlan, T.S. (1997) *Science*, **277** (5331), 1508–1510.

109 Kedar, R.P., Bourne, T.H., Powles, T.J., Collins, W.P., Ashley, S.E., Cosgrove, D.O., and Campbell, S. (1994) *Lancet*, **343** (8909), 1318–1321.

110 Langer, R. (1998) *Nature*, **392** (6679), 5–10.

111 Aronov, O., Horowitz, A.T., Gabizon, A., Fuertes, M.A., Perez, J.M., and Gibson, D. (2004) *Bioconjugate Chem.*, **15** (4), 814–823.

112 Pipkorn, R., Waldeck, W., and Braun, K. (2003) *J. Mol. Recognit.*, **16** (5), 240–247.

113 Kirin, S.I., Duebon, P., Weyhermueller, T., Bill, E., and Metzler-Nolte, N. (2005) *Inorg. Chem.*, **44** (15), 5405–5415.

114 Kirin, S.I., Ott, I., Gust, R., Mier, W., Weyhermueller, T., and Metzler-Nolte, N. (2008) *Angew. Chem. Int. Ed.*, **47** (5), 955–959.

115 Noor, F., Wuestholz, A., Kinscherf, R., and Metzler-Nolte, N. (2005) *Angew. Chem. Int. Ed.*, **44** (16), 2429–2432.

116 Mukhopadhyay, S., Barnes, C.M., Haskel, A., Short, S.M., Barnes, K.R., and Lippard, S.J. (2008) *Bioconjugate Chem.*, **19** (1), 39–49.

117 Tauro, J.R. and Gemeinhart, R.A. (2004) *Neuro-Oncol.*, **6** (4), 335–336.

118 Tauro, J.R. and Gemeinhart, R.A. (2005) *Mol. Pharm.*, **2** (5), 435–438.

119 Tauro, J.R. and Gemeinhart, R.A. (2005) *Bioconjugate Chem.*, **16** (5), 1133–1139.

120 Tauro, J.R., Lee, B.-S., Lateef, S.S., and Gemeinhart, R.A. (2008) *Peptides (Amsterdam)*, **29** (11), 1965–1973.

121 Kageyama, Y., Yamazaki, Y., and Okuno, H. (1998) *J. Inorg. Biochem.*, **70** (1), 25–32.

122 Sitlani, A. and Barton, J.K. (1994) *Biochemistry*, **33** (40), 12100–12108.

123 Sitlani, A., Dupureur, C.M., and Barton, J.K. (1993) *J. Am. Chem. Soc.*, **115** (26), 12589–12590.

124 Zeglis, B.M., Pierre, V.C., and Barton, J.K. (2007) *Chem. Commun.*, (44), 4565–4579.

125 Krotz, A.H., Kuo, L.Y., Shields, T.P., and Barton, J.K. (1993) *J. Am. Chem. Soc.*, **115** (10), 3877–3882.

126 Shields, T.P. and Barton, J.K. (1995) *Biochemistry*, **34** (46), 15037–15048.

127 Shields, T.P. and Barton, J.K. (1995) *Biochemistry*, **34** (46), 15049–15056.

128 Krotz, A.H., Hudson, B.P., and Barton, J.K. (1993) *J. Am. Chem. Soc.*, **115** (26), 12577–12578.

129 Perry, P.J. and Kelland, L.R. (1998) *Expert Opin. Ther. Pat.*, **8** (12), 1567–1586.

130 Harley, C.B., Futcher, A.B., and Greider, C.W. (1990) *Nature*, **345** (6274), 458–460.

131 Kim, N.W., Piatyszek, M.A., Prowse, K.R., Harley, C.B., West, M.D., Ho, P.L.C., Coviello, G.M., Wright, W.E., Weinrich, S.L., and Shay, J.W. (1994) *Science*, **266** (5193), 2011–2015.

132 Arola, A. and Vilar, R. (2008) *Curr. Top. Med. Chem.*, **8** (15), 1405–1415.

133 Arola-Arnal, A., Benet-Buchholz, J., Neidle, S., and Vilar, R. (2008) *Inorg. Chem.*, **47** (24), 11910–11919.

134 Reed, J.E., Arnal, A., Neidle, S., and Vilar, R. (2006) *J. Am. Chem. Soc.*, **128** (18), 5992–5993.

135 Reed, J.E., Arola, A., Vilar, R., and Neidle, S. (2005) *Clin. Cancer Res.*, **11** (24), 9051S–9051S.

136 Reed, J.E., Neidle, S., and Vilar, R. (2007) *Chem. Commun.*, (42), 4366–4368.

137 Reed, J.E., White, A.J.P., Neidle, S., and Vilar, R. (2009) *Dalton Trans.*, (14), 2558–2568.

138 Bagshawe, K.D., Springer, C.J., Searle, F., Antoniw, P., Sharma, S.K., Melton, R.G., and Sherwood, R.F. (1988) *Br. J. Cancer*, **58** (6), 700–703.

139 Niculescu-Duvaz, I. and Springer, C.J. (1997) *Adv. Drug Delivery Rev.*, **26** (2,3), 151–172.

140 Hanessian, S. and Wang, J.G. (1993) *Can. J. Chem.-Rev. Can. Chim.*, **71** (6), 896–906.

141 Tromp, R.A., van Boom, S., Timmers, C.M., van Zutphen, S., van der Marel, G.A., Overkleeft, H.S., van Boom, J.H., and Reedijk, J. (2004) *Bioorg. Med. Chem. Lett.*, **14** (16), 4273–4276.

142 Paschke, R., Paetz, C., Mueller, T., Schmoll, H.J., Mueller, H., Sorkau, E., and Sinn, E. (2003) *Curr. Med. Chem.*, **10** (19), 2033–2044.

143 Dhar, S., Gu, F.X., Langer, R., Farokhzad, O.C., and Lippard, S.J. (2008) *Proc. Natl. Acad. Sci. USA*, **105** (45), 17356–17361.

144 Kawakami, M. and Nakayama, J. (1997) *Cancer Res.*, **57** (12), 2321–2324.

145 Wright, G.L., Grob, B.M., Haley, C., Grossman, K., Newhall, K., Petrylak, D., Troyer, J., Konchuba, A., Schellhammer, P.F., and Moriarty, R. (1996) *Urology*, **48** (2), 326–334.

146 Al-Ejeh, F., Darby, J.M., Thierry, B., and Brown, M.P. (2009) *Nucl. Med. Biol.*, **36** (4), 395–402.

147 Payne, G. (2003) *Cancer Cell*, **3** (3), 207–212.

148 Gao, J., Liu, Y.G., Liu, R., and Zingaro, R.A. (2008), *Chem. Med. Chem.*, **3** (6), 954–962.

149 American Cancer Society (2007) *Breast Cancer Facts and Figures 2007–2008*, American Cancer Society, Inc., Atlanta.

3
Current Status and Mechanism of Action of Platinum-Based Anticancer Drugs

Shanta Dhar and Stephen J. Lippard

3.1
Introduction

3.1.1
Platinum Chemotherapy and Cancer

Chemotherapy, surgery, and radiation therapy are the main pillars of cancer treatment. The term "chemotherapy" refers to the use of any chemical agent to stop cancer cell proliferation. Chemotherapy has the ability to kill cancer cells at sites remote from the original cancer. Thus chemotherapy is referred to as systemic treatment. More than half of all people diagnosed with cancer receive chemotherapy. "Platinum chemotherapy" is the term used for cancer treatment where one of the chemotherapeutic drugs is a platinum derivative. The spectacular and first such platinum-based drug is cisplatin, *cis*-diamminedichloridoplatinum(II). Subsequently, the cisplatin relatives carboplatin and oxaliplatin were introduced to minimize side effects (Table 3.1). Platinum compounds have been the treatment of choice for ovarian, testicular, head and neck, and small cell lung cancer for the past 20 years.

3.1.2
Palette of Current Platinum Chemotherapeutic Drugs

During the last 30 years, over 700 FDA-approved drugs have entered into clinical practice. The success of cisplatin [1] has been the main impetus for the expansion of the family of platinum compounds. Carboplatin [2] and oxaliplatin [3] (Table 3.1) are registered worldwide and have been a major success in clinical practice. Nedaplatin [4] is used in Japan to treat head and neck, testicular, lung, ovarian, cervical, and non-small cell lung cancers. Heptaplatin [3, 4] is used in gastric cancer in South Korea. Lobaplatin [5] is approved in China for the treatment of chronic myelogenous leukemia, metastatic breast, and small cell lung

Bioinorganic Medicinal Chemistry. Edited by Enzo Alessio
Copyright © 2011 WILEY-VCH Verlag GmbH & Co. KGaA, Weinheim
ISBN: 978-3-527-32631-0

Table 3.1 List of platinum compounds.

Compound	Structure	Use	Current state
Cisplatin		Head and neck, testicular, lung, ovarian, cervical, and non-small cell lung cancers	FDA approved
Carboplatin		Head and neck, testicular, lung, ovarian, cervical, and non-small cell lung cancers	FDA approved
Oxaliplatin		Colon cancer	FDA approved
Nedaplatin		Head and neck, testicular, lung, ovarian, cervical, and non-small cell lung cancers	Phase II
Heptaplatin		Gastric, head and neck cancer, small cell lung cancer	Approved in South Korea
Lobaplatin		Chronic myelogenous leukemia (CML), metastatic breast, and small cell lung cancer, esophageal, ovarian cancers	Approved in China Phase II in USA
JM-11		Malignant disease	Abandoned
PAD		Leukemia	Failed in Phase I
Enloplatin		Refractory advanced ovarian carcinoma	Failed in Phase I
Zeniplatin		Ovarian cancer	Failed in Phase I
Cycloplatam		Ovarian and lung cancer	Failed in Phase I
Spiroplatin (TNO-6)		Ovarian cancer	Failed in Phase-II

Ormaplatin (Tetraplatin)		Melanoma, sarcoma, leukemia and breast cancer, refractory diseases	Failed in Phase I
Iproplatin (JM-9)		Small cell carcinoma of the lung, ovarian, metastatic breast, and head and neck cancer	Phase II and Phase III
Triplatin tetranitrate (BBR-3464)		Ovarian, small cell lung and gastric cancer	Failed in Phase II
Aroplatin (l-NDDP)		Colorectal and kidney cancer, malignant pleural mesothelioma	Phase II
Satraplatin (JM-216)		Prostate cancer	Failed in Phase III
Picoplatin		Non-small cell lung cancer	Phase III

cancer. These second-generation platinum drugs were developed to reduce the side effects generally shown by cisplatin, to enhance the therapeutic index, and for application against cisplatin-resistant tumors.

The clinical development of novel platinum compounds has been somewhat disappointing in view of the promise shown in preclinical studies. The vast majority of platinum compounds synthesized for cancer therapy have been abandoned because of low efficacy, high toxicity, and/or low water solubility. Included in this list (see Table 3.1) are JM-11, PAD, enloplatin [6], zeniplatin [7–10], cycloplatam [11], spiroplatin [12, 13], ormaplatin (tetraplatin) [14], iproplatin [15], the polynuclear platinum compound BBR-3464 [7], aroplatin [8], and other platinum conjugates. Although it is difficult to predict the clinical performance of a new platinum compound based solely on its geometry, structural features nonetheless provide important clues about its likely performance. Several platinum compounds are currently under clinical evaluation, including orally administered satraplatin [9] that showed promise against hormone refractory prostate cancer, the sterically hindered picoplatin [16] for small cell lung cancer, a liposomal cisplatin formulation, lipoplatin [10], as a first-line therapy in patients with non-small cell lung carcinoma (NSCLC), and a liposomal oxaliplatin, lipoxal [17]. Adverse side effects and low anticancer activity in Phase I and II clinical studies are the main reasons for the abandonment of platinum drugs. Of the two cisplatin liposomal formulations tested in the clinic, SPI-77 [12] failed in Phase II trials and was abandoned despite successful preclinical performance, whereas lipoplatin has progressed successfully through Phase III clinical trials in NSCLC with a response rate and stable disease

>70%. This result indicates that a formulation strategy, encapsulation of a platinum compound into tumor-targeted nanoparticles, could provide an attractive pathway for the development of clinically useful platinum compounds.

3.1.3
Early History of Cisplatin and Approved Platinum Drugs for the Clinic

The serendipitous discovery [13] of the anticancer properties of cisplatin and its clinical introduction in the 1970s represent a major landmark in the history of successful anticancer drugs. After the discovery of the biological activity of cisplatin, only two additional platinum compounds, carboplatin and oxaliplatin, have been approved by the FDA. Nedaplatin, lobaplatin, and heptaplatin are approved only in Japan, China, and South Korea, respectively. Cisplatin, carboplatin, oxaliplatin, and most other platinum compounds induce damage to tumors by apoptosis [14]. All these platinum drugs have characteristic nephrotoxicity and ototoxicity.

The present chapter focuses on nontraditional, strategically designed platinum(IV) complexes for targeted cancer therapy based on our knowledge of the mechanism of action of cisplatin.

3.2
Mechanism of Action of Cisplatin

The mechanism of cisplatin action is a multi-step process that includes (i) cisplatin accumulation, (ii) activation, and (iii) cellular processing.

3.2.1
Cisplatin Accumulation

The mechanism by which cisplatin enters cells is still under debate [18]. Originally, it was believed that cisplatin enters cells mainly by passive diffusion, being a neutral molecule. More recently, it was discovered that cisplatin might find its way into cells via active transport mediated by the plasma-membrane copper transporter Ctr1p present in yeast and mammals (Figure 3.1) [19]. Details about this active transport remain to be elucidated. Recent studies with Ctr1p−/− mouse embryonic fibroblasts exposed to 2 µM cisplatin or carboplatin revealed only 35% of platinum accumulation compared to that taken up by Ctr1p wild type cells, which supports such an active transport mechanism. Most likely there are multiple pathways by which the drug is internalized.

3.2.2
Cisplatin Activation

Cisplatin is administered to patients by intravenous injection into the bloodstream. The drop in Cl⁻ concentration as the drug enters the cytoplasm sets up a

[Cl-], ~100 mM
[Cl-], ~2-30 mM
Copper Transporter
Cisplatin
Passive Transport
Plasma membrane

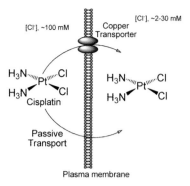

Figure 3.1 Cellular uptake of cisplatin by passive diffusion and via the copper influx transporter Ctr1.

complex pathway (Figure 3.2) for cisplatin activation. Several species form when water molecules enter the platinum coordination sphere, processes that essentially trap the activated form of cisplatin in the cell. The cationic, aquated forms of cisplatin can react with nuclear DNA, which contributes in a major way to the antitumor activity of cisplatin.

3.2.2.1 **Binding to DNA Targets**

There are significant consequences for the cell when cisplatin binds to nuclear DNA and forms covalent crosslinks with the nucleobases [20]. The 1,2-intrastrand d(GpG) crosslink is the major adduct, most likely responsible in large part for the ability of cisplatin to destroy cancer cells (Figure 3.3). Binding of cisplatin to DNA causes a significant distortion of the helical structure, which in turn results in inhibition of DNA replication and transcription. The platinated DNA adducts are recognized by different cellular proteins, including enzymes in DNA repair machinery, transcription factors, histones, and HMG-domain proteins [16, 21].

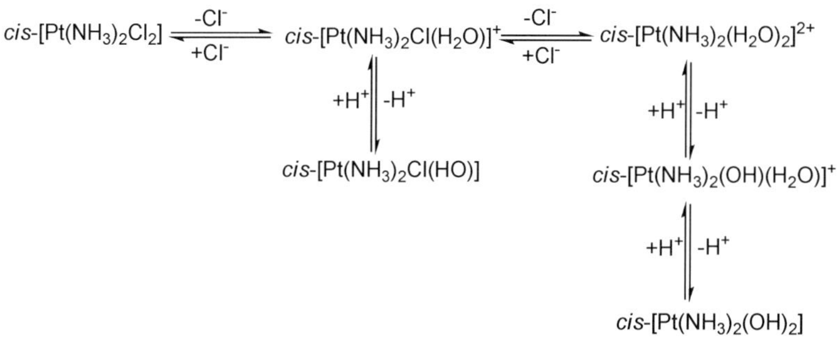

cis-[Pt(NH3)2Cl2]
-Cl-
+Cl-
cis-[Pt(NH3)2Cl(H2O)]+
-Cl-
+Cl-
cis-[Pt(NH3)2(H2O)2]2+
+H+
-H+
cis-[Pt(NH3)2Cl(HO)]
+H+
-H+
cis-[Pt(NH3)2(OH)(H2O)]+
+H+
-H+
cis-[Pt(NH3)2(OH)2]

Figure 3.2 Intracellular activation of cisplatin.

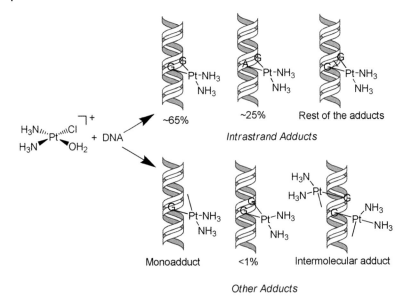

Figure 3.3 Different crosslinks formed by cisplatin.

3.2.2.2 Binding to Non-DNA Targets

Cysteine thiol groups in glutathione and metallothionein defend the cell against cisplatin [22]. Because of the high affinity of thiolate anions for Pt(II), after entering the cell a platinum complex may preferentially bind to sulfur atoms rather than to the bases of DNA [23]. Patients treated with cisplatin for the first time avoid this protective mechanism, but continuous exposure to the drug can build up resistance owing to increased levels of glutathione and metallothioneins [24]. The action of glutathione on cisplatin is catalyzed by glutathione S-transferases (GSTs), and the resulting complexes are exported from cells by the ATP-dependent glutathione S-conjugate export (GS-X) pump (Figure 3.4) [25].

3.2.3
Cellular Processing of Platinum-DNA Adducts

The therapeutic effect of cisplatin is due to the formation of adducts with nuclear DNA that inhibit DNA replication and/or transcription. The main mechanism for removing the intrastrand crosslinks is nucleotide excision repair (NER), but the efficacy of this process varies depending upon the nature of the adducts. NER in human cells depends on many factors, which include the XPA and RPA proteins [26], incision by structure-specific endonucleases, and repair DNA synthesis mediated by DNA polymerase (Figure 3.5). It is important to study the differential repair pathways of cisplatin–DNA intrastrand crosslinks to understand the intracellular processing of cisplatin and to design new platinum drug candidates. One potentially important factor is specific binding of high mobility group box

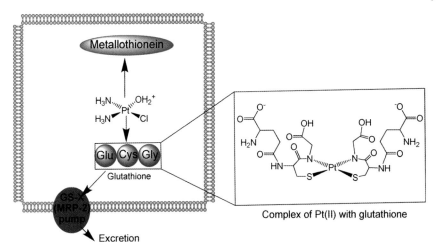

Figure 3.4   Cisplatin binding to glutathione and metallothionein.

(HMGB) proteins to 1,2-intrastrand cisplatin–DNA crosslinks, which shield these lesions from NER [27]. Signal-transduction pathways that control growth, differentiation, and stress responses, involving proteins such as ataxia telangiectasia and RAD3-related (ATR) [28], p53 [29], p73 [30], JUN amino-terminal kinase, and p38 mitogen activated protein kinase (MAPK14) [31], have also been implicated.

3.2.3.1   Cytotoxicity Associated with High Mobility Group (HMG) Proteins

HMG (high mobility group) domain proteins are non-histone chromosomal proteins that bind to specific structures in DNA or in chromatin with little or no sequence specificity [32]. There are two families of HMGB-domain proteins. The first contains two or more HMG domains, and includes the proteins HMGB1 and HMGB2, nucleolar RNA polymerase I transcription factor UBF, and mitochondrial transcription factor mtTF. In the second family, the proteins contain a single HMG domain and include tissue-specific transcription factors. Both families of proteins recognize the major 1,2-intrastrand crosslinks formed by cisplatin [27, 33, 34]. HMG-domain protein mediation of the cytotoxicity of cisplatin is the result of the recognition of DNA–cisplatin adducts by tissue-specific HMG proteins. Several HMG proteins, such as hSRY [35], are expressed in testis tissues and their presence might contribute to the efficacy of cisplatin in the treatment of testicular cancer. Binding of the HMG-domain proteins to cisplatin–DNA 1,2-intrastrand d-(GpG) crosslinks within nuclear DNA impairs the processing of DNA in tumor cells. The distortion caused by this adduct is well recognized by DNA-binding proteins containing HMG domains. The HMG protein forms a 1 : 1 complex with cisplatin–DNA adducts and acts as a protective shield against repair by NER. HMGB1 contains two tandem HMG domains, A and B, and a C-terminal acidic tail. The HMGB-1-induced inhibition of cisplatin–DNA adduct repair is accomplished through the acidic domain. A new member of HMGB family,

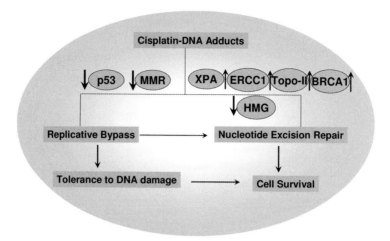

Figure 3.5 Proteins affect cytotoxicity of cisplatin.

HMGB4, was identified recently [36], which is preferentially expressed in the adult mouse testis and sperm cells. Sequence analysis reveals that a disulfide bond, which forms in HMGB1 between Cys23 and Cys45, cannot be formed in HMGB4 because of the absence of cysteine at position 23. The acidic C-terminal tail, which reduces the affinity of HMGB1 for DNA, is also absent in HMGB4. The fact that HMGB4 lacks the disulfide bond and an acidic tail would significantly improve its ability to shield cisplatin intrastrand d(GpG) crosslinks from NER and may contribute to the hypersensitivity of testicular cancer cells to treatment with cisplatin.

3.2.3.2 Cytotoxicity Associated with Non-HMG Proteins

Cisplatin adducts are recognized by mismatch repair proteins as well as ERCC-1 [37], an essential protein in NER (Figure 3.5). Human mismatch-repair protein, hMSH2 [38], also binds with modest specificity to DNA containing cisplatin adducts and displays selectivity for DNA adducts of therapeutically active platinum complexes. Similarly, the NER-related XPA gene [39], which contributes to enhanced repair, is overexpressed in cisplatin-resistant tumors. The sensitivity of testicular cancer to cisplatin has been related to a low expression of XPA and ERCC1/XPF [40]. Transcription-coupled nucleotide excision repair (TC-NER) is an important factor in the activity of cisplatin [41]. Both ERCC1 and XPA are involved in TC-NER. A gene that plays a key role in breast and ovarian cancer is BRCA1 [42]. Before the DNA repair machinery works on cisplatin–DNA crosslinks, these adducts are recognized by several specific proteins [16]. A futile attempt of MMR to repair cisplatin–DNA adducts leads to an apoptotic signal. The MMR complex consists of several proteins, including hMSH2, hMSH6, hMLH1, hMutL$_\alpha$, and hMutS$_\alpha$, with hMSH2 and hMutS$_\alpha$ involved directly in the recognition of cisplatin-d(GpG) intrastrand crosslinks.

The design of new platinum anticancer drug candidates can, in principle, benefit from this information by incorporating components that interfere with these processes or by using genes that provide improved platinum therapy. Of particular interest would be compounds that overcome cisplatin resistance.

3.3
Limitations of Current Platinum-Based Compounds: New Strategies

Despite its side effects, cisplatin-derived cancer therapy has been used successfully for three decades. Platinum-chemotherapy gives characteristic relief and modest improvement in survival. The unique pharmacological properties coupled with the side effects of cisplatin have led to the design of many analogs to broaden the spectrum of activity, reduce side effects, and overcome resistance. Although the cis configuration was initially identified as required for activity, trans-platinum complexes have shown significant antitumor activity in preclinical models. In addition to mononuclear platinum compounds, multinuclear platinum complexes are characterized by a different mode of interaction with DNA. One of the major limitations to the clinical efficacy of platinum compounds is drug resistance, and a most important feature of non conventional platinum compounds is their ability to overcome cellular resistance. The multifactorial nature of clinical resistance requires optimization of platinum-based therapy to include drug delivery approaches. The following discussion focuses on our recent studies to improve platinum therapy by introducing delivery systems that include single-walled carbon nanotubes (SWNTs), polymeric nanoparticles (NPs), and oligonucleotide-functionalized gold nanoparticles (DNA-Au NPs). We introduce a novel platinum(IV) compound, mitaplatin, which uses one of the unique pathways of cancer cell metabolism as a target for its selectivity towards cancer cells. We discuss the anticancer properties of these platinum constructs for their potential use in platinum-based chemotherapy.

3.4
Novel Concepts in the Development of Platinum Antitumor Drugs

The amount of platinum accumulated by cancer cells is an important factor that determines the efficacy of the drugs. Reduced cellular uptake or increased efflux is one reason for drug resistance [43]. A major goal is to develop platinum complexes that can overcome resistance by targeting them to cancer cells. Active and passive targeting of platinum compounds are attractive areas in the advancement of platinum-based drug development. For passive targeting, the vehicle for the drug exhibits prolonged circulation in blood. Active targeting results in higher therapeutic concentrations of the drug at the site of action. Normally, active targeting is achieved by using delivery systems that accumulate in cancer cells by a receptor-mediated mechanism. In passive targeting, the phenomenon known as the enhanced permeability and retention (EPR) effect [44] plays an important role. Low molecular weight platinum complexes that can attack a molecular pathway specific for cancer cells can also be targeted. We discuss different approaches of targeting to overcome resistance exhibited by conventional platinum drugs. The delivery of platinum-based compounds using nano-dimensional carriers of 100–150 nm, which include SWNTs [45], polymeric NPs [46], and gold NPs [47], makes it possible to reduce the general toxicity of chemotherapeutic drugs by decreasing free drug concentration in blood flow and by increasing passive transport and accumulation of nanocarriers in tumors due to enhanced permeability of defective capillary vessel walls. Platinum(IV) complexes provide an attractive alternative to Pt(II) compounds due to their inertness, which results in fewer side effects. Platinum(II)-based anticancer drugs are highly reactive, which leads to lower biological stability. Platinum(IV) complexes, also known as Pt(IV) prodrugs, are reduced in the intracellular environment to yield cytotoxic levels of Pt(II) species through reductive elimination of the axial ligands. In the current discussion, we focus on the use of Pt(IV) complexes as an attractive alternative to the existing portfolio of Pt(II) drugs.

3.4.1
Functionalized Single-Walled Carbon Nanotubes (SWNTs) as Vehicles for Delivery of Pt(IV)-Prodrugs

SWNTs [48–52] offer one of the most promising approaches as drug delivery systems because their physical dimensions mimic those of nucleic acids. Water-solubilized SWNTs have the ability to cross cell membranes by receptor-mediated endocytosis [51]. The surface of SWNTs can be easily modified by introducing several functionalities to attach therapeutic agents for delivery [48]. Well functionalized water soluble SWNTs with high hydrophilicity are non-toxic, even at high concentrations. The clearance of SWNTs from the blood compartment through a renal excretion route follows a first-order dependence without toxic side effects in animals. We have developed an efficient synthetic route to synthesize Pt(IV) compounds containing carboxylic acid functionalities from c,c,t-[Pt(NH$_3$)$_2$Cl$_2$(OH)$_2$] by reaction with acid anhydrides. Platinum(IV) compounds provide the opportunity to introduce ligands at the axial sites. We recently developed such an asymmetric Pt(IV) compound containing an alkoxy and a succinate group at the axial positions. We used non-covalently functionalized SWNTs having phospholipid tethered amines on their surface with a poly(ethylene glycol) (PEG) chain between the amine and phospholipid groups. This asymmetric Pt(IV) compound was tethered to SWNTs by coupling the carboxylate group from the succinate moiety via EDC/NHS promoted amide formation. This construct allowed us to successfully deliver cisplatin attached to SWNTs as a prodrug to human testicular cancer NTera-2 cells by endocytosis [53]. This method was extended by attaching cell-targeting moieties to the platinated SWNTs as "longboat" passengers to achieve high selectivity for cancer cells. In particular, we were able to successfully introduce a folic acid (FA) containing ligand at the axial site of a Pt(IV) center (Figure 3.6) [54]. Folic acid has the potential to target several types

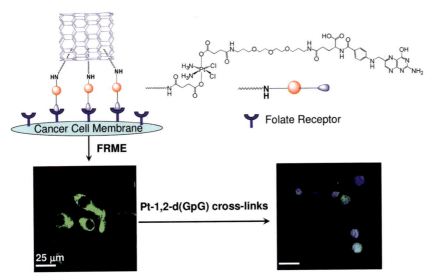

Figure 3.6 Folate receptor targeting Pt(IV) cargo for SWNT delivery vehicle.

of cancer cells because of its ability to interact with folate receptor (α-FR), a high-affinity membrane-anchored protein, through the α carboxylate group [55, 56].

The α-FR is overexpressed on a wide variety of human tumors, including those of ovarian, endometrial, breast, lung, renal, and colon origin. The highest frequency of α-FR overexpression ($>90\%$) occurs in ovarian carcinomas [56]. Our new construct was specifically delivered to folate receptor overexpressing KB and JAR cell lines by the folate receptor mediated endocytosis (FRME) pathway using SWNTs as longboat delivery system. Figure 3.6 displays this delivery as revealed by co-tethering a green fluorescent dye that subsequently appeared in the cytosol of FR-overexpressing KB cells. Most significantly, a cell-based MTT assay showed that the SWNT-Pt(IV)-folate construct was 8.6 times more efficient in killing FR($+$) KB cells than cisplatin. The cytotoxicity of SWNT-tethered Pt(IV) against FR($+$) vs. FR(-) cells demonstrated its ability to target tumor cells that overexpress the FR on their surface. Once inside the cell, cisplatin, formed upon reductive release from the longboat, enters the nucleus and reacts with its target nuclear DNA, as determined by platinum atomic absorption spectroscopic analysis of cell extracts. Formation of the major cisplatin 1,2-d(GpG) intrastrand crosslinks on the nuclear DNA was demonstrated by use of a fluorescent antibody specific for this adduct (Figure 3.6) [57].

3.4.2
Targeted Nanoparticles for Delivery of Cisplatin for Prostate Cancer

Controlled release polymeric nanoparticles (NPs) are effective nanocarriers [58–60]. Targeted uptake of therapeutic NPs in a cell, tissue, or disease-specific manner represents a potentially powerful technology. These NPs can encapsulate drugs and release them through surface or bulk erosion of the particles, diffusion of the drug, or swelling followed by diffusion of the drug. Surface engineering of the NPs offers the potential for functionalization with ligands such as peptides, antibodies, and nucleic acid aptamers, which can target their delivery. Patients with prostate cancer are clinically resistant to cisplatin-based chemotherapy. Targeted delivery of platinum drugs to prostate may offer a new treatment option to overcome this obstacle. We developed a new technology by targeting the prostate specific membrane antigen (PSMA) [61–63] overexpressed on the prostate cancer cells and used NPs as delivery vehicle to increase the effective dose of cisplatin to prostate cancer by means of a nontoxic platinum(IV) prodrug that is activated by intracellular reduction to release cytotoxic cisplatin [64]. We used poly(D,L-lactic-*co*-glycolic acid) (PLGA)–poly(ethylene glycol) (PEG) NPs with PSMA targeting aptamers (Apt) on the surface as a vehicle for the platinum(IV) compound c,t,c-[Pt(NH$_3$)$_2$(O$_2$CCH$_2$CH$_2$CH$_2$CH$_2$-CH$_3$)$_2$Cl$_2$] [65]. Use of the hydrophobic Pt(IV) compound (Figure 3.7) as the prodrug facilitated its encapsulation into pegylated PLGA NPs. Surface engineering with a PSMA targeting aptamer (Apt) allowed us to successfully target prostate cancer with an enormous boost in efficacy over that of free cisplatin, which was released as the cytotoxic warhead upon intracellular reduction [65].

3.4.3
Gold Nanoparticles as Delivery Vehicles for Platinum Compounds

We have devised a method of attaching and delivering platinum compounds using gold nanoparticles. Nanoparticles based on gold cores are promising candidates that provide many desirable features for drug-delivery systems [66–69]. Inherent features of gold nanoparticles are their core size, monodispersity, low toxicity, large surface-to-volume ratio, ease of fabrication, and ease of multifunctionalization. Drug loading can be achieved either by non-covalent interactions, such as with DNA through electrostatic interactions, or by covalent chemical conjugation. Human cells can

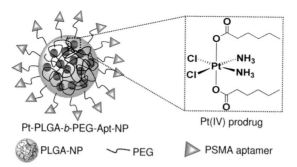

Pt-PLGA-*b*-PEG-Apt-NP

Pt(IV) prodrug

PLGA-NP   PEG   PSMA aptamer

Figure 3.7 Delivery of hydrophobic Pt(IV) prodrug by encapsulation into pegylated PLGA NP with surface modified with a PSMA targeting aptamer for prostate cancer application.

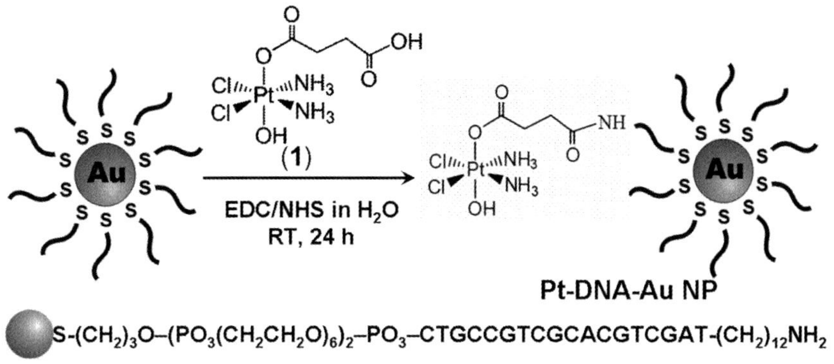

Figure 3.8 Attachment of a platinum(IV) compound to an amine-functionalized DNA-Au NP surface via amide linkages (Pt-DNA-Au NP).

take up gold nanoparticles without any adverse effects. In particular, we employed gold nanoparticles stabilized and labeled with oligonucleotides (DNA-Au NP) as a potential delivery system for platinum(IV) prodrug molecules [70]. We incorporated c,c,t-[Pt(NH$_3$)$_2$Cl$_2$(OH)(O$_2$CCH$_2$CH$_2$CO$_2$H)] (Figure 3.8), a platinum(IV) compound capable of being tethered to an amine-functionalized DNA-Au NP surface via amide linkages (Pt-DNA-Au NP). The use of this compound as the cargo allowed us to release a cytotoxic dose of cisplatin upon intracellular reduction. The Au NPs of 13 ± 1 nm used in this study were functionalized with thiolated oligonucleotides containing an 18-base recognition element of a specific RNA transcript via gold thiol bond formation. These particles are targeted to mRNA sequences coding for enhanced green fluorescent protein (EGFP) expressed in C166, a mouse endothelial cell line. Treatment of c,c,t-[Pt(NH$_3$)$_2$Cl$_2$(OH)(O$_2$CCH$_2$CH$_2$CO$_2$H)] with EDC and NHS afforded the N-succinimidyl ester, which readily formed amide linkages with the amines on the DNA-Au-NP surface (Figure 3.8).

We demonstrated that this Pt(IV) complex, which is otherwise inactive, can be made active against several cancer cell lines when attached to gold nanoparticles as delivery vehicles. The platinum-tethered gold nanoparticles are internalized through endocytosis. Pt-DNA-Au NPs are more active in several cell lines than cisplatin. Upon intracellular reduction, Pt-DNA-Au NP forms cisplatin 1,2-d(GpG) intrastrand crosslinks with nuclear DNA, as demonstrated by a monoclonal antibody assay.

3.4.4
Delivery of Pt(IV) Compounds Targeting Cancer Cell Metabolism

We have developed a Pt(IV) compound, c,t,c-[Pt(NH$_3$)$_2$(O$_2$CCHCl$_2$)$_2$Cl$_2$] (mitaplatin) (Figure 3.9), in which two dichloroacetate (DCA) units are appended to the axial sites of a six-coordinate Pt(IV) center [71]. The dichloroacetate ion is a kinase inhibitor [72]. It stimulates the activity of the mitochondrial enzyme pyruvate

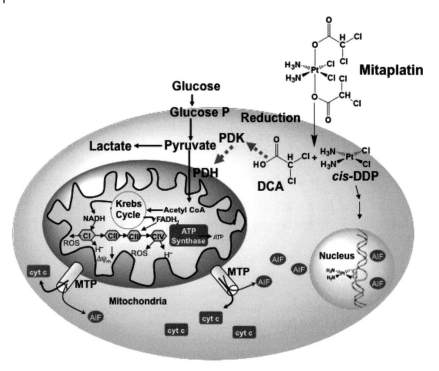

Figure 3.9 Dual targeted action of mitaplatin on cancer cells.

dehydrogenase (PDH) by inhibiting pyruvate dehydrogenase kinase (PDK). DCA thus shifts the cellular metabolism from glycolysis to glucose oxidation, which decreases the mitochondrial membrane potential, or $\Delta\psi_m$. This activity helps to open mitochondrial transition pores (MTPs), thus facilitating translocation of proapoptotic mediators like cytochrome c (cyt c) and apoptosis inducing factor (AIF), which stimulates apoptosis. Lactic acidosis is the common state of metabolism in cancer cells, which rely on glycolysis rather than glucose oxidation for their energy requirements. Healthy cells that become damaged are typically killed by apoptosis, which involves mitochondria, but this mechanism fails in cancer. Apoptotic resistance arises by hyperpolarization of the mitochondrial membrane, which prevents release of pro-apoptotic mediators from mitochondria to the cytoplasm. With the ability to decrease the $\Delta\psi_m$, DCA drives cancer cells to commit cellular suicide, or apoptosis. Unlike most other anticancer agents, DCA does not appear to have any deleterious effect on normal cells. Once inside the cell, DCA is released as the Pt(IV) center in mitaplatin is reduced to Pt(II) in the form of cisplatin. Thus mitaplatin attacks mitochondria as well as the nuclear DNA selectively in cancer cells. We established that the cytotoxicity of mitaplatin is comparable to that of cisplatin in several cancer cell lines. We investigated the *in vitro* selective killing of cancer cells by mitaplatin in a co-culture with normal fibroblasts and demonstrated that mitaplatin is also active against cisplatin resistant cancer cells. In the co-culture study, a mixture of one equivalent of cisplatin and two equivalents of DCA, the stoichiometric composition released upon intracellular mitaplatin reduction, killed both the normal fibroblasts and cancer cells. The reason for the inability of this mixture to selectively kill cancer cells, which differs from the results with mitaplatin, bears further investigation.

3.5
Concluding Remarks

The long trail leading from the discovery of the cytostatic properties of cisplatin to its activity as an anticancer drug to the development of new platinum constructs and incorporation into delivery vehicles demonstrates how fundamental knowledge of platinum chemistry and the molecular mechanism of a drug can eventually generate attractive new candidates for platinum-based cancer therapy. The goal of developing new platinum compounds with improved efficacy and tolerability is challenging. Efforts to find platinum complexes with improved profile will continue. Options under investigation include a focus on improving delivery mechanisms of existing platinum drugs. A valuable strategy for cancer therapy will be to link existing platinum compounds to targeting moieties to boost their uptake to cancer cells. Finally, the marriage of the inherent potency of cisplatin with factors that take advantage of specific cancer cell metabolic pathways, exemplified by mitaplatin, offers an additional powerful strategy that is only in its infancy.

Acknowledgment

Work reported from our laboratory in this chapter has been generously supported by the US National Cancer Institute.

References

1 Rosenberg, B. (1977) *Adv. Exp. Med. Biol.*, **91**, 129–150.

2 Canetta, R., Rozencweig, M., and Carter, S.K. (1985) *Cancer Treat. Rev.*, **12** (Suppl. A), 125–136.

3 Ahn, J.H., Kang, Y.K., Kim, T.W., Bahng, H., Chang, H.M., Kang, W.C., Kim, W.K., Lee, J.S., and Park, J.S. (2002) *Cancer Chemother. Pharmacol.*, **50**, 104–110.

4 Lee, W.S., Lee, G.W., Kim, H.W., Lee, O.J., Lee, Y.J., Ko, G.H., Lee, J.S., Jang, J.S., and Ha, W.S. (2005) *Cancer Res. Treat.*, **37**, 208–211.

5 Voegeli, R., Schumacher, W., Engel, J., Respondek, J., and Hilgard, P. (1990) *J. Cancer Res. Clin. Oncol.*, **116**, 439–442.

6 Kudelka, A.P., Siddik, Z.H., Tresukosol, D., Edwards, C.L., Freedman, R.S., Madden, T.L., Rastogi, R., Hord, M., Kim, E.E., Tornos, C., Mante, R., and Kavanagh, J.J. (1997) *Anticancer Drugs*, **8**, 649–656.

7 Manzotti, C., Pratesi, G., Menta, E., Di Domenico, R., Cavalletti, E., Fiebig, H. H., Kelland, L.R., Farrell, N., Polizzi, D., Supino, R., Pezzoni, G., and Zunino, F. (2000) *Clin. Cancer Res.*, **6**, 2626–2634.

8 Dragovich, T., Mendelson, D., Kurtin, S., Richardson, K., Von Hoff, D., and Hoos, A. (2006) *Cancer Chemother. Pharmacol.*, **58**, 759–764.

9 Bhargava, A. and Vaishampayan, U.N. (2009) *Expert Opin. Investig. Drugs*, **18**, 1787–1797.

10 Boulikas, T. (2009) *Expert Opin. Investig. Drugs*, **18**, 1197–1218.

11 Drees, M., Dengler, W.M., Hendriks, H.R., Kelland, L.R., and Fiebig, H.H. (1995) *Eur. J. Cancer*, **31A**, 356–361.

12 White, S.C., Lorigan, P., Margison, G.P., Margison, J.M., Martin, F., Thatcher, N., Anderson, H., and Ranson, M. (2006) *Br. J. Cancer*, **95**, 822–828.

13 Rosenberg, B., VanCamp, L., Trosko, J. E., and Mansour, V.H. (1969) *Nature*, **222**, 385–386.

14 Sedletska, Y., Giraud-Panis, M.J., and Malinge, J.M. (2005) *Curr. Med. Chem. Anticancer Agents*, **5**, 251–265.

15 Schroyens, W., Dodion, P., and Rozencweig, M. (1990) *J. Cancer Res. Clin. Oncol.*, **116**, 392–396.

16 Guggenheim, E.R., Xu, D., Zhang, C.X., Chang, P.V., and Lippard, S.J. (2009) *ChemBioChem.*, **10**, 141–157.

17 Stathopoulos, G.P., Boulikas, T., Kourvetaris, A., and Stathopoulos, J. (2006) *Anticancer Res.*, **26**, 1489–1493.

18 Gately, D.P. and Howell, S.B. (1993) *Br. J. Cancer*, **67**, 1171–1176.

19 Ishida, S., Lee, J., Thiele, D.J., and Herskowitz, I. (2002) *Proc. Natl. Acad. Sci. USA*, **99**, 14298–14302.

20 Wang, D. and Lippard, S.J. (2005) *Nat. Rev. Drug Discov.*, **4**, 307–320.

21 Jamieson, E.R. and Lippard, S.J. (1999) *Chem. Rev.*, **99**, 2467–2498.

22 Kraker, A., Schmidt, J., Krezoski, S., and Petering, D.H. (1985) *Biochem. Biophys. Res. Commun.*, **130**, 786–792.

23 Ishikawa, T. and Ali-Osman, F. (1993) *J. Biol. Chem.*, **268**, 20116–20125.

24 Schilder, R.J., Hall, L., Monks, A., Handel, L.M., Fornace, A.J. Jr, Ozols, R. F., Fojo, A.T., and Hamilton, T.C. (1990) *Int. J. Cancer*, **45**, 416–422.

25 Ishikawa, T., Wright, C.D., and Ishizuka, H. (1994) *J. Biol. Chem.*, **269**, 29085–29093.

26 Tapias, A., Auriol, J., Forget, D., Enzlin, J.H., Scharer, O.D., Coin, F., Coulombe, B., and Egly, J.M. (2004) *J. Biol. Chem.*, **279**, 19074–19083.

27 Reeves, R. and Adair, J.E. (2005) *DNA Repair*, **4**, 926–938.

28 Lewis, K.A., Lilly, K.K., Reynolds, E.A., Sullivan, W.P., Kaufmann, S.H., and Cliby, W.A. (2009) *Mol. Cancer Ther.*, **8**, 855–863.

29 Bhana, S., Hewer, A., Phillips, D.H., and Lloyd, D.R. (2008) *Mutagenesis*, **23**, 131–136.

30 Leong, C.O., Vidnovic, N., DeYoung, M. P., Sgroi, D., and Ellisen, L.W. (2007) *J. Clin. Invest.*, **117**, 1370–1380.

31 Givant-Horwitz, V., Davidson, B., Lazarovici, P., Schaefer, E., Nesland, J. M., Trope, C.G., and Reich, R. (2003) *Gynecol. Oncol.*, **91**, 160–172.

32 Bustin, M. and Reeves, R. (1996) *Prog. Nucleic Acid Res. Mol. Biol.*, **54**, 35–100.

33 Pasheva, E.A., Ugrinova, I., Spassovska, N.C., and Pashev, I.G. (2002) *Int. J. Biochem. Cell Biol.*, **34**, 87–92.

34 Huang, J.C., Zamble, D.B., Reardon, J. T., Lippard, S.J., and Sancar, A. (1994) *Proc. Natl. Acad. Sci. USA*, **91**, 10394–10398.

35 Trimmer, E.E., Zamble, D.B., Lippard, S.J., and Essigmann, J.M. (1998) *Biochemistry*, **37**, 352–362.

36 Catena, R., Escoffier, E., Caron, C., Khochbin, S., Martianov, I., and Davidson, I. (2009) *Biol. Reprod.*, **80**, 358–366.

37 Li, Q., Yu, J.J., Mu, C., Yunmbam, M.K., Slavsky, D., Cross, C.L., Bostick-Bruton, F., and Reed, E. (2000) *Anticancer Res.*, **20**, 645–652.

38 Mello, J.A., Acharya, S., Fishel, R., and Essigmann, J.M. (1996) *Chem. Biol.*, **3**, 579–589.

39 Weaver, D.A., Crawford, E.L., Warner, K.A., Elkhairi, F., Khuder, S.A., and Willey, J.C. (2005) *Mol. Cancer*, **4**, 18.

40 Welsh, C., Day, R., McGurk, C., Masters, J.R., Wood, R.D., and Koberle, B. (2004) *Int. J. Cancer*, **110**, 352–361.

41 Yang, L.Y., Jiang, H., and Rangel, K.M. (2003) *Int. J. Oncol.*, **22**, 683–689.

42 Moiseyenko, V.M., Protsenko, S.A., Brezhnev, N.V., Maximov, S.Y., Gershveld, E.D., Hudyakova, M.A., Lobeiko, O.S., Gergova, M.M., Krzhivitskiy, P.I., Semionov, I.I., Matsko, D.E., Iyevleva, A.G., Sokolenko, A.P., Sherina, N.Y., Kuligina, E., Suspitsin, E.N., Togo, A.V., and Imyanitov, E.N. (2010) *Cancer Genet. Cytogenet.*, **197**, 91–94.

43 Fuertes, M.A., Alonso, C., and Perez, J. M. (2003) *Chem. Rev.*, **103**, 645–662.

44 Maeda, H. (2001) *Adv. Enzyme Regul.*, **41**, 189–207.

45 Bianco, A., Kostarelos, K., Partidos, C. D., and Prato, M. (2005) *Chem. Commun.*, 571–577.

46 Brannon-Peppas, L. and Blanchette, J.O. (2004) *Adv. Drug Delivery Rev.*, **56**, 1649–1659.

47 Rosi, N.L., Giljohann, D.A., Thaxton, C. S., Lytton-Jean, A.K.R., Han, M.S., and Mirkin, C.A. (2006) *Science*, **312**, 1027–1030.

48 Kam, N.W.S., O'Connell, M., Wisdom, J. A., and Dai, H. (2005) *Proc. Natl. Acad. Sci. USA*, **102**, 11600–11605.

49 Kam, N.W.S., Liu, Z., and Dai, H. (2006) *Angew. Chem. Int. Ed.*, **45**, 577–581.

50 Kam, N.W.S., Liu, Z., and Dai, H. (2005) *J. Am. Chem. Soc.*, **127**, 12492–12493.

51 Kam, N.W.S., Jessop, T.C., Wender, P. A., and Dai, H. (2004) *J. Am. Chem. Soc.*, **126**, 6850–6851.

52 Kam, N.W.S. and Dai, H. (2005) *J. Am. Chem. Soc.*, **127**, 6021–6026.

53 Feazell, R.P., Nakayama-Ratchford, N., Dai, H., and Lippard, S.J. (2007) *J. Am. Chem. Soc.*, **129**, 8438–8439.

54 Dhar, S., Liu, Z., Thomale, J., Dai, H., and Lippard, S.J. (2008) *J. Am. Chem. Soc.*, **130**, 11467–11476.

55 Parker, N., Turk, M.J., Westrick, E., Lewis, J.D., Low, P.S., and Leamon, C.P. (2005) *Anal. Biochem.*, **338**, 284–293.

56 Campbell, I.G., Jones, T.A., Foulkes, W. D., and Trowsdale, J. (1991) *Cancer Res.*, **51**, 5329–5338.

57 Liedert, B., Pluim, D., Schellens, J., and Thomale, J. (2006) *Nucleic Acids Res.*, **34**, e47.

58 Zhang, L., Gu, F.X., Chan, J.M., Wang, A. Z., Langer, R.S., and Farokhzad, O.C. (2008) *Clin. Pharmacol. Ther.*, **83**, 761–769.

59 LaVan, D.A., McGuire, T., and Langer, R. (2003) *Nat. Biotechnol.*, **21**, 1184–1191.

60 Brigger, I., Dubernet, C., and Couvreur, P. (2002) *Adv. Drug Delivery Rev.*, **54**, 631–651.

61 Ghosh, A. and Heston, W.D.W. (2004) *J. Cell. Biochem.*, **91**, 528–539.

62 Farokhzad, O.C., Jon, S., Khademhosseini, A., Tran, T.-N.T., LaVan, D.A., and Langer, R. (2004) *Cancer Res.*, **64**, 7668–7672.

63 Farokhzad, O.C., Cheng, J., Teply, B.A., Sherifi, I., Jon, S., Kantoff, P.W., Richie, J.P., and Langer, R. (2006) *Proc. Natl. Acad. Sci. USA*, **103**, 6315–6320.

64 Mukhopadhyay, S., Barnes, C.M., Haskel, A., Short, S.M., Barnes, K.R., and Lippard, S.J. (2008) *Bioconjugate Chem.*, **19**, 39–49.

65 Dhar, S., Gu, F.X., Langer, R., Farokhzad, O.C., and Lippard, S.J. (2008) *Proc. Natl. Acad. Sci. USA*, **105**, 17356–17361.

66 Di Pasqua, A.J., Mishler, R.E., Ship, Y.-L., Dabrowiak, J.C., and Asefa, T. (2009) *Mater. Lett.*, **63**, 1876–1879.

67 Giljohann, D.A., Seferos, D.S., Patel, P.C., Millstone, J.E., Rosi, N.L., and Mirkin, C.A. (2007) *Nano Lett.*, **7**, 3818–3821.

68 Giljohann, D.A., Seferos, D.S., Prigodich, A.E., Patel, P.C., and Mirkin, C.A. (2009) *J. Am. Chem. Soc.*, **131**, 2072–2073.

69 Sood, P., Thurmond, K.B. II, Jacob, J.E., Waller, L.K., Silva, G.O., Stewart, D.R., and Nowotnik, D.P. (2006) *Bioconjugate Chem.*, **17**, 1270–1279.

70 Dhar, S., Daniel, W.L., Giljohann, D.A., Mirkin, C.A., and Lippard, S.J. (2009) *J. Am. Chem. Soc.*, **131**, 14652–14653.

71 Dhar, S. and Lippard, S.J. (2009) *Proc. Natl. Acad. Sci. USA*, **106**, 22199–22204.

72 Bonnet, S., Archer, S.L., Allalunis-Turner, J., Haromy, A., Beaulieu, C., Thompson, R., Lee, C.T., Lopaschuk, G.D., Puttagunta, L., Bonnet, S., Harry, G., Hashimoto, K., Porter, C.J., Andrade, M.A., Thebaud, B., and Michelakis, E.D. (2007) *Cancer Cell*, **11**, 37–51.

4

New Trends and Future Developments of Platinum-Based Antitumor Drugs

Xiaoyong Wang and Zijian Guo

4.1
Introduction

Platinum-based antitumor drugs have a strong impact on cancer chemotherapy and constitute a cornerstone for the treatment of various solid tumors such as genitourinary, colorectal, and non-small cell lung cancers [1–5]. One of the leading antitumor drugs, cisplatin, has been used for more than three decades in standard chemotherapy regimens either as single drug or in combination with other cyto-toxic agents or radiotherapy [6]. The efficacy of cisplatin, however, has been greatly hampered by drug resistance and severe side effects [7]. Many tumors display inherent resistance to cisplatin while others develop acquired resistance after in-itial treatment [8], and metastasis cancers lack response to it [9]. In addition, high systemic toxicities of cisplatin like nephrotoxicity, neurotoxicity, ototoxicity, and emetogenesis compel the patients to suffer from serious disorders or injuries [10–12].

The imperfection of cisplatin produced a forceful impetus to develop novel platinum-based antitumor drugs. Over the last 30 years, thousands of platinum compounds have been prepared and screened as potential antitumor drugs [13]. Among these candidates, five new drugs, that is, carboplatin, oxaliplatin, neda-platin, lobaplatin, and heptaplatin, have entered the clinical treatment (Figure 4.1), and about ten other complexes are undergoing clinical trials [14, 15]. Each of them possesses some properties that are not evinced by cisplatin. For example, neda-platin has a chelating leaving ligand and the mechanism of action is similar to that of cisplatin [16], but the agent appears to be less nephrotoxic and neurotoxic than cisplatin and carboplatin [17]. Oxaliplatin forms much lower levels of total DNA adducts and crosslinks, but its cytotoxicity is no less than that of cisplatin; in ad-dition, oxaliplatin can overcome cisplatin resistance in murine L1210 cells and colorectal cancer cells [18]. However, since most of these newcomers are structural congeners of cisplatin, with two ammine or amine donor groups and two anionic leaving groups in a cis geometry, some defects of cisplatin are inherited [19].

Bioinorganic Medicinal Chemistry. Edited by Enzo Alessio
Copyright © 2011 WILEY-VCH Verlag GmbH & Co. KGaA, Weinheim
ISBN: 978-3-527-32631-0

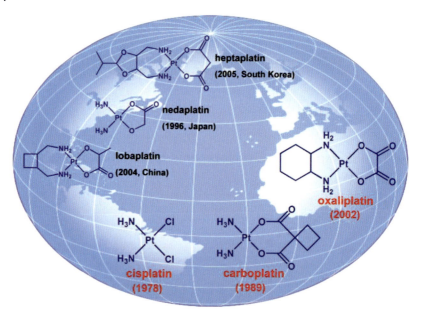

Figure 4.1 Platinum-based antitumor drugs used globally (cisplatin, carboplatin, oxaliplatin) or regionally (nedaplatin, lobaplatin, heptaplatin) in chemotherapy.

To overcome the resistance and improve the pharmacological properties of the currently used platinum drugs, continuing research for new platinum complexes is inevitable. Platinum complexes with totally different structures from those of cisplatin and its analogs provide many opportunities for finding antitumor drugs with different mechanisms of action [20, 21]. Such complexes may exhibit favorable properties like broader antitumor spectrum and distinctive cytotoxic profiles; they may also overcome the resistance pathways that have evolved to eliminate the cisplatin-like drugs. In view of these potential merits, platinum complexes that bind to DNA in a fundamentally different mode than cisplatin have evoked a particular interest in recent years [22, 23].

Since drug resistance is a major limitation to the clinical efficacy of platinum drugs, the most important feature of nonconventional platinum drugs should be the capability of overcoming cellular resistance. In fact, many newly designed platinum complexes exhibit a narrower range of resistance and improved antitumor properties in comparison to existing platinum drugs [24]. The antitumor mechanisms of platinum complexes, especially the information on DNA binding modes and DNA damage repair, are of great importance for the rational design of new compounds. For this reason, considerable attention has been paid to the cellular reactions associated with platinum drugs and molecular mechanisms underlying the resistance and sensitivity of different tumors. The knowledge acquired from these studies offers the prerequisite for the design and modification of platinum-based antitumor drugs.

This chapter focuses on the recent development of novel platinum complexes that may contribute to improve the efficacy of platinum-based chemotherapy. Special emphasis is given to the development of monofunctional platinum(II) complexes, multinuclear platinum(II) complexes, *trans*-platinum(II) complexes, and platinum(IV) complexes. The structure and mode of action of these complexes are fundamentally different from those of cisplatin. Since the pharmacological properties are crucial factors germane to the multifactorial nature of clinical resistance, drug delivery strategies and approaches aimed at optimizing the current platinum-based therapy are also included in this chapter. To help readers appreciate the originality behind the development, molecular alterations of tumor cells that are associated with resistance to platinum drugs will also be discussed briefly before the main theme. Since many valuable reviews and books on this topic have appeared over the years, the materials of this chapter have been sourced exclusively from the literatures after 2003.

4.2
Mechanisms of Action and Resistance

4.2.1
Mechanism of Action

The major mechanism of action for cisplatin involves intracellular activation through aquation to monoaqua species, and subsequent covalent binding to DNA purine bases, forming DNA adducts [25]. These adducts cause distortions in DNA, including unwinding and bending, which are recognized by several cellular proteins such as the DNA mismatch recognition protein hMSH2 and the DNA damage recognition protein HMG1 [26]. The DNA adducts impede cellular processes such as replication and transcription and, in some cases, trigger prolonged G2 phase cell cycle arrest; they also activate several signal transduction pathways that control cell growth, differentiation, and stress responses, involving proteins such as ATR (ataxia telangiectasia mutated and RAD3-related protein), p53, p73, JUN amino-terminal kinase (MAPK8), and p38 mitogen activated protein kinase (MAPK14) [27, 28]. The final cellular outcome is generally apoptotic cell death [29].

Most platinum–DNA bindings occur on the same DNA strand and involve bases adjacent to one another, namely, 1,2-d(GpG) (60–65%) and 1,2-d(ApG) (20–25%) intrastrand adducts or crosslinks. Other less frequently produced adducts are 1,3-d(GpNpG) intrastrand crosslinks (about 2%) and monofunctional adducts on guanines (about 2%). In addition, around 2% of adducts involve guanines on opposite DNA strands, that is, G–G interstrand crosslinks [14]. Nuclear proteins that uniquely recognize such interstrand crosslinks have been identified recently [30]. More information about the cellular uptake of platinum complexes and the interactions of cellular proteins with platinum–DNA adducts as well as the effects of these adducts on proteins that are involved in various DNA-related processes have been discussed comprehensively in the literature and elsewhere in this book [6, 9, 31, 32]. However, the pathways from platinum–DNA binding to apoptosis remain unclear; and the questions about how platinum drugs enter cells and how platinum–DNA damage initiates various cellular signaling pathways remain unsettled [33, 34].

4.2.2
Mechanism of Resistance

Multiple mechanisms have been proposed to elucidate the cellular resistance to cisplatin and its analogs in preclinical models. There are four representative mechanisms: (i) decreased drug accumulation or increased drug efflux; (ii) increased detoxification of the drug by thiol-containing molecules within the cells; (iii) enhanced repair and increased tolerance to DNA damage; and (iv) changes in molecular pathways involved in the regulation of cell survival or cell death [7, 35, 36]. These mechanisms are schematically shown in Figure 4.2 and will be expounded in the following sections.

4.2.2.1 Decreased Drug Accumulation
In tumor cells with acquired resistance to cisplatin, reduced platinum accumulation in comparison to the parental cells is a frequent observation. Thus, resistance to cisplatin is partly related to the cellular influx and efflux mechanisms of the drug.

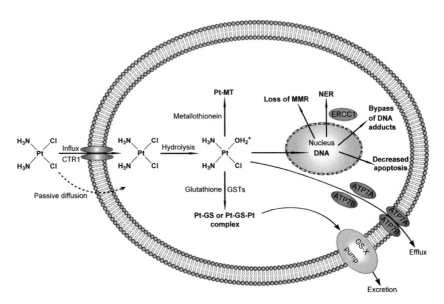

Figure 4.2 Key events of cellular resistance to platinum-based antitumor drugs; see text for details.

In principle, the influx and efflux of platinum drugs can regulate the accessibility to the target DNA and affect the pattern of cytotoxicity. In recent years, the copper transporter-1 (CTR1), a major plasma membrane transporter involved in copper homeostasis, has been shown to play a role in the influx of platinum drugs [37–39]. However, studies on drug-CTR1 interactions using model peptides show that cisplatin and its analogs lose all the ligands when bound to the Met rich sequences of the peptides [40, 41]; therefore, it is debatable whether platinum drugs are transported through such a pathway. Efflux proteins that are involved in copper transport, that is, the ATPases ATP7A and ATP7B, have also been shown to modulate the efflux of these drugs [42]. Nevertheless, as noted above, the underlying molecular mechanism by which platinum drugs enter cells remains poorly defined. Although different platinum drugs share some of the influx/efflux transporters, the molecular determinants of uptake and export are only in part overlapping. For example, contrary to cisplatin and carboplatin, the accumulation of oxaliplatin is not dependent on CTR1 at high concentrations [37]. In addition, cisplatin can enter cells through passive diffusion and facilitated diffusion. In cisplatin-resistant cells, platinum accumulation seems dependent on the lipophilicity of the drug, and an increased lipophilicity appears to favor the accumulation [43]. Therefore, the lipophilicity of a platinum drug may be a key factor to bypass the resistance.

4.2.2.2 Increased Detoxification

Plentiful evidence implies that increased cytoplasmic thiol-containing species are causative factors of resistance to cisplatin or carboplatin. Sulfur-containing biomolecules such as glutathione (GSH), metallothionein (MT), cysteine (Cys), and methionine (Met) are closely associated with the mechanism of resistance because of their high affinity for the Pt(II) ion. Platinum–sulfur interactions can significantly influence the cellular fate of platinum drugs [44]. The principal reaction products of platinum drugs with GSH have been identified as either mono-coordinated Pt-GS or bridged Pt-GS-Pt complexes (GS $=$ deprotonated GSH) [45–48]. Formation of these complexes would reduce the amount of intracellular platinum available for interaction with DNA and protect dividing cells from cisplatin toxicity. The conjugation of cisplatin with GSH might be catalyzed by glutathione *S*-transferases (GSTs). This process makes cisplatin more readily exported from cells by an ATP-dependent glutathione *S*-conjugate export (GS-X) pump [49]. In cell nucleus, GSH can quench platinum–DNA monoadducts before the formation of intrastrand crosslinks. In addition, some studies involving tumor biopsies from patients show that the GSH metabolic pathway is directly involved in the detoxification or inactivation of platinum drugs, and thus plays a role in the acquired and inherited resistance to these drugs [50]. Increased levels of the Cys-rich MTs have also been shown to induce resistance to cisplatin [51, 52].

4.2.2.3 Enhanced Repair and Increased Tolerance to DNA Damage

Enhanced DNA repair is another factor that may contribute to resistance to platinum drugs [27]. Many cisplatin-resistant cell lines from various tumor types have shown increased DNA-repair capacity in comparison to their sensitive counterparts [53]. Nucleotide excision repair (NER) is the major pathway to remove cisplatin lesions from DNA. In particular, the expression of the excision repair cross complementing-group 1 (ERCC1) protein has been correlated with resistance to platinum drugs [54]. For example, the suppression of ERCC1 expression by small interfering RNAs can decrease NER of cisplatin-induced DNA lesions and enhance cellular sensitivity to cisplatin [55]. In some cases, a polymorphism of ERCC1 might occur; this is associated with reduced translation of the gene and improved response to platinum-based chemotherapy [56].

Increased tolerance to DNA damage also plays a central role in platinum drug resistance, which can occur through loss of function of the mismatch repair (MMR) pathway [57]. Platinum drugs interfere with normal MMR activity and prevent a repair from being completed, and thereby lead to an apoptotic response. In this framework, the MLH1 and MSH2 genes seem to be particularly important. When MMR is deficient, cells can continue to proliferate despite DNA damage caused by platinum drugs, and consequently result in drug resistance [58]. Loss of this repair pathway leads to low-level resistance to cisplatin and carboplatin [28]. A clinical study shows that dysfunction of the MMR pathway through methylation of the hMLH1 gene after chemotherapy predicts poor survival for ovarian cancer patients [59]; while other data show no correlation with intrinsic resistance [60]. Another tolerance mechanism to platinum drugs involves enhanced replicative bypass, where cisplatin–DNA adducts can be bypassed by DNA polymerases β and η during a process termed translesion synthesis [61]. Oxaliplatin–GG adducts can be more efficiently bypassed by polymerases β and η than cisplatin–GG adducts [62]. Experiments indicate that inhibition of DNA polymerase η could increase the anticancer efficiency of cisplatin [63].

4.2.2.4 Reduced Apoptotic Response and Activation of Survival Pathways

Reduced apoptotic response has been accepted as a feature of tumor cells exhibiting intrinsic or acquired drug resistance. This feature is frequently associated with the tumor suppressor protein p53, which is a nuclear phosphoprotein involved in the control of cell cycle, DNA repair, and apoptosis [64]. The recognition of cisplatin–DNA adducts by the components of the NER system might mediate the signal transduction processes and hence activate the p53 pathway and cell cycle arrest, which allow DNA repair enhancing resistance to cisplatin [65, 66]. Nevertheless, the role of p53 in tumor cells response to cisplatin is yet ambiguous and depends on the tumor type or circumstance. A dual role may be seen for p53 after exposure to platinum-based treatment, either activating mechanisms that lead to apoptosis or launching processes directing to DNA repair and cell survival [67]. The tumor suppressors BRCA1 and BRCA2 may also be associated with acquired resistance to platinum-based chemotherapy. Recently, it has been shown that acquired resistance to cisplatin in BRCA1- or BRCA2-mutated tumors can be mediated by secondary intragenic mutations in BRCA1 or BRCA2 that restore the wild-type BRCA1 or BRCA2 reading frame. The results suggest that such secondary mutations can mediate resistance to platinum drugs [68, 69]. Apoptosis is activated mainly through the "intrinsic" pathway involving depolarization of the mitochondrial membrane potential. The Bcl-2 protein family consists of several key regulators of this process and includes pro- and antiapoptotic members [70]. Impaired apoptosis based on overexpression of Bcl-2 and Bcl-xL was shown to hamper the efficacy of clinically used platinum drugs [71]. Accordingly, the Bcl-2/ Bcl-xL inhibitor ABT-737 can sensitize cancer cells to carboplatin [72].

What we have discussed above mainly belongs to the "classical" resistance mechanisms. There are also several new molecular factors that have been linked to platinum resistance; however, their effect on platinum efficacy has been assessed only *in vitro* [73, 74]. Despite that, these potential resistance factors are of great interest since inhibitors of them are currently under development and may eventually prove useful as a means of reversing platinum resistance.

Overall, the mechanisms of tumor resistance to platinum drugs provide valuable insights for the rational design of platinum complexes to circumvent the resistance in patients. The development of novel platinum drugs has been pushed forward by the studies on cellular systems and the search for non-cross-resistant agents. In practice, the development of lead compounds to bypass any possible mechanism of resistance is regarded as a major advance. At this point, we turn to the major theme of this chapter and explain, when possible, how platinum complexes have been designed to overcome the problems associated with different mechanisms, based on the above knowledge.

4.3
Monofunctional Platinum(II) Complexes

Monofunctional platinum(II) complexes such as [PtCl(dien)]Cl (dien = diethylenetriamine) and [PtCl(NH$_3$)$_3$]Cl were considered to be biologically inactive in early studies. Furthermore, monodentate binding of such complexes to DNA was believed to be biologically insignificant because DNA conformation and downstream cellular processes are hardly affected. However, more recent studies show that some monofunctional Pt(II) complexes indeed exhibit potent antitumor effects. Interestingly, these active monofunctional complexes share a common feature with one or two considerably bulky amine ligands, as shown in the following examples.

Cationic monofunctional Pt(II) complex **1** was established long ago to possess significant antitumor activity in mouse tumor models. Recently it has been found that this compound is an excellent substrate for organic cation transporters SLC22A1 and SLC22A2, which are plentifully expressed in human colorectal cancers and are associated with the uptake of oxaliplatin. Unlike cisplatin or oxaliplatin, **1** binds to DNA monofunctionally, but blocks transcription nearly as efficiently as cisplatin. Although the DNA adducts of **1** can be removed by the NER apparatus, the process is significantly reduced relative to repair of cisplatin or oxaliplatin damage. These properties indicate that **1** merits consideration as a therapeutic option for treating colorectal and other cancers bearing appropriate cation transporters [75].

A series of monofunctional Pt(II) complexes of 8-aminoquinoline derivatives have been synthesized and tested against a wide range of tumor cell lines, including human liver (BEL-7402), colon (HCT-116), lung (SPC-A4), non-small-cell lung (A549), stomach (SGC-7901, MKN-28), epithelial ovary (HO-8910) cancer cell lines, and human or murine leukemia cell lines (MOLT-4, HL-60, P-388). Complex **2** is the most cytotoxic compound that exhibits significant activity against most of the cell lines. The IC_{50} value of **2** in HCT-116, SPC-A4, BEL-7402, and MOLT-4 cell lines is 0.38, 0.43, 0.43, 0.61 μM, respectively. Even at 6.6×10^{-7} mol L^{-1}, the inhibition rate to BEL-7402 can reach 75.1%, which is nearly six times higher than that of cisplatin [76, 77]. High lipophilicity can facilitate the passive uptake of drug molecules across the lipidic cell membrane and affect the activity of the drugs. In these complexes, the most cytotoxic compound (**2**) is also the most lipophilic complex. It has been shown that the introduction of bulky planar ligands maintains the cytotoxicity of Pt(II) complexes while significantly reducing the rate of deactivation by sulfur-containing molecules. For instance, sterically hindered complex **3** (ZD0473, picoplatin), a clinical developing candidate aimed at overcoming resistance mechanisms, has reduced susceptibility to inactivation by thiols as compared with cisplatin [78]. Therefore, a decrease in reactivity towards GSH is expected when a bulky 8-aminoquinolyl is introduced into the above complexes. Indeed, these complexes hardly react with GSH and hence they are likely to circumvent the resistance caused by detoxification mechanism. Although the exact mechanism of action remains to be elucidated, our preliminary studies show that the DNA binding mode of these complexes is radically different from that of existing platinum drugs because of the monofunctional nature.

Monofunctional complex **4** is characterized by a potent growth inhibitory activity with mean IC_{50} of 1.1 μM in human lung (NCI-H460, A549/ATCC), breast (MCF-7, MDA), colon (HCT-116, KM12, COLO205), and ovary (A2780, 41M, OVCAR-8, SKOV-3) cancer cell lines, whereas the mean IC_{50} of cisplatin in this panel is 3.8 μM. In particular, **4** is markedly more effective than cisplatin towards COLO205, HCT-116, KM12, MDA, and SKOV-3 cancer cells. Beside its remarkable activity towards the intrinsic cisplatin-resistant SKOV-3 cancer cells, **4** also exhibits inhibitory activity towards A2780cisR and 41McisR ovarian cancer cells with acquired resistance to cisplatin. The resistance factors (IC_{50} resistant line/IC_{50} parent line) in A2780cisR/A2780 and 41McisR/41M pairs are 0.6 and 0.7, respectively; by contrast, those for cisplatin are 16 and 4.5, respectively. The results indicate that the growth inhibitory profile of **4** is different from that of cisplatin and characterized by a major activity towards cisplatin-resistant tumor cells, thus pointing to a mechanism of action distinct from that of cisplatin [79].

4

Platinum–intercalator conjugates are hybrid complexes that interact with DNA through a dual binding mode, that is, intercalation plus platination [80]. Platinum–acridinylthiourea conjugates, represented by the prototype **5**, are examples [81]. Unlike cisplatin derivatives, **5** damages DNA by a unique mechanism involving monofunctional platinum binding to guanine (80%) or adenine (20%) and intercalation of the acridine moiety into the base pair step adjacent to the site of platination, instead of forming crosslinks [82–84]. More importantly, **5** appears to be the first monofunctional platinum-based agent that targets the minor groove of B-form DNA and leads to the platination of adenine-N3 situated there [85]. The structural perturbations in DNA produced by these adducts do not mimic those induced by cisplatin [86]. Furthermore, these DNA minor groove adducts can inhibit the association of human TATA binding protein (hTBP), which may be critical for transcription initiation [87]. Compound **5** shows a strong *in vitro* cytotoxicity against a broad range of tumor cell lines similar or superior to that of cisplatin, particularly against non-small-cell lung cancer (NSCLC) cell lines of different genetic backgrounds [88–90]. Unfortunately, the *in vitro* cytotoxicity did not translate into the *in vivo* inhibition of tumor growth. Hence, different modifications were made to seek more applicable candidates from the derivatives [91, 92]. Recently, it was found that the substitution of the thiourea with an amidine group could dramatically affect the chemical reactivity and biological activity of this type of conjugates. Complex **6** proved to be a more efficient DNA binder than **5** and induces adducts in sequences not targeted by the prototype. Complexes **6** and **7** inhibit NSCLC (H460) cells with IC_{50} values of 28 and 26 nM, respectively, and **7** slows tumor growth in a H460 mouse xenograft study by 40% when administered at a dose of 0.5 mg kg^{-1} [93]. By far, **7** is the first non-crosslinking platinum agent able to slow progression of NSCLC *in vivo*. Moreover, the reactivity of **6** with *N*-acetylcysteine is considerably reduced as compared to that of **5**, which may contribute to its superior potency *in vitro* and *in vivo* [94]. Nevertheless, the toxicity of these new hybrid agents is relatively high in animals, and therefore additional structural modifications are still needed.

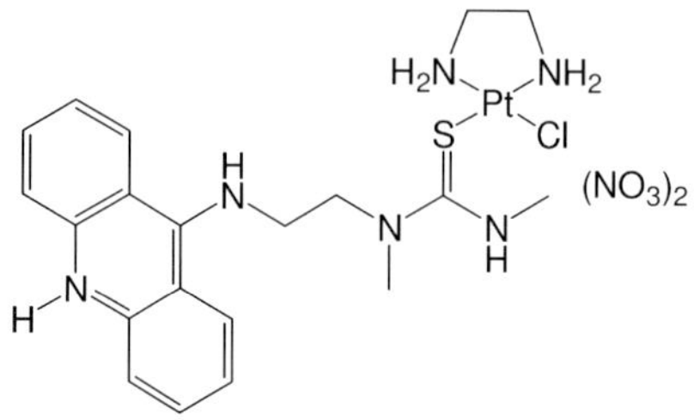

5

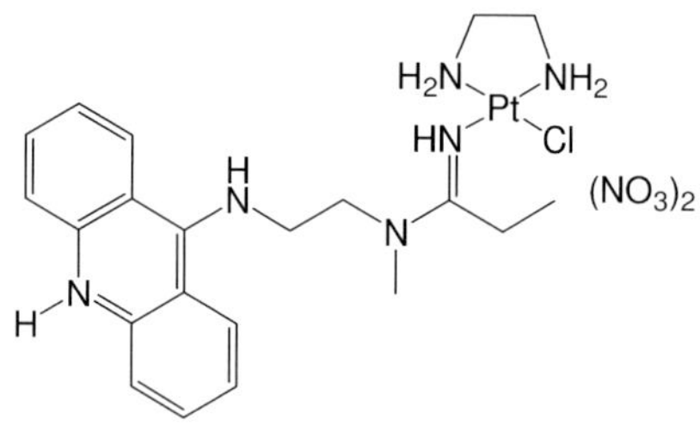

6

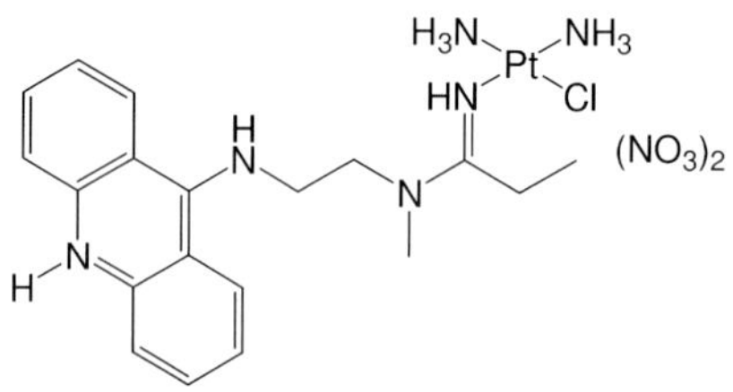

7

Another group of platinum–intercalator conjugates is exemplified by complexes **8** and **9**. These monofunctional Pt(II) complexes, like **5**–**7**, possess only one exchangeable chloride ligand, and display enhanced cytotoxicity as compared with cisplatin in human breast (MCF-7), human lung (A549), murine leukemia (P388), cisplatin-sensitive human ovarian (A2780), and cisplatin-resistant human ovarian (A2780cisR) cancer cell lines. In the MCF-7, A549 and P388 cell lines, the percentage of cell growth inhibition at 10 μM is in the order: cisplatin < **8** < **9**, indicating that the integration of the intercalator into a polyaza macrocyclic ligand helps to improve the efficiency of the complex. In addition, **8** and **9** are more potent against A2780cisR cells, and more effective in penetrating cell membranes than cisplatin [95].

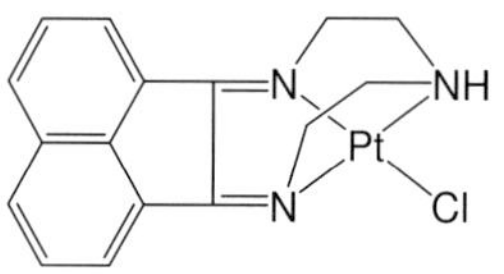

8 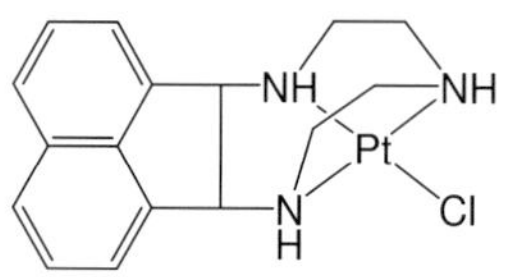　　　　**9**

The above examples demonstrate that monofunctional Pt(II) complexes represent a new class of antitumor agents that do not comply with the classic structure–activity relationships of platinum complexes but still possess high cytotoxic activity against tumor cells. A recent study performed in cell-free media indicated that enhancement of the bulkiness of the dien ligand in monofunctional $[Pt(NO_3)(dien)]^+$ by multiple methylation dramatically influences the biophysical properties and biochemical processes of DNA. This finding strongly supports the hypothesis that monodentate DNA binding of Pt(II) complexes could considerably affect the properties of DNA and consequently downstream cellular processes as a result of a large increase in the bulkiness of the non-leaving ligands [96]. In short, monofunctional Pt(II) complexes merit more intensive research, which may provide new opportunities to obtain promising antitumor drugs.

4.4
Trans-Platinum(II) Complexes

Platinum complexes of trans geometry were believed to be inactive previously because they are not able to form 1,2-intrastrand crosslinks, and thereby are unable to inhibit DNA replication and transcription effectively [97]. Moreover, *trans*-[PtCl$_2$(NH$_3$)$_2$] (transplatin) is kinetically more reactive than cisplatin, which may contribute to its deactivation by sulfur-containing molecules [98, 99]. However, later evidence revealed that different types of adducts such as 1,3-intrastrand and interstrand crosslinks can also be cytotoxic. Accordingly, in the last decade, large numbers of *trans*-platinum complexes have been prepared and many have shown significant antitumor activity in preclinical models [100]. Transplatin induces monoadducts that may be repaired or undergo further rearrangements to form DNA–DNA crosslinks as well as DNA–protein crosslinks; and it also forms interstrand crosslinks between guanine and cytosine of double-stranded DNA, but such crosslinks do not change the stability and structure of DNA significantly [101]. Interestingly, recent results indicate that transplatin is almost as cytotoxic as cisplatin toward HaCaT keratinocytes and A2780 ovarian cancer cells after irradiation by UVA light. Irradiation can activate both chloride ligands of transplatin, and greatly enhance formation of DNA interstrand crosslinks and DNA–protein crosslinks, which may largely account for the cytotoxicity of photoactivated transplatin [102].

Substitution of NH$_3$ in transplatin by a range of amines markedly enhances the cytotoxicity of this complex to a level equal to or better than that of cisplatin in many cases. The nature of the amine is a determinant for the properties of the bifunctional intrastrand and interstrand DNA adducts formed by transplatin derivatives. In fact, DNA adducts and cellular events induced by antitumor-active *trans*-platinum complexes are different from those of cisplatin and transplatin [103, 104]. The cytotoxicity for various series of *trans*-platinum complexes has been well summarized recently [105, 106]. In general, most complexes display cytotoxicity in the micromolar (1–20 μM) range and consistently display cytotoxicity in cisplatin-resistant cells. Enhancement of cytotoxicity by replacement of NH$_3$ is a general phenomenon. In this part, we discuss three different classes of mononuclear *trans*-platinum(II) complexes: those with imino ligands, aromatic amine ligands, and aliphatic amine ligands.

The replacement of ammine ligands by iminoether in transplatin potentiates significantly the cytotoxic activity, including activity against cisplatin-resistant tumor cells. For instance, complex **10** shows cytotoxicity comparable to that of cisplatin, and can circumvent the cisplatin resistance of A2780/cp8 cells. This complex forms mainly monofunctional adducts at guanine residues on DNA, which are not recognized by HMGB1 proteins and are readily removed from DNA by the NER system. These monofunctional adducts readily form crosslinks with proteins, which promote the termination of DNA polymerization by DNA polymerases *in vitro* and inhibit the removal of adducts from DNA by NER. The DNA–platinum–protein ternary crosslinks could persist longer than the non-crosslinked monofunctional adducts, which may potentiate the toxicity of **10** toward tumor cells. Thus, such ternary crosslinks may represent a unique feature of the mechanism underlying the antitumor effects of **10** [107].

10   11   12

For platinum–iminoether complexes, the antitumor activity could be affected by the ligand configuration. To avoid the isomerization between *(Z)-* and *(E)-*configurations of the iminoether ligands, cyclic ligands mimicking the stereochemistry of *(Z)-* and *(E)-*iminoethers were used to modify the structure of transplatin. In a panel of human tumor cell lines (ovary, colon, lung, and breast), **11** and **12** exhibit much higher cytotoxicity than transplatin and the corresponding platinum–iminoether complexes. More importantly, these complexes partially circumvent the cisplatin resistance of A2780cisR cells, and largely overcome the resistance of 41McisR cells. The interaction of DNA with **11** or **12** appears to be characterized by the formation of persistent monoadducts [108].

Substitution of acetonimine for one or two ammine ligands dramatically increases the antitumor activity of transplatin. For example, *trans* complexes **13** and **14** are more active than transplatin in a panel of human tumor cell lines representative of ovarian, colon, lung, and breast cancers; moreover, they are able to circumvent, at least partially, the cisplatin resistance of ovarian cell lines A2780cisR and 41McisR [109].

13   14

Mechanistic studies show that the cellular uptake of *trans*-platinum complexes containing at least one aromatic N-donor heterocycle is greater than that of transplatin, and these complexes are less reactive towards detoxifying biomolecules such as GSH. Complex **15** exhibits cytotoxic activity against MCF-7 and A2780 cancer cells. The complex induces DNA strand breakage and DNA–protein crosslinks in both cell lines; in addition, it interferes with topoisomerase I in the tumor cells, which is a function not evident in *cis*-platinum complexes. However, **15** is a poor apoptotic inducer in these cells [110]. Analyses of short DNA duplexes containing the single, site-specific monofunctional adduct of **15** show that the adduct inhibits DNA synthesis and creates a local conformational distortion similar to that produced in DNA by the 1,2-GG intrastrand crosslink of cisplatin. The monofunctional adduct of **15** can be recognized by HMGB1 domain proteins and removed by the NER system as happened to cisplatin [111]. Modification of DNA by **15** leads to mono- and bifunctional intra- or interstrand adducts in roughly equal proportions. Moreover, bifunctional crosslinks behave differently from the major intrastrand adduct of cisplatin, in that 1,3-intrastrand crosslinks of **15** are not recognized by HMG-domain proteins and are not removed by NER. These results suggest that the cytotoxicity of **15** may be a sum of contributions made by all the multiple DNA lesions [112].

15   **16**   **17**

Complexes **16** and **17** are also characterized by a planar aromatic heterocyclic amine ligand in the position trans to NH$_3$, and are more cytotoxic than cisplatin against the cisplatin-resistant A2780cisR and human leukemia cancer cell line HL-60 [113, 114]. Complex **17** forms stable intrastrand and interstrand crosslinks with DNA, which distort DNA conformation in a unique way and cause local DNA unwinding. DNA adducts of **17** also reduce the affinity of the tumor suppressor protein p53 for its consensus DNA sequence, which results in downstream effects related to p53 protein differing from those induced by transplatin. Further, **17** induces HL-60 cell death mainly through apoptosis (>80%) while cisplatin only reaches 50%. These differences may contribute to the distinct antitumor activity of **16** and **17**. However, the isomers of **17**, with 2-hydroxymethylpyridine or 3-hydroxymethyl-pyridine as ligand, are inactive [115, 116].

In recent years it has been proved that the replacement of halide ligands in various *trans*-platinum complexes by carboxylate ligands not only makes the complexes more water soluble but also enhances their stability toward hydrolysis. For example, *trans*-planar amine acetate complexes show increased aqueous solubility and slow hydrolysis rates that may lead to more desirable behavior *in vivo*. In addition, the biological profiles for these trans-complexes indicate a lack of cross-resistance in tumor cells resistant to cisplatin or oxaliplatin [117]. Complexes **18–20** are the first water-soluble cytotoxic *trans*-platinum(II) complexes containing a [N$_2$O$_2$] donor set with planar amines. These compounds are more cytotoxic against many cisplatin-resistant human ovarian cancer cell lines than against the parent cisplatin-sensitive cell lines (A2780, CH1, 41M) [118]. Water-soluble complexes **21** and **22** are highly cytotoxic against murine keratinocytes cell lines Pam 212/Pam 212-*ras* and human ovarian cancer cell lines 41M/41McisR. These complexes induce cell death through the apoptosis pathway and exhibit high levels of DNA platination in Pam 212-*ras* cells at concentrations five times lower than cisplatin. The amount of platinum bound to DNA relative to the platinum input is significantly higher for **21** and **22** than for cisplatin. The presence of two lipophilic picoline ligands in **21** and **22** may favor their transport through the cells and nuclear membranes and result in higher intracellular accumulation as compared with complexes having mixed picoline and ammine ligands [119]. Asymmetric *trans*-platinum(II) complexes **23** and **24** show high *in vitro* cytotoxicity in cisplatin-sensitive mouse leukemia L1210 cell line, and largely retain the activity in cis-platin-resistant L1210R cell line. However, the reactivity of the complexes towards DNA nucleobases is reduced significantly [120].

The aqueous solubility of complexes **25** and **26** is better than their dichlorido counterparts and is dependent on the steric hindrance of the heterocycle as well as the carboxylate group. As a result, the solubility decreases in the order **25a** > **25b** > **25c** > **26a** > **26b**. The hydrolysis rates of these carboxylate complexes follow the same pattern as the solubility. All the complexes exhibit cytotoxic activity in A2780 cells in the micromolar range. The most cytotoxic complex is **25a**, which is also the most soluble and hydrolabile compound of the series. The cytotoxicity of the carboxylate complexes follows the order **25a** > **25b** > **25c**, suggesting that the nature of the carboxylate leaving group is related to the cytotoxicity; while that of the acetate complexes follows the order of **25c** < **26a** < **26b**, suggesting that the steric hindrance of the methyl group can influence cytotoxicity (4-pic > 3-pic > 2-pic), with the more sterically hindered 2-pic complex being the most cytotoxic of the type [117].

Replacement of both ammine groups in transplatin by one planar amine and one aliphatic amine leads to complexes **27** and **28** that have a higher cytotoxicity against A2780/A2780cisR and CH1/CH1cisR cancer cells than cisplatin. The major merit of these complexes might be their potent cytotoxicity in cancer cell lines resistant to cisplatin. These complexes mainly form stable intrastrand crosslinks with DNA; however, they also generate interstrand crosslinks and mono-adducts. The reaction rate of **27** and **28** with DNA is somewhat faster than that of cisplatin and transplatin, and the DNA binding mode is also different as compared with transplatin and other analogs in which only one ammine group is replaced. In addition, the monofunctional DNA adducts of **27** and **28** are quenched by GSH to a considerably less extent than are the adducts of transplatin, which may potentiate their cytotoxic activity [121]. Similarly, asymmetric *trans*-platinum(II) complexes containing an isopropylamine ligand *trans* to an azole ligand (pyrazole, 1-methylimidazole, and 1-methylpyrazole) also possess cytotoxicity comparable to cisplatin against a panel of human tumor cell lines, and largely retain the activity in the cell line resistant to cisplatin (A2780cisR) [122].

27   **28**

The hydrophobic character of the ligands could help platinum complexes to cross cellular membranes and modify their intracellular accumulation. Generally, *trans*-platinum complexes with bulky ligands have lower water solubility than their cis counterparts, but this can be improved by using cyclic amine ligands. Complex **29** has significant cytotoxicity against A2780/A2780cisR, 41M/ 41McisR, and CH1/CH1cisR cell lines, especially the cisplatin-resistant A2780cisR cell line. Interestingly, the corresponding 2-, 3-, and 4-methylpiperidine complexes are less effective than **29** [123]. Unlike cisplatin, this complex forms mainly stable 1,3-GNG intrastrand crosslinks rather than 1,2-intrastrand crosslinks with DNA. In contrast with the 1,2-intrastrand crosslinks formed by cisplatin, the 1,3-GNG intrastrand crosslinks cannot be recognized by HMG1 proteins and hence cannot be removed efficiently from DNA by the NER system [124]. Complexes **30** and **31** also form 1,3-GNG intrastrand crosslinks that are more stable than the corresponding lesions induced by transplatin in double helical DNA. In addition, they induce more interstrand crosslinks with DNA in comparison to cis- and transplatin, which may be responsible for their high activity in tumor cells [125, 126].

29   **30**   **31**

Substitution of bulkier ligands for ammines in transplatin could produce complexes with higher *in vitro* antitumor activity, particularly towards cisplatin-resistant tumors and, in some cases, impart significant *in vivo* activity. Theoretically, the bulky ligands can retard substitution reactions of the chloride leaving groups, hence reducing the kinetic instability of transplatin. For example, *trans*-[PtCl$_2$(2-butylamine)(PPh$_3$)] is more active than cisplatin against the Pam 212-*ras* cells, which are cisplatin-resistant through overexpression of H-*ras* oncogene [127]. Platinum complexes with PPh$_3$ (**32**) or PMe$_2$Ph (**33**) in trans configuration to NH$_3$ or aliphatic amines show potent antitumor activity in ovarian carcinoma cell lines SKOV-3, CH1, CH1cisR, and Ewing's sarcoma cell line SR2910, and their IC$_{50}$ values are better than those for cisplatin and other *trans*-platinum complexes. Moreover, the toxicity in the normal cell line IMR90 is low and the intrinsic resistance to cisplatin in SKOV-3 and CH1cisR cell lines can be overcome. Complexes **32** and **33** induce apoptosis without G2/M and G1 accumulation, suggesting the mechanism of action is different from that of cisplatin. The presence of phosphine group in the complexes would enhance their lipophilicity and help to cross the cytoplasmic membrane, leading to more Pt binding to DNA. Meanwhile, the alteration in the amines also influences the cytotoxicity [128].

32   **33**

Trans-platinum(II) complex **34** has been assayed for its antiproliferative effect against the human colorectal adenocarcinoma HT29 cell line and the human non-small cell lung cancer A549 cell line as well as normal human peripheral blood lymphocytes. This complex demonstrates substantial cytotoxic activity against cancer cells and exerts weak influence on normal cells. Although the mechanism of action has not been elucidated, the presence of bulky substituents at the coordinated pyridines may contribute to the antiproliferative efficacy. Interestingly, **34** is more potent than its cis isomer [129]. *Trans*-platinum(II) complex **35**, bearing the even bulkier ligand oxadiazoline, also exhibits potent *in vitro* cytotoxicity in cisplatin-sensitive cell lines as well as cisplatin- and carboplatin-resistant cell lines, including ovarian cancer cell lines (PEO1, PEO1cisR, PEO1carboR, SKOV-3), colon cancer cell line (SW948), and testicular cancer cell line (N-TERA). This complex seems to differ from cisplatin and carboplatin in how it affects the cell cycle [130].

In complex **36**, replacement of one ammine of transplatin with 4-piperidinopiperidine significantly enhances the cytotoxicity. Studies with three pairs of cisplatin-sensitive and -resistant human ovarian cancer cell lines (A2780/A2780cisR, 41M/41McisR, CH1/CH1cisR) and colon cancer cells (C-26) showed that **36** is more potent than cisplatin in all the cisplatin-resistant cell lines and is nearly as cytotoxic as cisplatin against colon cancer cells. In the A2780 and A2780cisR tumor xenograft model mice, **36** is less efficacious than cisplatin; while in C-26 tumor-bearing mice, **36** has an efficacy comparable to that of cisplatin [131].

The replacement of one ammine group by 2-methylbutylamine or *sec*-butylamine ligand in transplatin (**37** and **38**) radically increases its activity in tumor cell lines sensitive (A2780, CH1) or resistant (A2780cisR, CH1cisR) to cisplatin. This replacement also markedly alters the DNA binding mode of transplatin and reduces the efficiency of repair systems to remove the Pt-DNA adducts of these complexes. Therefore, modification of the ineffective transplatin by aliphatic amine ligands may increase the efficiency to form DNA interstrand crosslinks and activate the *trans*-platinum(II) complexes to exert cytotoxic effect even on cisplatin-resistant cell lines [132].

The cytotoxic activity of the *trans*-platinum(II) complex **39** against A2780/A2780cisR, SKOV-3, and cisplatin-resistant human colon adenocarcinoma (SW480) cell lines was determined recently. This lipophilic complex is at least as cytotoxic as cisplatin towards these cells, while parent transplatin is markedly less potent than cisplatin. The DNA binding mode of **39** differs remarkably from that of ineffective transplatin. A notable feature of **39** is the capability to circumvent both acquired (A2780cisR) and intrinsic (SW480) cisplatin resistance. Consistent with the lipophilic character of the complex, its total accumulation in A2780 cells is considerably greater than that of cisplatin. In addition, the rate of reaction with GSH is lower than that of cisplatin and transplatin, which seems to be an important determinant of the cytotoxic effects of the complex [133].

$$H_3N\diagdown Pt \diagup Cl$$
$$Cl \diagup NH_2CH_2CH(CH_3)CH_2CH_3$$

37

$$H_3N\diagdown Pt \diagup Cl$$
$$Cl \diagup NH_2CH(CH_3)CH_2CH_3$$

38

$$Cl\diagdown Pt \diagup NH_3$$
$$H_2N \diagup \diagdown Cl$$

39

4.5
Multinuclear Platinum(II) Complexes

Multinuclear platinum complexes containing polyamine linkers of variable chain length represent a novel class of antitumor agents with pharmacological profiles different from those of currently used mononuclear platinum drugs. These complexes have more than one platinum centers with either cis or trans geometries and bind to DNA in a manner different from that of cisplatin. For instance, the BBR series of multinuclear complexes react with DNA more rapidly than cisplatin and produce characteristic long-range inter- and intrastrand crosslinked DNA adducts [134, 135]. It is generally believed that platinum complexes forming crosslinks different from cisplatin-DNA adducts would be recognized differently by cellular proteins and thereby have the potential to induce cell death via distinct pathways [136]. One of the first generation multinuclear Pt(II) complexes, **40** (BBR3464), even entered Phase II clinical evaluation in 2000. Unfortunately, this agent is not effective in patients with various cancers and is poorly tolerated by patients; thus, the trials have been stopped recently [137–140].

The second-generation multinuclear platinum complexes **41** (BBR3571) and **42** (BBR3610) contain two platinum units and a polyamine to replace the central platinum unit in **40**. These dinuclear complexes in general display activity profiles similar to those of **40**, with the cytotoxicity being dependent on the nature of the bridging polyamine. Complexes **41** and **42** are 20 and 250 times, respectively, more effective than cisplatin in LNZ308 and LN443 glioma cells in culture and animal models. In subcutaneous xenografts of U87MG glioma cells, **42** is more effective than **40**, **41**, and cisplatin, in that it significantly extends survival and nearly doubles the time needed for the tumor to reach a predetermined size. These complexes induce predominantly G2/M cell cycle arrest in glioma cells, while cisplatin induces apoptosis. This is the first direct evidence that the cellular response to multinuclear platinum complexes is distinct from that to cisplatin, although they seem to share the same signal transduction pathways [141]. In addition, **42** kills HCT-116, DLD1, SW480, and HT29 colon cancer cells more efficiently than **40**, cisplatin, or oxaliplatin. Compound **42** kills colon cancer cells via a caspase 8-dependent pathway, which can be enhanced by epidermal growth factor receptor (ERBB1) or phosphatidylinositol 3 kinase (PI3K) inhibitors [142]. Mechanistic studies with different phospholipids show that both non-covalent and covalent interactions are involved in the reactions of the positively charged platinum complexes with the negatively charged phospholipids. This suggests that the complexes interact not only with the phosphate head-group but also with the region of the fatty acid tail of liposomes and finally change the fluidity of the membrane. For this reason, their binding with the liposomes is significantly stronger than that of cisplatin [143].

On the other hand, the extraordinary cytotoxicity of the above complexes in the micro- to nanomolar range may lead to a relatively narrow therapeutic index and hence limit their clinical application. In an attempt to overcome this problem, blocked polyamine complexes that are less potent *in vitro* but are capable of releasing the blocking groups *in vivo* have been synthesized. Complexes **43** and **44** (BBR3537) are potential prodrugs for complex **41**. A time-dependent slow release of the active species can be achieved through spontaneous, pH dependent hydrolysis of the blocking groups. These blocked complexes are about two orders of magnitude less cytotoxic than the unprotected analogs in ovarian carcinoma cell lines sensitive and resistant to cisplatin, which confirms the importance of charge in dictating the cytotoxicity of this series. However, the cytotoxic properties of **43** and **44** do not arise exclusively from the spontaneous release of **41** at physiological pH. Inherent cytotoxicity and cell line specificity may also contribute to the activity. The DNA interstrand crosslinking efficiency of **43** and **44** is 90% and 74%, respectively, which is higher than that of **41** (33%) and cisplatin (6%) [144]. The micromolar cytotoxicity of **43** and **44** suggests that their rate of hydrolysis is suitable for a useful prodrug. Both complexes also display a reduced affinity for human serum albumin in comparison to **40** or **41**, suggesting that the rate of hydrolysis may be sufficient to bypass or diminish the possible deactivating reactions with plasma proteins [145].

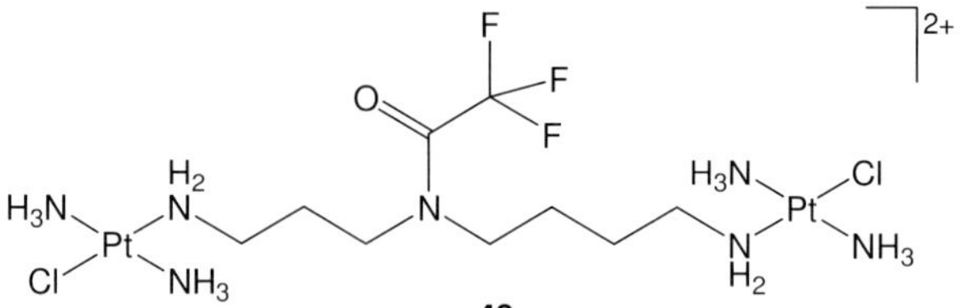

43

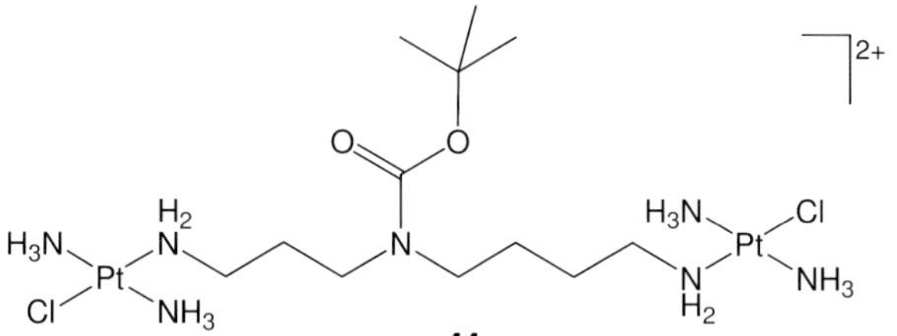

44

The dinuclear Pt(II) complex **45** was designed with the aim of combining the properties of fluorescent anthraquinone intercalators with those of platinum complexes. A major benefit of these complexes lies in the innate fluorescence of intercalators that allows the cellular processing of the complexes to be monitored via fluorescence microscopy. Complex **45** exhibits high cytotoxic activity against A2780 cells, with the complex with the shortest aminoalkyl chain being the most active. Different cell lines appear to process the complexes differently. In cisplatin-resistant A2780cisR cell line, the complexes are sequestered into lysosomes and display cross-resistance with cisplatin; in cisplatin-sensitive A2780 cell line, however, this is not observed. The cross-resistance may be due to the high levels of GSH present in A2780cisR cells, which could deactivate platinum drugs as mentioned above. Platinum accumulation in lysosomal vesicles may present a mechanism of resistance different from that of deactivation by GSH, which provides a plausible explanation for the decreased activity of these complexes in the A2780cisR cell line [146, 147]. Complex **45** also shows high cytotoxicity in the U2-OS human osteosarcoma cell line and its cisplatin-resistant U2-OS/Pt subline. Consistent with the lack of cross-resistance, the cellular processing of **45** is similar in cisplatin-resistant and -sensitive US-O2 cell lines. This is likely due to the formation of structurally different DNA-adducts that can evade the DNA repair mechanism responsible for removing cisplatin adducts. These complexes enter U2-OS cells quickly and accumulate in the nucleus, thereby reaching the biological target DNA, and are excreted from the cells via the Golgi apparatus. The cellular distribution of **45** in the pair of U2-OS and U2-OS/Pt cell lines is different from that in the pair of A2780 and A2780cisR cell lines, which most likely results from different resistance profiles in A2780cisR and U2-OS/Pt cells [148]. Based on the same conception, cytotoxic dinuclear platinum complexes of aliphatic diamines modified with a fluorogenic reporter (carboxyfluorescein diacetate) or a hapten (dinitrophenyl) have also been designed to investigate the cellular distribution and pathways of platinum complexes [149].

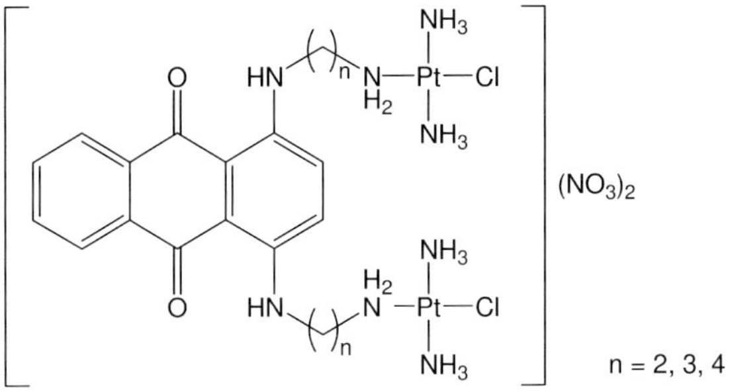

45

However, multinuclear platinum complexes with 1,1/t,t[1] structural motif are susceptible to decomposition in the presence of sulfur-containing nucleophiles. In other words, substitution of the chloride ligands by the sulfur donors could induce the loss of the di/polyamine linker because of the trans-influence [150]. This degradation decreases the bioavailability of the drugs and increases the amounts of toxic metabolites in the body. In contrast, complexes with 1,1/c,c geometry do not undergo breakdown upon reaction with sulfur nucleophiles. Studies on the reactions between 1,1/c,c dinuclear platinum complexes and sulfur-containing nucleophiles indicate that, for the thiol GSH, stable GS-bridged macrochelates are formed and [48], for the thioether *N*-acetyl-L-methionine, several dinuclear platinum compounds are generated and slow loss of NH_3 is observed [151]. Thus, 1,1/c,c *cis* geometry represents an intriguing template for the development of second-generation multinuclear platinum complexes. Because positive charge is critical to the cellular uptake, DNA binding, and high antitumor activity of the drugs, this feature should be retained in the new complexes.

Dinuclear platinum(II) complexes with 1,2-diaminocyclohexane as the carrier group, such as **46**, are dinuclear analogs of oxaliplatin. Complex **46** contains the main feature of BBR3610 (**42**) but with enhanced stability to metabolic deactivation. There is no labilization of the polyamine linker in the presence of sulfur-containing species at physiological pH, though metabolism reactions are somewhat dependent on the nature of the polyamine. The presence of the bidentate amine prevents the trans-influence while the bidentate carrier *per se* remains intact. This series of dinuclear complexes are expected to mimic the DNA-binding profile of BBR3464 while minimizing the bridge-deactivating reactions. Currently, the biological activity of these complexes is unknown [152].

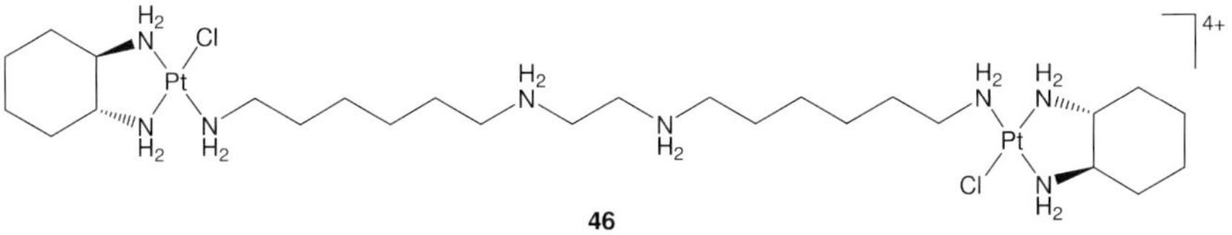

46

[1] "1,1/t,t" and "1,1/c,c": "1,1" refers to the number of leaving groups on each Pt center and "t,t" or "c,c" to the geometry relative to the linker group.

The 1,1/c,c type of dinuclear monofunctional Pt(II) complexes **47** and **48** are more cytotoxic than cisplatin against the A549 cell line, and so is **47** against the P-388 cell line at micromolar concentrations. In the A549 cell line, **47** exhibits more significant antitumor activity than **48**. The reaction of **47** with GSH proceeds very slowly and incompletely because phenyls in the complex greatly increase the steric hindrance around platinum centers. The diamine linker remains intact throughout the reaction. The results suggest that the steric hindrance of the linker cis to the leaving group is an important factor affecting the reaction mode of a multinuclear platinum complex with GSH, and the increase in steric hindrance and rigidity of the linker can inhibit the interaction. Complex **47** exhibits much higher DNA-binding ability than **48** and readily forms 1,3- and 1,4-intrastrand crosslinks with DNA oligonucleotides, while **48** preferentially forms 1,4-intrastrand crosslinks. These results demonstrate that the linker plays a critical role in controlling the DNA-binding and cytostatic abilities of dinuclear platinum complexes. The fine tuning of the linker may result in the variation in Pt…Pt distance and steric hindrance as well as lipophilicity of multinuclear platinum complexes, and thus provides a promising strategy to obtain platinum complexes with favorable DNA binding properties. The distinctive DNA crosslinks formed by these complexes may afford an opportunity to circumvent resistance pathways that have evolved to eliminate the cisplatin-DNA adduct [153, 154]. In another approach, two monofunctional diammineplatinum(II) moieties have been conjugated to the photodynamic therapeutic agent Si(IV)-phthalocyanine in a 1,1/c,c manner. The resulting dinuclear Pt(II) complexes **49** and **50** showed remarkable enhanced cytotoxicity against the human cervical cancer HeLa cell line under the irradiation of red light than in the dark [155].

In multinuclear platinum complexes, the active platinum centers are commonly linked by flexible aliphatic polyamines [156], such as in complexes **40–44, 46**, and the subsequent **51** and **52**, where the biogenic polyamines spermidine [$H_2N(CH_2)_3NH(CH_2)_4NH_2$] and spermine [$H_2N(CH_2)_3NH(CH_2)_4NH(CH_2)_3NH_2$], respectively, are used as bridging linkers. The latter two complexes display a rather high antiproliferative and cytotoxic activity toward HeLa cells and HSC-3 epithelial-type cells and their effect on healthy cells is reversible upon drug removal [157]. However, half-flexible polyamines like those in complexes **45, 47, 48**, and even rigid polyamines can also be used as linkers to bridge platinum centers in multinuclear complexes. For instance, the pyrazine-bridged dinuclear Pt(II) complex **53** exhibits cytotoxicity higher or comparable to cisplatin in both WIDR colon and IGROV ovarian cancer cell lines, and shows remarkable cytotoxicity in the cisplatin-resistant L1210 murine leukemia cell line [158, 159]. The pyrazole-bridged dinuclear Pt(II) complex **54**, where the co-bridging OH is the leaving group, possesses a cytotoxicity that is about 40 times higher than that of cisplatin against MCF-7 cells. It is also active against cisplatin-resistant cells *in vitro*. The rigid pyrazole keeps an appropriate distance between the two platinum centers for the binding of two adjacent guanines. In comparison with cisplatin, **54** is considerably slower in reacting with DNA and the major adduct, that is, the intrastrand 1,2-d (GG) crosslink, does not bend the double-helix significantly. The Watson–Crick base-pairing remains intact, and the melting temperature of DNA is unaffected by the crosslink. In addition, the helical twist is considerably reduced between the two platinated bases. As compared with the bend angle induced by the cisplatin 1,2-d (GG) crosslink (55–78°), **54** induces relatively minor structural perturbations upon the DNA double helix (15°), which may account for its observed lack of cross-resistance with cisplatin [160, 161].

The bridging linker in the trinuclear 3N-chelated monofunctional Pt(II) complex **55** is more rigid than aliphatic polyamines but more flexible than pure aromatic rings, which confers a favorable condition for DNA crosslinking. Complex **55** exhibits more potent cytotoxicity against the P-388 and A549 cell lines than cisplatin. It reacts with GSH to form mono- and disubstituted products such as $[Pt_3L(GS)Cl(OH)_2]^{2+}$ and $[Pt_3L(GS)_2(OH)_2]^{2+}$ but keeps the trinuclear skeleton stable for 24 h, which may ameliorate the biostability and bioavailability of the complex [162]. The combination of rigidity, flexibility, and distances between metals in **55** offers the possibility of forming both intra- and interstrand DNA crosslinks. The interaction between **55** and 18-mer duplex 5′-d(GAAGAAGTCA-CAAAATGT)-3′ · 5′-d(ACATTTTGTGACTTCTTC)-3′ indicates that **55** readily forms various DNA adducts, such as 1,3- and 1,4-intrastrand crosslinks, and the unprecedented interstrand crosslinked triadducts. These distinctive conformational features of triadducts and 1,3-intrastrand crosslinks could play important roles in biological processes such as protein recognition and NER. Molecular simulation shows that the 1,3-intrastrand crosslinks and interstrand triadducts of **55** bend DNA by 51–58° toward the minor groove, rather than toward the major groove, which is the key structural feature for 1,2-GG intrastrand adducts of cisplatin. These characteristics are valuable for Pt(II) antitumor complexes to overcome cisplatin resistance [163].

In contrast to the multinuclear Pt(II) complexes bridged by flexible aliphatic polyamines, multinuclear complexes **56** and **57** are greatly restricted in their flexibility. The neutral complex **56** presents three separate *cis*-PtCl$_2$ moieties, presumably acting in a bifunctional mode towards DNA. The cationic complex **57** contains both a monofunctional and a bifunctional Pt(II) moiety. The size and shape of these complexes enable them to behave as novel scaffolds for DNA binding. Both

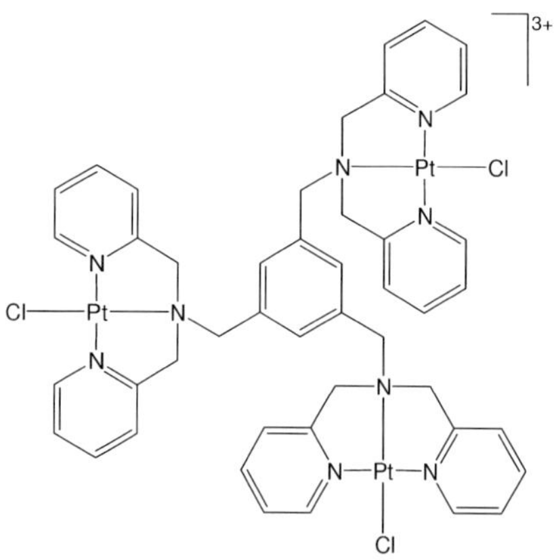

55

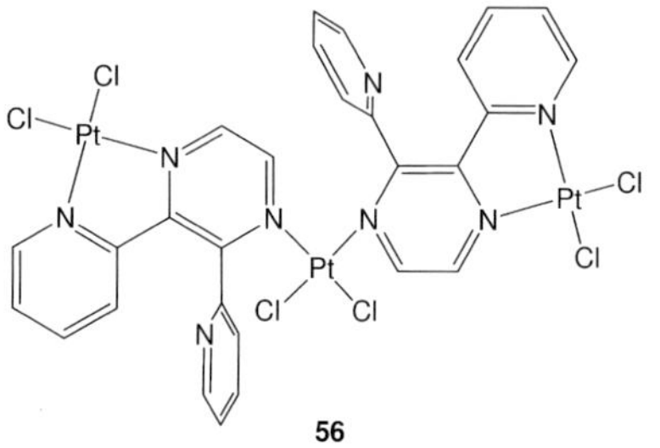

56

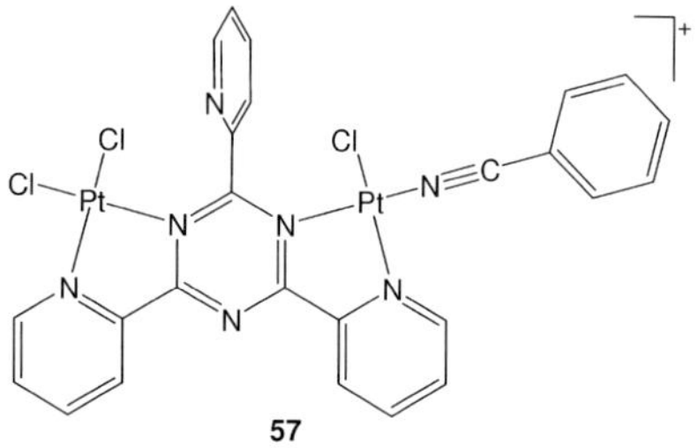

57

56 and 57 show cytotoxic activity against human hepatoma (HepG2) and human colon-rectal carcinoma (HT29) cell lines, but 56 seems less active than 57. HT29 cells are more sensitive than HepG2 cells to 56 and 57. The efficacy of the complexes, particularly that of 57, is markedly higher than that of cisplatin, and mitochondria are involved in apoptosis induced by the complexes in HT29 cells. These complexes are presumed to form unique types of Pt–DNA adducts[164].

Trinuclear complexes 58–60 are structural analogs of BBR3464 (40). The presence of one or two planar amine ligands at the central metal ion could introduce additional noncovalent interactions such as stacking with nucleobases in DNA. Complexes 58 and 59 are much more cytotoxic than cisplatin against human ovarian cancer cell lines (A2780, A2780cisR, and A2780ZD0473R). The resistance factors for 59 against the pairs of cell lines A2780/A2780cisR and A2780/A2780ZD0473R are 1.98 and 0.5,

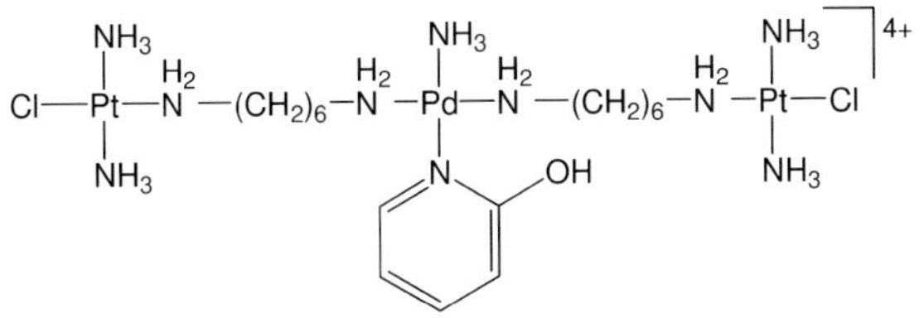

58

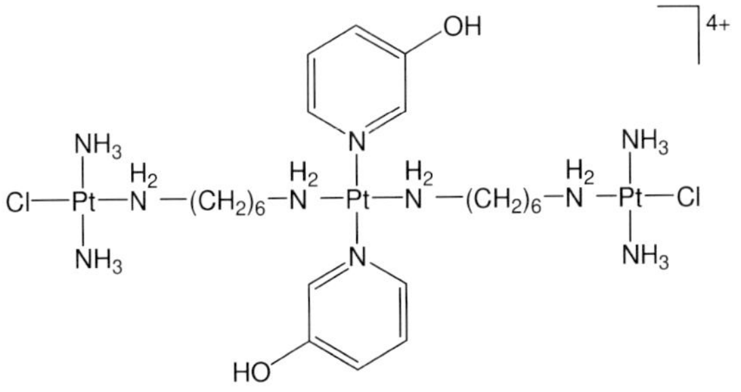

59

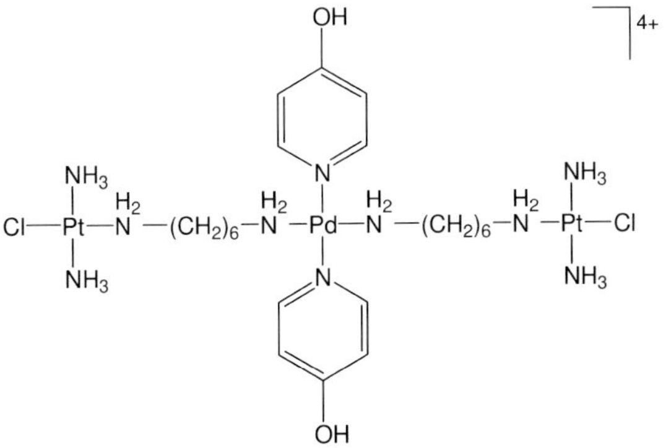

60

respectively, while the corresponding values for cisplatin are 12.9 and 3.0, respectively, indicating that **59** is able to overcome resistance in A2780cisR and A2780ZD0473R cell lines. Although **60** is less active than cisplatin against the parent cell line A2780, it is more active against the cisplatin-resistant cell line A2780cisR. Unlike cisplatin, complexes **58–60** are believed to form interstrand GG adducts that would cause global changes in DNA conformation. However, these complexes are likely to undergo unfavorable decomposition within the cells [165–167].

Non-covalent interactions such as major or minor groove binding and intercalation with DNA have been observed in nature and are of great importance [168]. Innovative trends in drug design of the second-generation multinuclear platinum complexes also focus on those characterized by non-covalent DNA binding [169]. Trinuclear Pt(II) complex **61** (TriplatinNC) is another structural analog of BBR3464, in which the chloride ligands are replaced by $NH_2(CH_2)_6NH_2$. Complex **61** displays a micromolar activity against human ovarian cancer cell lines and a greater cellular uptake than neutral cisplatin as well as other multinuclear platinum complexes [170]. Since this complex has no potential for covalent interaction and intercalation with DNA, it exhibits a peculiar binding mode that uses exclusively backbone functional groups. The mode was observed in the crystal structure of a double-stranded B-DNA dodecamer with **61**. As Figure 4.3 shows, the three Pt(II) units form bidentate NH…O…HN complexes with OP atoms through hydrogen bonds, motifs called "phosphate clamps." A series of such "phosphate clamps" with one strand of DNA results in "backbone tracking," and a combination of two interstrand clamps gives rise to "groove spanning." Both interactions may exist in solution. However, **61** does not bind the major or minor groove [171]. Owing to its high cytotoxic potency and novel mechanism of action, further studies are warranted before similar agents can reach the clinical setting.

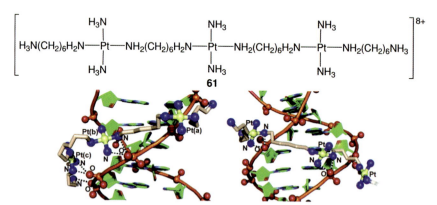

Figure 4.3 New DNA binding modes of platinum complexes through hydrogen bonds termed as "backbone tracking" (left) and "groove-spanning" (right). Both modes involve arrays of "phosphate clamps."
Adapted from Reference [171].

Besides conventional ligands such as derivatives of thiosemicarbazones [172], in recent years less common ligands or structural modes have also been adopted to construct novel multinuclear platinum complexes. Some of them display impressive antitumor activity in various tumor cell lines or animal models. For example, assays on human HeLa, HepG2, KB, and AGZY-83a tumor cell lines indicate that complexes **62** and **63** exhibit significant cytotoxic activity and specificity against these cells, especially against KB cells that are highly resistant to conventional chemotherapeutic agents. Complex **62** binds to DNA in both coordinative and intercalative mode, but shows less DNA binding affinity than complex **63** [173]. The tetranuclear polyimine dendrimer Pt(II) complex **64** strongly binds to human serum albumin by hydrophobic and electrostatic interactions. It shows a 20-fold higher cellular uptake and about 700-fold higher DNA binding than cisplatin. The complex crosses the cell membrane through a passive transport and the polyimine dendrimer seems to serve as a carrier for the shuttling of platinum into the cell nucleus. As a result, **64** exhibits strong cytotoxicity in MCF-7 cells and reduces the cell growth at 5 µM by 50% after an incubation of 150 h [174].

62   **63**

64

Complex **65** shows potent cytotoxicity against HL-60 (immature granulocyte leukemia), HCT-8 (colon carcinoma), MCF-7 (galactophore carcinoma), BGC-823 (gastric carcinoma), and EJ (bladder carcinoma) cell lines with IC_{50} values of 0.02, 1.70, 4.00, 0.98, and 1.02 µM, respectively. The LD_{50} of **65** is 815.3 mg kg^{-1}, which is significantly higher than that of cisplatin and carboplatin. In addition, **65** exhibits significant activity against A2780 and HCT-116 in nude mice at dose of

12 mg kg^{-1}, an activity similar to that of cisplatin at dose of 4 mg kg^{-1}. The results suggest that these new dinuclear platinum complexes could be promising candidates for anticancer drugs [175].

65

4.6
Platinum(IV) Complexes

Current clinical platinum drugs must be administered via intravenous infusion, which is inconvenient for both patients and care providers. The need for orally administrable platinum-based drugs elicited the design and synthesis of various platinum(IV) complexes. In Pt(IV) complexes, the octahedral geometry introduces two extra coordination sites, and the axial ligands provide an opportunity to modify the properties such as reduction potential, lipophilicity, and specificity of the complexes. Besides, some axial ligands themselves possess cytotoxicity upon release [176]. Pt(IV) complexes are kinetically inert to ligand substitution and *in vivo* reduction to Pt(II) analogs is critical for their reaction with the target DNA. For this reason, Pt(IV) complexes can be regarded as prodrugs and, as a prerequisite for activity, the Pt(II) species originated from them should be active [177]. Owing to the inertness, Pt(IV) complexes offer a potentiality for overcoming some of the limitations like clinical resistance and toxicities of platinum drugs. It has been demonstrated that Pt(IV) complexes are not subject to multicellular resistance [178].

Reduction potential is of primary importance for the design of active Pt(IV) complexes. Platinum(IV) complexes with low or intermediate reduction potentials can enter cells intact, and the proportion of the complexes reduced intracellularly is related to the reduction potentials of the complexes [179]. Aquation should not be an interfering factor in the design of Pt(IV) complexes. In fact, a [^{1}H,^{15}N] HSQC NMR investigation on ^{15}N-labeled Pt(IV) complexes with *trans*-diacetato or -dihydroxo ligands has revealed that the aquation of these complexes is extremely difficult, in that almost no reactions were observed even after several weeks in solution [180, 181]. The distribution of platinum within A2780 cells was shown to be similar to that in cells treated with cisplatin after 24 h [182], and the loss of the axial ligands on reduction has also been confirmed by *cis,trans,cis*-[PtCl$_2$(OC(O)CH$_2$Br)$_2$(NH$_3$)$_2$], which has a similar reduction potential and IC$_{50}$ to its *trans*-acetato analog [183, 184]. Platinum(IV) complexes do not lose cytotoxic activity in non-small cell lung cancer cell lines, suggesting that the hypoxic tumor microenvironment may favor the reduction of the complexes [185]. A study on the reduction of *cis,trans,cis*-[PtCl$_2$(OC(O)CH$_3$)$_2$(NH$_3$)$_2$] by extracts from different tumor cell lines has shown that the reduction rates are cell type dependent and may be influenced by cellular components with molecular weight larger than 3000 Da [186].

The development of Pt(IV) complexes has shown exciting prospects for practical application in therapy of different tumors. An outstanding example is satraplatin (JM216, **66**), which is a lipophilic compound that can be administered orally for its inertness. The rapid cellular uptake of this complex makes it possible to overcome accumulation defects in resistant cells [187]. *In vitro* studies have shown that **66** is active against several human cancer cell lines, including prostate, ovarian, cervical, and lung cancers, and it is also active against selected tumors resistant to cisplatin and taxanes [43, 188]. In preclinical studies, the antitumor activity of **66** is similar to that of cisplatin on various human tumor cell lines, and the toxicity is similar to that of carboplatin, with no nephrotoxicity, neurotoxicity, or ototoxicity, and therefore it is much better tolerated than cisplatin. Hematological and gastrointestinal toxicities are the predominant dose-limiting side effects [189]. Several Phase I clinical trials with **66** in combination with other chemotherapeutic agents have been launched. In several Phase II and III clinical trials, **66**, either alone or with prednisone, showed promising antineoplastic activity in hormone refractory prostate cancer [190, 191]. Currently, **66** is the only Pt(IV) complex undergoing clinical trials and its mechanistic studies are ongoing [192]. It has been demonstrated that the reduction of **66** can be mediated by heme proteins such as hemoglobin or cytochrome *c* in the presence of NADH *in vivo*. Under these conditions, **66** is reduced to *cis*-amminedichloro(cyclohexylamine)platinum(II) (JM 118) and hemoglobin is oxidized to methemoglobin [193]. Indeed, the major active metabolite found in plasma of patients treated with **66** is JM118. This metabolite binds to DNA to form intrastrand and interstrand crosslinks as cisplatin does. However, the formed adducts are not recognized by DNA mismatch repair proteins, which may play a role in overcoming resistance to other platinum drugs [194].

66   **67**

As mentioned earlier, the increase in lipophilicity of ligands often correlates positively with the enhanced cellular uptake of platinum complexes, which is beneficial for overcoming the cisplatin resistance. For example, Pt(IV) complex **67** (LA-12), with a bulky 1-adamantylamine ligand, is only four carbon atoms different from **66**, but it is highly lipophilic and has no cross-resistance with cisplatin. On a panel of cisplatin resistant cancer cell lines, **67** exhibits a high cytotoxic effect against leukemic, melanoma, and colorectal cancer cell lines and high efficiency to trigger apoptosis [195, 196]. IC_{50} values for **67** in A2780 and A2780cisR cell lines are 0.22 and 1.40 μM, respectively, which are significantly lower than those for cisplatin and **66** [197]. *In vivo* studies show that **67** is active against various tumor models such as murine plasmacytoma, prostate cancer, ovarian cancer, and colon cancer [198]. The cellular uptake of **67** in A2780 and A2780cisR cell lines is considerably greater than that of cisplatin and **66**, which may contribute to the increased cytotoxicity of **67** over cisplatin. DNA lesions of **67** are similar to those of cisplatin but they are less efficiently repaired and readily crosslink proteins. In contrast with the case of cisplatin, deactivation of **67** by GSH is less relevant to the resistance mechanism, although cisplatin and **67** react with GSH at a similar rate. The high efficacy of **67** against the A2780cisR cell line might result from an increased rate of intracellular reduction by GSH, which may represent a mechanism for circumvention of cisplatin resistance [199].

The cytotoxicity of complexes **68–71** against the cisplatin-sensitive CH1 and HeLa cell lines, and the cisplatin-resistant SKOV-3 and SW480 cell lines, is in the micromolar or even nanomolar range and is comparable to – or better than – those of cisplatin. The activity generally increases with the lipophilicity of the alcoholate moiety. For instance, in the series of ester derivatives **69**, the IC_{50} values remarkably decrease in all four cell lines with increasing chain length and lipophilicity of the alcoholate moiety, resulting in the following rank order of cytotoxicity: **69a** < **69b** < **69c** < **69d.** The cytotoxicity of **69b** nearly approaches that of cisplatin, while that of **69c** and **69d** consistently exceeds that of cisplatin. In CH1 and HeLa cancer cells, the IC_{50} values of **69d** are in the 10^{-8} M range, while those of cisplatin are in the 10^{-7} M range. However, increasing the overall

68

69

a. n = 0
b. n = 1
c. n = 2
d. n = 3

70

71

a. n = 0
b. n = 1

lipophilicity of the complex is not *per se* a guarantee for increased cytotoxicity. For example, the IC_{50} values of **70** are markedly increased in CH1 and HeLa, and are comparable to those of **69b** in SW480 and SKOV-3 cancer cells. Nevertheless, **71** exhibits promising IC_{50} values that are more potent than their counterparts **69a** and **69b**. DNA platination studies in SW480 cells revealed that a high platination capacity is in parallel to a high cytotoxic potential and vice versa. However, the resistance mechanisms of the cells can not be circumvented by these complexes [200, 201].

The *in vitro* cytotoxicity of Pt(IV) complexes **72** and **73** has been evaluated against human leukemia, and several human cancer cell lines. The IC_{50} values of **72** are lower than those of cisplatin. The *in vivo* antitumor activity of these complexes has been tested using mice bearing L1210 leukemia, L1210/cis-DDP leukemia, and B16 melanoma. The activity of complex **72a** is similar to that of cisplatin, but its toxicity seems to dwindle at effective drug concentrations. In cisplatin-resistant L1210/cis-DDP cancer cells, **73** shows the highest activity [202]. The HCT-116 cell line is particularly more sensitive to **72a** than to cisplatin, and the cell death is induced by apoptosis. ERK1/2 activation and the p53 pathway may play significant roles in mediating the **72a**-induced apoptosis in human colon cancer cells [203]. Again, the activity of these complexes may be associated with the enhanced intracellular accumulation because of the increase in lipophilicity.

72   **73**

Carboxylation of *trans*-dihydroxoplatinum(IV) complexes by anhydrides, cyclic anhydrides, isocyanates, pyrocarbonates, and carboxylic acid chlorides has led to numerous interesting Pt(IV) derivatives, and some of them even provide uncoordinated carboxylic acids for further derivatization [204]. These general reaction schemes allow large numbers of functional compounds to be conjugated to Pt(IV) complexes in a mode that does not disrupt the activity of either the ligand or the platinum center on reduction (*vide infra*) [205]. Here is an example in this context. As noted in Section 4.2.2.2, cytosolic glutathione *S*-transferase (GST) constitutes the main cellular defense against xenobiotics. Ethacrynic acid, a diuretic in clinical use, is an effective inhibitor of all GST isozymes. To overcome the platinum drug resistance, complex **74**, which can target GST enzymes in human cancer cells, was prepared by tethering ethacrynic acid to the *trans*-dihydroxoplatinum(IV) starting complex. Reduction of **74** in the cell results in the release of two equivalents of the potent GST inhibitor and one equivalent of the cytotoxic cisplatin. The cytotoxicity of **74** against the cisplatin-resistant breast

MCF-7 and T47D, lung A549, and colon HT29 human carcinoma cells is stronger and faster than that of cisplatin [206]. The aromatic carboxylate ligand was found to strongly influence the uptake and efficacy of *trans*-Pt(IV) complexes. For instance, complexes **75a–d** are 5–20-fold more cytotoxic than cisplatin against a panel of A549, HT29, MCF-7, and T47D carcinoma cell lines. Drug uptake is relatively high in cells treated with **75**, while that for cisplatin is much lower; **75b** and **75c** demonstrate the highest levels of penetration into cytosol. The efficacy of the complexes is closely correlated with the drug uptake, suggesting that drug uptake is an important factor in determining the effectiveness of **75** [207].

An increase in steric hindrance around the Pt(IV) center can stabilize the complex and enhance the antitumor efficacy, which is exemplified by all-*trans* Pt (IV) complex **76**. The antitumor and cellular pharmacological properties of **76** have been evaluated in the pairs of cisplatin-sensitive and -resistant A2780/A2780cisR, CH1/CH1cisR, and 41M/41McisR cell lines. The results indicate that **76** markedly circumvents cisplatin resistance in 41McisR and CH1cisR cell lines that are endowed with different mechanisms of resistance like decreased platinum accumulation and enhanced DNA repair or tolerance. At equitoxic concentrations, **76** induces a higher amount of apoptotic cells than cisplatin in CH1cisR cells, and the number of apoptotic cells correlates with the ability to form DNA interstrand crosslinks in CH1cisR cells. Moreover, this complex is able to inhibit the growth of CH1 carcinoma xenografts in mice. The protein-binding kinetics of **76** is distinctly slowed as compared with its corresponding *trans*-Pt(II) counterpart because of the improved stability of the complex [208].

Less often, some Pt(IV) complexes show activity towards novel chemotherapeutic targets. It is known that signal transducer and activator of transcription (STAT) proteins regulate many tumor cell processes such as proliferation, apoptosis, angiogenesis, and immune function. Consequently, STAT proteins, especially STAT3, are emerging as a promising molecular target for cancer therapy. Tumor cells relying on persistent STAT3 signaling are more sensitive to STAT3 inhibitors than normal cells; therefore, drugs inhibiting STAT3 may have specific effect on tumors but little effect on normal tissues [209]. Platinum(IV) complexes **77** (CPA-1) and **78** (CPA-7), particularly the latter, have recently been described in a series of patents as promising inhibitors of STAT3. Aside from inhibition of STAT3-ptyr705 phosphorylation, **78** can also inhibit STAT3 transcriptional activity [210].

Platinum(IV) complexes can be photoactivated to active antitumor agents directly at the tumor site by making the complexes photolabile with appropriate ligands. These complexes are inert and non-toxic to cells in the dark; however, upon irradiation at the tumor site, they undergo various photochemical reactions, including isomerization, substitution, and reduction. Such site-selective activation means that many unpleasant side effects and toxicity of conventional platinum-based drugs would be decreased and the therapeutic index would be increased consequently [211]. The photoactivation pathway of Pt(IV) complexes does not rely on oxygen, which is a significant advantage over the photosensitizers used in current photodynamic therapy. A possible application of such light-sensitive Pt(IV) prodrugs could be the treatment of localized cancers accessible to irradiation, such as bladder, lung, esophagus, and skin cancers [212].

Photolabile Pt(IV) complexes appear to kill cancer cells through a mechanism different from that of the classical platinum drugs, even though transcription mapping studies with treated plasmid DNA have shown that the platination sites of these complexes resemble those of cisplatin in dark conditions [213]. For example, Pt(IV) diazide complexes **79** and **80** are nontoxic in the dark, but demonstrate high cytotoxicity against human skin cells (HaCaT keratinocytes) upon irradiation ($\lambda =$ 366 nm), and **80** also exhibits cytotoxicity toward human bladder cancer cells on irradiation. No cross-resistance with cisplatin is observed. Photoactivation of the complexes results in a dramatic shrinking of the cancer cells, loss of adhesion, packing of nuclear material, and disintegration of nuclei. The rate of photolysis closely parallels that of DNA platination, indicating that the photolysis products interact directly and rapidly with DNA, which is a mechanism of action different from that of cisplatin. The mechanism of toxicity of **80** in HaCaT cells also differs markedly from that of transplatin, in that the DNA lesions formed after photoactivation are inaccessible to transplatin, suggesting that **80** is not simply a prodrug for transplatin [214, 215]. The photodecomposition of **79** induced by UVA light has been studied by multinuclear NMR spectroscopy and density functional theory (DFT) recently. The results illustrate that photoinduced reactions of Pt(IV) complexes can lead to novel reaction pathways, and therefore to new cytotoxic mechanisms in cancer cells [216–218].

79   **80**

Incorporation of a pyridine ligand into the Pt(IV) diazide complexes can greatly increase their potency. Complex **81** with azido ligands is very stable in the dark even in the presence of cellular reductant GSH, but it readily undergoes photoinduced ligand substitution and photoreduction. When **81** is photoactivated in cells, it is 13–80-fold more cytotoxic than cisplatin toward HaCaT keratinocytes and cisplatin-sensitive A2780 cells, and about 15-fold more cytotoxic toward cisplatin-resistant A2780cisR cells. Photoactivated **81** rapidly forms unusual *trans* azido/guanine, and then *trans* diguanine Pt^{II} adducts, which are probably mainly intrastrand crosslinks between two guanines separated by a third base. DNA interstrand and DNA–protein crosslinks also exist. DNA repair on plasmid DNA platinated by photoactivated **81** is markedly lower than those by cisplatin or transplatin because of the difference in DNA damage. Cell death is not solely dependent on activation of the caspase 3 pathway and, in contrast to cisplatin, p53 protein does not accumulate in cells after photosensitization. These remarkable properties make **81** a promising candidate for use in photoactivated cancer chemotherapy [219].

81

4.7
Delivery of Platinum Drugs

Since the discovery of cisplatin, medicinal chemists have been trying to improve the activity or resolve the clinical drawbacks of this drug through structural modifications [220]. Currently, lack of specificity is one of the major clinical problems for platinum-based drugs. The non-selective distribution of platinum drugs in normal and cancer cells can induce excessive systemic toxicity and limit the achievable drug dose within the tumor. In addition, rapid inactivation of the drug due to interactions with plasma and tissue proteins may lead to suboptimal treatment for the tumor. Therefore, targeting platinum antitumor drugs to specific tissues is an important issue in platinum-based chemotherapy [221].

To achieve targeting purpose and control systemic toxicity or side effects, it is required to harness and make use of molecular-level recognition events specific to the tissues concerned. The principal strategies used for the delivery and selective administration of platinum drugs include conjugation to biomolecules or polymers, encapsulation in macrocycles, nanotubes, proteins or nanocapsules, adsorption on ceramic materials, and prodrug techniques [222]. In these approaches, two main pathways are followed: (i) the enhanced permeability and retention (EPR) effect, which is caused by the increased angiogenesis and the enhanced production of permeability mediators as well as the impaired lymphatic drainage in tumor tissues, and (ii) the specific targeting of organs and receptors because of chemical modification of the drug [223]. Besides, targeting DNA directly by attaching the platinum moiety to a suitable carrier is also a common way for drug delivery. These strategies can decrease side effects or prevent non-DNA bindings through more localized and effective delivery of the drug to the biological targets. The following examples are some recent highlights in this area.

In clinical practice, cisplatin is usually administered intravenously as a short-term infusion. This yields a high drug concentration in the injection area and quick distribution in the rest of the body, leading to high local and systemic toxicity. Carbonated hydroxyapatite (HA) crystals are similar to the porous structure in bones and hence can be used to deliver cisplatin. This method has resulted in tumor inhibition and lower systemic toxicity. Cisplatin is adsorbed in the crystals and the adsorption depends on the physical and chemical properties of the HA crystals such as the composition, the morphology, the surface area or the size; while the release of the drug depends on temperature, chloride concentration in the medium and crystallinity of HA. Lower crystallinity leads to higher adsorption and slower release, and temperature slightly increases the drug release rate. *In vitro* cytotoxicity of the apatite/cisplatin conjugates in a K8 clonal murine osteosarcoma cell line shows that the drug activity is retained after adsorption onto the apatite crystals [224]. The shape of the HA crystals is important for the adsorption and desorption of cisplatin. When cisplatin is loaded into either plate-shaped or needle-shaped HA crystals, it is adsorbed better into the latter. Although the two crystal structures have similar Ca/P bulk ratios, the surface areas and Ca/P surface ratios are different. The lower amount of calcium in the surface of needle-shaped HA crystals allows easier loading of the positively charged aquated platinum species. However, cisplatin release is the same from HA crystals of both shapes [225].

Another novel strategy is the attachment of carboplatin analogs to a cysteine-binding molecule. In the albumin-binding complexes **82** and **83**, a maleimide group is attached through a spacer molecule to a modified cyclobutane-1, 1-dicarboxylic acid (CBDA) ligand as the platinum-complexing unit, and 1,2-diaminocyclohexane (DACH) or two NH_3 are incorporated as the amine ligands. The maleimide group is selected for thiol-binding owing to its specific and selective reaction with the cysteine-34 position of endogenous albumin. The platinum diammino moiety would be released through hydrolysis from the CBDA ligand as the active agent in the tumor cell. The aim of this approach is to bind the drug moiety to blood transport proteins and thereby localize it at tumor sites using the EPR effect. It was shown that around 50% of **82** or **83** were bound to human serum albumin (HSA) within the first minute of reaction and after 15 min only less than 10% of the free complexes remained. Complexes **82** and **83** showed a 5–8-fold decrease in activity against LXFL 529 human lung carcinoma and MaTu human breast carcinoma cell lines; however, they were more effective than carboplatin in reducing the tumor size of nude mice bearing MaTu human breast tumors [226].

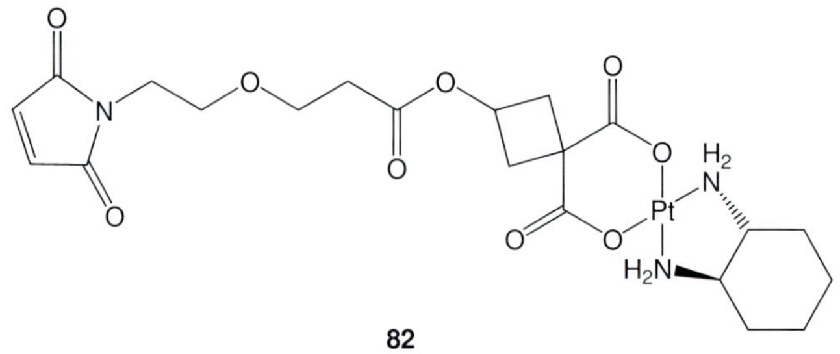

82

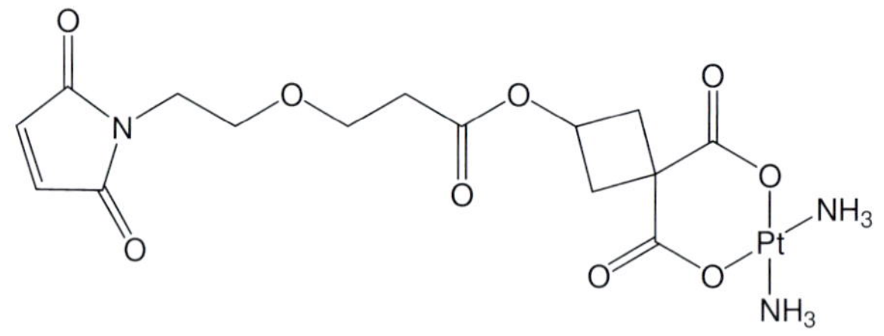

83

Tissues rich in estrogen receptors (ERs), such as breast and ovarian cancers, accumulate molecules that have high binding affinities for these receptors. Therefore, molecules that bind to the ER and have favorable cellular transport properties can be incorporated into platinum complexes to target these cancers. The notion that estradiol-linked platinum complexes might be selectively enriched in estrogen receptor-positive [ER(+)] tumor cells has led to the design and evaluation of different platinum complexes with affinity for the ER [227–229]. In complex **84**, a DNA damaging warhead, [Pt(ethylenediamine)Cl$_2$], is tethered through a linker to the steroid residue (in red). The ligand has 28% relative binding affinity for the ER as compared to 17β-estradiol. After covalent binding to a synthetic DNA duplex 16-mer, the affinity of **84** for the ER is still retained. Complex **84** shows higher cytotoxicity against the ER(+) ovarian cancer cell line CAOV3 than the control compound; it is also more toxic to the ER(+) line MCF-7 than to the ER(−) line MDA-MB231. These results indicate that both the presence of the estradiol moiety in **84** and the expression of the ER in target cells contribute to the enhanced activity [230].

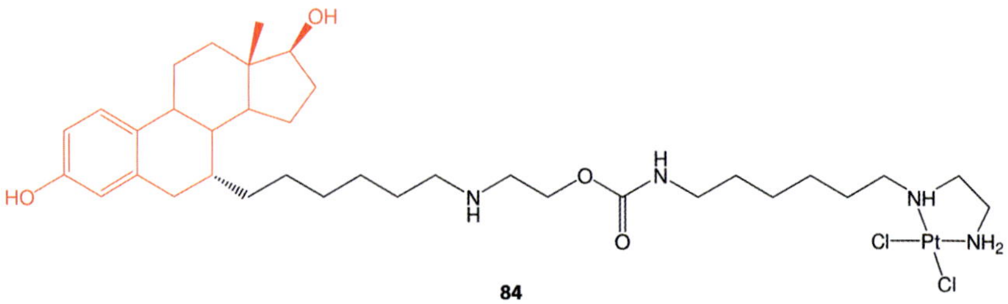

84

A series of estrogen-tethered Pt(IV) complexes (**85**) have been synthesized to target the platinum moiety to ER(+) breast cancer cells. In these molecules, estradiol (in red) is tethered to the terminal carboxylate groups of *cis,cis,trans*-diamminedichloro disuccinatoplatinum(IV) through polymethylene chains of varying lengths. Upon intracellular reduction, **85** is able to afford cisplatin and two equivalents of the modified estrogen, leading to the upregulation of HMGB1 in ER (+) MCF-7 cells. The upregulation of HMGB1 can shield cisplatin–DNA cross-links from repair and enhance cell death. The cytotoxicity of these complexes has been evaluated in ER(+) MCF-7 and ER(−) HCC-1937 human breast cancer cell lines. Complex **85c** is nearly twofold more cytotoxic in MCF-7 cells than in HCC-1937 cells [231]. This concurrent delivery of cisplatin and estrogen confers both DNA damage and HMGB1-induced repair shielding onto the same population of cells, and thus provides a novel strategy for the design of platinum complexes to target ER(+) malignancies.

a. n = 1
b. n = 2
c. n = 3
d. n = 4
e. n = 5

85

In recent years, carbon nanotubes have been explored for drug delivery because of their unique physical, chemical, and physiological properties. The structural stability of carbon nanotubes may prolong the circulation time and the bioavailability of the loaded drugs. Functionalized soluble single-walled carbon nanotubes (SWNTs) and single-walled carbon nanohorns (SWNHs) are two kinds of the most used nanotubes for the delivery of platinum-based drugs. SWNTs and SWNHs have plenty of inner spaces where the incorporation of the drugs is possible, and on the tube walls the drugs and various functional molecules can be physically adsorbed. More importantly, the edges of the tube holes have oxidized functional groups where further covalent modifications are feasible. For example, by combining the inertness of Pt(IV) complexes with the shuttling capacity of SWNTs, the Pt(IV) complex *cis,cis,trans*-[Pt(NH$_3$)$_2$Cl$_2$(OEt)(O$_2$CCH$_2$CH$_2$CO$_2$H)] has been tethered to SWNT through a covalent bond. Upon intracellular reduction of the conjugate, a lethal dose of cisplatin is released. On average, 65 Pt(IV) centers are attached to each SWNT and they enter the cell through endocytosis, leading to higher levels of platinum in the cell than the untethered complex or cisplatin. The SWNT–Pt(IV) conjugate shows a substantial increase in cytotoxicity (IC$_{50}$ = 0.02 µM) against the testicular carcinoma cell line NTera-2 with respect to that of the free complex as well as cisplatin (IC$_{50}$ = 0.05 µM) [232]. Further, such structure is functionalized by adding a folate derivative to the Pt(IV) moiety in the hope of

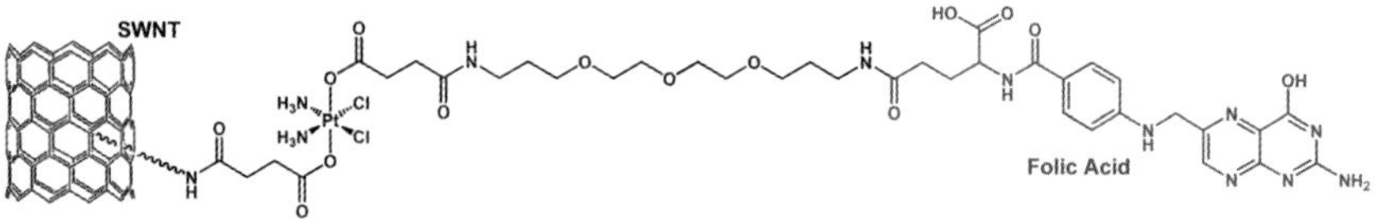

Figure 4.4 SWNT-tethered Pt(IV) prodrug with targeting property for folate receptor.

targeting human cells overexpressing the folate receptor (FR) (Figure 4.4). The conjugate indeed delivers the Pt(IV) pharmacophore selectively into the FR(+) cancer cells that overexpress the FR on their surface and releases cisplatin upon intracellular reduction of Pt(IV) to Pt(II). The IC_{50} values of the SWNT-tethered complex towards FR(+) human choriocarcinoma (JAR) and human nasopharyngeal carcinoma (KB) cell lines are 0.019 and 0.01 μM, respectively, which are significantly lower than those of cisplatin or the free complex. This construct is the first example in which both the targeting and delivery moieties have been incorporated into the same molecule [233].

More recently, the bioconjugates of Pt(II) complexes with SWNTs have been synthesized as well. Cisplatin and epidermal growth factor (EGF) have been attached to SWNTs to specifically target head and neck squamous carcinoma cells (HNSCC) that overexpress EGF receptors (EGFR) (Figure 4.5). The conjugates entered into the cell through EGFR directed endocytosis, as demonstrated by the lack of uptake when EGF was not attached or the EGFR was knocked out. The uptake of platinum observed in both *in vitro* and *in vivo* systems was higher for the targeted conjugates than the untargeted controls and the SWNTs were detected close to the nuclei. In short-term biodistribution studies, the presence of nanotube bioconjugates was observed in various vital organs of the mice, but in much smaller amounts than in the tumors. The EGFR-targeted bioconjugates kill cancer

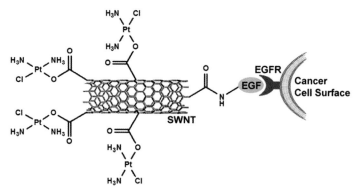

Figure 4.5 Schematic diagram of SWNT conjugated with cisplatin and EGF targeting the cell surface receptor EGFR on a single HNSCC cell.

cells more efficiently and inhibit tumor growth in mice more rapidly than cisplatin and the untargeted SWNT-cisplatin. This bioconjugate is the first targeted SWNT–drug construct showing selective antitumor activity *in vivo* [234].

SWNHs have a spherical structure between 80 and 100 nm assembled by several hundreds of SWNTs, a size that is adequate for drug delivery through vascular EPR. Cisplatin has been loaded into the SWNHs through a selective precipitation process in water. The amount of incorporated cisplatin is 46%, and the total released quantity of cisplatin is 100% over 48 h. Concurrently, *in vitro* antitumor efficiency of the drug-loaded SWNHs is 4–6 times greater than that of cisplatin; and *in vivo* antitumor activity against the growth of transplanted tumors in mice is also better than cisplatin and remains for a long time (25 days). Since cisplatin-SWNHs adhere to the cell surfaces *in vitro* and stay within the tumor tissues *in vivo*, the released cisplatin can realize high concentrations locally at the cells *in vitro* and in the tissues *in vivo* and hence can efficiently attack the tumor cells [235].

Encapsulation of a single drug molecule inside a synthetic macrocycle is a common strategy in the delivery of platinum-based antitumor drugs. Such systems have shown particular potential as protective delivery vehicles. In recent years, the most attractive macrocycles used for this purpose are cucurbit[*n*]urils (where n = 6–8). Cucurbit[*n*]urils contain two symmetrical hydrophilic carbonyl lined portals, capping a central hydrophobic cavity (Figure 4.6) [236]. These barrel shaped molecules could be used as molecular hosts for neutral and charged mono- and multinuclear platinum antitumor agents. Partial or full encapsulation within cucurbit[*n*]urils provides steric hindrance to drug degradation by peptides and proteins, and allows for the tuning of drug release rates, cytotoxicity, and toxicity [237, 238]. For example, the dinuclear platinum complex **86** has been included inside a cucurbit[7]uril macrocycle [239]. The cytotoxicity of **86** against the L1210 cell line and the corresponding cisplatin resistant L1210/DDP sub-line is not affected significantly by the encapsulation, but the reactivity at the platinum center is reduced at least threefold. Therefore, cucurbit[7]uril has the potential as a delivery vehicle for multinuclear platinum complexes [240, 241].

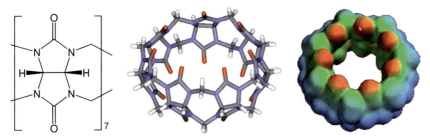

Figure 4.6 Chemical structure (left), X-ray crystal structure (middle), and electrostatic potential map (right) of cucurbit[7] uril.

Complex **87** is a newly developed DNA intercalator that displays cytotoxicity against a panel of human cancer cell lines, with some activity significantly higher, for example, up to 100-fold greater in the L1210 and A-427 cell lines, than that of cisplatin. Encapsulation of **87** by cucurbit[6]uril barely changes its cytotoxicity [242–245]. However, partial encapsulation of **87** by cucurbit[*n*]urils can drastically reduce the deactivation by GSH, thus protecting **87** from intracellular degradation [246]. The size of the cavity and the binding affinity are important for the effect on the cytotoxicity, in that small changes of macrocycle size could either decrease or increase the activity (up to 2.5-fold) of the platinum complexes. The decrease in activity may result from the protective effects of the macrocycles on the encapsulated complexes. In some cases, encapsulation by cucurbit[*n*]urils can significantly affect the cytotoxicity and limit the water solubility of platinum complexes [247]. For these reasons, several other macrocycles such as β-cyclo-dextrin and calix[4]arene are investigated as potential alternatives. For instance, encapsulation of **87** in these macrocycles increases its stability to GSH threefold, but shows no significant effect on the cytotoxicity against the LoVo human colorectal cancer cell line [248]. Similar result is also observed for **86** after such encapsulation [249].

The value of polymeric nanoparticles as sequential release carriers has become increasingly evident in the drug delivery for cancer therapy [250, 251]. In comparison with other delivery strategies, nanoparticle-based formulations possess many advantages, such as enhanced drug accumulation in tumor tissues, reduction in systemic toxicity, and surface-functionalization with targeting and passivating moieties. In addition, the nanotherapeutic approach allows protection of the loaded drug from the exterior environment, and thereby increases the blood circulation time of the active form before reaching its target. This strategy not only protects the drug *per se* but also shields the body from undesired side effects. Several nanoparticle-based anticancer drugs have been approved by the FDA, and there has been an intense interest in the development of nanoparticle formulations for effective delivery of platinum-based drugs to tumors [252, 253].

The prostate-specific membrane antigen (PSMA) is abundantly expressed in prostate cancer, its metastatic form, and the hormone-refractory form. Recently, a strategy to construct the Pt(IV)-encapsulated PSMA targeted nanoparticles has been explored, which includes the encapsulation of complex **88** in poly(D,L-lactic-*co*-glycolic acid)-poly(ethylene glycol)-functionalized polymers by nanoprecipitation and the subsequent conjugation of PSMA aptamers to the formed nanoparticles.

By using PSMA targeting aptamers on the surface of the nanoparticles, the Pt(IV) prodrug **88** is delivered specifically to the human PSMA-overexpressing (PSMA$^+$) LNCaP prostate cancer cells and internalized through endocytosis. Upon intracellular reduction of **88**, a lethal dose of cisplatin is released from the polymeric nanoparticles. Controlled release of **88** from the nanoparticles extends over 60 h. The nanoparticles loaded with **88** are highly cytotoxic to the LNCaP cells, having an IC$_{50}$ of 0.03 μM. Under the same conditions, free cisplatin has an IC$_{50}$ of 2.4 μM, which is 80-times less effective than the targeted nanoparticles [254].

A general strategy for the delivery of platinum-based drugs to cancer cells via their inclusion into nanoscale coordination polymers (NCPs) has been developed. NCPs have been constructed from Tb^{3+} ions and **89** simply by precipitating them from an aqueous solution via the addition of a poor solvent. The action of the metal ions is to form the polymerized metal–ligand networks. The platinum-based NCPs are stabilized with shells of amorphous silica to prevent rapid dissolution and to effectively control the release of the platinum species. The silica shells extend the release half-time to 5.5 or 9 h, depending on the size of the coating (2 or 7 nm). These rates of release would allow sufficient time for the platinum-based NCPs to circulate throughout the body and accumulate in tumor tissue. To enhance the cellular uptake of NCPs *in vitro*, silyl-derived c(RGDfK) is grafted onto the surface. This small cyclic peptide sequence could enhance the binding affinity for the $\alpha_v\beta_3$ integrin that is upregulated in many angiogenic cancers such as HT29. As a result, the IC$_{50}$ values of the c(RGDfK)-targeted NCPs are lower than that of cisplatin for this cell line. Cytotoxicity against the MCF-7 cell line, which does not overexpress the $\alpha_v\beta_3$ integrin, is similar to that of cisplatin [255]. These targeted NCPs may be internalized via receptor-mediated endocytosis.

Polymer-based drug carriers could protect platinum complexes from degradation and accomplish controlled release in the delivery, which would substantially improve the efficacy of platinum-based chemotherapy. Polymeric systems are expected to show preferential accumulation of the platinum drugs in the tumor cells because of the EPR effect and the leaky neovasculature [256]. Different delivery strategies have been proposed on the basis of polymeric molecules. Polymer–platinum conjugates are the most studied systems thus far: they are formed between a polymer with suitable metal-binding groups and a platinum drug moiety through coordination bonds. Polymers such as poly(amino acids), poly(amidoamine) dendrimers, and poly(N-(2-hydroxypropyl)methacrylamide) (PHPMA) are generally used as carriers because they contain either inherent ligating groups or pendant or terminal ligating groups. Cleavable linking groups can be used to provide potential for tissue or tumor specificity under desired conditions [257, 258]. By far, PHPMA is one of the most successful polymers that have been used to construct platinum conjugates. Two of such conjugates, **90** (AP5280) and **91** (AP5346), have entered clinical trials. In **90**, the *cis*-diammine-platinum(II) moiety is coordinated to an aminomalonic acid terminal group linked by a Gly-Phe-Leu-Gly-peptide spacer to a *N*-(2-hydroxypropyl)methacrylamide (HPMA) copolymer [259]. Complex **91** is an analog of **90** in which the diamino-cyclohexaneplatinum(II) moiety is bound to a polymer with triglycine side chains and an aminomalonic acid terminal group [260]. Both polymer–platinum conjugates contain pH sensitive peptide side chains to which the active fragment of cisplatin (for **90**) or oxaliplatin (for **91**) is bound. The platinum drug is not active until it reaches the tumor tissues where the environment is more acidic than that of normal tissues. In Phase I trials, however, **90** presented dose limiting toxicity of nausea and vomiting; **91** on the other hand has progressed through Phase I trials and a Phase II study in patients with recurrent ovarian cancer has been completed under the commercial name of Prolindac^{TM} [261–263].

90   91

In the above examples, approaches used for drug protection and controlled release are based on systems with non-physiological carriers. Recently, a strategy based on a natural protein has been developed for the targeting delivery of platinum drugs. The native iron-storage protein ferritin (Ft) is a promising vehicle for targeted drug delivery since the binding sites and endocytosis of Ft have been identified in tumor cells. Ft can be easily demineralized into apoferritin (AFt), a hollow protein cage with internal and external diameters of 8 and 12 nm, respectively. This protein cage can be employed to deliver platinum drugs, which may enhance the drug selectivity for cell surfaces that express Ft receptors.

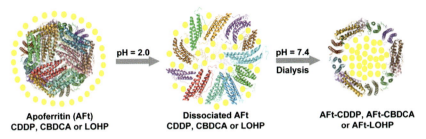

Scheme 4.1 Schematic illustration of the pH-mediated encapsulation of cisplatin (CDDP), carboplatin (CBDCA), or oxaliplatin (LOHP) by apoferritin (AFt) via an unfolding–refolding process.

Cisplatin, carboplatin, and oxaliplatin have been successfully encapsulated in the cavity of AFt [264]. The encapsulation was achieved through manipulating the pH-dependent unfolding–refolding process of AFt at pH 2.0 and 7.4, respectively, in saturated drug solution (Scheme 4.1). The structural integrity of the protein shell remains intact after encapsulation and hence the potential recognition nature should not be affected. *In vitro* assays on the rat pheochromocytoma cell line (PC12) showed that AFt–cisplatin inhibits the cells in a slow but sustaining mode and the cellular uptake of platinum is enhanced by AFt [265]. These protein coated drugs are expected to improve the toxicity profiles of the naked ones and finally to overcome the detrimental effects of platinum-based drugs.

4.8
Concluding Remarks and Future Perspectives

This chapter presents the latest scenario in the area of platinum-based antitumor drugs. The mechanism of tumor resistance to platinum drugs and diverse tactics in overcoming the resistance are highlighted. Platinum complexes with DNA damage mode radically different from that of cisplatin may evade the cellular DNA repair machinery and inhibit tumor cells through different mechanisms. Structural changes could substantially modulate the DNA binding mode and the DNA damage process and, as a result, the biological efficacy of platinum complexes. Therefore, the design of new functional ligands remains the main strategy to tune the biological behavior of platinum complexes. Tailored multifunctional ligands offer many exciting possibilities to prepare platinum complexes with improved physicochemical and biological properties such as targeting potential. In addition, they can play an integral role in modulating the systemic toxicity of the complexes. Undoubtedly, rational design will become a standard procedure in the future development of platinum-based antitumor drugs. In this context, the increasing knowledge of the biochemical and cellular processes underlying the antitumor activity of platinum complexes not only builds up the basis for the rational design of novel complexes but also speeds up their development in clinical tests. Moreover, an understanding of the molecular determinants of cellular response to platinum complexes helps to identify new cellular targets for the development of novel drugs. In this respect, a recent proteomic study in rat cochlear cells has led to the identification of novel cisplatin-modulated proteins that are involved in apoptosis, cell survival, or progression through the cell cycle [266]. In addition, a novel mechanism for the cisplatin-induced neuropathy has been characterized in a mouse model using monoclonal antibodies specific for the cisplatin–DNA adducts [267]. Such discoveries reveal new determinants of drug response that should be taken into account in future drug design. Likewise, the development of innovative delivery systems such as nanomaterials is expected to significantly improve the therapeutic index of platinum-based antitumor drugs in the near future [268].

Enormous progress has been made in the last decade or so in the field of platinum-based antitumor drugs. The advance foreshows an even brighter prospect in the coming years. On this basis, an appreciable improvement in efficacy will prove to be a reachable objective for platinum-based chemotherapy.

Acknowledgments

We thank the National Natural Science Foundation of China for financial support (grant numbers 30870554, 20631020).

References

1 Abu-Surrah, A.S. and Kettunen, M. (2006) *Curr. Med. Chem.*, **13**, 1337–1357.
2 Ho, Y.-P., Au-Yeung, S.C.F., and To, K.W. (2003) *Med. Res. Rev.*, **23**, 633–655.
3 Reedijk, J. (2003) *Proc. Natl. Acad. Sci. USA*, **100**, 3611–3616.
4 Boulikas, T. and Vougiouka, M. (2004) *Oncol. Rep.*, **11**, 559–595.
5 Muggia, F. (2009) *Gynecol. Oncol.*, **112**, 275–281.
6 Barnes, K.R. and Lippard, S.J. (2004) *Metal Ions Biol. Syst.*, **42**, 143–177.
7 Rabik, C.A. and Dolan, M.E. (2007) *Cancer Treat. Rev.*, **33**, 9–23.
8 Galanski, M., Jakupec, M.A., and Keppler, B.K. (2005) *Curr. Med. Chem.*, **12**, 2075–2094.
9 Wang, D. and Lippard, S.J. (2005) *Nat. Rev. Drug Discov.*, **4**, 307–320.

10 Argyriou, A.A., Polychronopoulos, P., Iconomou, G., Chroni, E., and Kalofonos, H.P. (2008). *Cancer Treat. Rev.*, **34**, 368–377.
11 Barabas, K., Milner, R., Lurie, D., and Adin, C. (2008) *Vet. Comp. Oncol.*, **6**, 1–18.
12 McWhinney, S.R., Goldberg, R.M., and McLeod, H.L. (2009) *Mol. Cancer Ther.*, **8**, 10–16.
13 Zhang, C.X., and Lippard, S.J. (2003). *Curr. Opin. Chem. Biol.*, **7**, 481–489.
14 Fuertes, M.A., Alonso, C., and Pérez, J.M. (2003). *Chem. Rev.*, **103**, 645–662.
15 Montaña, Á.M. and Batalla, C. (2009) *Curr. Med. Chem.*, **16**, 2235–2260.
16 Koshiyama, M., Kinezaki, M., Uchida, T., and Sumitomo, M. (2005) *Anticancer Res.*, **25**, 4499–4502.

17 Hanada, K., Asano, K., Nishimura, T., Chimata, T., Matsuo, Y., Tsuchiya, M., and Ogata, H. (2008) *J. Pharm. Pharmacol.*, **60**, 317–322.

18 Graham, J., Muhsin, M., and Kirkpatrick, P. (2004) *Nat. Rev. Drug Discov.*, **3**, 11–12.

19 Hartmann, J.T. and Lipp, H.P. (2003) *Expert Opin. Pharmacother.*, **4**, 889–901.

20 Bruijnincx, P.C.A. and Sadler, P.J. (2008) *Curr. Opin. Chem. Biol.*, **12**, 197–206.

21 Lovejoy, K.S. and Lippard, S.J. (2009) *Dalton Trans.*, 10651–10659.

22 Sun, R.W.-Y., Ma, D.-L., Wong, E.L.-M., and Che, C.-M. (2007) *Dalton Trans.*, 4884–4892.

23 Wang, X.Y. and Guo, Z.J. (2008) *Dalton Trans.*, 1521–1532.

24 Cossa, G., Gatti, L., Zunino, F., and Perego, P. (2009) *Curr. Med. Chem.*, **16**, 2355–2365.

25 Cepeda, V., Fuertes, M.A., Castilla, J., Alonso, C., Quevedo, C., and Pérez, J. M. (2007) *Anti-Cancer Agents Med. Chem.*, **7**, 3–18.

26 Chaney, S.G., Campbell, S.L., Temple, B., Bassett, E., Wu, Y., and Faldu, M. (2004) *J. Inorg. Biochem.*, **98**, 1551–1559.

27 Siddik, Z.H. (2003) *Oncogene*, **22**, 7265–7279.

28 Kelland, L. (2007) *Nat. Rev. Cancer*, **7**, 573–584.

29 Rebillard, A., Lagadic-Gossmann, D., and Dimanche-Boitrel, M.-T. (2008) *Curr. Med. Chem.*, **15**, 2656–2663.

30 Zhu, G. and Lippard, S.J. (2009) *Biochemistry*, **48**, 4916–4925.

31 Jung, Y. and Lippard, S.J. (2007) *Chem. Rev.*, **107**, 1387–1407.

32 Arnesano, F. and Natile, G. (2009) *Coord. Chem. Rev.*, **253**, 2070–2081.

33 Klein, A.V. and Hambley, T.W. (2009) *Chem. Rev.*, **109**, 4911–4920.

34 Gibson, D. (2009) *Dalton Trans.*, 10681–10689.

35 Benedetti, V., Perego, P., Beretta, G.L., Corna, E., Tinelli, S., Righetti, S.C., Leone, R., Apostoli, P., Lanzi, C., and Zunino, F. (2008) *Mol. Cancer Ther.*, **7**, 679–687.

36 Stordal, B., Pavlakis, N., and Davey, R. (2007) *Cancer Treat. Rev.*, **33**, 347–357.

37 Holzer, A.K., Manorek, G.H., and Howell, S.B. (2006) *Mol. Pharmacol.*, **70**, 1390–1394.

38 Holzer, A.K. and Howell, S.B. (2006) *Cancer Res.*, **66**, 10944–10952.

39 Kuo, M.T., Chen, H.H.W., Song, I.-S., Savaraj, N., and Ishikawa, T. (2007) *Cancer Metastasis Rev.*, **26**, 71–83.

40 Wu, Z.Y., Liu, Q., Liang, X., Yang, X. L., Wang, N.Y., Wang, X.H., Sun, H.Z., Lu, Y., and Guo, Z.J. (2009) *J. Biol. Inorg. Chem.*, **14**, 1313–1323.

41 Arnesano, F., Scintilla, S., and Natile, G. (2007) *Angew. Chem. Int. Ed.*, **46**, 9062–9064.

42 Safaei, R., Holzer, A.K., Katano, K., Samimi, G., and Howell, S.B. (2004) *J. Inorg. Biochem.*, **98**, 1607–1613.

43 Martelli, L., Di Mario, F., Ragazzi, E., Apostoli, P., Leone, R., Perego, P., and Fumagalli, G. (2006) *Biochem. Pharmacol.*, **72**, 693–700.

44 Heffeter, P., Jungwirth, U., Jakupec, M., Hartinger, C., Galanski, M., Elbling, L., Micksche, M., Keppler, B., and Berger, W. (2008) *Drug Resist. Update*, **11**, 1–16.

45 Liu, Q., Wei, H.Y., Lin, J., Zhu, L.G., and Guo, Z.J. (2004) *J. Inorg. Biochem.*, **98**, 702–712.

46 Wei, H.Y., Liu, Q., Lin, J., Jiang, P.J., and Guo, Z.J. (2004) *Inorg. Chem. Commun.*, **7**, 792–794.

47 Wei, H.Y., Wang, X.Y., Liu, Q., Mei, Y. H., Lu, Y., and Guo, Z.J. (2005) *Inorg. Chem.*, **44**, 6077–6081.

48 Oehlsen, M.E., Hegmans, A., Qu, Y., and Farrell, N. (2005) *Inorg. Chem.*, **44**, 3004–3006.

49 Wang, X.Y. and Guo, Z.J. (2007) *Anti-Cancer Agents Med. Chem.*, **7**, 19–34.

50 Yang, P., Ebbert, J.O., Sun, Z., and Weinshilboum, R.M. (2006) *J. Clin. Oncol.*, **24**, 1761–1769.

51 Wei, H.Y., Wang, X.Y., and Guo, Z.J. (2005) *Platinum–sulfur interactions and new platinum-based drug design*, in *Metal Compounds in Cancer Chemotherapy* (eds J.M. Pérez, M.A. Fuertes, and C. Alonso), Research Signpost, Kerala, India, pp. 241–267.

52 Knipp, M. (2009) *Curr. Med. Chem.*, **16**, 522–537.

53 Yu, J.J. (2009) *Curr. Drug Ther.*, **4**, 19–28.

54 Gossage, L. and Madhusudan, S. (2007) *Cancer Treat. Rev.*, **33**, 565–577.

55 Chang, I.-Y., Kim, M.-H., Kim, H.B., Lee, D.Y., Kim, S.-H., Kim, H.-Y., and You, H.J. (2005) *Biochem. Biophys. Res. Commun.*, **327**, 225–233.

56 Reed, E. (2005) *Clin. Cancer Res.*, **11**, 6100–6102.

57 O'Brien, V. and Brown, R. (2006) *Carcinogenesis*, **27**, 682–692.

58 Martin, L.P., Hamilton, T.C., and Schidler, R.J. (2008) *Clin. Cancer Res.*, **14**, 1291–1295.

59 Gifford, G., Paul, J., Vasey, P.A., Kaye, S.B., and Brown, R. (2004) *Clin. Cancer Res.*, **10**, 4420–4426.

60 Helleman, J., van Staveren, I.L., Dinjens, W.N.M., van Kuijk, P.F., Ritstier, K., Ewing, P.C., van der Burg, M.E.L., Stoter, G., and Berns, E.M.J.J. (2006) *BMC Cancer*, **6**, 201.

61 Alt, A., Lammens, K., Chiocchini, C., Lammens, A., Pieck, J.C., Kuch, D., Hopfner, K.-P., and Carell, T. (2007) *Science*, **318**, 967–970.

62 Chaney, S.G., Campbell, S.L., Bassett, E. and Wu, Y. (2005) *Crit. Rev. Oncol. Hematol.*, **53**, 3–11.

63 Albertella, M.R., Green, C.M., Lehmann, A.R., and O'Connor, M.J. (2005) *Cancer Res.*, **65**, 9799–9806.

64 Yu, Q. (2006) *Drug Resist. Update*, **9**, 19–25.

65 Wang, G., Dombkowski, A., Chuang, L., and Xu, X.X.S. (2004) *Cell Res.*, **14**, 303–314.

66 Wang, G., Chuang, L., Zhang, X., Colton, S., Dombkowski, A., Reiners, J., Diakiw, A., and Xu, X.S. (2004) *Nucleic Acids Res.*, **32**, 2231–2240.

67 Brabec, V. and Kasparkova, J. (2005) *Drug Resist. Update*, **8**, 131–146.

68 Sakai, W., Swisher, E.M., Karlan, B.Y., Agarwal, M.K., Higgins, J., Friedman, C., Villegas, E., Jacquemont, C., Farrugia, D.J., Couch, F.J., Urban, N., and Taniguchi, T. (2008) *Nature*, **451**, 1116–1120.

69 Swisher, E.M., Sakai, W., Karlan, B.Y., Wurz, K., Urban, N., and Taniguchi, T. (2008) *Cancer Res.*, **68**, 2581–2586.

70 Adams, J.M. and Cory, S. (2007) *Oncogene*, **26**, 1324–1337.

71 Hayward, R.L., Macpherson, J.S., Cummings, J., Monia, B.P., Smyth, J. F., and Jodrell, D.I. (2004) *Mol. Cancer Ther.*, **3**, 169–178.

72 Witham, J., Valenti, M.R., De-Haven-Brandon, A.K., Vidot, S., Eccles, S.A., Kaye, S.B., and Richardson, A. (2007) *Clin. Cancer Res.*, **13**, 7191–7198.

73 Stewart, D.J. (2007) *Crit. Rev. Oncol. Hematol.*, **63**, 12–31.

74 Centerwall, C.R., Kerwood, D.J., Goodisman, J., Toms, B.B., and Dabrowiak, J.C. (2008) *J. Inorg. Biochem.*, **102**, 1044–1049.

75 Lovejoy, K.S., Todd, R.C., Zhang, S., McCormick, M.S., D'Aquino, J.A., Reardon, J.T., Sancar, A., Giacomini, K.M., and Lippard, S.J. (2008) *Proc. Natl. Acad. Sci. USA.*, **105**, 8902–8907.

76 Zhang, J.Y., Wang, X.Y., Tu, C., Lin, J., Ding, J., Lin, L.P., Wang, Z.M., He, C., Yan, C.H., You, X.Z., and Guo, Z.J. (2003) *J. Med. Chem.*, **46**, 3502–3507.

77 Gao, X.L., Wang, X.Y., Ding, J., Lin, L. P., Li, Y.Z., and Guo, Z.J. (2006) *Inorg. Chem. Commun.*, **9**, 722–726.

78 Gelmon, K.A., Stewart, D., Chi, K.N., Chia, S., Cripps, C., Huan, S., Janke, S., Ayers, D., Fry, D., Shabbits, J.A., Walsh, W., McIntosh, L., and Seymour, L.K. (2004) *Ann. Oncol.*, **15**, 1115–1122.

79 Margiotta, N., Natile, G., Capitelli, F., Fanizzi, F.P., Boccarelli, A., De Rinaldis, P., Giordano, D., and Coluccia, M. (2006) *J. Inorg. Biochem.*, **100**, 1849–1857.

80 Baruah, H., Barry, C.G., and Bierbach, U. (2004) *Curr. Top. Med. Chem.*, **4**, 1537–1549.

81 Guddneppanavar, R. and Bierbach, U. (2007) *Anti-Cancer Agents Med. Chem.*, **7**, 125–138.

82 Barry, C.G., Baruah, H., and Bierbach, U. (2003) *J. Am. Chem. Soc.*, **125**, 9629–9637.

83 Baruah, H. and Bierbach, U. (2003) *Nucleic Acids Res.*, **31**, 4138–4146.

84 Budiman, M.E., Alexander, R.W., and Bierbach, U. (2004) *Biochemistry*, **43**, 8560–8567.

85 Barry, C.G., Day, C.S., and Bierbach, U. (2005) *J. Am. Chem. Soc.*, **127**, 1160–1169.

86 Baruah, H., Wright, M.W., and Bierbach, U. (2005) *Biochemistry*, **44**, 6059–6070.

87 Budiman, M.E., Bierbach, U., and Alexander, R.W. (2005) *Biochemistry*, **44**, 11262–11268.

88 Hess, S.M., Anderson, J.G., and Bierbach, U. (2005) *Bioorg. Med. Chem. Lett.*, **15**, 443–446.

89 Guddneppanavar, R., Saluta, G., Kucera, G.L., and Bierbach, U. (2006) *J. Med. Chem.*, **49**, 3204–3214.

90 Hess, S.M., Mounce, A.M., Sequeira, R.C., Augustus, T.M., Ackley, M.C., and Bierbach, U. (2005) *Cancer Chemother. Pharmacol.*, **56**, 337–343.

91 Ackley, M.C., Barry, C.G., Mounce, A. M., Farmer, M.C., Springer, B.-E., Day, C.S., Wright, M.W., Berners-Price, S.J., Hess, S.M., and Bierbach, U. (2004) *J. Biol. Inorg. Chem.*, **9**, 453–461.

92 Guddneppanavar, R., Choudhury, J.R., Kheradi, A.R., Steen, B.D., Saluta, G., Kucera, G.L., Day, C.S., and Bierbach, U. (2007) *J. Med. Chem.*, **50**, 2259–2263.

93 Ma, Z., Choudhury, J.R., Wright, M. W., Day, C.S., Saluta, G., Kucera, G.L., and Bierbach, U. (2008) *J. Med. Chem.*, **51**, 7574–7580.

94 Ma, Z., Rao, L., and Bierbach, U. (2009) *J. Med. Chem.*, **52**, 3424–3427.

95 Gao, J., Woolley, F.R., and Zingaro, R. A. (2005) *J. Med. Chem.*, **48**, 7192–7197.

96 Nováková, O., Malina, J., Kašpárková, J., Halámiková, A., Bernard, V., Intini, F., Natile, G., and Brabec, V. (2009) *Chem. Eur. J.*, **15**, 6211–6221.

97 Kalinowska-Lis, U., Ochocki, J., and Matlawska-Wasowska, K. (2008) *Coord. Chem. Rev.*, **252**, 1328–1345.

98 Jakupec, M.A., Galanski, M., and Keppler, B.K. (2003) *Rev. Physiol. Biochem. Pharmacol.*, **146**, 1–53.

99 Marchán, V., Pedroso, E., and Grandas, A. (2004) *Chem. Eur. J.*, **10**, 5369–5375.

100 Coluccia, M. and Natile, G. (2007) *Anti-Cancer Agents Med. Chem.*, **7**, 111–123.

101 Brabec, V. and Kasparkova, J. (2005) *DNA interactions of platinum anticancer drugs. Recent advances and mechanisms of action*, in *Metal Compounds in Cancer Chemotherapy* (eds J.M. Pérez, M.A. Fuertes, and C. Alonso), Research Signpost, Kerala, India, pp. 187–281.

102 Heringova, P., Woods, J., Mackay, F.S., Kasparkova, J., Sadler, P.J., and Brabec, V. (2006) *J. Med. Chem.*, **49**, 7792–7798.

103 Boccarelli, A., Giordano, D., Natile, G., and Coluccia, M. (2006) *Biochem. Pharmacol.*, **72**, 280–292.

104 Suchánková, T., Vojtíšková, M., Reedijk, J., Brabec, V., and Kašpárková, J. (2009) *J. Biol. Inorg. Chem.*, **14**, 75–87.

105 Aris, S.M. and Farrell, N.P. (2009) *Eur. J. Inorg. Chem.*, 1293–1302.

106 Natile, G. and Coluccia, M. (2004) *Metal Ions Biol. Syst.*, **42**, 209–250.

107 Novakova, O., Kasparkova, J., Malina, J., Natile, G., and Brabec, V. (2003) *Nucleic Acids Res.*, **31**, 6450–6460.

108 Intini, F.P., Boccarelli, A., Francia, V. C., Pacifico, C., Sivo, M.F., Natile, G., Giordano, D., De Rinaldis, P., and Coluccia, M. (2004) *J. Biol. Inorg. Chem.*, **9**, 768–780.

109 Boccarelli, A., Intini, F.P., Sasanelli, R., Sivo, M.F., Coluccia, M., and Natile, G. (2006) *J. Med. Chem.*, **49**, 829–837.

110 Farrell, N., Povirk, L.F., Dange, Y., DeMasters, G., Gupta, M.S., Kohlhagen, G., Khan, Q.A., Pommier, Y., and Gewirtz, D.A. (2004) *Biochem. Pharmacol.*, **68**, 857–866.

111 Kasparkova, J., Novakova, O., Farrell, N., and Brabec, V. (2003) *Biochemistry*, **42**, 792–800.

112 Marini, V., Christofis, P., Novakova, O., Kasparkova, J., Farrell, N., and Brabec, V. (2005) *Nucleic Acids Res.*, **33**, 5819–5828.

113 Huq, F., Daghriri, H., Yu, J.Q., Beale, P., and Fisher, K. (2004) *Eur. J. Med. Chem.*, **39**, 691–697.

114 Martínez, A., Lorenzo, J., Prieto, M.J., de Llorens, R., Font-Bardia, M., Solans, X., Avilés, F.X., and Moreno, V. (2005) *ChemBioChem.*, **6**, 2068–2077.

115 Stehlíková, K., Kašpárková, J., Nováková, O., Martínez, A., Moreno,

V., and Brabec, V. (2006) *FEBS J.*, **273**, 301–314.

116 Martínez, A., Lorenzo, J., Prieto, M.J., Font-Bardia, M., Solans, X., Avilés, F. X., and Moreno, V. (2007) *Bioorg. Med. Chem.*, **15**, 969–979.

117 Bulluss, G.H., Knott, K.M., Ma, E.S.F., Aris, S.M., Alvarado, E., and Farrell, N. (2006) *Inorg. Chem.*, **45**, 5733–5735.

118 Ma, E.S.F., Bates, W.D., Edmunds, A., Kelland, L.R., Fojo, T., and Farrell, N. (2005) *J. Med. Chem.*, **48**, 5651–5654.

119 Quiroga, A.G., Pérez, J.M., Alonso, C., Navarro-Ranninger, C., and Farrell, N. (2006) *J. Med. Chem.*, **49**, 224–231.

120 van Zutphen, S., Pantoja, E., Soriano, R., Soro, C., Tooke, D.M., Spek, A.L., den Dulk, J., Brouwer, J., and Reedijk, J. (2006) *Dalton Trans.*, 1020–1023.

121 Ramos-Lima, F.J., Vrána, O., Quiroga, A.G., Navarro-Ranninger, C., Halámiková, A., Rybníčková, H., Hejmalová, L., and Brabec, V. (2006) *J. Med. Chem.*, **49**, 2640–2651.

122 Pantoja, E., Gallipoli, A., van Zutphen, S., Komeda, S., Reddy, D., Jaganyi, D., Lutz, M., Tooke, D.M., Spek, A.L., Navarro-Ranninger, C., and Reedijk, J. (2006) *J. Inorg. Biochem.*, **100**, 1955–1964.

123 Jawbry, S., Freikman, I., Najajreh, Y., Perez, J.M., and Gibson, D. (2005) *J. Inorg. Biochem.*, **99**, 1983–1991.

124 Kasparkova, J., Novakova, O., Marini, V., Najajreh, Y., Gibson, D., Perez, J.-M., and Brabec, V. (2003) *J. Biol. Chem.*, **278**, 47516–47525.

125 Kasparkova, J., Marini, V., Najajreh, Y., Gibson, D., and Brabec, V. (2003) *Biochemistry*, **42**, 6321–6332.

126 Najajreh, Y., Kasparkova, J., Marini, V., Gibson, D., and Brabec, V. (2005) *J. Biol. Inorg. Chem.*, **10**, 722–731.

127 Ramos-Lima, F.J., Quiroga, A.G., Pérez, J.M., Font-Bardía, M., Solans, X., and Navarro-Ranninger, C. (2003) *Eur. J. Inorg. Chem.*, 1591–1598.

128 Ramos-Lima, F.J., Quiroga, A.G., García-Serrelde, B., Blanco, F., Carnero, A., and Navarro-Ranninger, C. (2007) *J. Med. Chem.*, **50**, 2194–2199.

129 Kalinowska, U., Matławska, K., Chęcińska, L., Domagała, M., Kontek,

R., Osiecka, R., and Ochocki, J. (2005) *J. Inorg. Biochem.*, **99**, 2024–2031.

130 Coley, H.M., Sarju, J., and Wagner, G. (2008) *J. Med. Chem.*, **51**, 135–141.

131 Najajreh, Y., Khazanov, E., Jawbry, S., Ardeli-Tzaraf, Y., Perez, J.M., Kasparkova, J., Brabec, V., Barenholz, Y., and Gibson, D. (2006) *J. Med. Chem.*, **49**, 4665–4673.

132 Prokop, R., Kasparkova, J., Novakova, O., Marini, V., Pizarro, A.M., Navarro-Ranninger, C., and Brabec, V. (2004) *Biochem. Pharmacol.*, **67**, 1097–1109.

133 Halámiková, A., Heringová, P., Kašpárková, J., Intini, F.P., Natile, G., Nemirovski, A., Gibson, D., and Brabec, V. (2008) *J. Inorg. Biochem.*, **102**, 1077–1089.

134 Farrell, N. (2004) *Metal Ions Biol. Syst.*, **42**, 251–296.

135 Qu, Y., Tran, M.-C., and Farrell, N.P. (2009) *J. Biol. Inorg. Chem.*, **14**, 969–977.

136 Wheate, N.J. and Collins, J.G. (2003) *Coord. Chem. Rev.*, **241**, 133–145.

137 Gourley, C., Cassidy, J., Edwards, C., Samuel, L., Bisset, D., Camboni, G., Young, A., Boyle, D., and Jodrell, D. (2004) *Cancer Chemother. Pharmacol.*, **53**, 95–101.

138 Jodrell, D.I., Evans, T.R.J., Steward, W., Cameron, D., Prendiville, J., Aschele, C., Noberasco, C., Lind, M., Carmichael, J., Dobbs, N., Camboni, G., Gatti, B., and De Braud, F. (2004) *Eur. J. Cancer*, **40**, 1872–1877.

139 Hensing, T.A., Hanna, N.H., Gillenwater, H.H., Camboni, M.G., Allievi, C., and Socinski, M.A. (2006) *Anti-Cancer Drugs*, **17**, 697–704.

140 Jakupec, M.A., Galanski, M., Arion, V. B., Hartinger, C.G., and Keppler, B.K. (2008) *Dalton Trans.*, 183–194.

141 Billecke, C., Finniss, S., Tahash, L., Miller, C., Mikkelsen, T., Farrell, N.P., and Bögler, O. (2006) *Neuro-Oncol.*, **8**, 215–226.

142 Mitchell, C., Kabolizadeh, P., Ryan, J., Roberts, J.D., Yacoub, A., Curiel, D.T., Fisher, P.B., Hagan, M.P., Farrell, N.P., Grant, S., and Dent, P. (2007) *Mol. Pharmacol.*, **72**, 704–714.

143 Liu, Q., Qu, Y., van Antwerpen, R., and Farrell, N. (2006) *Biochemistry*, **45**, 4248–4256.

144 Hegmans, A., Kasparkova, J., Vrana, O., Kelland, L.R., Brabec, V., and Farrell, N.P. (2008) *J. Med. Chem.*, **51**, 2254–2260.

145 Montero, E.I., Benedetti, B.T., Mangrum, J.B., Oehlsen, M.J., Qu, Y., and Farrell, N.P. (2007) *Dalton Trans.*, 4938–4942.

146 Jansen, B.A.J., Wielaard, P., Kalayda, G.V., Ferrari, M., Molenaar, C., Tanke, H.J., Brouwer, J., and Reedijk, J. (2004) *J. Biol. Inorg. Chem.*, **9**, 403–413.

147 Kalayda, G.V., Jansen, B.A.J., Molenaar, C., Wielaard, P., Tanke, H.J., and Reedijk, J. (2004) *J. Biol. Inorg. Chem.*, **9**, 414–422.

148 Kalayda, G.V., Jansen, B.A.J., Wielaard, P., Tanke, H.J., and Reedijk, J. (2005) *J. Biol. Inorg. Chem.*, **10**, 305–315.

149 Kalayda, G.V., Zhang, G.F., Abraham, T., Tanke, H.J., and Reedijk, J. (2005) *J. Med. Chem.*, **48**, 5191–5202.

150 Oehlsen, M.E., Qu, Y., and Farrell, N. (2003) *Inorg. Chem.*, **42**, 5498–5506.

151 Oehlsen, M.E., Hegmans, A., Qu, Y., and Farrell, N. (2005) *J. Biol. Inorg. Chem.*, **10**, 433–442.

152 Williams, J.W., Qu, Y., Bulluss, G.H., Alvorado, E., and Farrell, N.P. (2007) *Inorg. Chem.*, **46**, 5820–5822.

153 Fan, D.M., Yang, X.L., Wang, X.Y., Zhang, S.C., Mao, J.F., Ding, J., Lin, L.P., and Guo, Z.J. (2007) *J. Biol. Inorg. Chem.*, **12**, 655–665.

154 Zhu, J.H., Lin, M.X., Fan, D.M., Wu, Z.Y., Chen, Y.C., Zhang, J.F., Lu, Y., and Guo, Z.J. (2009) *Dalton Trans.*, 10889–10895.

155 Mao, J.F., Zhang, Y.M., Zhu, J.H., Zhang, C.L., and Guo, Z.J. (2009) *Chem. Commun.*, 908–910.

156 Cesar, E.T., de Almeida, M.V., Fontes, A.P.S., Maia, E.C.P., Garnier-Suillerot, A., Couri, M.R.C., and Felício, E. de C. A. (2003) *J. Inorg. Biochem.*, **95**, 297–305.

157 Teixeira, L.J., Seabra, M., Reis, E., da Cruz, M.T.G., de Lima, M.C.P., Pereira, E., Miranda, M.A., and Marques, M.P.M. (2004) *J. Med. Chem.*, **47**, 2917–2925.

158 Kalayda, G.V., Komeda, S., Ikeda, K., Sato, T., Chikuma, M., and Reedijk, J. (2003) *Eur. J. Inorg. Chem.*, 4347–4355.

159 Komeda, S., Kalayda, G.V., Lutz, M., Spek, A.L., Yamanaka, Y., Sato, T., Chikuma, M., and Reedijk, J. (2003) *J. Med. Chem.*, **46**, 1210–1219.

160 Komeda, S., Bombard, S., Perrier, S., Reedijk, J., and Kozelka, J. (2003) *J. Inorg. Biochem.*, **96**, 357–366.

161 Teletchéa, S., Komeda, S., Teuben, J.-M., Elizondo-Riojas, M.-A., Reedijk, J., and Kozelka, J. (2006) *Chem. Eur. J.*, **12**, 3741–3753.

162 Zhao, Y.M., He, W.J., Shi, P.F., Zhu, J. H., Qiu, L., Lin, L.P., and Guo, Z.J. (2006) *Dalton Trans.*, 2617–2619.

163 Zhu, J.H., Zhao, Y.M., Zhu, Y.Y., Wu, Z.Y., Lin, M.X., He, W.J., Wang, Y., Chen, G.J., Dong, L., Zhang, J.F., Lu, Y., and Guo, Z.J. (2009) *Chem. Eur. J.*, **15**, 5245–5253.

164 Rubino, S., Portanova, P., Albanese, A., Calvaruso, G., Orecchio, S., Fontana, G., and Stocco, G.C. (2007) *J. Inorg. Biochem.*, **101**, 1473–1482.

165 Cheng, H., Huq, F., Beale, P., and Fisher, K. (2006) *Eur. J. Med. Chem.*, **41**, 896–903.

166 Tayyem, H., Huq, F., Yu, J.Q., Beale, P., and Fisher, K. (2008) *ChemMedChem*, **3**, 145–151.

167 Huq, F., Tayyem, H., Yu, J.Q., Beale, P., and Fisher, K. (2009) *Med. Chem.*, **5**, 372–381.

168 Hannon, M.J. (2007) *Chem. Soc. Rev.*, **36**, 280–295.

169 Harris, A.L., Ryan, J.J., and Farrell, N. (2006) *Mol. Pharmacol.*, **69**, 666–672.

170 Harris, A.L., Yang, X., Hegmans, A., Povirk, L., Ryan, J.J., Kelland, L., and Farrell, N.P. (2005) *Inorg. Chem.*, **44**, 9598–9600.

171 Komeda, S., Moulaei, T., Woods, K.K., Chikuma, M., Farrell, N.P., and Williams, L.D. (2006) *J. Am. Chem. Soc.*, **128**, 16092–16103.

172 Quiroga, A.G. and Ranninger, C.N. (2004) *Coord. Chem. Rev.*, **248**, 119–133.

173 Gao, E., Zhu, M., Yin, H., Liu, L., Wu, Q., and Sun, Y. (2008) *J. Inorg. Biochem.*, **102**, 1958–1964.

174 Kapp, T., Dullin, A., and Gust, R. (2006) *J. Med. Chem.*, **49**, 1182–1190.

175 Zhang, J.C., Liu, L., Gong, Y.Q., Zheng, X.M., Yang, M.S., Cui, J.R., and Shen, S.G. (2009) *Eur. J. Med. Chem.*, **44**, 2322–2327.

176 Hall, M.D., Mellor, H.R., Callaghan, R., and Hambley, T.W. (2007) *J. Med. Chem.*, **50**, 3403–3411.

177 Hall, M.D., Dolman, R.C., and Hambley, T.W. (2004) *Metal Ions Biol. Syst.*, **42**, 297–322.

178 Hall, M.D., Martin, C., Ferguson, D.J. P., Phillips, R.M., Hambley, T.W., and Callaghan, R. (2004) *Biochem. Pharmacol.*, **67**, 17–30.

179 Hall, M.D., Foran, G.J., Zhang, M., Beale, P.J., and Hambley, T.W. (2003) *J. Am. Chem. Soc.*, **125**, 7524–7525.

180 Berners-Price, S.J., Ronconi, L., and Sadler, P.J. (2006) *Prog. Nucl. Magn. Reson. Spectrosc.*, **49**, 65–98.

181 Davies, M.S., Hall, M.D., Berners-Price, S.J., and Hambley, T.W. (2008) *Inorg. Chem.*, **47**, 7673–7680.

182 Hall, M.D., Dillon, C.T., Zhang, M., Beale, P., Cai, Z., Lai, B., Stampfl, A.P. J., and Hambley, T.W. (2003) *J. Biol. Inorg. Chem.*, **8**, 726–732.

183 Hall, M.D., Amjadi, S., Zhang, M., Beale, P.J., and Hambley, T.W. (2004) *J. Inorg. Biochem.*, **98**, 1614–1624.

184 Hall, M.D., Alderden, R.A., Zhang, M., Beale, P.J., Cai, Z., Lai, B., Stampfl, A. P.J., and Hambley, T.W. (2006) *J. Struct. Biol.*, **155**, 38–44.

185 Mellor, H.R., Snelling, S., Hall, M.D., Modok, S., Jaffar, M., Hambley, T.W., and Callaghan, R. (2005) *Biochem. Pharmacol.*, **70**, 1137–1146.

186 Nemirovski, A., Kasherman, Y., Tzaraf, Y., and Gibson, D. (2007) *J. Med. Chem.*, **50**, 5554–5556.

187 Kelland, L. (2007) *Expert. Opin. Investig. Drugs*, **16**, 1009–1021.

188 Wosikowski, K., Lamphere, L., Unteregger, G., Jung, V., Kaplan, F., Xu, J.P., Rattel, B., and Caligiuri, M. (2007) *Cancer. Chemother. Pharmacol.*, **60**, 589–600.

189 Vouillamoz-Lorenz, S., Buclin, T., Lejeune, F., Bauer, J., Leyvraz, S., and Decosterd, L.A. (2003) *Anticancer Res.*, **23**, 2757–2765.

190 Sternberg, C.N., Whelan, P., Hetherington, J., Paluchowska, B., Slee, P.H., Vekemans, K., Van Erps, P., Theodore, C., Koriakine, O., Oliver, T., Lebwohl, D., Debois, M., Zurlo, A., and Collette, L. (2005) *Oncology*, **68**, 2–9.

191 Latif, T., Wood, L., Connell, C., Smith, D.C., Vaughn, D., Lebwohl, D., and Peereboom, D. (2005) *Invest. New Drugs*, **23**, 79–84.

192 Choy, H. (2006) *Expert Rev. Anticancer Ther.*, **6**, 973–982.

193 Carr, J.L., Tingle, M.D., and McKeage, M.J. (2006) *Cancer Chemother. Pharmacol.*, **57**, 483–490.

194 Choy, H., Park, C., and Yao, M. (2008) *Clin. Cancer Res.*, **14**, 1633–1638.

195 Žák, F., Turánek, J., Kroutil, A., Sova, P., Mistr, A., Poulová, A., Mikolin, P., Žák, Z., Kašná, A., Záluská, D., Neča, J., šindlerová, L., and Kozubík, A. (2004) *J. Med. Chem.*, **47**, 761–763.

196 Turánek, J., Kasná, A., Záluská, D., Neca, J., Kvardová, V., Knötigová, P., Horváth, V., Sindlerová, L., Kozubík, A., Sova, P., Kroutil, A., Žák, F., and Mistr, A. (2004) *Anti-Cancer Drugs*, **15**, 537–543.

197 Kozubík, A., Horváth, V., švihálková-šindlerová, L., Souček, K., Hofmanová, J., Sova, P., Kroutil, A., Žák, F., Mistr, A., and Turánek, J. (2005) *Biochem. Pharmacol.*, **69**, 373–383.

198 Sova, P., Mistr, A., Kroutil, A., Zak, F., Pouckova, P., and Zadinova, M. (2006) *Anti-Cancer Drugs*, **17**, 201–206.

199 Kašpárková, J., Nováková, O., Vrána, O., Intini, F., Natile, G., and Brabec, V. (2006) *Mol. Pharmacol.*, **70**, 1708–1719.

200 Reithofer, M.R., Valiahdi, S.M., Jakupec, M.A., Arion, V.B., Egger, A., Galanski, M., and Keppler, B.K. (2007) *J. Med. Chem.*, **50**, 6692–6699.

201 Reithofer, M.R., Schwarzinger, A., Valiahdi, S.M., Galanski, M., Jakupec, M.A., and Keppler, B.K. (2008) *J. Inorg. Biochem.*, **102**, 2072–2077.

202 Kwon, Y.-E., Whang, K.-J., Park, Y.-J., and Kim, K.H. (2003) *Bioorg. Med. Chem.*, **11**, 1669–1676.

203 Kwon, Y.-E. and Kim, K.H. (2006) *Anti-Cancer Drugs*, **17**, 553–558.

204 Reithofer, M., Galanski, M., Roller, A., and Keppler, B.K. (2006) *Eur. J. Inorg. Chem.*, 2612–2617.

205 Dyson, P.J. and Sava, G. (2006) *Dalton Trans.*, 1929–1933.

206 Ang, W.H., Khalaila, I., Allardyce, C.S., Juillerat-Jeanneret, L., and Dyson, P.J. (2005) *J. Am. Chem. Soc.*, **127**, 1382–1383.

207 Ang, W.H., Pilet, S., Scopelliti, R., Bussy, F., Juillerat-Jeanneret, L., and Dyson, P.J. (2005) *J. Med. Chem.*, **48**, 8060–8069.

208 Pérez, J.M., Kelland, L.R., Montero, E. I., Boxall, F.E., Fuertes, M.A., Alonso, C., and Navarro-Ranninger, C. (2003) *Mol. Pharmacol.*, **63**, 933–944.

209 Yu, H. and Jove, R. (2004) *Nat. Rev. Cancer*, **4**, 97–105.

210 Littlefield, S.L. and Baird, M.C. (2008) *Inorg. Chem.*, **47**, 2798–2804.

211 Bednarski, P.J., Mackay, F.S., and Sadler, P.J. (2007) *Anti-Cancer Agents Med. Chem.*, **7**, 75–93.

212 Ronconi, L. and Sadler, P.J. (2007) *Coord. Chem. Rev.*, **251**, 1633–1648.

213 Kašpárková, J., Mackay, F.S., Brabec, V., and Sadler, P.J. (2003) *J. Biol. Inorg. Chem.*, **8**, 741–745.

214 Bednarski, P.J., Grünert, R., Zielzki, M., Wellner, A., Mackay, F.S., and Sadler, P.J. (2006) *Chem. Biol.*, **13**, 61–67.

215 Mackay, F.S., Woods, J.A., Moseley, H., Ferguson, J., Dawson, A., Parsons, S., and Sadler, P.J. (2006) *Chem. Eur. J.*, **12**, 3155–3161.

216 Phillips, H.I.A., Ronconi, L., and Sadler, P.J. (2009) *Chem. Eur. J.*, **15**, 1588–1596.

217 Ronconi, L. and Sadler, P.J. (2008) *Chem. Commun.*, 235–237.

218 Salassa, L., Phillips, H.I.A., and Sadler, P.J. (2009) *Phys. Chem. Chem. Phys.*, **11**, 10311–10316.

219 Mackay, F.S., Woods, J.A., Heringová, P., Kašpárková, J., Pizarro, A.M., Moggach, S.A., Parsons, S., Brabec, V., and Sadler, P.J. (2007) *Proc. Natl. Acad. Sci. USA*, **104**, 20743–20748.

220 Hannon, M.J. (2007) *Pure Appl. Chem.*, **79**, 2243–2261.

221 van Zutphen, S. and Reedijk, J. (2005) *Coord. Chem. Rev.*, **249**, 2845–2853.

222 Sanchez-Cano, C. and Hannon, M.J. (2009) *Dalton Trans.*, 10702–10711.

223 Galanski, M. and Keppler, B.K. (2007) *Anti-Cancer Agents Med. Chem.*, **7**, 55–73.

224 Barroug, A., Kuhn, L.T., Gerstenfeld, L. C., and Glimcher, M.J. (2004) *J. Orthop. Res.*, **22**, 703–708.

225 Palazzo, B., Iafisco, M., Laforgia, M., Margiotta, N., Natile, G., Bianchi, C.L., Walsh, D., Mann, S., and Roveri, N. (2007) *Adv. Funct. Mater.*, **17**, 2180–2188.

226 Warnecke, A., Fichtner, I., Garmann, D., Jaehde, U., and Kratz, F. (2004) *Bioconjugate Chem.*, **15**, 1349–1359.

227 Descôteaux, C., Provencher-Mandeville, J., Mathieu, I., Perron, V., Mandal, S. K., Asselin, É., and Bérubé, G. (2003) *Bioorg. Med. Chem. Lett.*, **13**, 3927–3931.

228 Cassino, C., Gabano, E., Ravera, M., Cravotto, G., Palmisano, G., Vessières, A., Jaouen, G., Mundwiler, S., Alberto, R., and Osella, D. (2004) *Inorg. Chim. Acta*, **357**, 2157–2166.

229 Gust, R., Niebler, K., and Schönenberger, H. (2005) *J. Med. Chem.*, **48**, 7132–7144.

230 Kim, E., Rye, P.T., Essigmann, J.M., and Croy, R.G. (2009) *J. Inorg. Biochem.*, **103**, 256–261.

231 Barnes, K.R., Kutikov, A., and Lippard, S.J. (2004) *Chem. Biol.*, **11**, 557–564.

232 Feazell, R.P., Nakayama-Ratchford, N., Dai, H., and Lippard, S.J. (2007) *J. Am. Chem. Soc.*, **129**, 8438–8439.

233 Dhar, S., Liu, Z., Thomale, J., Dai, H., and Lippard, S.J. (2008) *J. Am. Chem. Soc.*, **130**, 11467–11476.

234 Bhirde, A.A., Patel, V., Gavard, J., Zhang, G., Sousa, A.A., Masedunskas, A., Leapman, R.D., Weigert, R., Gutkind, J.S., and Rusling, J.F. (2009) *ACS Nano*, **3**, 307–316.

235 Ajima, K., Murakami, T., Mizoguchi, Y., Tsuchida, K., Ichihashi, T., Iijima, S., and Yudasaka, M. (2008) *ACS Nano*, **2**, 2057–2064.

236 Lagona, J., Mukhopadhyay, P., Chakrabarti, S., and Isaacs, L. (2005) *Angew. Chem., Int. Ed.*, **44**, 4844–4870.

237 Wheate, N.J. (2008) *J. Inorg. Biochem.*, **102**, 2060–2066.

238 Jeon, Y.J., Kim, S.-Y., Ko, Y.H., Sakamoto, S., Yamaguchi, K., and Kim, K. (2005) *Org. Biomol. Chem.*, **3**, 2122–2125.

239 Kennedy, A.R., Florence, A.J., McInnes, F.J., and Wheate, N.J. (2009) *Dalton Trans.*, 7695–7700.

240 Wheate, N.J., Day, A.I., Blanch, R.J., Arnold, A.P., Cullinane, C., and Collins, J.G. (2004) *Chem. Commun.*, 1424–1425.

241 Wheate, N.J., Buck, D.P., Day, A.I., and Collins, J.G. (2006) *Dalton Trans.*, 451–458.

242 Krause-Heuer, A.M., Grünert, R., Kühne, S., Buczkowska, M., Wheate, N.J., Le Pevelen, D.D., Boag, L.R., Fisher, D.M., Kasparkova, J., Malina, J., Bednarski, P.J., Brabec, V., and Aldrich-Wright, J.R. (2009) *J. Med. Chem.*, **52**, 5474–5484.

243 Wheate, N.J., Taleb, R.I., Krause-Heuer, A.M., Cook, R.L., Wang, S., Higgins, V.J., and Aldrich-Wright, J.R. (2007) *Dalton Trans.*, 5055–5064.

244 Wheate, N.J., Brodie, C.R., Collins, J.G., Kemp, S., Aldrich-Wright, J.R. (2007) *Mini Rev. Med. Chem.*, **7**, 627–648.

245 Fisher, D.M., Bednarski, P.J., Grünert, R., Turner, P., Fenton, R.R., and Aldrich-Wright, J.R. (2007) *ChemMedChem*, **2**, 488–495.

246 Kemp, S., Wheate, N.J., Pisani, M.J., and Aldrich-Wright, J.R. (2008) *J. Med. Chem.*, **51**, 2787–2794.

247 Kemp, S., Wheate, N.J., Wang, S., Collins, J.G., Ralph, S.F., Day, A.I., Higgins, V.J., and Aldrich-Wright, J.R. (2007) *J. Biol. Inorg. Chem.*, **12**, 969–979.

248 Krause-Heuer, A.M., Wheate, N.J., Tilby, M.J., Pearson, D.G., Ottley, C.J., and Aldrich-Wright, J.R. (2008) *Inorg. Chem.*, **47**, 6880–6888.

249 Wheate, N.J., Abbott, G.M., Tate, R.J., Clements, C.J., Edrada-Ebel, R., and Johnston, B.F. (2009) *J. Inorg. Biochem.*, **103**, 448–454.

250 Ferrari, M. (2005) *Nat. Rev. Cancer*, **5**, 161–171.

251 Cho, K., Wang, X., Nie, S., Chen, Z., and Shin, D.M. (2008) *Clin. Cancer Res.*, **14**, 1310–1316.

252 Peer, D., Karp, J.M., Hong, S., Farokhzad, O.C., Margalit, R., and Langer, R. (2007) *Nat. Nanotechnol.*, **2**, 751–760.

253 Zhang, L., Gu, F.X., Chan, J.M., Wang, A.Z., Langer, R.S., and Farokhzad, O. C. (2008) *Clin. Pharmacol. Ther.*, **83**, 761–769.

254 Dhar, S., Gu, F.X., Langer, R., Farokhzad, O.C., and Lippard, S.J. (2008) *Proc. Natl. Acad. Sci. USA*, **105**, 17356–17361.

255 Rieter, W.J., Pott, K.M., Taylor, K.M.L., and Lin, W. (2008) *J. Am. Chem. Soc.*, **130**, 11584–11585.

256 Haag, R. and Kratz, F. (2006) *Angew. Chem. Int. Ed.*, **45**, 1198–1215.

257 Haxton, K.J. and Burt, H.M. (2009) *J. Pharm. Sci.*, **98**, 2299–2316.

258 Tomalia, D.A., Reyna, L.A., and Svenson, S. (2007) *Biochem. Soc. Trans.*, **35**, 61–67.

259 Lin, X., Zhang, Q., Rice, J.R., Stewart, D.R., Nowotnik, D.P., and Howell, S.B. (2004) *Eur. J. Cancer*, **40**, 291–297.

260 Sood, P., Thurmond, K.B. II, Jacob, J. E., Waller, L.K., Silva, G.O., Stewart, D. R., and Nowotnik, D.P. (2006) *Bioconjugate Chem.*, **17**, 1270–1279.

261 Rademaker-Lakhai, J.M., Terret, C., Howell, S.B., Baud, C.M., de Boer, R. F., Pluim, D., Beijnen, J.H., Schellens, J.H.M., and Droz, J.-P. (2004) *Clin. Cancer Res.*, **10**, 3386–3395.

262 Rice, J.R., Gerberich, J.L., Nowotnik, D. P., and Howell, S.B. (2006) *Clin. Cancer Res.*, **12**, 2248–2254.

263 Campone, M., Rademaker-Lakhai, J.M., Bennouna, J., Howell, S.B., Nowotnik, D.P., Beijnen, J.H., and Schellens, J.H. M. (2007) *Cancer Chemother. Pharmacol.*, **60**, 523–533.

264 Yang, Z., Wang, X.Y., Diao, H.J., Zhang, J.F., Li, H.Y., Sun, H.Z., and

Guo, Z.J. (2007) *Chem. Commun.*, 3453–3455.

265 Xing, R.M., Wang, X.Y., Zhang, C.L., Zhang, Y.M., Wang, Q., Yang, Z., and Guo, Z.J. (2009) *J. Inorg. Biochem.*, **103**, 1039–1044.

266 Jamesdaniel, S., Ding, D., Kermany, M. H., Davidson, B.A., Knight, P.R. III, Salvi, R., and Coling, D.E. (2008) *J. Proteome Res.*, **7**, 3516–3524.

267 Dzagnidze, A., Katsarava, Z., Makhalova, J., Liedert, B., Yoon, M.-S., Kaube, H., Limmroth, V., and Thomale, J. (2007) *J. Neurosci.*, **27**, 9451–9457.

268 Xu, C.J., Wang, B.D., and Sun, S.H. (2009) *J. Am. Chem. Soc.*, **131**, 4216–4217.

5
Ruthenium and Other Non-platinum Anticancer Compounds

Ioannis Bratsos, Teresa Gianferrara, Enzo Alessio, Christian G. Hartinger, Michael A. Jakupec, and Bernhard K. Keppler

5.1
Introduction

Chapters 3 and 4 of this book describe in detail the many positive aspects, and also the drawbacks, of cisplatin and of the other clinically used Pt chemotherapeutics. The major limitations of such drugs are a narrow range of activity (they are scarcely active against several malignancies with high social incidence), severe toxic side-effects, and intrinsic or acquired tumor resistance observed during treatment.

Compounds of almost all metals of the periodic table have been investigated for *in vitro* anticancer activity against cancer cell lines – the most widely used screening method – and in many cases some activity has been reported. However, most often *in vitro* cytotoxicity has not been paralleled by *in vivo* therapeutic activity, and only a few of the many metal compounds investigated *in vitro* have shown some realistic follow-up. Therefore, given that cytotoxic activity against cancer cells should not be mistaken for anticancer activity, this chapter will treat exclusively those non-platinum compounds, or classes of compounds, that have shown the most promising results in convincing tumor models or have even entered clinical trials. They include derivatives of titanium, gallium, iron, osmium, germanium, arsenic, lanthanides, and, above all, ruthenium. Gold compounds are treated separately in Chapter 7.

It is worth stressing that the primary goal of developing non-platinum metal anticancer drugs is to find activity against tumors that are resistant to Pt drugs. A secondary goal is that activity should not be accompanied by severe toxicity, that is, that the compounds have a good tolerability and a large therapeutic window (i.e., a large range of effective dosage before the onset of severe adverse effects). In principle, non-platinum compounds may be expected to have anticancer activity and toxic side-effects markedly different from those of Pt drugs for several obvious reasons: their different coordination geometries, binding preferences, and ligand-exchange rates are likely to lead to different mechanism(s) of action and, as a consequence, to different biological properties [1].

Bioinorganic Medicinal Chemistry. Edited by Enzo Alessio
Copyright © 2011 WILEY-VCH Verlag GmbH & Co. KGaA, Weinheim
ISBN: 978-3-527-32631-0

5.2
Ruthenium Anticancer Compounds

Among the several metal compounds that have been investigated for anticancer activity, those of ruthenium occupy a prevalent position [2–5]. In the last 35 years, basically three main classes of active Ru compounds (i.e., compounds that have demonstrated effectiveness *in vivo* against animal models including transplanted human tumors) have been discovered: Ru-dmso compounds, "Keppler-type" Ru(III) complexes, and organometallic Ru(II)-arene compounds. Other Ru compounds, such as the group of [RuCl$_2$(2-phenylazopyridine)$_2$] isomers developed by Reedijk and coworkers, were found to possess significant cytotoxicity *in vitro* against cancer cells [6] but were not tested *in vivo*, and for this reason will not be treated here.

In general, the most promising Ru compounds developed so far show some similarities with the established Pt drugs. They are typically considered as *prodrugs* that are activated by aquation (i.e., replacement of one or more ligands by water molecules), possibly preceded by reduction/oxidation. In other words, as for Pt drugs, activity is believed to derive mainly from the direct coordination of one (or more) of their metabolites to the biological target(s) (*functional compounds*) [7]. Still, the biological target(s) and mechanism of action of ruthenium compounds – even of the most successful and thoroughly investigated ones – are largely unknown. A possible explanation for the success of Ru compounds derives from their ligand exchange kinetics: similarly to Pt, Ru is a relatively inert metal, and ligand exchange kinetics are typically within the same timescale as cellular division processes [8].

In recent years, however, an increasing number of coordinatively saturated and inert ruthenium compounds (i.e., *structural compounds*) have shown interesting biological activity *in vitro*, such as interactions with DNA [9] or selective enzyme inhibition [10] and in some cases also cytotoxic activity and induction of apoptosis in cancer cells [9, 11, 12]. Such ruthenium compounds are still at a preliminary stage of development, and for this reason they are not treated in detail in this chapter. Nevertheless, they are likely to be the basis for important future developments. Examples of promising *structural compounds* are reported in Chapter 12.

5.2.1
Chemical Features of Ruthenium Compounds

Under physiological conditions ruthenium has two main oxidation states accessible: Ru(II) (d^6, diamagnetic) and Ru(III) (d^5, paramagnetic). Ruthenium(IV) compounds are also possible, but require several acido, oxo, or sulfido ligands for stabilization. The Ru ion is typically six-coordinate with octahedral geometry, whereas Pt(II) is square-planar. Similarly to Pt(II), Ru(II/III) ions show affinity for nitrogen and sulfur ligands. Ruthenium(II) compounds are air-stable only if they bear good π-acceptor ligands. Ligand exchange typically occurs through a dissociative mechanism with kinetics that, broadly speaking, are similar to those of Pt(II) species. It is also generally accepted that Ru(II) complexes are less inert than the corresponding Ru(III) species. For this reason, the so-called "activation-by-reduction" hypothesis was advanced in the pioneering work by Clarke, who was among the first to investigate ruthenium compounds as potential anticancer agents [2, 3]. To explain the activity of coordinatively saturated and very inert Ru(III) chlorido-ammine compounds, such as *fac*-[RuCl$_3$(NH$_3$)$_3$] and *cis*-[RuCl$_2$(NH$_3$)$_4$]Cl, Clarke and coworkers suggested that aquation was preceded by *in vivo* reduction to the less inert Ru(II) species. The latter would release the chloride anions and thus generate the aquated species that, since the aqua ligand is typically labile, are capable of coordinating to the biological target(s). Abundant cellular molecules, such as glutathione, might act as reducing agents. In principle, this activation pathway might offer some kind of selectivity against solid tumors, which are more hypoxic because of insufficient vascularization and thus more reducing environments compared to normal tissues and should facilitate the reduction of Ru(III) to Ru(II) [and/or disfavor the possible O$_2$ re-oxidation to Ru(III) species]. This fascinating hypothesis has been later applied to almost all Ru(III) species, even though in most cases not adequately supported by experimental evidence. Some general features should not be forgotten [13]: (i) The Ru(III) reduction potential depends on the ligand environment. This implies that it can be tuned by an appropriate choice of the ligands to obtain biologically accessible values. On the other hand, the parent Ru(III) complex and its metabolites may have significantly different potentials. (ii) The kinetics of aquation of Ru(II/III) compounds depend strongly on the nature of the ligands and on the net charge. (iii) As remarked above, structural Ru compounds were also shown to possess *in vitro* activity. In other words it is possible that some Ru(III) compounds without exchangeable ligands are active *per se* and need no activation.

Another feature often invoked as a source of potential selectivity towards cancer cells and for explaining the low toxicity of many Ru compounds (compared to Pt compounds) is the capability of ruthenium to mimic iron in binding to many biological molecules, including serum proteins (e.g., transferrin and albumin). A transferrin-mediated uptake is often proposed for some Ru compounds. However, aside from the obvious similarities between Fe and Ru ions (e.g., size, charge, coordination geometry), their coordination chemistry is quite different in terms of thermodynamics and kinetics. In addition, whereas it is conceivable that the Ru$^{3+/2+}$ ions might effectively compete for the binding sites of iron proteins and enzymes (e.g., apotransferrin), it is less easily understood how such competition might occur for ruthenium species with residual ligands and different net charge. In other words, speciation is of paramount importance in determining the affinity of ruthenium compounds for the metal binding sites of biomolecules. Finally, a mimetic behavior might actually lead to the opposite result by affecting iron pathways that are crucial in cell metabolism (see below Section 5.3 for Ga^{3+}).

5.2.2
Trans-[tetrachloridobis(1*H*-indazole)ruthenate(III)] Complexes and their Development

Based on the evaluation of a plethora of Ru(III) compounds with heterocyclic N-ligands (L), *trans*-[RuCl$_4$L$_2$]$^-$ ("Keppler type") complexes were identified as the most promising candidates for clinical development (for recent reviews see References [14, 15]). In particular, KP418 (L = imidazole, Him, Figure 5.1c) and KP1019 (L = 1*H*-indazole, Figure 5.1a) were active against primary tumors and metastases in animal models, for example, in an autochthonous colorectal carcinoma of the rat, which resembles colon cancer of humans [16]. KP1019 was investigated *in vitro* against more than 50 primary explanted human tumors, a model with high predictivity for the clinical situation [17], showing activity in more than 80% of the specimens.

A multitude of biological studies were conducted to identify intracellular processes relevant for the mode of action of this class of compounds. For both KP1019 and its sodium salt analogue KP1339 (Figure 5.1b), the induction of apoptosis in the colorectal cancer cell lines SW480 and HT29 was driven predominantly by the intrinsic mitochondrial pathway. In addition, formation of reactive oxygen species and DNA-strand breaks were observed, which both can be prevented by addition of *N*-acetylcysteine (NAC). Especially noteworthy was the low degree of acquired resistance of KB-3-1 epithelial cells after treatment with KP1019 for more than a year, which was shown to be unrelated to a reduced drug accumulation [18].

KP1019 entered clinical Phase I trials in 2004, and promising results were obtained: five out of six evaluable patients experienced stabilization of their disease, and treatment was accompanied only by very moderate toxicity, which might also be related to the high degree of binding to serum proteins and probably to the selective activation in the tumor tissue [14, 15]. However, the relatively low solubility of the drug limited the total administrable dose to 600 mg per patient (administered twice a week for 3 weeks). For this reason KP1339, which is about 35 times more soluble, was selected for further development.

The development of KP1019 was accompanied by extensive studies on chemical reactivity and mode of action. The compound is sufficiently stable in aqueous solution for intravenous administration, which is an important requirement for

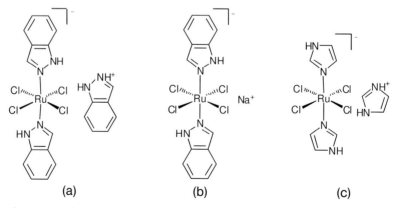

Figure 5.1 Schematic structures of the most widely investigated "Keppler type" Ru(III) complexes KP1019 (a), KP1339 (b), and KP418 (c).

clinical evaluation. However, the kinetics of hydrolysis are strongly dependent on the temperature and the pH of the solution. Similarly to other structurally related compounds, KP1019 in water hydrolyzes a chloride to give the respective mono-aqua complex, which was characterized by X-ray diffraction analysis. In the test tube in buffered solutions at pH ~ 7, KP1019 and KP1339 also hydrolyze and form insoluble species, most probably of polynuclear nature [15].

Like many other transition metal complexes, KP1019 and analogs exhibit high affinity to donor atoms present in both proteins and DNA. After intravenous administration, plasma proteins are the first potential targets. Human serum albumin and transferrin are highly abundant plasma proteins, and the reactions with them in the test tube were demonstrated to proceed quickly and with high affinity by liquid chromatography/mass spectrometry (LC/MS) and capillary electrophoresis (CE)/MS methods [19]. The data suggested that transferrin is the kinetically preferred binding partner of KP1019, but albumin is thermodynamically favored. When the methodology was applied to clinical blood serum samples, albumin was identified as the serum component loaded to the highest degree with ruthenium [14]. The reaction with proteins after intravenous administration appears to be highly relevant for the mode of action and the low toxicity of these drug candidates (see also Section 5.2.1): their transport via transferrin or albumin could result in an enrichment of the drug in the tumor, considering that transferrin receptors are expressed to a higher degree at the surface of tumor cells (due to the iron requirement of quickly growing tumors) and that macromolecules are accumulated as a result of the enhanced permeability and retention effect.

Once inside the cell, the compounds are thought to be activated by reduction (compare Section 5.2.1), followed by chloride dissociation [2]. They also show affinity to DNA bases [20], although DNA is not necessarily considered the primary target. The role of redox processes seems to be important for the selectivity of Ru(III) compounds, provided that their redox potential is physiologically accessible. In the case of KP1019, naturally abundant reducing agents such as ascorbic acid (11–79 µM in the blood plasma) and glutathione (0.5–10 mM in the cell) may cause a reduction of the complex to the corresponding Ru(II) species. As shown by NMR and CE experiments, this process strongly influences the reactivity to the DNA model compound GMP. At low glutathione-to-KP1019 molar ratios the reaction with GMP is enhanced as compared to incubation mixtures without the reducing agents. However, at a higher excess of glutathione the formation of GMP adducts is prevented, most probably because of saturation of the Ru coordination sphere with glutathione or because of quick precipitation due to fast hydrolysis [20].

Many other studies were conducted to investigate the binding of KP1019 to DNA and DNA model compounds. Guanine was identified by CE studies as the primary binding site on DNA, followed by adenine, if single nucleotides were reacted with the Ru complex [19]. Gel electrophoretic analysis of reaction mixtures with plasmid DNA and with a linear radioactively labeled dsDNA fragment revealed that KP1019 can untwist (though to a lower degree than cisplatin) and bend DNA. In addition, a lower interstrand crosslinking efficiency was observed as compared to clinically established drugs. Studies on the influence of inhibitors of the DNA

repair systems on the cytotoxic activity suggests involvement of both the nucleotide excision repair and base excision repair systems, which does not conform with
results obtained for cisplatin. Furthermore, KP1019 was demonstrated to form
DNA–DNA and DNA–protein crosslinks in tumor cells, though to a lesser extent
than cisplatin.

In general, KP1019 showed a behavior quite different from that of cisplatin,
which makes this Ru(III) compound a promising drug candidate with a mode of
action different from established anticancer chemotherapeutics [14, 15].

After the development of these Ru(III) complexes, the more inert Os(III) analogs were synthesized, and in some cases a comparable or even higher *in vitro*
antiproliferative activity was found [21, 22]. However, no clear-cut relationships
between reduction potentials and *in vitro* anticancer potency were found. Most
likely additional factors, such as cellular uptake and interactions with unknown
molecular targets, are of importance for understanding the activity of the Os
compounds [21].

5.2.3
Ruthenium-dmso Compounds: The Development of NAMI-A

Ruthenium compounds containing dimethyl sulfoxide (dmso) ligands have been
investigated since the early 1980s. The motivations for choosing dmso were the
following: dmso is an ambidentate ligand that can bind to Ru either through
the sulfur (dmso-S) or through the oxygen atom (dmso-O). The preferred binding
mode of dmso to Ru ions depends on both steric and electronic factors [23]. Most
importantly, dmso-S is a moderately good π-acceptor ligand and stabilizes Ru(II);
as a consequence, Ru(III)-dmso-S compounds are easily reduced to the corresponding Ru(II) species, whereas Ru(II)-dmso-S compounds are air-stable. In
general, dmso is also a relatively good leaving ligand, and therefore Ru-dmso
compounds are expected to generate reactive aquated metabolites rather easily.
Finally, dmso is expected to impart to its complexes a reasonably good solubility in
water and, by virtue of its intrinsic capability to diffuse easily through phospho-
lipidic membranes, also a good cell permeability.

Initially, after establishing that *cis*-[RuCl$_2$(dmso)$_4$] (Figure 5.2a) is active *in vivo*
only at unreasonably high dosages (800 mg kg^{-1}), the investigation focused on
trans-[RuCl$_2$(dmso)$_4$] (Figure 5.2b). This compound, despite its negligible *in vitro*
cytotoxicity, showed *in vivo* antitumor activity remarkably different from cisplatin
against several murine tumor models. In comparative experiments, cisplatin
was more active in reducing the primary tumor, whereas *trans*-[RuCl$_2$(dmso)$_4$] was
more effective in reducing the number and weight of spontaneous metastases
derived from the primary tumor. This specific antimetastatic activity, which clearly
emerged here for the first time, was later found as a hallmark for other Ru-dmso
complexes (see below).

In the 1990s, following the results on KP418 (Figure 5.1c), the structurally similar
Ru(III) complex [H$_2$im]*trans*-[RuCl$_4$(Him)(dmso)] (NAMI-A, Figure 5.2c) was prepared. NAMI-A was found to be highly effective in inhibiting the development and

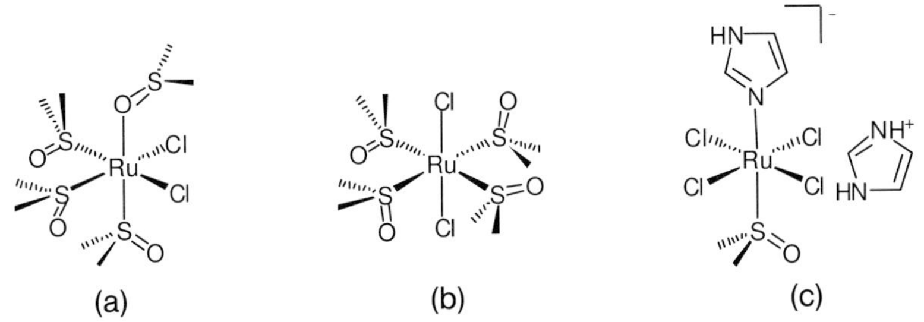

Figure 5.2 Schematic structures of *cis*-[RuCl$_2$(dmso)$_4$] (a), *trans*-[RuCl$_2$(dmso)$_4$] (b), and NAMI-A (c).

growth of pulmonary metastases in many *in vivo* models of solid tumors, including human non-small cell lung cancer xenotransplanted into nude mice. In this latter case, treatment with NAMI-A induced a statistically significant reduction of lung metastases growth, regardless of whether the treatment was done prior to or after surgical removal of the primary tumor. Conversely, the activity of NAMI-A towards the primary tumor and *in vitro* cytotoxicity are both negligible when compared to those of cisplatin [24, 25].

It is worth stressing here that the finding of drugs with specific activity against metastases is an extremely important goal. In fact, metastases of solid tumors, because of their multiple locations and low accessibility to surgery and/or radio-therapy, typically need to be addressed by chemotherapy. However, they are often poorly responsive to chemotherapy and thus, more than the primary tumor, re-present the leading cause of cancer death [26].

NAMI-A has favorable biological and chemical features – good antimetastatic activity, low general toxicity, facile and reproducible preparation, good stability as a solid – that prompted its further development, and in 1999, after extensive pre-clinical tests, it became the first ruthenium compound to be investigated on hu-man beings. NAMI-A successfully accomplished Phase I clinical trials on 24 patients, showing good tolerability over a wide range of sub-toxic doses without any unexpected toxicity. Hematological toxicity was negligible. At high dose levels mild renal dysfunction, completely reversible, was observed together with other drug-related toxicities such as nausea and vomiting [27]. The recommended dose for further clinical testing was established at 300 mg m^{-2}.

In fall 2008 NAMI-A entered a therapeutic combination Phase I/II clinical trial in patients affected by non-small cell lung carcinoma.

5.2.3.1 Chemical Features of NAMI-A
NAMI-A is structurally similar to KP418 with a dmso-S that formally replaces one of the axial imidazole ligands. However, this similarity is deceptive because the presence of dmso-S in the coordination sphere of Ru(III) strongly affects its chemical properties. First of all, by virtue of the π-acceptor property of dmso-S, the reduction potential of NAMI-A is dramatically more positive than that of KP418 (+235 and −275 mV vs. NHE, respectively), that is, NAMI-A is reduced much

more easily to Ru(II). In addition, NAMI-A is remarkably less inert than KP418: at 37 °C and physiological pH 7.4 (phosphate buffer), the complex undergoes stepwise dissociation of two chlorides within a few minutes, a process that is believed to be responsible for its activation [24, 25, 28]. The first step is catalyzed by ruthenium(II) species; in fact, its rate is considerably enhanced by the addition of traces of biological reductants, such as ascorbic acid. For comparison, under the same conditions, KP418 is remarkably more inert [29]. The aquation processes of NAMI-A are, however, strongly pH-dependent. NAMI-A is administered by infusion at room temperature and slightly acidic pH: under such conditions it is remarkably stable and aquation is almost completely suppressed (only slow dissociation of dmso occurs).

In vivo the presence of biological ligands is expected to affect heavily the behavior of NAMI-A. The results of Phase I trials, supported by *in vitro* studies with albumin, showed that in the blood NAMI-A is highly bound to plasma proteins. Most importantly, at physiological pH stoichiometric amounts of common biological reducing agents (e.g., ascorbic acid and glutathione) rapidly and quantitatively reduce NAMI-A to the corresponding dianionic Ru(II) species *trans*-$[RuCl_4(Him)(dmso)]^{2-}$, NAMI-AR, which then undergoes rapid stepwise aquation with rates comparable to those of NAMI-A [30, 31]. Neither the loss of the neutral ligands nor reoxidation by atmospheric oxygen were observed. Metabolites of NAMI-A that still bear the dmso-S are expected to have reduction potentials similar to that of the parent compound and thus to be easily reducible. Owing to the abundance of natural electron donors (Section 5.2.2), reduction of NAMI-A is likely to occur *in vivo*. Nevertheless, notably, at physiological pH activation of NAMI-A may easily occur through aquation. That is, for NAMI-A activation by reduction is a conceivable pathway, but is not a prerequisite to explain its activity. Interestingly NAMI-AR, obtained by reduction of NAMI-A with ascorbic acid in a test tube immediately prior to administration, was even more effective than the parent compound against metastases in animal tumor models [30].

The different biological activities found for KP418 and NAMI-A strongly suggest that the two similar compounds generate different active metabolites. In other words, the antimetastatic activity of NAMI-A is most likely strictly related to the presence of dmso in the coordination sphere of its active metabolite(s). This consideration is consistent with the strong influence of dmso on the chemical behavior of Ru(III) species.

5.2.3.2 NAMI-A-Type Complexes

Over the years, in systematic structure–activity relationship investigations, a remarkable number of NAMI-A-type complexes of the general formula [X]*trans*-$[RuCl_4(L)(dmso)]$ ($X^+ = Na^+$, $NH_4{}^+$, LH^+; L = NH_3 or N-heterocycle), including the dinuclear species [Na]$_2$*trans*-$[\{RuCl_4(dmso)\}_2(\mu\text{-}L)]$ (L = ditopic linker such as pyrazine, pyrimidine or 4,4′-bipyridine), have been prepared and tested. In addition, neutral compounds of the general formula *mer*-$[RuCl_3(L)(dmso)_2]$ and nitrosyl compounds such as $[H_2im]$*trans*-$[RuCl_4(Him)(NO)]$ were investigated [24, 25, 32].

In general, many of the NAMI-A-type complexes, including dinuclear compounds, showed *in vivo* effectiveness against metastasis growth similar to that of NAMI-A and in some cases were even slightly superior. However, NAMI-A still had the best combination of biological activity and chemical features (ease of synthesis, purity, stability). Best alternatives were the mononuclear NAMI-A-type compounds with N-ligands of comparable size but lower basicity than imidazole (e.g., L = thiazole or oxazole).

Surprisingly, Os analogs of NAMI-A-type compounds exhibited a tenfold increased *in vitro* cytotoxic activity compared to NAMI-A against the two human carcinoma cell lines HT-29 (colon) and SK-BR-3 (breast) [22].

5.2.3.3 Mechanism of Action of NAMI-A

The mechanism of action of NAMI-A is still largely unknown. The *in vitro* data and the mild general host toxicity suggest that its activity is unrelated to direct tumor cytotoxicity (see above) [24, 25, 32]. NAMI-A is capable of interacting with DNA in cell-free medium. However, contrary to Pt drugs and similarly to KP1019, nuclear DNA does not seem to be its primary target. *In vitro* and *in vivo* studies showed that it also binds strongly to the plasma proteins albumin and transferrin [33, 34].

As already mentioned, *in vivo* experiments, typically performed on tumors that metastasize in the lung, showed that NAMI-A has apparently a strong selectivity for metastatic cells over the other tumor cells. The cell population in the primary tumor is heterogeneous and only some cell subtypes possess metastatic ability. It has been suggested that at the primary tumor site NAMI-A selectively eliminates only the metastasizing cells, which represent a small fraction of the total number of tumor cells, hence explaining its modest activity at this site. On the other hand, only metastatic cells are present in the lung, because they have a clonal origin: as a consequence, pronounced reduction of lung metastases is consistent with the selectivity of NAMI-A for this kind of cell. In addition, NAMI-A metabolites reach a significant concentration in the lung (estimated about 0.2–0.4 mM by atomic absorption measurements), where they bind strongly to collagens of the extracellular matrix.

In vitro experiments evidenced several features of NAMI-A that occur at subtoxic dosages and that might be relevant to its *in vivo* activity: (i) it stops cell proliferation at the pre-mitotic G_2–M phase; (ii) at 0.001–0.05 mM concentrations (depending on the cell line) it interacts with actins on the cell surface, thus increasing tumor cell adhesion to the culture substrate and strongly inhibiting cell migration; (iii) at 0.05–0.1 mM concentrations it reduces the spontaneous invasion of matrigel by tumor cells; (iv) at doses compatible with anti-metastatic activity it affects several functions of endothelial cells, thus inhibiting angiogenesis; (v) acts as NO scavenger; (vi) inhibits matrix metallo proteinases (MMPs) at mM concentrations; (vii) induces caspase activation and thus apoptosis; and (viii) at doses compatible with induction of apoptosis it down-regulates the extracellular signal-regulated kinase (ERK1/2), thus inhibiting the mitogen activated protein kinase (MAPK) signaling pathway.

The abundance of biological effects suggests that NAMI-A acts through multiple mechanisms, both at the cell membrane and intracellularly.

5.2.4
Half-Sandwich Ru-Organometallics

In recent years, entirely new classes of organometallic Ru(II)-arene compounds, developed by the groups of Sadler and Dyson, were found to have promising anticancer activity *in vitro* and *in vivo*. Representative examples are $[(\eta^6\text{-biphenyl})$Ru(en)Cl][PF$_6$] (RM175, en = ethane-1,2-diamine) and $[(\eta^6\text{-}p\text{-cymene})$RuCl$_2$(pta)] (RAPTA-C, pta = 1,3,5-triaza-7-phosphatricyclo[3.3.1.1]decane), (Figure 5.3). The geometry of these half-sandwich compounds can be described as pseudo-tetrahedral (piano-stool geometry), assuming that the six-electron donor arene ligand occupies one coordination position, or as octahedral, assuming that it occupies three facial coordination positions. The $+2$ oxidation state of ruthenium is stabilized by the arenes, which are considered as both π-donor and π-acceptor ligands towards ruthenium. Ru–arene bonds are generally stable towards hydrolysis. However, there are quite a few examples showing that the arene can be cleaved off, and this reaction might be an important part in the mode of action of such compounds.

5.2.4.1 Piano-Stool Ru-Arene Compounds

Organometallic Ru(II) compounds of the type $[(\eta^6\text{-arene})$Ru(chel)X], where chel is a neutral or mono-anionic *N,N*-, *N,O*-, or *O,O*-chelating ligand (e.g., en, bipy, picolinate, 8-hydroxyquinolate, acetylacetonate, maltolate) and X is typically a halide, have been extensively investigated by Sadler and coworkers (see also Chapter 1). These half-sandwich compounds are either neutral or positively charged (typically isolated as PF$_6$ salts), depending on the nature of the chel ligand. Extensive structure–activity relationship investigations showed that the most active are those with chel = en and X = Cl. Indeed, $[(\eta^6\text{-arene})$Ru(en)Cl][PF$_6$] compounds (Figure 5.3) showed promising anticancer activity, both *in vitro* against human cancer cell lines, including the cisplatin-resistant variant A2780cis, and *in vivo* with significant growth delay against both A2780 and A2780cis xenografts [35, 36]. Cytotoxicity increases with the hydrophobicity of the arene ligands. Thus, when evaluated against the human ovarian cancer cells A2780, compounds with arene = *p*-cymene or biphenyl have IC$_{50}$ values in the range of 6–9 μM, that is, similar to that of the established antitumor

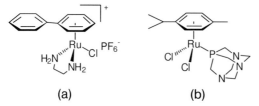

(a)   (b)

Figure 5.3 Schematic structures of the half-sandwich organometallic Ru(II) compounds RM175 (a) and RAPTA-C (b).

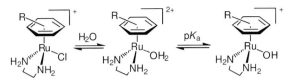

Scheme 5.1 Activation of Ru(II) piano-stool compounds occurs through aquation: the concentration of the active Ru$-$OH$_2$ species depends on Cl$^-$ concentration and pH.

drug carboplatin, whereas compounds with arene = tetrahydronaphthalene are equipotent with cisplatin (IC$_{50}$ = 0.6 µM).

Activation of piano-stool compounds is believed to involve rapid hydrolysis of the Ru$-$Cl bond, thus generating an active monofunctional Ru$-$OH$_2$ metabolite (Scheme 5.1). Aquation is largely suppressed at extracellular chloride concentration ([Cl$^-$] about 0.1 M) but becomes possible inside the cells, where [Cl$^-$] is much lower (4$-$25 mM). The pK_a values of the [(η^6-arene)Ru(en)(H$_2$O)]$^{2+}$ aqua species are typically between 7 and 8, and thus at physiological pH the Ru$-$OH$_2$ species largely prevails over the less reactive Ru$-$OH species.

For this type of compounds nuclear DNA is believed to be the main target. Their reactivity to DNA was tested against nucleotides and a DNA 14-mer, and the preferential formation of monofunctional adducts via the *N*7 atom of guanine residues was observed. In cell-free media, these adducts distort the DNA duplex, as shown by several biophysical techniques [37]. Both [(η^6-arene)Ru(en)]$-$nucleobase and $-$nucleoside adducts have been isolated and characterized by X-ray crystallography. According to solution and solid-state evidence, the main Ru$-$*N*7(guanine) coordination bond is assisted by stereospecific hydrogen bonding between the C6=O of guanine and the NH of en, and π$-$π stacking between the aromatic ligand and the nucleobase.

The interactions between the [(η^6-arene)Ru(en)X] complexes and several biologically relevant molecules and potential targets (e.g., cytochrome *c*, the amino acids histidine, cysteine and methionine, and the tripeptide glutathione) have also been investigated [35, 38–40]. Overall, the results suggest that, in the cell, DNA and RNA are the preferred targets.

In conclusion, these piano-stool compounds appear to target DNA similar to cisplatin, but the formation of monofunctional adducts (preferentially at guanine sites) and the non-covalent interactions (i.e., intercalation and H-bonding) are believed to be fundamental for DNA recognition and, as a consequence, anticancer activity.

Using the radioactive ^{106}Ru isotope it was possible to establish that, within 15 min after injection of the radiolabeled piano-stool compound in the rat, ruthenium distributes in all tissues, with higher levels in the liver and kidneys [41].

Cell biological studies have revealed cross-resistance with adriamycin but not with cisplatin, which was also confirmed *in vivo*. At a molecular level induction of p53 and p21/WAF1 in a concentration-dependent manner and of Bax were demonstrated. The cell cycle is blocked at G1 and G2 phases in a p53- and p21/WAF1-dependent manner, and topoisomerase I/II inhibition appears not to be of relevance in the mode of action [38].

Recently, Os(II)-arene analogs of the Ru(II) piano-stool compounds have also been investigated by Sadler and coworkers. Compounds with N,N or O,O chelates were found to be scarcely cytotoxic either because aquation was too slow or because of formation of inert hydroxo-bridged dimers under physiologically relevant conditions. Conversely, Os-arene compounds with anionic N,O chelating ligands such as $[(\eta^6\text{-}p\text{-cymene})Os(pico)Cl]$ (pico = picolinate) were found to be active towards both A2780 and A549 cancer cell lines, whereas the corresponding Ru compounds are inactive. These results demonstrate that biological activity of the piano-stool compounds depends on a delicate balance of the chemical properties of all components, both the ligands and the metal [42].

5.2.4.2 **RAPTA Compounds**

RAPTA compounds of the general formula $[(\eta^6\text{-arene})Ru(X)_2(pta)]$ (X = Cl or chelating dicarboxylate, Figure 5.3b) are characterized by the presence of the monodentate phosphane ligand pta. The choice of pta was originally motivated by the rationale that it would be preferentially protonated in the slightly acidic environment of solid tumors (where the pH can be as low as 5.5), thereby increasing the uptake of the complex that would remain trapped inside tumor cells. Even though this hypothesis was later demonstrated to be inconsistent (the pK_a of coordinated pta is too low, about 3) [43], the prototype of this class of organometallic half-sandwich compounds, RAPTA-C, and, more recently, the closely similar toluene derivative $[(\eta^6\text{-toluene})RuCl_2(pta)]$ (RAPTA-T), were subject of thorough *in vitro* and *in vivo* investigations [43, 44]. *In vivo*, in the MCa mammary carcinoma model, both RAPTA-C and RAPTA-T inhibit lung metastases formation (though to a lower extent than NAMI-A) without affecting the primary tumor significantly. *In vitro* RAPTA-T is virtually devoid of cytotoxicity, but it interacts with extracellular matrix components, thus inhibiting some steps typical of the metastatic process. Interestingly, the effects were more pronounced against the highly invasive MDA-MB-231 cells than against the non-invasive MCF-7 or the non-tumorigenic HBL-100 cells [44]. Overall, RAPTA-T was found to have *in vivo* and *in vitro* behavior quite similar to NAMI-A, which is surprising, given the structural differences.

In vivo activation of RAPTA compounds is believed to occur through hydrolysis of the chloride ligands, the extent of which depends on the amount of chloride present in solution and the pH. In some cases loss of the arene was observed, whereas the pta ligand is bound very tightly. RAPTA derivatives in which chelating dicarboxylate ligands replace the chlorides and that are more inert towards aquation display nevertheless an *in vitro* activity very similar to RAPTA-C.

The RAPTA chemical frame has been the subject of detailed structure–activity investigations, in which the π-bonded arene, pta and the anionic ligands have been systematically derivatized or changed [45]. The prototypes RAPTA-C and RAPTA-T proved to be the most active representatives of the series. Changes leading to an increase of cytotoxicity were accompanied by a larger increase of general toxicity.

More recently RAPTA-like compounds have been investigated also in a more targeted chemotherapeutic approach [4, 46–48]. However, these new compounds, in which the RAPTA scaffold is functionalized to have multiple modes of action, are at an early stage of development.

5.2.4.3 Other Half-Sandwich Ru Compounds

Given the promising results obtained with the piano-stool and RAPTA compounds, the half-sandwich Ru(II) scaffold has been explored along different pathways in the search of novel anticancer compounds.

For example, water-soluble dinuclear Ru(II)-arene compounds with maltol-derived linkers between $[(\eta^6\text{-}p\text{-cymene})RuCl]$ units (Figure 5.4a) were reported to exhibit promising cytotoxic effects in human cancer cell lines [49]. Notably, these compounds are potent DNA duplex and DNA–protein crosslinkers [50]. Dyson and coworkers also described diruthenium-arene compounds in which two $[(\eta^6\text{-}p\text{-cymene})RuCl_2]$ moieties are connected via a ferrocene-modified bis-pyridine linker [51].

Alessio and coworkers demonstrated that in the active piano-stool compounds replacement of the arene moiety with a neutral face-capping six-electron donor ligand (fcl), such as 1,4,7-trithiacyclononane ([9]aneS3, Figure 5.4b), leads to coordination compounds that maintain a reasonable cytotoxicity *in vitro* [52]. In other words, half-sandwich Ru(II) coordination compounds of the type $[Ru(fcl)(en)Cl]^+$ represent an entirely new class of compounds that deserve a thorough investigation.

With the intention to use multinuclear cytotoxic organometallics in combination with photosensitizers for photodynamic cancer chemotherapy, half-sandwich Ru–arene moieties were attached to porphyrin rings [53]. In the dark, these compounds were not cytotoxic, whereas upon irradiation with visible light they exhibited high cytotoxicity in the low µM range. Accumulation of the Ru-porphyrin conjugates in the cytoplasm of melanoma cells was observed by fluorescence microscopy. Recently, coordinatively saturated and inert half-sandwich Ru(II)-arene compounds,

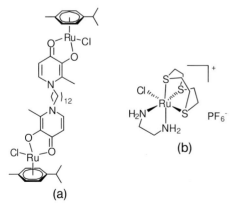

(a)

(b)

Figure 5.4 A dinuclear half-sandwich organometallic compound (a) and $[Ru([9]aneS3)(en)Cl]^+$ (b).

which are not expected to set free any coordination position, were also found to have *in vitro* activity against cancer cell lines, either through the catalytic production of ROS [54] or through kinase inhibition [55].

5.3
From Gallium Nitrate to Oral Gallium Complexes

Gallium salts are endowed with various biological properties potentially useful for therapeutic applications, with inhibition of bone resorption and cellular proliferation as the most relevant [56–58]. Gallium nitrate is approved as an alternative to bisphosphonates for treatment of cancer-associated hypercalcemia, a severe complication of malignancies affecting bone turnover [59]. On the other hand, efforts to establish gallium nitrate as an anticancer drug have ceased because of a subtle dilemma concerning tolerability and mode of administration, despite clinical evidence for activity in lymphoma and cancer of the bladder, prostate, ovary, and cervix [60].

The applicability of Ga(III) compounds strongly depends on suitable ligands to prevent, or at least to slow down, hydrolysis. For this reason, citrate has been used to stabilize solutions of gallium nitrate for clinical application. Currently, complexation with ligands that improve intestinal absorption is being pursued as an approach to orally applicable gallium drugs. The first of these compounds to enter clinical studies was tris(maltolato)gallium(III) (Figure 5.5a), containing three bidentate O,O donor ligands known to facilitate iron absorption [61]. A set of more lipophilic N,O donor ligands was employed in the case of tris(8-quinolinolato) gallium(III), KP46 (Figure 5.5b), which showed promising signs of preclinical activity in primary explanted melanoma *in vitro* [62] and of clinical activity in renal cell cancer [63]. The high lipophilicity of KP46 has implications for pharmacokinetics and biodistribution, but might also affect its molecular mode of action compared to other Ga(III) compounds.

The anticancer activity of Ga compounds appears to be related to the similarity of the Ga^{3+} and Fe^{3+} ions in terms of charge, size (Fe^{3+} 0.65 Å; Ga^{3+} 0.62 Å), and Lewis acidity [64]. Transport of Ga into the cell is believed to utilize iron uptake routes [65], but since intracellular iron metabolism involves redox processes,

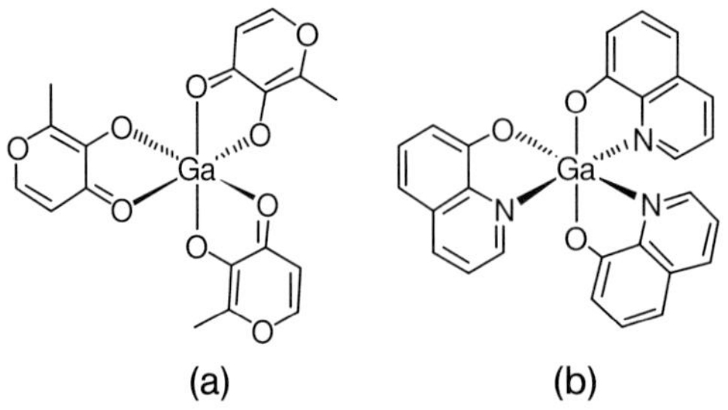

(a)   (b)

Figure 5.5 Schematic structures of gallium maltolate (a) and KP46 (b).

trafficking in the cell probably differs [66]. Furthermore, Ga competes with Fe for the same binding sites in enzymes such as ribonucleotide reductase, which is inhibited by the binding of this redox-inactive ion. As a consequence, binding of Ga^{3+} to the iron-binding site of the R2 subunit results in destabilization of the tyrosyl radical, which is essential for catalytic activity of this enzyme [67]. The inhibition of ribonucleotide reductase results in depletion of dNTP pools in pro-liferating cells, impaired DNA synthesis, cell cycle perturbations, and apoptosis through the mitochondrial pathway [68, 69].

5.4
Titanium Anticancer Compounds

5.4.1
Budotitane: The First Transition Metal Complex in Clinical Trials in the Post-cisplatin Era

Budotitane, or *cis*-diethoxybis(1-phenylbutane-1,3-dionato-$\kappa^2 O^1, O^2$)titanium(IV) (Figure 5.6a), was among the first non-platinum metal-based anticancer agents studied in clinical trials, which commenced in 1986 [70]. The compound was se-lected from more than 200 structurally similar β-diketonato compounds based on efficacy in mice bearing subcutaneous sarcoma-180 and other transplantable tu-mor models and proved to be more active than 5-fluorouracil in chemically in-duced autochthonous colorectal tumors in the rat [71]. Budotitane was actually obtained as a mixture of isomers and, owing to the presence of two leaving groups, the formation of several aquated species was observed under equilibrium condi-tions. Because of its low solubility in aqueous solution, the compound was used in clinical trials as a co-precipitate with the emulsifier Cremophor EL and 1,2-propyleneglycol. With co-administration of mannitol to prevent nephrotoxicity, cardiac arrhythmia caused by volume effects was identified as the dose-limiting toxicity [72]. Further development was stopped because the formulation did not allow appropriate analytical characterization of the drug.

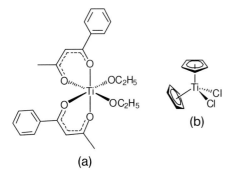

Figure 5.6 Schematic structures of budotitane (a) and titanocene dichloride (b).

5.4.2
Titanocene Dichloride and Related Compounds

Titanocene dichloride, $[Ti(\eta^5\text{-}C_5H_5)_2Cl_2]$ (Figure 5.6b), was the first organometallic transition metal compound to be investigated clinically as an anticancer agent. It contains a *cis*-dichloride motif as cisplatin and was selected from several early transition metal cyclopentadienyl complexes as the best candidate for further development [73]. The chemistry of the hard Ti(IV) ion is different from that of Pt(II): for example, cisplatin binds preferentially to the N7 of guanine in DNA, whereas titanocene dichloride exhibits higher affinity for the phosphate backbone [74]. $[Ti(\eta^5\text{-}C_5H_5)_2Cl_2]$ hydrolyzes quickly in water, yielding a solvated Ti(IV) ion with high affinity for transferrin [75]. As for Ga(III), selective transport of titanium ions via the transferrin route appears plausible. *In vitro*, titanocene dichloride is active in cisplatin-resistant cancer cells [76]. It entered clinical trials in 1993, revealing nephrotoxicity as dose-limiting side effect. In Phase II studies as single-agent therapy no advantage over other treatment regimens was observed, and the trials were thus abandoned [77].

Current research on titanocene compounds focuses on improved stability in aqueous solution by using chelating *ansa* systems (Figure 5.7). These compounds were evaluated against a renal cancer cell line, and *in vitro* activity was found at a concentration one order of magnitude lower than that of titanocene dichloride. However, the presence of isomers appears problematic for clinical development.

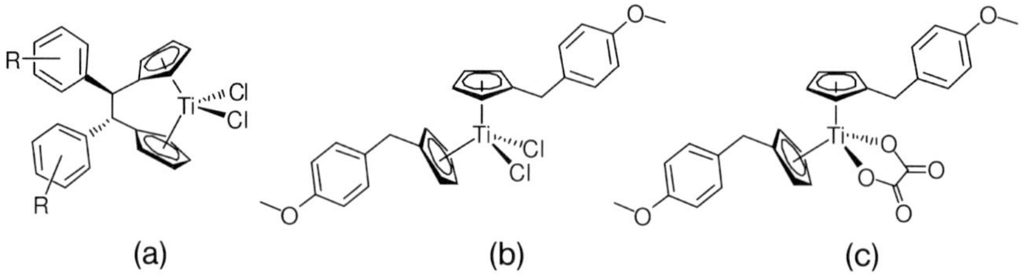

(a) (b) (c)

Figure 5.7 Schematic structures of an *ansa*-titanocene (a), titanocene Y (b) and its oxalato derivative (c).

Furthermore, the functionalized titanocene dichloride compound dichloridobis (*p*-methoxybenzyl)cyclopentadienyltitanium(IV), also called titanocene Y (Figure 5.7b), was defined as a new lead structure because of its *in vivo* activity in mice bearing Ehrlich ascites tumor. The oxalato derivative of titanocene Y (Figure 5.7c), prepared with the aim of improving the stability, yielded promising results in an initial *in vitro* screening [78, 79].

5.5
Ferrocene-Derived Anticancer Agents

Several ferrocene and ferrocinium compounds have been studied for their activity in cancer cells *in vitro*. The compounds developed to the most advanced stage by

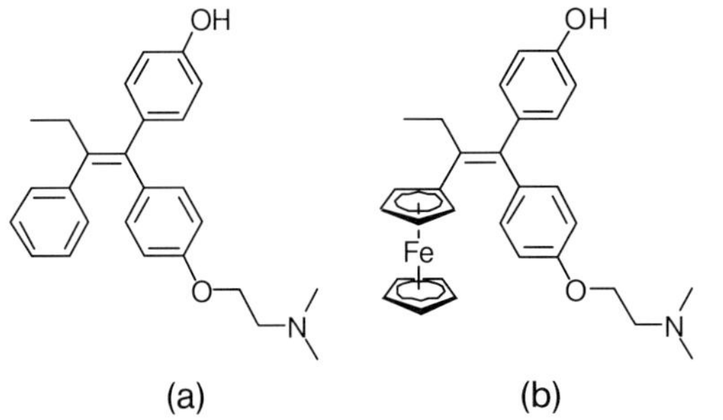

Figure 5.8 Schematic structures of hydroxytamoxifen (a) and a ferrocifen compound (b).

Jaouen and coworkers are the ferrocifens, in which a phenyl ring of the anticancer drug tamoxifen is substituted with a ferrocenyl moiety [80]. Tamoxifen (through its active metabolite hydroxytamoxifen; Figure 5.8a) targets the estrogen receptor, which is present in hormone-dependent (ER+) tumors. For example, estrogens are responsible for the growth induction of about 66% of breast cancers.

Ferrocifens are more lipophilic than tamoxifen and can cross cell membranes more easily. They exhibit a stronger cytotoxic effect, probably due to the combination of binding to the estrogen receptor and additional cytotoxicity induced by the redox-active ferrocene moiety itself. This would explain why they are also active in hormone-independent (ER-) breast cancer cells, in which hydroxytamoxifen and ferrocene are inactive.

The extended π-system plays an important role in the mode of action of these compounds, and a correlation of anticancer activity with the electron transfer capacity was observed [81]. The mechanism of action has been proposed to involve a series of redox processes that originate from the oxidation of the ferrocene fragment and eventually lead to the generation of ROS. Notably, ruthenocene compounds structurally similar to ferrocifens exhibit very high affinity to the estrogen receptor. However, in contrast to the iron compounds ($Fe^{II} \leftrightarrow Fe^{III}$), the ruthenium analogs (irreversible oxidation $Ru^{II} \rightarrow Ru^{III}$) behave as antiestrogens with activity against ER+ human breast adenocarcinoma cells, but not against ER- cell lines [80, 81].

5.6
The Main Group Organometallics Spirogermanium and Germanium-132

The germanium compound *N*-(3-dimethylaminopropyl)-2-aza-8,8-diethyl-8-germaspiro-4,5-decane dihydrochloride or spirogermanium (Figure 5.9a) was shown to inhibit DNA, RNA, and protein synthesis and to be cytotoxic in different cell cultures. Moreover, activity in mammary and prostatic carcinoma in rats was demonstrated [82]. A Phase I study initiated in the early 1980s revealed that treatment was accompanied by dose-limiting neurotoxicity [83]. The compound was further developed in a series of Phase II studies, in which objective responses in malignant lymphomas, ovarian, breast, and prostatic cancer were observed [84].

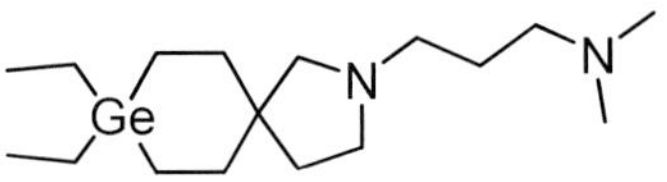

$$[(GeCH_2CH_2COOH)_2O_3]_n$$

(a)   (b)

Figure 5.9 Schematic structures of spirogermanium (a) and germanium-132 (b).

Germanium-132 (carboxyethylgermanium sesquioxide, Figure 5.9b) has shown activity in preclinical experiments in a wide spectrum of tumors [85], such as Lewis lung carcinoma and bladder tumors. It is thought that the compound stimulates the immune system, including the activation of T-lymphocytes and macrophages, an increase in NK cell activity and induction of interferon secretion. In clinical studies, activity in unresectable lung cancer patients was observed [86].

Even though these germanium compounds have been the subject of many patents and were investigated in the clinic, not much is known (or at least published) about their chemical behavior. Similarly, the reasons for abandoning their clinical investigation were not clearly stated.

5.7
Arsenic in Cancer Chemotherapy

Arsenic trioxide (As_2O_3) had been used in a potassium bicarbonate-containing formulation (potassium arsenite, "Fowler's solution") for treatment of Hodgkin's disease and leukemia (in particular chronic myelogenous leukemia) in the late nineteenth and early twentieth century but had been abandoned because of chronic toxicity. Its antileukemic activity was reappraised in the late twentieth century based on its long history in Chinese traditional medicine [87]. Studies revealed remarkable therapeutic activity in acute promyelocytic leukemia (APL) in comparatively low and well-tolerable doses given daily by intravenous infusion, prompting drug approval for second-line therapy of this disease [88].

The mechanism of action of arsenic trioxide involves the induction of partial differentiation of promyelocytes and the induction of apoptosis. The former effect results from degradation of the PML-RARα protein, which is the product of a fusion gene resulting from a chromosome translocation specific for APL and central to the pathogenesis of this disease in blocking myeloid differentiation [89]. Additionally, arsenic trioxide may restore the antiproliferative and proapoptotic functions of the PML gene [90]. However, other mechanisms such as induction of oxidative stress by inhibition of cellular antioxidant defense systems (glutathione, thioredoxin) [91] and interference with various signal transduction pathways [89] are likely to be involved in the apoptotic response of APL cells to arsenic trioxide.

Darinaparsin (S-dimethylarsinoglutathione, Figure 5.10a) is a putative bio-transformation product of arsenic trioxide with high cytotoxicity in cancer cells [92]. This compound, which is potentially suited for both intravenous and oral administration, is currently being studied in clinical trials in various malignancies, in

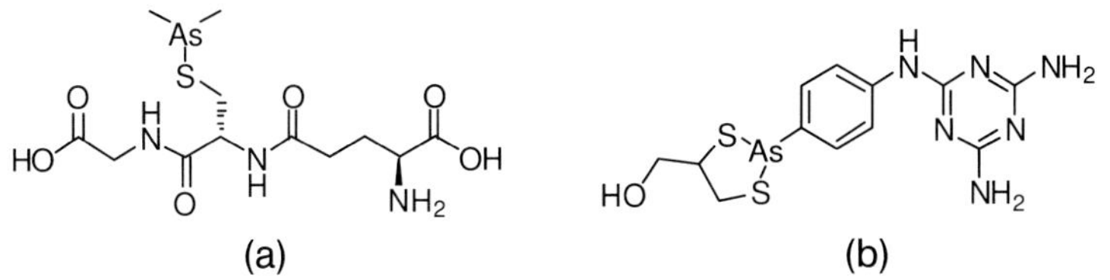

Figure 5.10 Schematic structures of darinaparsin (b) and melarsoprol (b).

particular hepatocellular carcinoma, multiple myeloma, and lymphoma [93]. Melarsoprol (2-[4-[(4,6-diamino-1,3,5-triazin-2-yl)amino]phenyl]-1,3,2-dithiarsolane-4-methanol, Figure 5.10b), an organoarsenic drug for treatment of trypanosomiasis, has also been clinically studied in hematological malignancies, but was abandoned because of cerebral toxicity [94]. Efforts are now being made to develop a formulation with limited access to the central nervous system [95].

Similar to the Ge compounds mentioned above, very little is reported in the literature about the chemistry of these As compounds.

5.8
Overcoming the Resistance of Tumors to Anticancer Agents by Rare Earth Element Compounds

Several lanthanide (Ln) compounds have been tested for the treatment of cancer and as anti-emetics due to favorable pharmacological properties [96]. In particular, the Gd-texaphyrin complex motexafin gadolinium (Figure 5.11a) is at an advanced stage of clinical development for the treatment of non-small cell lung cancer [97]. In addition, some Ln compounds found application as medications for burn wound treatment and as phosphate binders for the treatment of hyperphosphatemia.

The mode of action of lanthanides is related to their similarity to calcium: Ln^{3+} ions exhibit high affinity to Ca^{2+} binding sites in biomolecules because of their similar ionic radii, but have a higher charge [97, 98]. Therefore, Ln compounds are able to inhibit the calcium fluxes required for cell cycle regulation, but they can also substitute for Ca^{2+} and other metal ions such as Mg^{2+}, Fe^{3+}, and Mn^{2+} in proteins, leading to the inhibition of their functions [99, 100]. The tumor-inhibiting activity of La^{3+} is considerably enhanced by complexation with various ligands, such as phenanthroline derivates [101, 102].

The La^{3+} complex tris(1,10-phenanthroline)tri(thiocyanato-κN)lanthanum(III) (KP772, Figure 5.11b) exerts potent activity in a wide range of tumor cell lines *in vitro* and in a colon carcinoma xenograft model *in vivo*, with an efficacy comparable to cisplatin and methotrexate [103]. Notably, multidrug-resistant cell models are hypersensitive to this compound, and long-term treatment of KBC-1 cells with KP772 leads to a progressive loss of the multidrug-resistance protein-1 (MRP1), resulting in restoration of drug sensitivity, apparently without a direct interaction with MRP1. Complementary studies showed that exposure of cells to subtoxic,

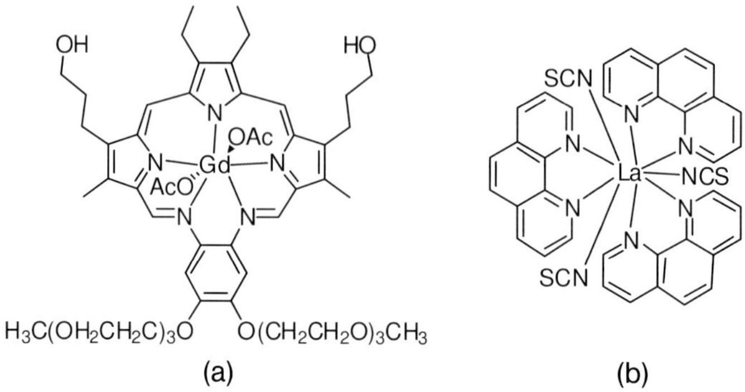

Figure 5.11 Schematic structures of motexafin gadolinium (a) and KP772 (b).

stepwise increasing, KP772 concentrations does not lead to acquired resistance [104]. KP772 is therefore expected to be active against multidrug-resistant tumors, making it an intriguing investigational drug for future clinical development.

5.9
Conclusions

The results reported in this chapter support conclusions of opposite signs. On the optimistic side, it is apparent that many and diverse non-platinum metal compounds have shown promising anticancer activity and that the pharmacological potential of the broad spectrum of metals is still far from being fully recognized. In other words, the *in vitro* and *in vivo* results collected over several decades for non-platinum metal compounds, despite being sparse and inhomogeneous, confirm that there is no reason to assume that only platinum complexes can be successfully applied in cancer therapy. Preclinical studies of ruthenium, gallium, titanium, and other metal lead-compounds revealed activity profiles different from that of cisplatin. Even though we are only now beginning to gain an understanding of the cellular and molecular effects of these drug candidates, they demonstrated that metal compounds are much more versatile than previously envisaged and that they can be fine-tuned almost at will by an appropriate choice of the metal, its oxidation state, and of the ligands. In most compounds the coordination of the metal center to some biological target seems to be necessary for anticancer activity, and in this regard the kinetics of activation through dissociation of some ligands are equally important as the kinetic and thermodynamic binding parameters of the metal center. Very often, additional non-covalent interactions are essential for increasing activity and target recognition (i.e., selectivity). However, a growing number of metal compounds that are inert and coordinatively saturated, and thus can give only non-coordinative interactions with their targets, are being investigated and show interesting biological properties.

On the other hand, it has to be recognized that there are no clear-cut guidelines for the rational design of new metal-anticancer compounds. Most of the current lead-compounds were found through serendipity or through intuition coupled with systematic investigations. In addition, in the case of ruthenium, the most well-studied metal in this field beside platinum, there are yet no general structure–activity relationships that might be used as guidelines for the development of new active compounds. For example, the reason for the selective activity of some Ru compounds against metastases is not clearly understood yet, in particular when considering that NAMI and RAPTA compounds are structurally so different.

One major drawback in the quest for chemotherapeutics in general is the lack of biological models with high predictivity, which would allow screening at an early development level for the best drug candidates for further development. As mentioned in the Introduction (Section 5.1), experience has demonstrated that metal compounds with too low cytotoxicity to meet the criteria for activity in cell line screens can still be effective agents *in vivo*.

Finally, we believe that this field might benefit from a careful re-examination of the wealth of chemical and biological results collected over the years by many different groups in the light of the recently acquired knowledge of cellular pathways. This process might produce new understanding and ideas for the design of novel compounds following non-canonical guidelines. In addition, it might lead to the biological re-investigation of compounds that have been abandoned or to the study of classes of metal compounds that have so far been neglected because they were considered devoid of features previously deemed essential for anticancer activity.

Acknowledgments

The group of Trieste gratefully acknowledges Regione FVG, Fondazione CRTrieste, Fondo Trieste, and Fondazione Beneficentia Stiftung that over the years have generously supported the research on anticancer Ru compounds. BASF Italia Srl is also gratefully acknowledged for a donation of hydrated ruthenium chloride. C. G. H., M. A. J., and B. K. K. thank the University of Vienna, the "Hochschuljubiläumsstiftung Vienna," the "Johanna Mahlke geb. Obermann-Stiftung," the FFG – Austrian Research Promotion Agency (811591), the Austrian Council for Research and Technology Development (IS526001) and the Austrian Science Fund for financial support. We are grateful to COST D39 and CM0902.

References

1 For a very recent update on current frontier research, see the *Dalton Transactions* themed issue on Metal Anticancer Compounds. (2009) *Dalton Trans.*, 10629–10936.

2 Clarke, M.J., Zhu, F., and Frasca, D.R. (1999) *Chem. Rev.*, **99**, 2511–2533.

3 Clarke, M.J. (2003) *Coord. Chem. Rev.*, **236**, 209–233.

4 Ang, W.H. and Dyson, P.J. (2006) *Eur. J. Inorg. Chem.*, 4003–4018.

5 Levina, A., Mitra, A., and Lay, P.A. (2009) *Metallomics*, **1** (6), 458–470.

6 Velders, A.H., Kooijman, H., Spek, A.L., Haasnoot, J.G., De Vos, D., and Reedijk, J. (2000) *Inorg. Chem.*, **39**, 2966–2967.

7 Gianferrara, T., Bratsos, I., and Alessio, E. (2009) *Dalton Trans.*, 7588–7598.

8 Reedijk, J. (2003) *Proc. Natl. Acad. Sci. USA*, **100**, 3611–3616.

9 Pascu, G.I., Hotze, A.C.G., Sanchez-Cano, C., Kariuki, B.M., and Hannon, M.J. (2007) *Angew. Chem. Int. Ed.*, **46**, 4374–4378.

10 Debreczeni, J.É., Bullock, A.N., Atilla, G.E., Williams, D.S., Bregman, H., Knapp, S., and Meggers, E. (2006) *Angew. Chem.*, **45**, 1580–1585.

11 Smalley, K.S.M., Contractor, R., Haass, N.K., Kulp, A.N., Atilla-Gokcumen, G.E., Williams, D.S., Bregman, H., Flaherty, K.T., Soengas, M.S., Meggers, E., and Herlyn, M. (2007) *Cancer Res.*, **67**, 209–217.

12 Schatzschneider, U., Niesel, J., Ott, I., Gust, R., Alborzinia, H., and Woelfl, S. (2008) *ChemMedChem.*, **3**, 1104–1109.

13 Reisner, E., Arion, V.B., Keppler, B.K., and Pombeiro, A.J.L. (2008) *Inorg. Chim. Acta*, **361**, 1569–1583.

14 Hartinger, C.G., Jakupec, M.A., Zorbas-Seifried, S., Groessl, M., Egger, A., Berger, W., Zorbas, H., Dyson, P.J., and Keppler, B.K. (2008) *Chem. Biodiversity*, **5**, 2140–2155.

15 Hartinger, C.G., Zorbas-Seifried, S., Jakupec, M.A., Kynast, B., Zorbas, H., and Keppler, B.K. (2006) *J. Inorg. Biochem.*, **100**, 891–904.

16 Berger, M.R., Garzon, F.T., Keppler, B.K., and Schmaehl, D. (1989) *Anticancer Res.*, **9**, 761–765.

17 Pieper, T., Borsky, K., and Keppler, B.K. (1999) *Top. Biol. Inorg. Chem.*, **1**, 171–199.

18 Heffeter, P., Pongratz, M., Steiner, E., Chiba, P., Jakupec, M.A., Elbling, L., Marian, B., Körner, W., Sevelda, F., Micksche, M., Keppler, B.K., and Berger, W. (2005) *J. Pharmacol. Exp. Ther.*, **312**, 281–289.

19 Hartinger, C.G., and Keppler, B.K. (2007) *Electrophoresis*, **28**, 3436–3446.

20 Schluga, P., Hartinger, C.G., Egger, A., Reisner, E., Galanski, M., Jakupec, M.A., and Keppler, B.K. (2006) *Dalton Trans.*, 1796–1802.

21 Stepanenko, I.N., Krokhin, A.A., John, R.O., Roller, A., Arion, V.B., Jakupec, M.A., and Keppler, B.K. (2008) *Inorg. Chem.*, **47**, 7338–7347.

22 Egger, A., Cebrian-Losantos, B., Stepanenko, I.N., Krokhin, A.A., Eichinger, R., Jakupec, M.A., Arion, V.B., and Keppler, B.K. (2008) *Chem. Biodiversity*, **5**, 1588–1593.

23 Alessio, E. (2004) *Chem. Rev.*, **104**, 4203–4242.

24 Alessio, E., Mestroni, G., Bergamo, A., and Sava, G. (2004) in *Metal Ions in Biological Systems* Vol. 42, (eds A. Sigel and H. Sigel), Marcel Dekker, New York, pp. 323–351.

25 Bratsos, I., Jedner, S., Gianferrara, T., and Alessio, E. (2007) *Chimia*, **61**, 692–697.

26 Bergamo, A. and Sava, G. (2007) *Dalton Trans.*, 1267–1272.

27 Rademaker-Lakhai, J.M., Van Den Bongard, D., Pluim, D., Beijnen, J.H., and Schellens, J.H.M. (2004) *Clin. Cancer Res.*, **10**, 3717–3727.

28 Bacac, M., Hotze, A.C.G., van der Schilden, K., Haasnoot, J.G., Pacor, S., Alessio, E., Sava, G., and Reedijk, J. (2004) *J. Inorg. Biochem.*, **98**, 402–412.

29 Ni Dhubhghaill, O.M., Hagen, W.R., Keppler, B.K., Lipponer, K.-G., and Sadler, P.J. (1994) *J. Chem. Soc., Dalton Trans.*, 3305–3310.

30 Sava, G., Bergamo, A., Zorzet, S., Gava, B., Casarsa, C., Cocchietto, M., Furlani, A., Scarcia, V., Serli, B., Iengo, E., Alessio, E., and Mestroni, G. (2002) *Eur. J. Cancer*, **38**, 427–435.

31 Brindell, M., Piotrowska, D., Shoukry, A.A., Stochel, G., and Eldik, R. (2007) *J. Biol. Inorg. Chem.*, **12**, 809–818.

32 Alessio, E., Mestroni, G., Bergamo, A., and Sava, G. (2004) *Curr. Top. Med. Chem.*, **4**, 1525–1535.

33 Groessl, M., Reisner, E., Hartinger, C.G., Eichinger, R., Semenova, O., Timerbaev, A.R., Jakupec, M.A., Arion, V.B., and Keppler, B.K. (2007) *J. Med. Chem.*, **50**, 2185–2193.

34 Brindell, M., Stawoska, I., Supel, J., Skoczowski, A., Stochel, G., and Eldik, R. (2008) *J. Biol. Inorg. Chem.*, **13**, 909–918.

35 Peacock, A.F.A. and Sadler, P.J. (2008) *Chem. Asian J.*, **3**, 1890–1899.

36 Yan, Y.K., Melchart, M., Habtemariam, A., and Sadler, P.J. (2005) *Chem. Commun.*, 4764–4776.

37 Novakova, O., Chen, H., Vrana, O., Rodger, A., Sadler, P.J., and Brabec, V. (2003) *Biochemistry*, **42**, 11544–11554.

38 Dougan, S.J. and Sadler, P.J. (2007) *Chimia*, **61**, 704–715.

39 Wang, F., Bella, J., Parkinson, J.A., and Sadler, P.J. (2005) *J. Biol. Inorg. Chem.*, **10**, 147–155.

40 Wang, F., Xu, J., Habtemariam, A., Bella, J., and Sadler, P.J. (2005) *J. Am. Chem. Soc.*, **127**, 17734–17743.

41 Hoeschele, J.D., Habtemariam, A., Muir, J., and Sadler, P.J. (2007) *Dalton Trans.*, 4974–4979.

42 Peacock, A.F.A., Parsons, S., and Sadler, P.J. (2007) *J. Am. Chem. Soc.*, **129**, 3348–3357.

43 Scolaro, C., Bergamo, A., Brescacin, L., Delfino, R., Cocchietto, M., Laurenczy, G., Geldbach, T.J., Sava, G., and Dyson, P.J. (2005) *J. Med. Chem.*, **48**, 4161–4171.

44 Bergamo, A., Masi, A., Dyson, P.J., and Sava, G. (2008) *Int. J. Oncol.*, **33**, 1281–1289.

45 Dyson, P.J. (2007) *Chimia*, **61**, 698–703.

46 Ang, W.H., De Luca, A., Chapuis-Bernasconi, C., Juillerat-Jeanneret, L., Lo Bello, M., and Dyson, P.J. (2007) *ChemMedChem.*, **2**, 1799–1806.

47 Vock, C.A., Ang, W.H., Scolaro, C., Phillips, A.D., Lagopoulos, L., Juillerat-Jeanneret, L., Sava, G., Scopelliti, R., and Dyson, P.J. (2007) *J. Med. Chem.*, **50**, 2166–2175.

48 Ang, W.H., Parker, L.J., De Luca, A., Juillerat-Jeanneret, L., Morton, C.J., Lo Bello, M., Parker, M.W., and Dyson, P.J. (2009) *Angew. Chem. Int. Ed.*, **48**, 3854–3857.

49 Mendoza-Ferri, M.G., Hartinger, C.G., Eichinger, R.E., Stolyarova, N., Jakupec, M.A., Nazarov, A.A., Severin, K., and Keppler, B.K. (2008) *Organometallics*, **27**, 2405–2407.

50 Nováková, O., Nazarov, A.A., Hartinger, C.G., Keppler, B.K., and Brabec, V. (2009) *Biochem. Pharmacol.*, **77**, 364–374.

51 Auzias, M., Therrien, B., Suess-Fink, G., Stepnicka, P., Ang, W.H., and Dyson, P.J. (2008) *Inorg. Chem.*, **47**, 578–583.

52 Serli, B., Zangrando, E., Gianferrara, T., Scolaro, C., Dyson, P.J., Bergamo, A., and Alessio, E. (2005) *Eur. J. Inorg. Chem.*, 3423–3434.

53 Schmitt, F., Govindaswamy, P., Suess-Fink, G., Ang, W.H., Dyson, P.J., Juillerat-Jeanneret, L., and Therrien, B. (2008) *J. Med. Chem.*, **51**, 1811–1816.

54 Dougan, S.J., Habtemariam, A., McHale, S.E., Parsons, S., and Sadler, P.J. (2008) *Proc. Natl. Acad. Sci. USA*, **105**, 11628–11633.

55 Meggers, E., Atilla-Gokcumen, G.E., Gruendler, K., Frias, C., and Prokop, A. (2009) *Dalton Trans.*, 10882–10888.

56 Bernstein, L.R. (1998) *Pharmacol. Rev.*, **50**, 665–682.

57 Collery, P., Keppler, B., Madoulet, C., and Desoize, B. (2002) *Crit. Rev. Oncol. Hematol.*, **42**, 283–296.

58 Chitambar, C.R. (2004) *Curr. Opin. Oncol.*, **16**, 547–552.

59 Leyland-Jones, B. (2003) *Semin. Oncol.*, **30**, 13–19.

60 Jakupec, M.A. and Keppler, B.K. (2004) *Curr. Top. Med. Chem.*, **4**, 1575–1583.

61 Chitambar, C.R., Purpi, D.P., Woodliff, J., Yang, M., and Wereley, J.P. (2007) *J. Pharmacol. Exp. Ther.*, **322**, 1228–1236.

62 Valiahdi, S.M., Heffeter, P., Jakupec, M.A., Marculescu, R., Berger, W., Rappersberger, K., and Keppler, B.K. (2009) *Melanoma Res.*, **19** (5), 283–293.

63 Hofheinz, R.D., Dittrich, C., Jakupec, M.A., Drescher, A., Jaehde, U., Gneist, M., Graf von Keyserlingk, N., Keppler, B.K., and Hochhaus, A. (2005) *Int. J. Clin. Pharmacol. Ther.*, **43**, 590–591.

64 Weaver, K.D., Heymann, J.J., Mehta, A., Roulhac, P.L., Anderson, D.S., Nowalk, A.J., Adhikari, P., Mietzner, T.A., Fitzgerald, M.C., and Crumbliss, A.L. (2008) *J. Biol. Inorg. Chem.*, **13**, 887–898.

65 Gray, D.J., Burns, R., Brunswick, P., Le Huray, J., Chan, W., Mast, M., Allamneni, K., and Sreedharan, S. (2005) *Spec. Publ. R. Soc. Chem.*, **301**, 43–58.

66 Davies, N.P., Rahmanto, Y.S., Chitambar, C.R., and Richardson, D.R. (2006) *J. Pharmacol. Exp. Ther.*, **317**, 153–162.

67 Narasimhan, J., Antholine, W.E., and Chitambar, C.R. (1992) *Biochem. Pharmacol.*, **44**, 2403–2408.

68 Hedley, D.W., Tripp, E.H., Slowiaczek, P., and Mann, G.J. (1988) *Cancer Res.*, **48**, 3014–3018.

69 Chitambar, C.R., Wereley, J.P., and Matsuyama, S. (2006) *Mol. Cancer Ther.*, **5**, 2834–2843.

70 Heim, M.E., Flechtner, H., and Keppler, B.K. (1989) *Prog. Clin. Biochem.*, **10**, 217–223.

71 Keppler, B.K., Friesen, C., Vongerichten, H., and Vogel, E. (1993) in *Metal Complexes in Cancer Chemotherapy* (ed. B.K. Keppler), VCH, Weinheim, pp. 297–323.

72 Schilling, T., Keppler, K.B., Heim, M. E., Niebch, G., Dietzfelbinger, H., Rastetter, J., and Hanauske, A.R. (1996) *Invest. New Drugs*, **13**, 327–332.

73 Köpf-Maier, P. (1999) *Anticancer Res.*, **19**, 493–504.

74 Guo, M., Guo, Z., and Sadler, P.J. (2001) *J. Biol. Inorg. Chem.*, **6**, 698–707.

75 Guo, M., Sun, H., McArdle, H.J., Gambling, L., and Sadler, P.J. (2000) *Biochemistry*, **39**, 10023–10033.

76 Tshuva, E.Y. and Ashenhurst, J.A. (2009) *Eur. J. Inorg. Chem.*, **2009**, 2203–2218.

77 Abeysinghe, P.M. and Harding, M.M. (2007) *Dalton Trans.*, 3474–3482.

78 Claffey, J., Hogan, M., Müller-Bunz, H., Pampillón, C., and Tacke, M. (2008) *ChemMedChem.*, **3**, 729–731.

79 Strohfeldt, K. and Tacke, M. (2008) *Chem. Soc. Rev.*, **37**, 1174–1187.

80 Nguyen, A., Vessieres, A., Hillard, E.A., Top, S., Pigeon, P., and Jaouen, G. (2007) *Chimia*, **61**, 716–724.

81 Hillard, E.A., Pigeon, P., Vessieres, A., Amatore, C., and Jaouen, G. (2007) *Dalton Trans.*, 5073–5081.

82 Slavik, M., Blanc, O., and Davis, J. (1983) *Invest. New Drugs*, **1**, 225–234.

83 Slavik, M., Elias, L., Mrema, J., and Saiers, J.H. (1982) *Drugs Exp. Clin. Res.*, **8**, 379–385.

84 Heim, M.E. (1993) in *Metal Complexes in Cancer Chemotherapy* (ed. B.K. Keppler), VCH, Weinheim, pp. 9–24.

85 Goodman, S. (1988) *Med. Hypotheses*, **26**, 207–215.

86 Keppler, B.K. (1993) *Metal Complexes in Cancer Chemotherapy*, VCH, Weinheim, p. 434.

87 Wang, Z.-Y. (2001) *Cancer Chemother. Pharmacol.*, **48**, S72–S76.

88 Dombret, H., Fenaux, P., Soignet, S.L., and Tallman, M.S. (2002) *Semin. Hematol.*, **39**, 8–13.

89 Miller, W.H. Jr, Schipper, H.M., Lee, J. S., Singer, J., and Waxman, S. (2002) *Cancer Res.*, **62**, 3893–3903.

90 Hayakawa, F. and Privalsky, M.L. (2004) *Cancer Cell*, **5**, 389–401.

91 Lu, J., Chew, E.-H., and Holmgren, A. (2007) *Proc. Natl. Acad. Sci. USA*, **104**, 12288–12293.

92 Hirano, S. and Kobayashi, Y. (2006) *Toxicology*, **227**, 45–52.

93 Quintas-Cardama, A., Verstovsek, S., Freireich, E., Kantarjian, H., Chen, Y. W., and Zingaro, R. (2008) *Anti-Cancer Agents Med. Chem.*, **8**, 904–909.

94 Soignet, S.L., Tong, W.P., Hirschfeld, S., and Warrell, R.P. Jr (1999) *Cancer Chemother. Pharmacol.*, **44**, 417–421.

95 Ben Zirar, S., Astier, A., Muchow, M., and Gibaud, S. (2008) *Eur. J. Pharm. Biopharm.*, **70**, 649–656.

96 Jakupec, M.A., Unfried, P., and Keppler, B.K. (2005) *Rev. Physiol. Biochem. Pharmacol.*, **153**, 101–111.

97 Fricker, S.P. (2006) *Chem. Soc. Rev.*, **35**, 524–533.

98 Evans, C.H. (1990) in *Biochemistry of the Elements* (ed. E. Frieden), Plenum Press, New York, Vol. **8**.

99 Anghileri, L.J. (1979) *Eur. J. Cancer*, **15**, 1459–1462.

100 Sato, T., Hashizume, M., Hotta, Y., and Okahata, Y. (1998) *BioMetals*, **11**, 107–112.

101 Wang, Z.M., Lin, H.K., Zhu, S.R., Liu, T.F., Zhou, Z.F., and Chen, Y.T. (2000) *Anti-Cancer Drug Des.*, **15**, 405–411.

102 Krishnamurti, C., Saryan, L.A., and Petering, D.H. (1980) *Cancer Res.*, **40**, 4092–4099.

103 Heffeter, P., Jakupec, M.A., Koerner, W., Wild, S., Von Keyserlingk, N.G., Elbling, L., Zorbas, H., Korynevska, A., Knasmueller, S., Sutterluety, H., Micksche, M., Keppler, B.K., and Berger, W. (2006) *Biochem. Pharmacol.*, **71**, 426–440.

104 Heffeter, P., Jakupec, M.A., Koerner, W., Chiba, P., Pirker, C., Dornetshuber, R., Elbling, L., Sutterluety, H., Micksche, M., Keppler, B.K., and Berger, W. (2007) *Biochem. Pharmacol.*, **73**, 1873–1886.

6
The Challenge of Establishing Reliable Screening Tests for Selecting Anticancer Metal Compounds

Angela Boccarelli, Alessandra Pannunzio, and Mauro Coluccia

6.1
Introduction

Cisplatin (**1**, *cis*-[PtCl$_2$(NH$_3$)$_2$])[1] is one of the most used anticancer agents, and with carboplatin (**2**, *cis*-[Pt(1,1-cyclobutanedicarboxylate)(NH$_3$)$_2$]) and oxaliplatin (**3**, *cis*-[Pt(oxalate)(*R,R*-1,2-diaminocyclohexane)]) is present in a large number (>50%) of chemotherapy regimens. Soon after the serendipitous discovery of the antitumor activity of cisplatin in 1969 [1], much of the early research effort was carried out on murine tumors, aimed at identifying the structure–activity relationships of platinum compounds [2, 3]. In parallel and subsequently, mechanistic investigations performed at first in bacteria and then in mammalian cells led to the identification of DNA as the pharmacological target of cisplatin [4]. The extraordinary and long-term research activity focused on metal-based anticancer compounds was inspired by the great efficacy of cisplatin in testicular cancer treatment, where cure rates of around 80% are obtained [5]; this induced the hope of repeating that clinical success in other solid tumors, exploiting also compounds of metals different from platinum. At the same time, the research effort was supported by the great versatility of metals and metal compounds, whose properties may be variously modulated depending upon the metal itself, its oxidation state, number and type of ligands, and coordination geometry of the complex.

In addition to cisplatin, two platinum drugs are at present in clinical use: carboplatin and oxaliplatin, approved by FDA in 1989 and 2002, respectively [6]. Carboplatin has the same indications of cisplatin, but a different toxicity profile. In contrast, oxaliplatin has a spectrum of activity different from that of cisplatin. Oxaliplatin is indicated indeed in colorectal cancer, a malignancy quite refractory to cisplatin or carboplatin, and has no indication in germ cell tumors or squamous cell carcinomas. Therefore, oxaliplatin satisfies the requirements for novel compounds with a spectrum of activity different from cisplatin; however, notably, the impact of oxaliplatin on colorectal cancer is hardly comparable to that of cisplatin on germ

Schematic structures of the labeled compounds can be found in Chapters 1–5 (e.g. Table 3.1).

Bioinorganic Medicinal Chemistry. Edited by Enzo Alessio
Copyright © 2011 WILEY-VCH Verlag GmbH & Co. KGaA, Weinheim
ISBN: 978-3-527-32631-0

cell tumors. Other platinum compounds are currently under clinical trials [7, 8], and so are ruthenium [9] and gallium [10] compounds. At present, there are also many metal-based drug candidates in the preclinical stage; numerous reviews have been published recently, giving broad coverage of this field [11–15]. As far as platinum compounds are concerned, many strategies are being developed to improve tumor targeting and reduce toxicity [16–22], as well as to modify the DNA interaction properties [23, 24] and/or the biological targets [25, 26]. The ruthenium compounds under investigation are characterized by a wide diversity of modes of action. For example, the interaction with proteins of the extracellular matrix and cell surface is likely involved in the antitumor activity of NAMI-A (**4**, imidazolium [*trans*-tetrachloro(dimethyl sulfoxide) imidazoleruthenate(III)]) [27]; the transferrin pathway and the interaction with cellular DNA and mitochondria appear to be involved in the mechanism of action of the KP1019 complex (**5**, indazolium [*trans*-tetrachlorobis(1H-indazole)ruthenate(III)]) [28]; the inhibition of protein kinases has been reported for the organometallic complex DW1/2 (**6**) [29], and the inhibition of enzymes implicated in tumor progression (thioredoxin reductase and cathepsin B) has been reported for ruthenium-arene compounds [30]; more recently, a ruthenium-derived organometallic compound capable of inducing the endoplasmic reticulum stress pathway has also been described [31]. The pharmacological action of gallium nitrate, $Ga(NO_3)_3$, which has already shown antitumor activity in clinical trials [32, 33], appears to be associated with the interference with iron homeostasis, including the inhibition of the ribonucleotide reductase enzyme [34], even though other mechanisms of cytotoxicity may be involved [35]. Moreover, gallium complexes acting as proteasome inhibitors have been reported recently [36]. Gold compounds are investigated with respect to the ability of inducing mitochondrial damage [37, 38] and proteasome inhibition [39]. Finally, inhibition of macromolecular synthesis, and mitochondrial and cell membrane damage, have been implicated in cellular effects of organotin compounds [40].

The reported examples show that several distinct molecular and cellular targets may be involved in the pharmacological action of the compounds at present under investigation. In some cases, the targets are different from DNA but are in any event implicated in replicative functions, therefore justifying early screening assays based on tumor cell proliferation inhibition. In other cases, the targets are not directly related to replicative functions, thus suggesting the necessity of different screening assays.

Herein, we illustrate the main *in vitro* screening systems available for the development of metal-based anticancer drugs. In particular, the role of screening assays based on tumor cell growth inhibition and cell death is discussed. Subsequently, the impact of genomics and proteomics on metal-based drug development is illustrated by some examples of both established and candidate anticancer compounds. Historically, the design of metal-based anticancer drugs, after an initial phase of empirical research, has always been inspired by mechanistic considerations that have evolved in parallel with scientific and technological advancements in basic sciences and cancer research. A typical (and exciting) feature of metal-based drug research is the contribution of different disciplines (from bioinorganic chemistry and biology to pharmacology and oncology), each of which has determined specific and important breakthroughs in the course of time. The most recent advancements in understanding the mechanistic properties of anticancer drugs have come from genomics and proteomics, and the application of such technologies for screening purposes constitutes a present major challenge of anticancer drug discovery and development.

6.2
Tumor Cell Growth Inhibition and Cell Death Screening Assays

Over the last few decades, a fundamental goal of cancer research has been the identification of therapies capable of inducing selective cancer cell death. In general, killing as many tumor cells as possible in most cases allows an improvement of survival and quality of life of patients, but complete eradication unfortunately can be obtained only in few cases (e.g., in testicular cancer, Hodgkin's lymphoma, and acute myeloid leukemia). The great efficacy of cisplatin in testicular cancer treatment fostered a growing interest in metal-based drugs, and over a period of 40 years thousands of compounds have been synthesized and investigated as potential anticancer agents.

Metal-based anticancer drugs are generally classified as DNA-damaging cytotoxic agents, and in all programs of research and development of candidate anticancer compounds the ability of determining both tumor cell death and DNA damage is almost always systematically investigated. The use of human tumor cell lines for the screening of anticancer agents, including the metal-based compounds, gained ground in the late 1980s when the National Cancer Institute (NCI) selected 60 cell lines representing nine human tumor types (leukemia, non-small-cell lung, colon, central nervous system, melanoma, ovarian, renal, prostate, and breast). The NCI-60 panel of cancer cell lines was established as a tool for a disease-oriented *in vitro* anticancer drug screening [41]. To date, tens of thousands of compounds, including over 1100 metal or metalloid containing compounds, have been screened in the NCI-60 panel for evidence of the ability to inhibit the growth of human tumor cell lines (http://dtp.nci.nih.gov/docs/cancer/cancer_data.html). Data concerning the activity of a given compound in a cancer cell panel, such as the NCI-60 or the JFCR-45 [42], generate a typical profile of cellular response, for example, the "mean graph"; through computational techniques, such as COMPARE, the pattern of activity of a new compound can be compared with the patterns of previously screened compounds, and this analysis is used to define compounds with similar mechanisms of action [43]. A major goal of this screening system is to recognize compounds characterized by a pattern of activity (and presumed mechanism of action) different from those of known classes. For example, the platinum compounds tested against the NCI-60 panel have been analyzed by Fojo and colleagues; 107 compounds have been organized into distinct groups on the basis of their activity profiles and structural features, thus identifying the most promising compounds for further experimentation [44]. In platinum-based drug discovery programs specifically aimed at developing drugs capable of circumventing cisplatin/carboplatin resistance, Hills and colleagues established human ovarian carcinoma cell panels showing a close *vitro/vivo* correlation in cisplatin sensitivity/response [45]. By using this ovarian screening system, several promising compounds were identified [7], including satraplatin (**7,** *cis,trans, cis*-[PtCl$_2$(acetate)$_2$(NH$_3$)(cyclohexylamine)]) and picoplatin (**8,** *cis*-[PtCl$_2$(NH$_3$)(2-methylpyridine)]), at present investigated in Phase II/III clinical trials. Oxaliplatin, too, is characterized by an activity profile different from that of cisplatin in the NCI-60 panel (Figure 6.1), even though the drug was originally selected and investigated in *in vivo* murine systems [46, 47]. Panels of tumor cells representative of common human tumors and including cisplatin-resistant cells have also been used to screen platinum compounds with *trans* geometry [48, 49]. On the whole, tumor cell panels have been successful at identifying platinum-based compounds characterized, *in vitro*, by a spectrum of activity different from cisplatin and/or effective against cisplatin-resistant tumors.

As far as methodological aspects are concerned, in the early investigations of activity on tumor cells, the ability to inhibit cell proliferation is usually analyzed; the parameter most frequently defined is the compound concentration resulting in a 50% reduction in the cell number increase compared to control cells, that is, the 50% growth inhibitory concentration (GI50). The wide use of cell growth inhibition assays in the screening of potential anticancer agents is connected to the fact that anticancer drug development is still largely focused on cellular processes involved in cell proliferation. As tumor cells are characterized by proliferative activity, the inhibition of proliferation is considered an operative definition of cell death, and in this sense is used in the so-called cytotoxicity assays. To determine the treatment-induced inhibition of proliferation, the tumor cell number (in microtiter plates) is calculated by spectrophotometric techniques adapted for automation and large-scale applications. The Sulforhodamine B (SRB) assay is a widely used method that relies on the ability of the SRB dye to bind to proteins, thus providing a measurement of cellular protein content and the determination of cell density [50, 51]. Tetrazolium dye-based assays (MTT, XTT) are other popular systems for measuring cell density [52, 53]. Unlike SRB, which does not distinguish between viable and dead cells, the tetrazolium assays are based upon viable cell metabolic activity to convert tetrazolium into formazan dye. In any case, it is important to note that the GI50 parameter is a measure of cell proliferation, and not of cell kill. Moreover, even when the long-term fate of proliferating cells is measured by using clonogenic assays [54], these methods do not distinguish cell death from long-lasting cell-cycle arrest [55].

The identification of dead cells after treatment with metal compounds, and in general after a toxic damage, is not a difficult task and should be considered from the initial development stage. In *in vitro* systems, a cell should be considered dead when it has lost the integrity of the plasma membrane (thus incorporating exclusion dyes, such as propidium iodide or trypan blue) and/or the whole cell has undergone fragmentation into discrete fragments not incorporating exclusion dyes (apoptotic bodies). These morphological features can be detected by

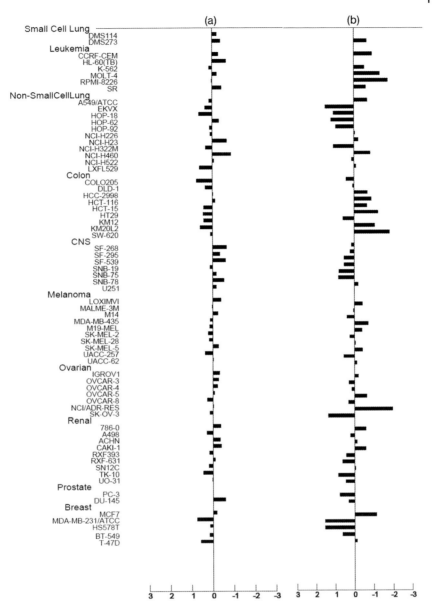

Figure 6.1 Mean GI50 graphs showing the different activity profile of cisplatin (a) and oxaliplatin (b) in the NCI-60 tumor cell line panel (data obtained from the NCI web site at http://dtp.nci.nih.gov/webdata.html). The zero value represents the mean GI50 concentration across all cell lines to cisplatin and oxaliplatin (9.5 and 2.8 μM, respectively). Using a logarithmic scale, the sensitivity or resistance of each cell line is represented by deflections to the right or left, respectively, from the mean. The mean graphs format highlights the differences of sensitivity patterns for distinct cancer cell subpanels, for example, showing the sensitivity of colon cancer cells to oxaliplatin.

microscopy- and/or flow cytometry-based techniques, which provide also important clues of the underlying mechanisms. Cell death is classically considered to be consequent to regulated and unregulated mechanisms [56]. Apoptosis is the best known regulated mode of cell death, naturally occurring in multicellular organisms, and is characterized by cell shrinkage, chromatin condensation, and nuclear and cell fragmentation; importantly, apoptosis is often disrupted in cancer cells [57]. In contrast, unregulated cell death, often called necrosis, is an unwarranted (pathologic) form of cell death characterized by a gain in cell volume, swelling of organelles, plasma membrane rupture, and loss of intracellular contents. However, evidence is accumulating that necrosis may also be regulated by specific signal transduction pathways [58], and the term "necroptosis" has been suggested to differentiate regulated from accidental necrosis [59]. In some cases, investigated mainly in *in vitro* systems, dying cells can show a morphology called autophagic cell death [60], in which wide portions of cytoplasm are sequestered and degraded inside the autophagosomes, in the absence of chromatin condensation. Notably, the breakdown of cellular components in the autophagosomes may provide a cell with nutrients, ensuring that vital processes can continue, for example, after nutrient starvation [61] and/or toxic damage [62, 63]. This suggests that autophagy is basically a survival mechanism. However, a persistent autophagy can progressively weaken the cell, eventually leading to its death.

In vitro studies on effects of anticancer drugs on tumor cells have allowed investigators to identify all three major modes of cell death: apoptosis, necrosis, and autophagy. Since these events represent the end result of treatment-induced perturbations of the interconnected metabolic pathways regulating cell death-survival homeostasis, it is not surprising to find examples in which cell death shows mixed morphological features (e.g., signs of both apoptosis and necrosis), thus witnessing the interdependence of cell death mechanisms [64]. Characterizing the tumor cell death mode induced by anticancer drugs is also important as far as the clearance mechanisms of dead cells are concerned. Apoptosis is generally considered immune-silent or immunosuppressive, whereas necrosis initiates the immune response. The insights from the last decade increasingly support the relevance of cell death pathways, removal of dying and dead cells, and immune system response in anticancer therapy [65].

The investigation of tumor cell death induced by metal-based compounds is important, particularly in a mechanism-driven development program. The biochemical aspects of different cell death modes can be detected by several techniques (immunofluorescence microscopy, colorimetric/fluorogenic assays, flow cytometry, immunoblotting, etc.) [66, 67]. On the whole, the available techniques, in several cases adapted for commercially available high content assays, can give rather specific information on cellular demise processes involved in apoptosis (activation of caspases, mitochondrial transmembrane permeabilization, DNA fragmentation, phosphatidylserine exposure), necrosis (leakage of intracellular proteins, ATP measurement, activation of non-caspase proteases), and autophagy (LC3 protein conversion, Beclin-1 dissociation).

Before leaving cell death modes behind, we add two additional considerations. Cell death is a dynamic process, whose features may appear even after the usual time intervals (1–3 days) of short-term assays; therefore, when investigating cell death mechanisms, a wide range of doses, schedules, and times should be examined. Moreover, the biological features of tumor cells used in the assay must be carefully considered: cell death–survival homeostasis of tumor cells is affected by tumor-specific genetic and epigenetic mechanisms, which support the adaptive features of cancer cells. This implies that cellular effects of anticancer drugs depend both on the intrinsic pharmacodynamic properties of the compounds being tested and on tumor cell specific context [68].

6.3
Metal-Based Anticancer Compounds and Gene Expression Microarray Technologies

The tumor biology early notion that an increased cell proliferation represents the basic difference between tumor and normal cells has long since been discarded in favor of a more complex notion of dysregulation of numerous and interconnected cell processes. Because of genetic mutations, epigenetic changes, and microenvironment-dependent events, tumor cells show abnormalities of functions relating to growth, division, differentiation, senescence, apoptosis, DNA repair, cell signaling, and tissue architecture. As overall consequence, tumor cells adaptively acquire the ability to survive and reproduce in defiance of normal restraints, and to invade and colonize anatomical districts normally reserved for other cells. A fundamental contribution to the present understanding of cancer molecular and cellular biology comes from gene expression microarray technologies. Developed in the 1990s, DNA microarrays can monitor RNA products of thousands of genes simultaneously. Figure 6.2 shows an example of a 22 000-probe oligonucleotide microarray.

Transcriptional profiling provides unprecedented and comprehensive information on which genes are switched on (or off) when tumor cells grow, proliferate, differentiate, or respond to molecular signals as well as to therapeutic agents. An indication of the increasing application of microarrays in cancer can be appreciated when interrogating PubMed database for "microarray + cancer." The search produces 131 citations from 1995 to 2000, whereas more than 11 000 citations are at present reported. Gene expression analyses are also facilitating the contemporary anticancer drug development, being usable in most stages of the process, including target identification and validation, mechanism of action studies, and identification of pharmacodynamic endpoints. In the current literature, numerous applications of transcriptional profiling with anticancer agents can be found [69]. Herein, some examples are provided to illustrate how the transcriptional profiling approach is being used for investigating the pharmacological action of metal-based anticancer drugs. In more detail, the use of basal or

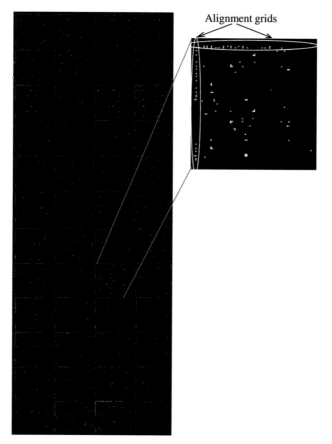

Alignment grids

Figure 6.2 Pseudo-color image of a competitive hybridization experiment on a 70mer oligonucleotide microarray. The competitive hybridization was performed by using cyanine 3-labeled cDNA (green), prepared from mRNA isolated from cisplatin-sensitive A2780 ovarian cancer cells, and cyanine 5-labeled cDNA (red), prepared from mRNA isolated from cisplatin-resistant A2780cisR cells. A representative (magnified) subgrid is also shown on the right. The scanner generates separate data images for green (532 nm) and red (635 nm) wavelengths. By overlaying the red and green data images, the differential expression of various genes is visualized. Green spots indicate expression in the A2780 sample, while red spots indicate expression in the A2780cisR sample. Yellow spots indicate genes that are expressed in both samples. The fluorescence intensity of each color at each spot indicates the level of expression of that gene in each sample.

constitutive gene expression profiling to understand and predict drug sensitivity and resistance is described; likewise, the value of determining changes in gene expression in response to treatment is illustrated with examples concerning both established and candidate metal-based drugs.

6.3.1
**Basal Transcription Profiling and Sensitivity/Resistance
to Metal-Based Anticancer Drugs**

This strategy is generally applied to a panel of tumor cell lines, in which measures of compound activity (e.g., growth inhibition, cell death, or other cellular parameters) are correlated with gene expression profiles prior to drug exposure.

A comprehensive review of metal compounds with potential anticancer activity investigated in the NCI-60 screen has been published by Huang and colleagues [70]. More than 1100 compounds containing transition metals, main group metals, or metalloids were examined and clustered according to similarities of their activity profiles (GI50) in the NCI-60 panel. The compounds in the clusters were subsequently analyzed according to their structural features and reactivity. Along with literature support, similarities in chemical features and reactivity were used to assign the compounds to four broad classes of mechanism of action, according to preference for binding to sulfhydryl groups, chelation, generation of reactive oxygen species, and production of lipophilic ions. Finally, the genomic features uniquely associated to specific cytotoxic responses were analyzed, with the aim of identifying putative genes or biological pathways affected by metal compounds. Several examples of cytotoxicity/gene expression correlations were reported, as in the case of metal compounds with high affinity for sulfhydryl groups and cysteine- or glutathione-dependent protein gene expression profiles. On the whole, the correlations between cytotoxic responses and gene expression profiles provided additional support for mechanism of action assignment, and contributed to identify subsets of compounds specifically active towards certain cancer cell sub-panels. The NCI database with gene expression profiles of NCI-60 tumor cells was explored also by Vekris and colleagues, to identify the genes whose expression was positively or negatively correlated to the sensitivity of tumor cells to cisplatin, carboplatin, oxaliplatin, or tetraplatin. Interestingly, important similarities were noticed between cisplatin and carboplatin, on the one hand, and tetraplatin and oxaliplatin on the other hand [71].

Metal compound sensitivity-basal transcription profiling relationships have also been investigated in cell lines not included in the NCI-60 panel, representative of tumors for which platinum-based chemotherapy is usually performed (e.g., head and neck tumors), as well as in distinct series and pairs of tumor cells sensitive and resistant/refractory to platinum drugs. Most studies have been focused on clinically used platinum compounds, and some examples are reported below. The cytotoxic activity of platinum drugs has been correlated to gene expression profiles of a series of head and neck, esophageal, gastric, testicular, ovarian, and colorectal cancer cell lines [72–87]. Gene expression profiles correlating with cisplatin sensitivity of oral squamous cell carcinoma pairs of cisplatin-sensitive and- resistant cell lines have been reported by different groups [72–76], which provided useful models for identifying the genes associated with cisplatin resistance. Gosepath and colleagues showed that cisplatin resistance of the oral cancer cell line Cal27 is associated with a decreased expression of DKK1, an inhibitor of WNT signaling, and that resistance could be partially reversed by DKK1 overexpression [77]. More recently, Yamano and colleagues [78] identified differentially expressed genes between cisplatin-sensitive and -resistant head and neck squamous cell carcinoma (HNSCC) cell lines; importantly, it was shown that siRNA-directed silencing of some selected genes resulted in enhanced susceptibility to cisplatin, and that a positive immunohistochemical staining of the same genes correlated with resistance to cisplatin-based chemotherapy in HNSCC patients. The cytotoxic activity of cisplatin was correlated with basal expression profiles of esophageal cancer cell lines and normal esophageal epithelial cells [79], chemoresistant gastric cancer cell lines and parental cells [80], and cisplatin-resistant and -sensitive testicular germ cell tumor (TGCT) cell lines [81], thus providing comprehensive gene information of different tumor histotypes usually treated with platinum drugs. In particular, the genomic differences between three TGCT parental cell lines and their cisplatin-resistant derivatives were investigated by single nucleotide polymorphism (SNP) microarray analysis and fluorescence *in situ* hybridization. Two chromosomal regions (6q26-27 and 10p14) with copy number changes were identified across cisplatin-resistant cells, thus suggesting the localization of cisplatin chemoresistance genes [81].

Transcription profiling data of 14 human ovarian cancer cell lines (derived from patients who were either untreated or treated with platinum-based chemotherapy) were correlated to cell proliferation rate as well as to cisplatin, carboplatin, oxaliplatin, or picoplatin sensitivity by Roberts and colleagues [82]. Stat1 gene expression was associated with decreased sensitivity to cisplatin and picoplatin; importantly, it was shown that resistance could be induced in sensitive cells by Stat1 transfection. Cheng and colleagues used basal transcriptional profiling to identify genes differentially expressed in multiple pairs of cisplatin-sensitive and - resistant human ovarian carcinoma cell lines. For the first time, many genes involved in cell surface interactions and trafficking pathways were associated with cisplatin resistance [83]. Basal transcriptional profiling of ovarian cancer cells intrinsically characterized by different sensitivity to platinum drugs, such as ovarian clear cell or serous adenocarcinoma cell lines, has also been investigated. In cisplatin-refractory clear cells, Saga and colleagues identified six highly expressed genes, including glutathione peroxidase 3 (GPX3), and showed also that RNAi suppression of GPX3 increased cisplatin sensitivity [84]. In ovarian clear cells, Oishi and colleagues reported a higher expression of galectin-3, a protein involved in cell interactions, and confirmed the observation by immunohistochemistry of clear cell and serous adenocarcinomas [85].

Oxaliplatin is an effective chemotherapeutic agent for the treatment of colorectal cancer, and several apoptosis and DNA damage repair genes correlated with response to oxaliplatin were identified through transcriptional profiling of several colorectal cancer cell lines [86]. Genes correlated with oxaliplatin or cisplatin cytotoxicity in six human colorectal cancer cell lines were investigated also by Meynard and colleagues; protein synthesis, cell energetics, and response to oxidative stress were identified as the major functions involved in oxaliplatin activity [87].

Finally, basal transcription profiling of gallium-resistant and gallium-sensitive human T-lymphoblastic leukemia/lymphoma CCRF-CEM cells has been performed recently, showing a role for metal-responsive transcription factor-1, metallothionein-2A, and zinc transporter-1 in modulating the antineoplastic activity of gallium nitrate [88].

To sum up, the above examples demonstrate that the correlations between basal transcription profiling and drug activity may contribute to the validation of candidate target functions. Moreover, such correlations contribute to the understanding of sensitivity/resistance mechanisms, and to the identification of activity biomarkers. Knowledge of factors that may affect the efficacy of anticancer drugs is of utmost importance because such information may be used to identify patients more likely to benefit from treatment. This is a major task of cancer chemotherapy, since clinical resistance at present can be determined only retrospectively.

6.3.2
Profiling Gene Expression Alterations in Response to Treatment with Metal-Based Anticancer Drugs

Gene expression profiling of cells in response to anticancer drug treatment provides a valuable means to identify potential pharmacodynamic markers, and to investigate the cellular mechanism of action. For example, gene expression profiling was used by Le Fevre and colleagues to distinguish among anticancer drugs according to their presumed mode of action. In this study, human lymphoblastoid TK6 cells were exposed to different concentrations of DNA-reactive (alkylating, crosslinking, and oxidative agents) or non-DNA-reactive anticancer drugs (mitotic spindle interacting agents, antimetabolites, and topoisomerase inhibitors), and gene expression was examined at the end of treatment or after a recovery period. Gene expression analysis allowed the drugs to be classified according to their presumed mechanism of action, and a molecular signature of 28 genes mainly involved in cell cycle and signal transduction pathways was identified [89]. Given that metal-based anticancer compounds are generally considered as DNA-damaging agents, a distinction between DNA-reactive and non-DNA-reactive compounds could be useful information from the early screening onwards of metal-based anticancer drug candidates. In the following, some examples of the different types of studies focused on transcriptional profile modifications induced by both established [90–94] and candidate metal-based drugs [95–101] are reported.

The effect of cisplatin upon global gene expression of testicular germ cell tumor (TGCT) cells has been investigated by Duale and colleagues [90], to provide a general picture of the unusual sensitivity of TGCT cells to platinum agents. New functional classes of genes and pathways not previously known to be associated with cisplatin-induced cellular response were identified (terminal growth arrest, and senescent-like arrest), along with microRNA target genes that may play important functional role in determining cisplatin sensitivity. The effects of oxaliplatin on HCT116 human colon carcinoma cells at the gene expression level (and also at protein level) were compared to those of cisplatin to investigate the mechanisms underlying the higher efficacy of oxaliplatin in colon carcinoma by Voland and colleagues [91]. Some genes equally affected by both drugs were identified (e.g., Bax and Fas), whereas other genes were found to be conversely regulated. In particular, oxaliplatin down-regulated several cell cycle-related genes involved in DNA replication and G2-M progression, thus suggesting a different profile of cellular response, in any event converging toward a common cell death fate. A time–course analysis of transcriptional modifications associated with the development of cisplatin resistance was proposed by Whiteside and colleagues [92]. In this study, changes in gene expression of NCI-H226 and NCI-H2170 lung cancer cells treated weekly with minimal inhibitory concentrations of cisplatin were systematically analyzed; the time–course method contributed to identify genes directly rather than passively involved in resistance development, and genes never before associated with cisplatin resistance occurrence were identified, including interferon-inducible protein 9–27 (IIP 9–27), stromelysin-3 (Str-3), ERBB-3, and macrophage inhibitory cytokine 1 (MIC1). More recently, time–course and concentration–effect experiments were performed by Brun and colleagues, to investigate the effects of cisplatin and oxaliplatin on gene expression of A2780 ovarian cancer cells across multiple time-points and drug concentrations. A model for fitting simultaneously time profiles and concentration effects to gene expression data was used to identify genes with different expression patterns between the two platinum drugs [93].

Gene expression modifications induced by cisplatin have been investigated also in spheroid cultures of ovarian cancer cells [94]. Multicellular spheroids may provide a better *in vitro* approximation of solid tumors [102], in particular for studying early occurrence of acquired resistance, a phenomenon that likely represents an adaptive multicellular mechanism rather than the result of genetic changes [103]. Consistent with this hypothesis, the major functional gene categories that were specifically upregulated in cisplatin-treated ovarian cancer spheroids included cellular assembly and organization, cell death and DNA replication, recombination, and repair.

Gene expression modifications induced by candidate metal-based anticancer drugs have been also actively investigated, with the aim to provide a mechanistic basis for their development [95–101]. Transcriptional modifications induced by the multinuclear platinum complex BBR3464 (**9**, [{$trans$-PtCl(NH$_3$)$_2$}$_2${μ-$trans$-Pt(NH$_3$)$_2$(H$_2$N(CH$_2$)$_6$NH$_2$)$_2$}]$^{4+}$) on a pair of cisplatin-sensitive (A431) and resistant (A431/Pt) human cervix squamous cell carcinoma cells have been investigated by Gatti and colleagues [95]. The results showed a differential pattern of transcription profile between sensitive (upregulation of cell cycle and growth regulators, tumor suppressors, and signal transduction genes) and resistant (upregulation of apoptosis and DNA damage genes) cells.

The differential processing of antitumor-active $trans$-Z (**10**, $trans$-[PtCl$_2${Z-HN=C(OCH$_3$)CH$_3$}(NH$_3$)]) and antitumor-inactive transplatin (**11**, $trans$-[PtCl$_2$(NH$_3$)$_2$]) by SKOV-3 ovarian cancer cells treated with equi-cytotoxic doses (IC$_{50}$) of the two compounds has been investigated by Boccarelli and colleagues [96]. The time–course analysis of gene expression modifications showed that phenotypic effects of transplatin were driven by an initial and transient upregulation of some genes related to cell-cycle checkpoint and arrest networks, whereas the more dramatic phenotypic effects of *trans-Z* were associated with a persistent upregulation of more numerous genes involved in cell-cycle checkpoint and arrest networks, and in genome stability and DNA repair. The effect of transplatin on gene expression of human foreskin fibroblasts has been compared to that of cisplatin by Galea and Murray [97]. Many genes responded similarly to treatment with both compounds, but a differential expression of DNA damage response genes was induced by cisplatin.

The novel *trans-R,R*-diaminocyclohexane (DACH) derivative [Pt(*R,R*-DACH) (DMC)], characterized by the presence of the dimethylcantharidin (DMC) ligand, has been compared to oxaliplatin on HCT116 colon cancer cells. The DMC ligand is expected to be released from the parent complex, thus determining an additional DNA damage-dependent cytotoxic effect. The analysis of treatment-induced transcriptional modifications allowed the identification of genes specifically affected by the DMC ligand, including base excision, homologous recombination, and mismatch repair genes, thus confirming the mechanistic basis for the marked efficacy of [Pt(*R,R*-DACH)(DMC)] [98].

Platinum(IV) complexes are thought to function as prodrugs for anticancer Pt(II) drugs [99]. Olszewski and colleagues compared transcriptional profile modifications induced by the orally applicable Pt(IV) compound oxoplatin [**12**, *cis,trans,cis*-diammine-dihydroxido-dichlorido-platinum(IV)] with those of cisplatin in H526 small cell lung cancer cells [100]. Oxoplatin and cisplatin showed a partial overlap of gene expression modifications, but 80% of affected genes were different for the two compounds, thus suggesting distinct mechanistic properties. Gene expression profiling of SiSo cervical adenocarcinoma cells resistant to either cisplatin or oxoplatin was also investigated, highlighting marked global differences between the two compounds [101]. On the whole, these results suggest a different mechanism of action for oxoplatin, and question its role as inactive prodrug of cisplatin.

In summary, microarray technology may greatly facilitate the analysis of the mechanisms of action of metal-based drugs. As shown by the reported examples, gene expression profiles of cells in response to metal-based compounds can readily be compared with untreated control cells, to reveal those genes that have undergone a change in response to drug treatment. For screening purposes, when structurally similar compounds are compared, the identification of gene expression responses unique to antitumor-active compounds can greatly improve the design and development of lead agents. Importantly, the global effects of treatment, including the toxicity effects, can be analyzed by transcriptional profiling, thus allowing one to identify the various pharmacological actions of the metal compounds under investigation.

6.4
Metal-Based Anticancer Compounds and the Proteomic Approach

Scientists have long focused on the characterization of platinum–DNA interactions, exploiting advanced analytical and molecular biology techniques. However, protein-bound platinum represents the most part of administered drug in the case of cisplatin, both in the extracellular medium (up to 98% of cisplatin interacts with albumin) [104] and into the cytoplasm [105], where the binding to intracellular S-donors is markedly favored [106]. In the case of cisplatin, it is generally believed that the reactions occurring with cytosolic components account for the broad spectrum of side-effects [107], and contribute to resistance [108]. Consequently, the investigation of platinum interactions with biological molecules different from DNA is an important task of both toxicity and resistance investigations [109]. In addition, cytotoxicity mechanisms alternative (or additional) to DNA binding are increasingly considered for platinum compounds, based upon interactions with the plasma membrane [110] or with regulatory proteins [111]. Interestingly, anti-tumor platinum compounds that do not interact with cellular DNA have also been reported recently [112].

In this general context, as well as in a more specific mechanistic-driven drug development program, the relevance of the identification of protein targets of metal-based compounds is undoubtable. The most recent findings in identification of binding proteins and target sites of platinum, gold, arsenic, and ruthenium anticancer compounds have been recently reviewed by Sun and colleagues [113], and the potential of proteomic approaches in mechanistic investigations by Wang and Chiu [114].

Similarly to the genomic investigations mentioned in the previous paragraphs, some studies aimed at identifying the basal proteomic profile associated with drug sensitivity/resistance of cancer cells are mentioned below [115–120]; likewise, some examples of proteomic profile changes induced by treatment are provided [121–124].

By using the isotope-coded affinity tags combined with tandem mass spectrometry technique (ICAT/MS/MS), Stewart and colleagues compared the proteomes of cisplatin-sensitive IGROV1 ovarian cancer cells and cisplatin-resistant IG-ROV1/CP cells; 121 differentially expressed proteins between the two cell lines were identified, including proteins overexpressed in resistant cells (cell recognition molecule CASPR3, S100 protein family members, junction adhesion molecule Claudin 4, and CDC42-binding protein kinase beta) and in sensitive cells (hepatocyte growth factor inhibitor 1B, and programmed cell death 6-interacting protein) [115]. The proteomic pattern of cisplatin-resistant IGROV1-R10 cells was compared to that of IGROV1 by Le Moguen and colleagues. Among the 40 proteins identified using the MALDI-TOF mass spectrometry technique, cytokeratins 8 and 18 and aldehyde dehydrogenase 1 were found to be overexpressed in IG-ROV1-R10, whereas annexin IV was down-regulated [116].

Yan and colleagues compared the proteomes of cisplatin-sensitive SKOV3 and A2780 ovarian cancer cells to those of corresponding resistant sublines, by using two-dimensional gel electrophoresis and MALDI-TOF mass spectrometry. Out of 57 differentially expressed proteins, five proteins (annexin A3, destrin, cofilin 1, Glutathione-S-transferase omega 1, and cytosolic $NADP^+$-dependent isocitrate dehydrogenase) were found to be co-instantaneously significant [117]. More recently, membrane-associated glycoproteins of ovarian cancer cells sensitive (A2780) and resistant (A2780cis) to cisplatin have been investigated by liquid chromatography and mass-spectrometry. Six overexpressed proteins were identified in resistant cells, and one of them, namely the CD70 cytokine, was revealed also in ovarian tumors by immunohistochemistry, suggesting a potential role as resistance marker [118].

Differentially expressed proteins have been investigated also in cisplatin-sensitive and- resistant breast cancer cell lines (MCF7) by MALDI-TOF mass spectrometry, and some candidates for further validation in clinical samples were identified [119]. Finally, through a proteomic approach, a link between decreased pyruvate kinase M2 expression and oxaliplatin resistance in human HT29 colorectal cancer cell lines as well as in patients has been identified recently by Martinez-Balibrea and colleagues [120].

Proteomic profile changes associated with cisplatin treatment of IGROV1 cells were investigated by Le Moguen and colleagues by electrophoresis coupled to MALDI-TOF mass spectrometry. The kinetic analysis of IGROV1 cell behavior following cisplatin treatment revealed time and/or concentration-dependent modifications in protein expression, including events likely associated with cell cycle blockade (decreased amino-acid and nucleotide synthesis) and resistance development (enhanced glycolysis and increased proliferating potential) [121]. The differential protein expression associated with cisplatin treatment of HeLa cervical carcinoma cells has been investigated by Yim and colleagues. Interestingly, proteomic data confirmed the ability of cisplatin to induce apoptosis through the activation of both death receptor-mediated and mitochondria-mediated pathways [122]. A human nasopharyngeal carcinoma cell line (SUNE1) was used by Wang and colleagues to compare the proteomic profile modifications induced by cisplatin and a by a gold(III) porphyrin complex. Several clusters of altered proteins were identified, including cellular structure and stress-related chaperone proteins, proteins involved in reactive oxygen species metabolism, translation factors, proteins that mediate cell proliferation or differentiation, and proteins participating in the internal degradation systems [123]. Finally, the global changes in protein expression levels after oxaliplatin treatment in three colon cancer cell lines (HT29, SW620, and LoVo) have been investigated recently for the first time by Yao and colleagues. Through electrophoresis coupled to MALDI-TOF mass spectrometry, 21 differentially expressed proteins common to the three cell lines were identified, including proteins associated with apoptosis, signal transduction, transcription and translation, cell structural organization, and metabolic processes [124].

On the whole, the investigations of basal proteomic profiles as well as of treatment-induced proteomic modifications demonstrate how global proteomic profiling approaches can be used to identify putative candidates that can be further evaluated individually. Important clues on proteins potentially involved in modulating tumor cell response to metal-based drugs can be identified, even though further investigations are needed to understand their mechanistic role as well as the potential use as therapeutic targets and/or markers of treatment response.

6.5
Concluding Remarks

A survey of the recent literature illustrates that metal-based anticancer agents constitute a heterogeneous class of compounds from both chemical and mechanistic points of view. The heterogeneity of mechanisms of action, in particular, can determine some perplexity in choosing the *in vitro* primary screening system; this chapter aims to contribute to the understanding of currently available assays, and therefore to a rational choice of the most appropriate experimental systems.

6.5.1
Cytotoxic Metal-Based Anticancer Agents

Metal-based anticancer agents can damage DNA and associated functions as well as other biochemical processes involved in proliferation and survival/death homeostasis of tumor cells. In such cases, a cell growth inhibition assay in a panel of tumor cell lines allows the identification (i) of the most active compound in a series of structurally similar compounds and/or (ii) of the compound(s) whose activity profile is different from that of reference compounds. However, a cell growth inhibition assay does not give information about the target(s); in contrast, a knowledge of the treatment-induced cell death mode would provide more information on the mechanism of action, thus contributing more significantly to the development of selected compounds. Importantly, cell death is a complex phenomenon; specific experimental methods, discussed in Section 6.2, must be used with tumor cells, which are themselves a research tool according to their genetic and genomic features. The importance of a detailed knowledge of tumor cell lines is also emphasized in specific research projects, such as the Cancer Cell Line Project (http://www.sanger.ac.uk/genetics/CGP/CellLines), focused on the characterization of currently used cancer cell lines.

6.5.2
Non-Cytotoxic Metal-Based Anticancer Agents

In some cases, tumor-inhibiting metal compounds have been reported whose pharmacological action does not appear to depend upon killing of tumor cells; rather, other malignant phenotype properties appear inhibited. For example, the antitumor activity of the ruthenium complex NAMI-A is associated with its anti-metastatic [125] and antiangiogenic [126] effects. In this case, it is difficult to think that a tumor cell growth inhibition or a tumor cell death assay could be useful as primary screen of compounds of the same mechanistic class, and more appropriate systems should be selected. In this regard, invasion, metastasis, or angiogenesis are very complex, multistep and multicomponent (the tumor microenvironment) processes, and no single *in vitro* assay exists that can accurately model them. However, different steps of the above processes can be assayed using standard cell-based techniques; for example, endothelial cell proliferation and migration, tube formation and so on in the case of angiogenesis [127]. However, also in these cases the screening assays are not informative as far as target(s) and molecular mechanisms are concerned, and the results can hardly orient or re-orient the chemical design and subsequent developments.

6.5.3
Tumor-Targeted Metal-Based Anticancer Agents

Tumor targeting of metal-based anticancer agents is also an active field of research, and poses specific screening problems, according to the targeting modality. In particular, when microenvironment features (e.g., hypoxia, pH) are chosen for the activation of metal compounds in tumor microenvironment-specific conditions, the available experimental systems are poorly suited for screening purposes. Even though *in vitro* three-dimensional tumor microenvironment models are rapidly evolving research tools [128], the use of animal models is at present almost unavoidable.

6.5.4
Towards a Global Mechanistic-Based Screening Approach

It has been demonstrated that metal-based compounds can target specific enzymes [29, 30]; in such cases, they can be considered and developed as targeted agents, using biochemical systems as primary screen [129]. For the most part of research programs on metal-based anticancer compounds, however, cell-based assays are used as primary screening systems, and the choice of both tumor cells and cell-based assay depends upon the hypotheses on the mechanism of action of the compounds under investigation. However, even when an appropriate cell panel is used, and a well-defined endpoint is measured, a cell-based assay is characterized by high hit rates, because a complex biological process such as proliferation, cell death, invasion, or angiogenesis has many potential targets that, in addition, can be shared by interconnected biological pathways. On the other hand, metal-based compounds have heterogeneous mechanisms of action, and the same compound can interact with different targets; therefore, it seems reasonable to suppose that a class of compounds sharing a common overall structure may have multiple cellular targets, and that the effect on the endpoint measured in the primary screen depends upon both chemical features of the compounds and cellular context. For such reasons, the unique criterion of effectiveness towards any phenotypic feature investigated in the primary cell-based screen may not be sufficient to address effectively the subsequent development, which includes also the choice of an appropriate *in vivo* model. To this end, mechanistic information on the same cell-based assay would be more useful, representing also a proof-of-concept of the working hypothesis. In this regard, genomic and proteomic technologies, such as DNA microarray and protein mass spectrometry, allow a global measurement of mRNA levels and of thousands of proteins, respectively, thus providing a detailed assessment of the biological state of a cell. These techniques do not allow a direct identification of molecular targets but, importantly, they allow the identification of cellular pathways affected by the treatment, thus providing a preliminary evaluation of the putative mode of action and, at the same time, a global view of the cellular response to treatment. Currently, the cost of DNA microarray for global profiling is affordable, in particular when the number of agents under investigation is relatively low, as usually happens for metal-based compounds. As to proteomic approach by mass spectrometry, the current limitations are represented by the instrumentation high cost and protein detection limits in complex mixtures. However, the analysis of cellular phenotype through multiparameter profiling techniques is increasingly important in the drug discovery general context [130], and its association with target identification approaches [131] is currently being investigated to improve selection of lead compounds and optimization in the anticancer drug development.

As far as metal-based anticancer compounds are concerned, their importance in cancer chemotherapy is undisputable, considering the clinical relevance of platinum drugs, the potential success of metal compounds at present in clinical trials, the mechanistic heterogeneity of metal compounds, and the enormous knowledge patrimony on metal-based inorganic and organometallic compounds acquired in the last decades. The most recent findings indicate that multiparameter profiling allows the identification of regulatory pathways affected by the treatment and helps the understanding of the mechanism of action, thus representing a novel and important methodological breakthrough. We believe that an earlier integration of transcriptional profiling and proteomic analyses can markedly improve screening efficacy of metal-based compounds, facilitating in particular their subsequent preclinical development.

Acknowledgments

This work was supported by grants from Regione Puglia (Progetto Strategico Biotecnoter) and the University of Bari (Ricerche Finanziate dall'Università).

References

1 Rosenberg, B., Van Camp, L., Trosko, J.E., and Mansour, V.H. (1969) *Nature*, **222**, 385–386.

2 Connors, T.A., Cleare, M.J., and Harrap, K.R. (1979) *Cancer Treat. Rep.*, **63**, 1499–502.

3 Macquet, J.P. and Butour, J.L. (1983) *J. Natl. Cancer Inst.*, **70**, 899–905.

4 Siddik, Z.H. (2003) *Oncogene*, **22**, 7265–7279.

5 Einhorn, L.H. (2002) *Proc. Natl. Acad. Sci. USA*, **99**, 4592–4595.

6 Muggia, F.M. and Fojo, T.J. (2004) *J. Chemother.*, **16**, 77–82.

7 Kelland, L.R. (2007) *Nat. Rev. Cancer*, **7**, 573–584.

8 Montaña, A.M. and Batalla, C. (2009) *Curr. Med. Chem.*, **16**, 2235–2260.

9 Levina, A., Mitra, A., and Lay, P.A. (2009) *Metallomics*, **1**, 458–470.

10 Keppler, B.K. (2004) *Curr. Top. Med. Chem.*, **4**, 1575–1583.

11 Hambley, T.H. (2007) *Dalton Trans.*, 4929–4937.

12 Jakupec, M.A., Galanski, M., Arion, V. B., Hartinger, C.G., and Keppler, B.K. (2008) *Dalton Trans.*, 183–194.

13 Bruijnincx, P.C. and Sadler, P.J. (2008) *Curr. Opin. Chem. Biol.*, **12**, 197–206.

14 Gianferrara, T., Bratsos, I., and Alessio, E. (2009) *Dalton Trans.*, 7588–7598.

15 Chen, D., Milacic, V., Frezza, M., and Dou, Q.P. (2009) *Curr. Pharm. Des.*, **15**, 777–791.

16 Rice, J.R., Gerberich, J.L., Nowotnik, D. P., and Howell, S.B. (2006) *Clin. Cancer Res.*, **12**, 2248–2254.

17 Feazell, R.P., Nakayama-Ratchford, N., Dai, H., and Lippard, S.J. (2007) *J. Am. Chem. Soc.*, **129**, 8438–8439.

18 Dhar, S., Gu, F.X., Langer, R., Farokhzad, O.C., and Lippard, S.J. (2008) *Proc. Natl. Acad. Sci. USA*, **105**, 17356–17361.

19 Dhar, S., Liu, Z., Thomale, J., Dai, H., and Lippard, S.J. (2008) *J. Am. Chem. Soc.*, **130**, 11467–11476.

20 Rieter, W.J., Pott, K.M., Taylor, K.M.L., and Lin, W. (2008) *J. Am. Chem. Soc.*, **130**, 11584–11585.

21 MacDiarmid, J.A., Madrid-Weiss, J., Amaro-Mugridge, N.B., Phillips, L., and Brahmbhatt, H. (2007) *Cell Cycle*, **6**, 2099–2105.

22 Ang, W.H., Khalaila, I., Allardyce, C.S., Juillerat-Jeanneret, L., and Dyson, P.J. (2005) *J. Am. Chem. Soc.*, **127**, 1382–1383.

23 Coluccia, M. and Natile, G. (2007) *Anti-Cancer Agents Med. Chem.*, **7**, 111–123.

24 Farrell, N.P. (2004) *Semin. Oncol.*, **31**, 1–9.

25 Sasanelli, R., Boccarelli, A., Giordano, D., Laforgia, M., Arnesano, F., Natile, G., Cardellicchio, C., Capozzi, M.A., and Coluccia, M. (2007) *J. Med. Chem.*, **50**, 3434–3441.

26 Arnesano, F., Boccarelli, A., Cornacchia, D., Nushi, F., Sasanelli, R., Coluccia, M., and Natile, G. (2009) *J. Med. Chem.*, **52**, 7847–7855.

27 Dyson, P.J. and Sava, G. (2006) *Dalton Trans.*, 1929–1933.

28 Hartinger, C.G., Zorbas-Seifried, S., Jakupec, M.A., Kynast, B., Zorbas, H., and Keppler, B.K. (2006) *J. Inorg. Biochem.*, **100**, 891–904.

29 Smalley, K.S., Contractor, R., Haass, N. K., Kulp, A.N., Atilla-Gokcumen, G.E., Williams, D.S., Bregman, H., Flaherty, K.T., Soengas, M.S., Meggers, E., and Herlyn, M. (2007) *Cancer Res.*, **67**, 209–217.

30 Casini, A., Gabbiani, C., Sorrentino, F., Rigobello, M.P., Bindoli, A., Geldbach, T.J., Marrone, A., Re, N., Hartinger, C. G., Dyson, P.J., and Messori, L. (2008) *J. Med. Chem.*, **51**, 6773–6781.

31 Meng, X., Leyva, M.L., Jenny, M., Gross, I., Benosman, S., Fricker, B., Harlepp, S., Hébraud, P., Boos, A., Wlosik, P., Bischoff, P., Sirlin, C., Pfeffer, M., Loeffler, J.P., and Gaiddon, C. (2009) *Cancer Res.*, **69**, 5458–5466.

32 Straus, D.J. (2003) *Semin. Oncol.*, **30**, 25–33.

33 Chitambar, C.R. (2004) *Curr. Opin. Oncol.*, **16**, 547–552.

34 Narasimhan, J., Antholine, W.E., and Chitambar, C.R. (1992) *Biochem. Pharmacol.*, **44**, 2403–2408.

35 Collery, P., Keppler, B., Madoulet, C., and Desoize, B. (2002) *Crit. Rev. Oncol. Hematol.*, **42**, 283–296.

36 Chen, D., Frezza, M., Shakya, R., Cui, Q.C., Milacic, V., Verani, C.N., and Dou, Q.P. (2007) *Cancer Res.*, **67**, 9258–9265.

37 Rigobello, M.P., Scutari, G., Folda, A., and Bindoli, A. (2004) *Biochem. Pharmacol.*, **67**, 689–696.

38 Saggioro, D., Rigobello, M.P., Paloschi, L., Folda, A., Moggach, S.A., Parsons, S., Ronconi, L., Fregona, D., and Bindoli, A. (2007) *Chem Biol.*, **14**, 1128–1139.

39 Milacic, V., Chen, D., Ronconi, L., Landis-Piwowar, K.R., Fregona, D., and Dou, Q.P. (2006) *Cancer Res.*, **66**, 10478–10486.

40 Alama, A., Tasso, B., Novelli, F., and Sparatore, F. (2009) *Drug Discov.Today*, **14**, 500–508.

41 Shoemaker, R.H. (2006) *Nat. Rev. Cancer*, **6**, 813–823.

42 Nakatsu, N., Yoshida, Y., Yamazaki, K., Nakamura, T., Dan, S., Fukui, Y., and Yamori, T. (2005) *Mol. Cancer Ther.*, **4**, 399–412.

43 Paull, K.D., Shoemaker, R.H., Hodes, L., Monks, A., Scudeiro, D.A., Rubinstein,

L., Plowman, J., and Boyd, M.R. (1989) *J. Natl. Cancer Inst.*, **81**, 1088–1092.

44 Fojo, T., Farrell, N., Ortuzar, W., Tanimura, H., Weinstein, J., and Myers, T.G. (2005) *Crit. Rev. Oncol./ Hematol.*, **53**, 25–34.

45 Hills, C.A., Kelland, L.R., Abel, G., Siracky, J., Wilson, A.P., and Harrap, K.R. (1989) *Br. J. Cancer*, **59**, 527–534.

46 Rixe, O., Ortuzar, W., Alvarez, M., Parker, R., Reed, E., Paull, K., and Fojo, T. (1996) *Biochem. Pharmacol.*, **52**, 1855–1865.

47 Mathé, G., Kidani, Y., Noji, M., Maral, R., Bourut, C., and Chenu, E. (1985) *Cancer Lett.*, **27**, 135–143.

48 Farrell, N., Kelland., L.R., Roberts, J.D., and Van Beusichem, M. (1992) *Cancer Res.*, **52**, 5065–5072.

49 Natile, G. and Coluccia, M. (2004) *Met. Ions Biol. Syst.*, **42**, 209–250.

50 Skehan, P., Storeng, R., Scudiero, D., Monks, A., McMahon, J., Vistica, D., Warren, J.T., Bokesch, H., Kenney, S., and Boyd, M.R. (1990) *J. Natl. Cancer Inst.*, **82**, 1107–1112.

51 Vichai, V. and Kirtikara, K. (2006) *Nat. Protoc.*, **1**, 1112–1116.

52 Alley, M.C., Scudiero, D.A., Monks, A., Hursey, M.L., Czerwinski, M.J., Fine, D.L., Abbott, B.J., Mayo, J.G., Shoemaker, R.H., and Boyd, M.R. (1988) *Cancer Res.*, **48**, 589–601.

53 Scudiero, D.A., Shoemaker, R.H., Paull, K.D., Monks, A., Tierney, S., Nofziger, T.H., Currens, M.J., Seniff, D., and Boyd, M.R. (1988) *Cancer Res.*, **48**, 4827–4833.

54 Boyer, M.J. and Tannock, I.F. (1998) in *The Basic Science of Oncology*, 3rd edn (eds I.F. Tannock and R.F. Hill), McGraw-Hill, New York, pp. 350–369.

55 Kroemer, G., Galluzzi, L., Vandenabeele, P., Abrams, J., Alnemri, E.S., Baehrecke, E.H. *et al.* (2009) *Cell Death Differ.*, **16**, 3–11.

56 Degterev, A. and Yuan, J. (2008) *Nat. Rev. Mol. Cell Biol.*, **9**, 378–390.

57 Cotter, T.G. (2009) *Nat. Rev. Cancer*, **9**, 501–507.

58 Festjens, N., Vanden Berghe, T., and Vandenabeele, P. (2006) *Biochim. Biophys. Acta*, **1757**, 1371–1387.

59 Hitomi, J., Christofferson, D.E., Ng, A., Yao, J., Degterev, A., Xavier, R.J., and Yuan J. (2008) *Cell*, **135**, 1311–1323.

60 Kroemer, G. and Levine, B. (2008) *Nat. Rev. Mol. Cell Biol.*, **9**, 1004–1010.

61 Yorimitsu, T. and Klionsky, D.J. (2005) *Cell Death Differ.*, **12**, 1542–1552.

62 Scherz-Shouval, R., Shvets, E., Fass, E., Shorer, H., Gil, L., and Elazar, Z. (2007) *EMBO J.*, **26**, 1749–1760.

63 Katayama, M., Kawaguchi, T., Berger, M.S., and Pieper, R.O. (2007) *Cell Death Differ.*, **14**, 548–558.

64 Amaravadi, R.K. and Thompson, C.B. (2007) *Clin. Cancer Res.*, **13**, 7271–7279.

65 Krysko, D.V. and Vandenabeele, P. (2008) *Cell Death Differ.*, **15**, 29–38.

66 Galluzzi, L., Aaronson, S.A., Abrams, J., Alnemri, E.S., Andrews, D.W., Baehrecke, E.H. *et al.* (2009) *Cell Death Differ.*, **16**, 1093–1107.

67 Klionsky, D.J., Abeliovich, H., Agostinis, P., Agrawal, D. K., Aliev, G., Askew, D.S. *et al.* (2008) *Autophagy*, **4**, 151–175.

68 Rixe, O. and Fojo, T. (2007) *Clin. Cancer Res.*, **13**, 7280–7287.

69 Clarke, P.A., te Poele, R., and Workman, P. (2004) *Eur. J. Cancer*, **40**, 2560–2591.

70 Huang, R., Wallqvist, A., and Covell, D.A. (2005) *Biochem. Pharmacol.*, **69**, 1009–1039.

71 Vekris, A., Meynard, D., Haaz, M.C., Bayssas, M., Bonnet, J., and Robert, J. (2004) *Cancer Res.*, **64**, 356–362.

72 Akervall, J., Guo, X., Qian, C.N., Schoumans, J., Leeser, B., Kort, E., Cole, A., Resau, J., Bradford, C., Carey, T., Wennerberg, J., Anderson, H., Tennvall, J., and Teh, B.T. (2004) *Clin. Cancer Res.*, **10**, 8204–8213.

73 Nakamura, M., Nakatani, K., Uzawa, K., Ono, K., Uesugi, H., Ogawara, K., Shiiba, M., Bukawa, H., Yokoe, H., Wada, T., Fujita, S., and Tanzawa, H. (2005) *Oncol. Rep.*, **14**, 1281–1286.

74 Zhang, P., Zhang, Z., Zhou, X., Qiu, W., Chen, F., and Chen, W. (2006) *BMC Cancer*, **6**, 224.

75 Negoro, K., Yamano, Y., Fushimi, K., Saito, K., Nakatani, K., Shiiba, M., Yokoe, H., Bukawa, H., Uzawa, K.,

Wada, T., Tanzawa, H., and Fujita, S. (2007) *Int. J. Oncol.*, **30**, 1325–1332.

76 Ansell, A., Jerhammar, F., Ceder, R., Grafström, R., Grénman., R., and Roberg, K. (2009) *Oral Oncol.*, **45**, 866–871.

77 Gosepath, E.M., Eckstein, N., Hamacher, A., Servan, K., von Jonquieres, G., Lage, H., Györffy, B., Royer, H.D., and Kassack, M.U. (2008) *Int. J. Cancer*, **123**, 2013–2019.

78 Yamano, Y., Uzawa, K., Saito, K., Nakashima, D., Kasamatsu, A., Koike, H., Kouzu, Y., Shinozuka, K., Nakatani, K., Negoro, K., Fujita, S., and Tanzawa, H. (2010) *Int. J. Cancer*, **126**, 437–449.

79 Takashima, N., Ishiguro, H., Kuwabara, Y., Kimura, M., Mitui, A., Mori, Y., Mori, R., Tomoda, K., Hamaguchi, M., Ogawa, R., Katada, T., Harada, K., and Fujii, Y. (2008) *Dis. Esophagus*, **21**, 230–235.

80 Kang, H.C., Kim, I.J., Park, J.H., Shin, Y., Ku, J.L., Jung, M.S., Yoo, B.C., Kim, H.K., and Park, J.G. (2004) *Clin. Cancer Res.*, **10**, 272–284.

81 Noel, E.E., Perry, J., Chaplin, T., Mao, X., Cazier, J.B., Joel, S.P., Oliver, R.T., Young, B.D., and Lu, Y.J. (2008) *Genes Chromosomes Cancer*, **47**, 604–613.

82 Roberts, D., Schick, J., Conway, S., Biade, S., Laub, P.B., Stevenson, J.P., Hamilton, T.C., O'Dwyer, P.J., and Johnson, S.W. (2005) *Br. J. Cancer*, **92**, 1149–1158.

83 Cheng, T.C., Manorek, G., Samimi, G., Lin, X., Berry, C.C., and Howell, S.B. (2006) *Cancer Chemother. Pharmacol.*, **58**, 384–395.

84 Saga, Y., Ohwada, M., Suzuki, M., Konno, R., Kigawa, J., Ueno, S., and Mano, H. (2008) *Oncol. Rep.*, **20**, 1299–1303.

85 Oishi, T., Itamochi, H., Kigawa, J., Kanamori, Y., Shimada, M., Takahashi, M., Shimogai, R., Kawaguchi, W., Sato, S., and Terakawa, N. (2007) *Int. J. Gynecol. Cancer*, **17**, 1040–1046.

86 Arango, D., Wilson, A.J., Shi, Q., Corner, G.A., Arañes, M.J., Nicholas, C., Lesser, M., Mariadason, J.M., and

Augenlicht, L.H. (2004) *Br. J. Cancer*, **91**, 1931–1946.

87 Meynard, D., Le Morvan, V., Bonnet, J., and Robert, J. (2007) *Oncol. Rep.*, **17**, 1213–1221.

88 Yang, M., Kroft, S.H., and Chitambar, C.R. (2007) *Mol. Cancer Ther.*, **6**, 633–643.

89 Le Fevre, A.C., Boitier, E., Marchandeau, J.P., Sarasin, A., and Thybaud, V. (2007) *Mutat. Res.*, **619**, 16–29.

90 Duale, N., Lindeman, B., Komada, M., Olsen, A.K., Andreassen, A., Soderlund, E.J., and Brunborg, G. (2007) *Mol. Cancer*, **6**, 53.

91 Voland, C., Bord, A., Péleraux, A., Pénarier, G., Carrière, D., Galiègue, S., Cvitkovic, E., Jbilo, O., and Casellas, P. (2006) *Mol. Cancer Ther.*, **5**, 2149–2157.

92 Whiteside, M.A., Chen, D.T., Desmond, R.A., Abdulkadir, S.A., and Johanning, G.L. (2004) *Oncogene*, **23**, 744–752.

93 Brun, Y.F., Varma, R., Hector, S.M., Pendyala, L., Tummala, R., and Greco, W.R. (2008) *Cancer Genomics Proteomics*, **5**, 43–53.

94 L'Espérance, S., Bachvarova, M., Tetu, B., Mes-Masson, A.M., and Bachvarov, D. (2008) *BMC Genomics*, **9**, 99.

95 Gatti, L., Beretta, G.L., Carenini, N., Corna, E., Zunino, F., and Perego, P. (2004) *Cell Mol. Life Sci.*, **61**, 973–981.

96 Boccarelli, A., Giordano, D., Natile, G., and Coluccia, M. (2006) *Biochem. Pharmacol.*, **72**, 280–292.

97 Galea, A.M. and Murray, V. (2008) *Cancer Inform.*, **6**, 315–355.

98 Pang, S.K., Yu, C.W., Guan, H., Au-Yeung, S.C., and Ho, Y.P. (2008) *Oncol. Rep.*, **20**, 1269–1276.

99 Hall, M.D. and Hambley, T.W. (2002) *Coord. Chem. Rev.*, **232**, 49–67.

100 Olszewski, U., Ach, F., Ulsperger, E., Baumgartner, G., Zeillinger, R., Bednarski, P., and Hamilton, G. (2009) *Met. Based Drugs*, 2009:348916.

101 Hamberger, J., Liebeke, M., Kaiser, M., Bracht, K., Olszewski, U., Zeillinger, R., Hamilton, G., Braun, D., and

Bednarski, P.J. (2009) *Anti-Cancer Drugs*, **20**, 559–572.

102 Sutherland, R. (1988) *Science*, **240**, 177–184.

103 Graham, C.H., Kobayashi, H., Stankiewicz, K.S., Man, S., Kapitain, S. J., and Kerbel, R.S. (1994) *J. Natl. Cancer Inst.*, **86**, 975–982.

104 Reedijk, J. (2008) *Macromol. Symp.*, **270**, 193–201.

105 Kraker, A., Schmidt, J., Krezoski, S., and Petering, D.H. (1985) *Biochem. Biophys. Res. Commun.*, **130**, 786–792.

106 Alderlen, R.A., Hall, M.D., and Hambley, T.W. (2006) *J. Chem. Educ.*, **83**, 728–734.

107 Reedijk, J. (1999) *Chem. Rev.*, **99**, 2499–2510.

108 Boulikas, T. and Vougiouka, M. (2003) *Oncol. Rep.*, **10**, 1663–1682.

109 Knipp, M. (2009) *Curr. Med. Chem.*, **16**, 522–537.

110 Rebillard, A., Lagadic-Gossmann, D., and Dimanche-Boitrel, M.T. (2008) *Curr. Med. Chem.*, **15**, 2656–2663.

111 Sheikh-Hamad, D. (2008) *Am. J. Physiol. Renal. Physiol.*, **295**, 42–43.

112 Bose, R.N., Maurmann, L., Mishur, R. J., Yasui, L., Gupta, S., Grayburn, W.S., Hofstetter, H., and Salley, T. (2008) *Proc. Natl. Acad. Sci. USA*, **105**, 18314–18319.

113 Sun, X., Tsang C., and Sun, H. (2009) *Metallomics*, **1**, 25–31.

114 Wang, Y. and Chiu, J.F. (2008) *Metal-Based Drugs*, 2008:716329.

115 Stewart, J.J., White, J.T., Yan, X., Collins, S., Drescher, C.W., Urban, N. D., Hood, L., and Lin, B. (2006) *Mol. Cell Proteomics*, **5**, 433–443.

116 Le Moguen, K., Lincet, H., Deslandes, E., Hubert-Roux, M., Lange, C., Poulain, L., Gauduchon, P., and Baudin, B. (2006) *Proteomics*, **6**, 5183–5192.

117 Yan, X.D., Pan, L.Y., Yuan, Y., Lang, J. H., and Mao, N. (2007) *Proteome Res.*, **6**, 772–780.

118 Aggarwal, S., He, T., Fitzhugh, W., Rosenthal, K., Field, B., Heidbrink, J., Mesmer, D., Ruben, S.M., and Moore, P.A. (2009) *Gynecol. Oncol.*, **115**, 430–437.

119 Smith, L., Welham, K.J., Watson, M.B., Drew, P.J., Lind, M.J., and Cawkwell, L. (2007) *Oncol. Res.*, **16**, 497–506.

120 Martinez-Balibrea, E., Plasencia, C., Ginés, A., Martinez-Cardús, A., Musulén, E., Aguilera, R., Manzano, J. L., Neamati, N., and Abad, A. (2009) *Mol. Cancer Ther.*, **8**, 771–778.

121 Le Moguen, K., Lincet, H., Marcelo, P., Lemoisson, E., Heutte, N., Duval, M., Poulain, L., Vinh, J., Gauduchon, P., and Baudin, B. (2007) *Proteomics*, **7**, 4090–4101.

122 Yim, E.K., Lee K.H., Kim, C.J., and Park, J.S. (2006) *Int. J. Gynecol. Cancer*, **16**, 690–697.

123 Wang, Y., He, Q.Y., Che, C.M., and Chiu, J.F. (2006) *Proteomics*, **6**, 131–142.

124 Yao, Y., Jia, X.Y., Tian, H.Y., Jiang, Y. X., Xu, G.J., Qian, Q.J., and Zhao, F.K. (2009) *Biochim. Biophys. Acta*, **1794**, 1433–1440.

125 Bergamo, A. and Sava, G. (2007) *Dalton Trans.*, 1267–1272.

126 Vacca, A., Bruno, M., Boccarelli, A., Coluccia, M., Ribatti, D., Bergamo, A., Garbisa, S., Sartor, L., and Sava, G. (2002) *Br. J. Cancer*, **86**, 993–998.

127 Taraboletti, G. and Giavazzi, R. (2004) *Eur. J. Cancer*, **40**, 881–889.

128 Smalley, K.S.M., Lioni, M., Noma, K., Haass, N.K., and Herlyn, M. (2008) *Expert Opin. Drug Discov.*, **3**, 1–10.

129 Hait, W.N. (2009) *Cancer Res.*, **69**, 1263–1267.

130 Feng, Y., Mitchison, T.J., Bender, A., Young, D.W., and Tallarico, J.A. (2009) *Nat. Rev. Drug Discovery*, **8**, 567–578.

131 Terstappen, G.C., Schlüpen, C., Raggiaschi, R., and Gaviraghi, G. (2007) *Nat. Rev. Drug Discovery*, **6**, 891–903.

7
Gold-Based Therapeutic Agents: A New Perspective

Susan J. Berners-Price

7.1
Introduction

7.1.1
An Historical Perspective

The longstanding clinical use of gold drugs in rheumatoid arthritis, as well as the current interest in finding new medical applications for gold-based therapeutic agents, follow a trend that dates back into history, perhaps as far back as 2500 BC, which marks the earliest application of gold as a therapeutic agent in China. This fascinating topic has been reviewed both for medicinal uses of gold in western cultures (before 1900) [1] and in ancient Chinese medicine [2].

Modern interest in the medicinal use of gold compounds stems from Robert Koch's discovery in 1890 of the anti-tubercular activity of gold cyanide *in vitro*. While this compound was too toxic for clinical use, the "gold decade" (1925–1935) saw extensive use of intravenously administered Au(I) thiolate salts for the treatment of tuberculosis, despite a lack of experimental evidence for anti-tubercular benefits of the gold treatment [3]. Observations by Landé [4] that gold therapy brought about significant reductions in joint pain in a group of non-tubercular patients led the French physician Jacques Forestier to initiate a study of gold compounds for treatment of rheumatoid arthritis (RA) [5]. The results of a six-year trial were published in 1935 [6]. Over the years conflicting reports emerged concerning the efficacy of gold therapy in rheumatoid arthritis, until the Empire Rheumatism Council published the results of a very large well-controlled, double-bind trial in 1960, which concluded that gold drugs do have a beneficial effect [7]. Various Au(I) thiolate drugs (Figure 7.1), which were first introduced in the 1920s, are still in clinical use today (see, for example, References [11, 12]), and are included in the class of disease-modifying antirheumatic drugs (DMARDs), which have the principal effect of retarding the progression of the disease. The topic of

Bioinorganic Medicinal Chemistry. Edited by Enzo Alessio
Copyright © 2011 WILEY-VCH Verlag GmbH & Co. KGaA, Weinheim
ISBN: 978-3-527-32631-0

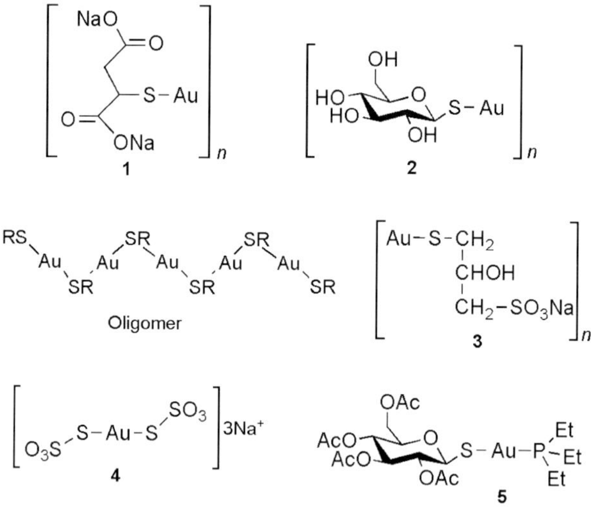

Figure 7.1 Structures of gold(I) drugs used for the treatment of rheumatoid arthritis: (**1**) sodium aurothiomalate (Myocrisin), (**2**) aurothioglucose (Solganol), (**3**) sodium aurothiopropanolsulfonate (Allochrysine), (**4**) sodium aurothiosulfate (Sanochrysin), and (**5**) tetraacetyl-β-D-thioglucose gold(I) triethylphosphine (auranofin). The 1 : 1 Au-S drugs are polymers. Crystal structures have been published for **1** [8], **4** [9], and **5** [10].

chrysotherapy (the treatment of RA by gold drugs) has been reviewed more recently [13].

During the 1970s and early 1980s two major advances occurred. The first was the development by Sutton and coworkers of an orally active Au(I) phosphine compound (auranofin, Figure 7.1) for the treatment of rheumatoid arthritis, which was approved for clinical use in 1985 [14, 15]. The nature of the phosphine ligand influenced the extent of oral absorption and the maximum absorption and activity occurred for Et$_3$P. Initially, Et$_3$PAuCl was selected for clinical trial but was abandoned in favor of the tetracetylated thioglucose derivative, which was better tolerated. Early indications from clinical trials [16, 17] were that auranofin offered significant advantages over the traditional injectable Au(I) thiolates, having comparable efficacy and only mild side-effects. However, auranofin later proved to be less effective than the injectable gold drugs [18] and oral gold now seems to be rarely used clinically [11]. Nevertheless, there is much current interest in the medicinal chemistry of auranofin, for reasons that will be discussed in detail below.

The second advance was an understanding of the biological chemistry of gold drugs from the application of sophisticated techniques that emerged at that time (e.g., NMR, EXAFS, ^{197}Au Mössbauer spectroscopy). The work of Sadler, Shaw, Elder, and others provided the first investigations of the solid-state and solution structures of Au(I) thiolate drugs, as well as biologically relevant ligand-exchange reactions (particularly with protein thiols), which underpin ongoing investigations of their mode of action. This important work has been discussed in detail in several earlier reviews [19–25].

In the mid-1980s the first reports of the anticancer activity of gold compounds appeared. Auranofin was found to be cytotoxic to tumor cells *in vitro* [26], and this led to the identification of other Au(I) phosphines with a broader spectrum of antitumor activity, in particular the bis-chelated Au(I) diphosphine complex [Au (dppe)$_2$]Cl (where dppe = Ph$_2$P(CH$_2$)$_2$PPh$_2$, Figure 7.2) [27]. Around this time Au (III) complexes were first investigated as potential antitumor agents, with the idea that square planar Au(III) compounds, being d^8 and hence isoelectronic with Pt (II), could mimic the activity of cisplatin. However, only in the 1990s were promising results reported for several classes of Au(III) antitumor compounds [28]. Several reviews published in the late 1990s summarized the state of the art for gold-based therapeutic agents at that time [24, 25, 29, 30], but many significant developments have occurred more recently, including new insight into molecular targets and likely modes of action.

7.1.2
Current Interest in Gold-Based Drugs

Over the past few years there has been a resurgence of interest in the medicinal chemistry of gold compounds, particularly as anticancer agents [31–35]. These developments for gold-based drugs reflect those seen in other areas of medicinal inorganic chemistry, where there is increasing realization that the unique properties of metal ions can be exploited in the *design* of new drugs that have different mechanisms of action to existing drugs and/or a more targeted, cancer-cell specific approach [36–41].

The unique chemistry of gold, particularly the high affinity for protein thiols and selenols, has provided a stimulus for much of the recent research, due to the broad range of novel disease targets that have emerged recently that involve cysteine or

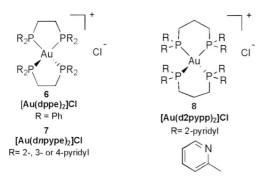

Figure 7.2 Examples of Au(I) diphosphine antitumor compounds.

selenocysteine residues. Cysteine proteases have been implicated in the patho-physiology of several diseases, including inflammatory airway diseases, bone and joint disorders, parasitic diseases and cancer [42] and the cathepsins B, K, and S [42–46] have been the subject of recent attention, along with tyrosine phosphatases [47], which are implicated in several disease states.

A particular focus of interest has been the thioredoxin system [48], which plays a key role in regulating the overall intracellular redox balance. Thioredoxin reductase (TrxR) has been implicated in several chronic diseases such as certain cancers, rheumatoid arthritis, and Sjögren's syndrome [49] and many emerging cancer therapies use TrxR as a target for drug development [50, 51]. Gold(I) complexes are the most effective and selective inhibitors of purified mammalian TrxR found to date [49, 51, 52], with auranofin being particularly potent [52]; the inhibition has been attributed to Au(I) binding to the -Cys-Sec- redox active center (Sec = selenocysteine) [51, 52]. Anti-arthritic gold(I) drugs are known to interact with other selenoenzymes such as glu-tathione peroxidase [53], and notably the activity of glutathione reductase (closely related to TrxR but lacking the Sec residue) is inhibited at 1000-fold higher con-centrations [49]. Recent studies have shown that different classes of cytotoxic gold compounds [both Au(I) and Au(III)] are potent inhibitors of TrxR. A unifying me-chanism has been proposed that involves inhibition of mitochondrial TrxR by these gold compounds that ultimately leads to cell death [48]. Moreover, it has been shown that certain Au(I) phosphine and N-heterocyclic carbene (NHC) complexes are *selec-tively* toxic to cancer cells and not to normal cells and the mechanism may depend on the ability to selectively target mitochondrial TrxR in cancer cells [54, 55].

The potential application of gold-drugs against major tropical diseases has re-ceived recent attention [56] and is an area of growing importance due to the variety of thiol and selenol proteins that have been validated as drug targets. Similarly, targeting selenium metabolism with gold-based drugs offers a new avenue for antimicrobial development against selenium-dependent pathogens [57, 58].

These new developments in gold-based therapeutic agents are the focus of this chapter. Another area of emerging interest, not included here, is the potential application of gold nanoparticles for cellular imaging, diagnostic, or therapeutic purposes. This topic has been the subject of several recent reviews [59–61].

7.2
Biological Chemistry of Gold

There are several excellent reviews where the biological chemistry of gold has been discussed in detail [19, 24, 25] and so only a brief overview is given here. While various different oxidation states are known for gold, studies on gold-based therapeutic agents have been restricted to compounds in the two common oxidation states of $+1$ and $+3$.

7.2.1
Gold(I) Oxidation State

Gold(I) $(5d^{10})$, being a large ion with a low charge, is a "soft" Lewis acid and hence forms it most stable complexes with "soft" ligands such as CN, S-donors (RS^H,

R_2S, and $S_2O_3^{2H}$), P-donors (PR_3), and Se ligands. In the absence of stabilization by "soft" ligands, disproportionation into metallic gold and gold(III) readily occurs in aqueous solution:

$$3Au(I) \rightarrow 2Au(0) + Au(III)$$

Gold(I) has a much higher affinity for thiolate S (cysteine) compared to thioether S (methionine) and a low affinity for N and O ligands. Hence antiarthritic gold(I) drugs contain thiolate and phosphine ligands, the biological chemistry is dominated by ligand exchange reactions with "soft" cysteine and selenocysteine binding sites on proteins, and DNA is not a target for gold(I) antitumor compounds. The highest affinity is for thiols with the lowest pK_a values. Consequently, in blood, most of the circulating Au from antiarthritic drugs is bound to the cysteine-34 of serum albumin ($pK_a \sim 5$) and transcription factors (Jun, Fos, NF-κB), which have cysteine residues flanked by basic lysine and arginine residues, are likely targets [62]. Gold(I) has a particularly high affinity for selenocysteine residues (e.g., in glutathione peroxidase and thioredoxin reductase) because Se is more polarizable (hence "softer") than S and the pK_a of selenocysteine (~ 5.2 [63]) is much lower than that of cysteine (8.5 [64]).

Linear, two-coordination is most common for Au(I) but higher (three and four) coordination numbers are known. Relativistic effects increase the 6s–6p energy gap of gold, which enhances the stability of the two-coordinate geometry compared to the lighter elements Ag(I) and Cu(I) [25]. Notably, tetrahedral four-coordination can be imposed by the use of bidentate phosphine ligands, and bis-chelated Au(I) diphosphine complexes such as $[Au(dppe)_2]^+$ have a high thermodynamic stability [65]. The formation of chelate rings can contribute to the driving force of unusual reactions [66]. For example, in the presence of thiols (SR) and blood plasma, bridged digold complexes RSAu(dppe)AuSR [linear two coordinate Au(I)] convert into the tetrahedral complex $[Au(dppe)_2]^+$ via the reaction [67]:

$$2[(AuSR)_2(dppe)] + 2RS^- \rightleftharpoons [Au(dppe)_2]^+ + 3[Au(SR)_2]^-$$

For linear two-coordinate Au(I) compounds thiolate ligand exchange reactions are facile [64], occurring via an associative mechanism and a three-coordinate transition state. Hence, following administration of gold antiarthritic drugs, Au is readily transported by serum albumin and a thiol shuttle mechanism [68] is probably responsible for the transport of Et_3PAu^+ across cell membranes to key thiol/selenol protein target sites. In contrast, bis-chelated Au(I) diphosphine complexes such as $[Au(dppe)_2]^+$ are stable in the presence of thiols and in blood plasma [27], because ligand exchange reactions must occur by a ring-opening mechanism.

Small mononuclear Au(I) complexes show a pronounced tendency to self-associate and short sub-van der Waals Au· · ·Au distances of about 3.05 Å indicate the presence of an attractive (aurophilic) interaction [69] with a bond energy comparable to that of standard hydrogen bonds. Recent theoretical/computational studies indicate that "aurophilicity" results primarily from dispersion forces reinforced by relativistic effects [70, 71]. Compounds displaying these short Au· · ·Au distances are often luminescent [72] and this native luminescence has been exploited recently to determine the intracellular distribution of a dinuclear Au(I)-NHC complex using fluorescence microscopy [73].

7.2.2
Gold(III) Oxidation State

Gold(III) is a d^8 metal ion, isoelectronic with Pt(II), and its complexes are generally four-coordinate and square planar. Ligand substitution reactions are likely to occur via five-coordinate intermediates and are faster for Au(III) than Pt(II) [74], but slower than for Au(I) [19]. However, while various ligand types form stable complexes with this oxidation state (Section 7.4.2), the biological chemistry is dominated by the strong oxidizing properties. Thus, *in vivo* most Au(III) compounds will be reduced to Au(I) or Au(0), driven by naturally occurring reductants such as thiols (cysteine), thioethers (methionine), and protein disulfides [24, 25, 75]. On the other hand, while the biological environment is strongly reducing, Au(I) compounds can be converted into Au(III) by strong oxidants, such as hypochlorite, which is produced in inflammatory situations during the oxidative burst. The immunological toxic side effects of antiarthritic gold(I) drugs are attributed to oxidation to Au(III) and subsequent interaction with proteins [75] (Section 7.3.1).

7.3
Gold Antiarthritic Drugs

7.3.1
Structural Chemistry and Biotransformation Reactions

The chemistry and pharmacology of gold(I) antiarthritic drugs were comprehensively reviewed in 1999 by Shaw [24, 25] and more recently by Messori and Marcon [76]. A few pertinent points are highlighted here.

Whereas the orally active complex auranofin is a crystalline monomeric complex [10], the injectable Au(I) thiolate complexes, such as aurothiomalate, are polymers with thiolate S bridging linear Au(I) ions. The crystal structure of aurothiomalate (Myocrisin) was determined only relatively recently [8] and shows linear S-Au-S units arranged into double-helical chains, in good agreement with the chain and cyclic structures indicated in early EXAFS and WAXS studies [77, 78].

After administration, these linear gold(I) complexes rapidly undergo ligand exchange reactions so that the administered drugs are unlikely to be the pharmacologically active species. For auranofin the phosphine ligand confers membrane solubility and affects the pharmacological profile, including uptake into cells. While release of the phosphine does not occur readily in most model

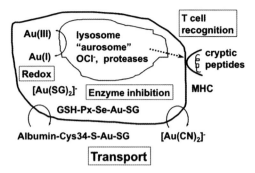

Figure 7.3 Biotransformations of gold antiarthritic drugs, from the article by Sadler and Guo [81]. Gold accumulates in the lysosomes of cells, forming gold rich deposits known as aurosomes. Oxidation of Au(I) to Au(III) can occur, due to the production of hypochlorite by the lysosomal enzyme myeloperoxidase during the oxidative burst in inflamed sites. The formation of Au(III) in lysosomes could lead to the modification of "self proteins," which are degraded and transported to the cell surface. The presentation of these "cryptic" peptides at the cell surface, bound to the major histocompatibility complex (MHC) protein, could lead to T cell recognition and triggering of the immune response, accounting for the toxic effects of chrysotherapy [24, 87]. Enzyme inhibition includes the Se enzyme glutathione peroxidase (GSH-Px).
Reproduced from Reference [81] with permission.

reactions, studies with ^{195}Au, ^{35}S, and ^{32}P-radiolabeled auranofin in dogs have shown that the ^{35}S and ^{32}P are excreted more rapidly than ^{195}Au [79], and Et$_3$PO has been identified in the urine of auranofin-treated patients [16]. On binding to albumin the acetylthioglucose ligand is substituted first and the phosphine ligand is liberated slowly (with formation of Et$_3$PO) driven by the liberated acetylthioglucose ligand and thiol ligands such as glutathione (GSH) [80]. Once the phosphine is released, the products of auranofin metabolism could be similar to those of Au(I) thiolates. Understanding the mechanism of action of gold antiarthritic drugs is made difficult by the complicated biotransformation reactions that ensue (Figure 7.3 [81]). Albumin can transfer Au(I) into cells (via a thiol shuttle mechanism [68]) and the metabolite [Au(SG)$_2$]$^-$ can be excreted from cells and the Au(I) transferred back to albumin [82]. [Au(CN)$_2$]$^-$ is the major metabolite identified in the urine of patients treated with either injectable Au(I)-thiolate drugs or auranofin [83] and may play a key role in the pharmacology. The neutrophil enzyme myeloperoxidase converts aurothiomalate into [Au(CN)$_2$]$^-$ through the oxidation of thiocyanate [84]. [Au(CN)$_2$]$^-$ readily enters cells and can inhibit the oxidative burst of white blood cells, and thus may alleviate secondary effects of the chronic inflammation in the joints of RA patients. Under the oxidative conditions that exist in inflamed joints, oxidation of Au(I) to Au(III) can occur and some of the immunological side effects (gold-induced dermatitis) observed in chrysotherapy are attributable to the production of Au(III) metabolites [85–87]. Hypochlorite (produced by the enzyme myeloperoxidase during the oxidative burst in inflamed sites) has been shown to oxidize Au(I) in aurothiomalate, auranofin, and [Au(CN)$_2$]$^-$ to Au(III) [24, 25]. The operation of a redox cycle [with Au(III) species reduced back to Au(I) by biologically occurring reductants] has also been proposed [24, 25, 87].

7.3.2
Mode of Action

Rheumatoid arthritis (RA) is a chronic inflammatory disease characterized by the migration of activated phagocytes and leukocytes into synovial tissue, which causes progressive destruction of cartilage bone and joint swelling. Evidence suggests that gold drugs have multiple modes of action in this complex disease [24], and an overriding theme is the interaction with protein cysteine (or seleno-cysteine) residues. More recent studies on the mechanism of action of DMARDs, including gold drugs, have focused on their effects on macrophage signal transduction and the induction of proinflammatory cytokines (see References [88, 89] for reviews). Cytokines are low molecular weight peptides, proteins, or glycoproteins participating in intracellular signaling and are important mediators in many inflammatory diseases. Of particular importance in RA are tumor necrosis factor (TNF-α) and interleukin-1 (IL-1). Gold drugs have been shown to play a role in each of the different phases of the immune reaction. At the initiation stage gold is taken up by macrophages and inhibits antigen processing. Peptide antigens containing cysteine and methionine residues are especially important [87, 89]. Gold accumulates in the lysosomes of synovial cells and macrophages, forming gold laden deposits known as aurosomes. EXAFS measurements have shown that the gold in aurosomes is in the form S–Au(I)–S [22]. At the effector level, gold drugs inhibit degradative enzymes such as collagenase. Many of the degradative enzymes in the lysosome are cysteine dependent and of particular interest are the cathepsins, which are implicated in inflammation and joint destruction. They play a role in antigen processing and presentation and have been implicated in auto-immune disorders [43]. Recent studies have focused on understanding the mechanism of inhibition of cathepsin B by auranofin and in tuning the potency by alteration of the phosphine ligand [43–45]. Cathepsins K and S have been shown to play central roles in the inflammatory and erosive components of RA, and a recent study [46] shows efficient inhibition of both these cathepsins by auranofin and aurothiomalate; a crystal structure of a cathepsin K/aurothiomalate complex shows linear S–Au–S coordination with Au bound to the active site cysteine residue and a thiomalate ligand still coordinated [46].

At the transcription level gold(I) drugs downregulate a range of proin-flammatory genes by inhibiting transcriptional activities of the NF-κB and AP-1 (Jun/Fos) [62] transcription factors. AP-1 controls the expression of genes for collagenase and the cytokine IL-2, and NF-κB controls transcription of other inflammatory mediators, including TNF-α, IL-1, and IL-6. It was suggested that these transcription factors would be attractive targets for gold(I) drugs because they have conserved lysine-cysteine-arginine sequences in which the thiol pK_a of the cysteine residues is lowered by the positive charge of the flanking basic amino acid residues [62]. NF-κB activation is a complex process that can be triggered by many agents. The potential targets for gold drugs (via crucial cysteine/selenocysteine residues) include NF-κB itself [90], IκB kinase (thus preventing dissociation of NF-κB from the inhibitory protein IκB) [91, 92], and TrxR [93]. Gold drugs have been shown also to activate transcription factor Nrf2/small Maf, which leads to the up-regulation of antioxidative stress genes, whose products contribute to the scavenging of reactive oxygen species and exhibit anti-inflammatory effects [94].

Gold drugs also act at the T-cell level [89], and have been shown to inhibit os-teoclast bone resorption [95], recently attributed to the inhibition of the cathepsins [43, 46, 96]. RA patients have elevated levels of copper that can be correlated to the severity of the disease. Gold drugs could interfere with copper homeostasis by binding to Cu(I) responsive transcription factors and other Cu(I) transport proteins [97].

7.4
Gold Complexes as Anticancer Agents

7.4.1
Gold(I) Compounds

Analysis of the literature to date indicates that gold(I) antitumor compounds can be broadly divided into two distinct classes based on coordination chemistry, lipophilic-cationic properties, and propensity to undergo ligand exchange reactions with biological thiols and selenols [31, 39]. The two classes are (i) neutral, linear, two-coordinate complexes, such as auranofin and (ii) lipophilic cationic complexes such as [Au(dppe)$_2$]$^+$ and dinuclear Au(I) NHC complexes. For both classes tumor cell mitochondria are likely targets [98], with apoptosis induced by alteration of the thiol redox balance [31, 39, 48].

7.4.1.1 Auranofin and Related Compounds
Auranofin, in common with a large variety of other linear, two-coordinate Au(I) phosphine complexes, has been shown to inhibit the growth of cultured tumor cells *in vitro* [26, 34, 66, 99–104]. The cytotoxic activity of auranofin against HeLa cancer cells was first reported in 1979 by Lorber and coworkers [105], and subsequent studies showed that auranofin increased the survival times of mice with P388 leukemia [106]. Mirabelli and coworkers carried out an extensive study of the antitumor activity of auranofin against 15 tumor models in mice [26] and found that it was active only in ip (intraperitoneal) P388 leukemia, and required ip administration for activity. A comprehensive structure–activity study [99] of the *in*

vitro cytotoxic activity and *in vivo* antitumor activity of a series of 63 linear Au(I) complexes of type LAuX (where L is generally – but not exclusively – a phosphine ligand) showed that *in vivo* antitumor activity (against P388 leukemia) was optimized for complexes with both phosphine and thiosugar ligands, and variation of the phosphine substituent modulated the antitumor activity. Since that time, cytotoxic activity has been reported for various other linear Au(I) phosphine complexes incorporating S-ligands such as thionucleobases and dithiocarbamates (for comprehensive reviews see References [34, 100]), sulfanylpropenoates [101], and, in recent studies, bioactive vitamin K3 [102], azacoumarin [103], and naphthalimide [104] derivatives.

Extensive early mechanistic studies provided evidence that both auranofin and Et$_3$PAuCl affect mitochondrial function (reviewed in Reference [98]) and these results have been reinterpreted more recently to be consistent with an antitumor mechanism involving the induction of mitochondrial-dependent apoptosis pathways [98]. The recent work of Bindoli and coworkers [48, 107–109] has shed light on the mechanism and it is proposed that apoptosis induction occurs due to alteration of the thiol redox balance in cells, as a result of TrxR inhibition (see below, Section 7.4.3). Early studies on Au(I) compounds suggested that the phosphine ligands were a necessary requirement for antitumor activity [66], but in the last few years antitumor activity has been reported for linear Au(I) complexes with N-heterocyclic carbene ([55] Section 7.4.1.3), cyclodiphosphazene [110], and phosphole [111] ligands in place of phosphines. The role of the ligand is likely to be related to cellular uptake. Notably, polymeric Au(I) thiolates do not readily enter cells and were found to exhibit very low cytotoxicity to B16 melanoma cells *in vitro* and were totally inactive against ip P388 leukemia [99]. Moreover, aurothiomalate – in contrast to auranofin – is poorly effective in inhibiting TrxR and inducing apoptosis in Jurkat T cells [112]. A recent study, however, has shown that both aurothiomalate and aurothioglucose exhibit potent antitumor effects in *in vitro* and *in vivo* pre-clinical models of non-small cell lung cancer [113]. The mechanism has been attributed to selective targeting of Cys-69 within the PB1 domain of protein kinase C$_\iota$ [114].

7.4.1.2 Tetrahedral Gold(I) Diphosphines and Related Compounds

Early studies on auranofin showed that the cytotoxicity to cultured tumor cells *in vitro* was significantly reduced when the culture medium contained serum proteins [26] and the high reactivity toward protein thiols (a characteristic feature of linear two-coordinate Au(I) complexes) limits its antitumor activity *in vivo*. While a range of auranofin analogs have shown promising cytotoxic activity *in vitro* (see above), similar reactions are likely to limit their application as anticancer agents. It was this aim of reducing the high thiol reactivity that led to early investigations of Au(I) complexes with chelated diphosphines, and to the development of [Au(dppe)$_2$]Cl, shown to exhibit significant antitumor activity against a range of tumor models in mice [27]. Structure–activity relationships have been evaluated for a wide range of diphosphine ligands and their metal complexes [66, 115, 116]. For complexes of the type [Au(R$_2$P(CH$_2$)nPR$_2$)$_2$]Cl, highest activity was found where R $=$ R^1 $=$ phenyl and $n = 2,3$ or *cis*-CH $=$ CH. In general, activity was reduced, or lost altogether, when the phenyl substituents on the phosphine were replaced by other substituents. In contrast to auranofin, $[Au(dppe)_2]^+$ retains its structural integrity in the presence of thiols and in human plasma [27, 67]. Moreover, structure–activity relationships for the series of linear digold complexes $ClAu(Ph_2P(CH_2)_nPPh_2)AuCl$ ($n = 1-6$) and $XAu(dppe)AuX$ [X = e.g., Cl, Br, OAc, SMan, SGlu, SGlu(Ac)$_4$, SMan(Ac)$_4$] [115] showed that antitumor activity was related to whether they could undergo ring closure reactions to form bis-chelated $[Au(P-P)_2]^+$ species *in vivo*, by reaction with thiols [66, 67].

The clinical development of $[Au(dppe)_2]^+$ was halted following preclinical toxicological studies in dogs and rabbits that identified severe toxicities to heart, liver, and lung attributable to mitochondrial dysfunction [117–120]. The behavior of $[Au(dppe)_2]^+$, and related compounds, is consistent with that of the class of antitumor agents known as delocalized lipophilic cations (DLCs) [121], which accumulate in the mitochondria of tumor cells driven by the elevated mitochondrial membrane potential that is a characteristic feature of cancer cells [31, 39, 122]. The high lipophilicity of $[Au(dppe)_2]^+$ results in its non-selective concentration into mitochondria, causing general membrane permeabilization. Replacing the phenyl substituents with pyridyl groups (with the N atom in either the 2, 3, or 4 position in the ring), provided a series of compounds of type $[Au(dnpype)_2]Cl$ (Figure 7.2) that are structurally similar to $[Au(dppe)_2]Cl$ and exhibit a hydrophilic–lipophilic character spanning a very large range [123, 124]. Studies carried out in isolated rat hepatocytes and a panel of cisplatin-resistant human ovarian carcinoma cell lines showed a general increase in cytotoxic potency (and decrease in selectivity) with increasing lipophilicity [123, 124]. Evaluation of the *in vivo* antitumor activity in colon 38 tumors in mice showed that whereas the most lipophilic and hydrophilic complexes had no significant tumor growth delay, the 2-pyridyl complex with intermediate lipophilicity showed significant antitumor activity that correlated with highest drug concentrations in plasma and tumor tissue [123]. This compound has been shown to accumulate preferentially in the mitochondrial fractions of cancer cells, driven by the mitochondrial membrane potential [124]. To further fine-tune the hydrophilic–lipophilic balance in the optimal range, the related compound [Au(d2pypp)$_2$]$^+$ (Figure 7.2), with the propyl-bridged 2-pyridyl-phosphine ligand (d2pypp), was designed with the idea of combining the features of the two distinct classes of Au(I) phosphines, that is, retaining the lipophilic cationic properties of the tetrahedral bis-chelated complexes that allow accumulation into mitochondria but enhancing the reactivity toward protein thiols/selenols that underlies the inhibition of TrxR by auranofin [54, 125]. Ligand exchange reactions will be more facile, compared to $[Au(dppe)_2]^+$ and its pyridylphosphine analogs, due to the increased chelate ring size. Recent findings have shown that $[Au(d2pypp)_2]^+$ is selectively toxic to breast cancer cells but not to normal breast cells (Figure 7.4), accumulates in the mitochondria of cells driven by the high membrane potential, and selectively induces apoptosis of breast cancer cells but not of normal breast cells [54]. The same concentrations of $[Au(dppe)_2]^+$ are non-selectively toxic to both cells lines (Figure 7.4) by causing necrosis, not apoptosis. Furthermore, the activities of both thioredoxin (Trx) and thioredoxin reductase (TrxR) are inhibited by $[Au(d2pypp)_2]^+$, being more pronounced in breast cancer cells compared to normal cells. These findings indicate that mitochondria and the thioredoxin system may be the critical targets responsible for the selective toxicity seen in the breast cancer cell line; the results are consistent with the proposed model illustrated in Figure 7.4.

Within this class of lipophilic cationic Au(I) phosphine complexes is the mixed gold phosphine compound $[Au(dppp)(PPh_3)Cl]$, which has recently been shown to exhibit potent cytotoxicity in a range of cancer cell lines *in vitro* [126, 127], with the inhibition of melanoma cell growth shown to involve induction of mitochondria-mediated apoptosis [127]. While the complex is neutral, it decomposes in solution to give several products, including $[Au(dppp)_2]^+$ [128], and so the pharmacologically active species may be positively charged complexes. Also of interest is the hydrophilic four-coordinate complex $[Au(P(CH_2OH)_3)_4]Cl$, which has been shown to be cytotoxic *in vitro* against several human tumor cell lines and a mouse tumor model *in vivo* [129], and to be well tolerated in pharmacokinetic studies in dogs [130]. In addition, interestingly, in a recent study of a series of neutral $[Au(PPh_3)(Hxspa)]$ complexes with sulfanylpropenoate ligands (Hxspa), deprotonation of the carboxylate group produced cationic complexes of type $[Au(PPh_3)(xspa)]^+$, with a modified profile of cytotoxic activity [131].

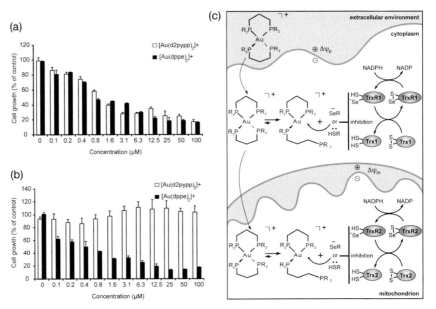

Figure 7.4 (a) Inhibition of cell growth by $[Au(d2pypp)_2]Cl$ (**8**) and $[Au(dppe)_2]Cl$ (**6**) in MDA-468 breast cancer cells (a) and normal breast cells (b); (c) proposed mechanism of uptake and activity of $[Au(d2pypp)_2]^+$ in cells. Trx, thioredoxin, TrxR, thioredoxin reductase. Reprinted from Reference [54] with permission from Elsevier.

7.4.1.3 Gold(I) N-Heterocyclic Carbene Compounds

Many studies have shown that N-heterocyclic carbenes (NHCs) have similar properties to phosphines in the way they interact with metals, including gold. An attractive feature of NHC chemistry is the relative ease with which a series of structurally similar complexes with varying lipophilicity can be prepared from simple imidazolium salt precursors. For a family of linear, cationic Au(I) NHC complexes $[(R_2Im)_2Au]^+$ (Figure 7.5), the lipophilicity is fine-tuned by incorporation of different functional groups and log P values vary across the series within a wide range, from log $P = -1.09$ (R = Me) to 1.73 (R = cyclohexyl) [132]. An initial study showed their ability to induce cyclosporin A-sensitive swelling in isolated rat liver mitochondria correlated with lipophilicity, indicating that these compounds could potentially target mitochondrial cell death pathways [132]. More recently it was shown that the complexes are selectively toxic to two highly tumorigenic breast cancer cell lines and not to normal breast cells, and that the degree of selectivity and potency are optimized by modification of the substituent (Figure 7.5) [55]. Model studies with cysteine (Cys) and selenocysteine (Sec) showed that release of the NHC ligands occurs by two-step ligand exchange reactions and, at physiological pH, the rate constants for the reactions with Sec are 20- to 80-fold higher than those with Cys. Consistent with this result, the lead compound $[(iPr_2Im)_2Au]^+$ was shown to accumulate in mitochondria of cancer cells, to cause cell death through a mitochondrial apoptotic pathway, and (in treated cells) to inhibit the activity of TrxR but not of the closely related and Se-free enzyme glutathione reductase (Figure 7.5).

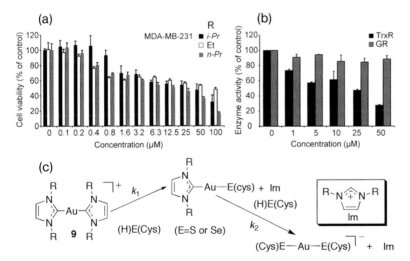

Figure 7.5 (a) Selective cytotoxicity of a series of Au(I) N-heterocyclic complexes $[(R_2Im)_2Au]^+$ (**9**, R = *i*-Pr, *n*-Pr, and Et) in MDA-MB-231 breast cancer cells. (b) Selective inhibition of intracellular thioredoxin reductase (TrxR) compared to glutathione reductase (GR) activity by $[(i-Pr_2Im)_2Au]^+$. The mechanism is proposed to occur by a two-step reaction (c) involving displacement of the NHC ligands by the selenocysteine residues of TrxR, as shown in model reactions with Cys and Sec.

Adapted with permission from Reference [55]. Copyright 2008 American Chemical Society.

7.4.2
Gold(III) Compounds

Prior to the mid-1990s there were only a few known Au(III) complexes with antitumor activity, including some dimethyl Au(III) compounds with modest activity in ip P388 leukemia in mice [133]. In general, Au(III) complexes are not very stable under physiological conditions because of their high reduction potential and fast rate of hydrolysis. However, in recent times a range of strategies have been used to stabilize the Au(III) oxidation state and various different classes of Au(III) compounds have been shown to have significant antitumor properties. Some examples are shown in Figure 7.6 (see References [28, 32, 35] for recent reviews). Parish, Buckley, and coworkers first investigated [AuCl$_2$(damp)] (**10**, where damp = 2-[(dimethylamino)methyl]phenyl) [134] as a cisplatin analog, in which the Au(III) oxidation state is stabilized by coordination of the σ-bonded aryl group in a five-membered chelate ring. This compound showed differential toxicity in a panel of human tumor cell lines [134] and acetate and malonate derivatives were found to be moderately active *in vivo* against human carcinoma xenografts [135].

Messori and coworkers have reported various different classes of cytotoxic Au(III) complexes in which the reduction potential of the metal center is lowered by the use of simple chelating polyamines (e.g., [Au(en)$_2$]Cl$_3$ and [Au(dien)Cl]Cl$_2$ (**11**) [136]) and ligands based on the 2,2-bipyridyl motif such as [Au(bipy)(OH)$_2$]PF$_6$ and the cyclometalated complex [Au(bipy^{dmb}-H)(OH)]PF$_6$ [**13**, bipy^{dmb} = 6-(1,1-dimethylbenzyl)2,2′-bipyridine] [137, 138, 147]. More recently they have reported dinuclear bipyridyl gold(III) oxo compounds having a common Au$_2$O$_2$ motif (e.g., **15** [139, 140]). A recent paper investigates the activity of 13 of these compounds against a panel of human tumor cell lines and various DNA-independent molecular mechanisms are indicated, based on analysis with the Compare algorithm [148]. The results indicate that there is a delicate balance between reactivity (redox properties and stability in aqueous solution) and cytotoxicity. Thus [Au(cyclam)]$^{3+}$ is very stable toward both reduction and ligand substitution reactions [136] and is not cytotoxic (IC$_{50}$ values > 100μM [136]). The reactivity is fine-tuned by the ligands and a single C-Au(III) bond confers great redox stability on the Au(III) center. The dinuclear gold(III) oxo complexes undergo reduction in the presence of GSH and ascorbic acid at physiologically relevant concentrations [140] and the compound [Au$_2$(μ-O)$_2$(6,6′-Me$_2$bipy)$_2$](PF$_6$)$_2$ (**13**), which is most easily reduced, was ranked highest in terms of tumor selectivity and cytotoxic potency [148]. Ranford and coworkers have reported a series of cytotoxic Au(III) 2-phenylpyridine complexes with carboxylate [141] and thiolate ligands (e.g., **16** [149]) and Guo and coworkers have described cytotoxic Au(III) complexes with aminoquinoline (e.g., **17** [142]) and 1,4,7-triazacylononane (TACN) (**18**) ligands [143], as well as several Au (III) terpyridine derivatives [150] that extend the earlier study on [Au(terpy)Cl]Cl$_2$ [136]. These complexes are not reduced by GSH and there is a correlation between DNA binding affinity and cytotoxicity suggesting that, in contrast to most Au(III) complexes, DNA is a possible target.

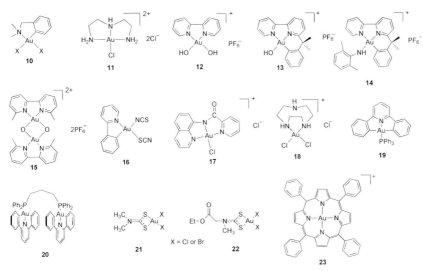

Figure 7.6 Examples of Au(III) compounds with antitumor activity. [AuX$_2$(damp)] (**10** X = Cl, acetate, malonate [134, 135]); [Au(dien)Cl]Cl$_2$ (**11** [136]); [Au(bipy)(OH)$_2$]PF$_6$, (**12**) and [Au(bipy^{dmb}-H)(OH)]PF$_6$ (**13**) [137]; [Au(bipy^{dmb}-H)(2,6-xylidine-H)]PF$_6$ (**14** [138]); [Au$_2$(μ-O)$_2$(6,6-Me$_2$bipy)$_2$](PF6)$_2$ (**15** [139, 140]); [Au(ppy)X$_2$] (X = carboxylate [141] or thiolate, e.g., SCN (**16**) [141]); [Au(Quinpy)Cl]Cl (**17** [142]); [Au(TACN)Cl$_2$]Cl (**18** [143]); [Au(C^N^C) (PPh$_3$)]$^+$ (**19**) and [Au$_2$(C^N^C)$_2$(μ-dppp)]$^{2+}$ (**20**) [144]; [Au(DMDT)X$_2$] (**21**) and [Au(ESDT) X$_2$] (**22**) [145]; and Au(III) *meso*-tetraphenylporphyrin (TPP) (**23**) [146]. While all complexes have been shown to be cytotoxic to cancer cell lines *in vitro*, only **10**, **22**, and **23** have been shown also to have antitumor activity *in vivo*.

Che and coworkers have recently described a series of interesting cyclometalated Au(III) compounds [Au$_m$(C^N^C)$_m$L]$^{n+}$ (*m* = 1–3; *n* = 0–3; HC^N^CH = 2,6-diphenylpyridine) that contain various N-donor or phosphine ligands (L). These complexes exhibit potent cytotoxicity against a panel of cancer cell lines including a cisplatin resistant variant [144], and the Au(III) oxidation state has been shown to be very stable under physiological conditions. The complexes appear to act by various different mechanisms dependent on the nature of the auxiliary ligand, which influences the DNA-binding affinity. A series of complexes where the [Au(C^N^C) L]$^+$ unit is ligated to various mono- and bi-dentate phosphine ligands {e.g., [Au(C^N^C)(PPh$_3$)]$^+$ (**19**) and [Au$_2$(C^N^C)$_2$(μ-dppp)]$^{2+}$ (**20**)} show a cytotoxicity at least ten-fold higher than other Au(III) analogs and react only weakly with DNA.

Despite the large variety of Au(III) complexes that have shown cytotoxicity to cancer cell lines *in vitro*, very few have been shown to demonstrate anticancer activity *in vivo*. Since the investigations on the [AuX$_2$(damp)] complexes (see above), reports of *in vivo* anticancer activity have only been documented for two other classes of Au(III) compounds that require special attention. The first is a series of Au(III) dithiocarbamate complexes reported by Fregona and coworkers

[145, 151–153]. The complexes [Au(DMDT)X$_2$] (**21**) and [Au(ESDT)X$_2$] (**22**) (where DMDT = *N,N*-dimethylthiocarbamate and ESDT = ethylsarcosinedithiocarbamate; X = Cl, Br) are more cytotoxic *in vitro* than cisplatin (including in human tumor cell lines intrinsically resistant to cisplatin) [145]. A representative compound of this series [Au(DMDT)Br$_2$] was shown to significantly inhibit the growth of MDA-MB-231 breast cancer xenografts in nude mice [152]. The second is a series of Au(III) porphyrins reported by Che and coworkers [154–160, 146] that show potent *in vitro* anticancer properties toward a range of human cancer cell lines with some selectivity for cancer cells over normal cells. The prototypical compound [Au(III)(TPP)]Cl (**23**, Figure 7.6), which has been studied most extensively, exhibits promising *in vivo* activity against hepatocellular carcinoma [156] and nasopharyngeal carcinoma [160]. The porphyrinato ligand markedly stabilizes the Au(III) ion against reduction so that it is not reduced by biological reductants such as GSH and ascorbic acid [154]. It is proposed that the compound interacts with biomolecular targets through non-covalent interactions as no reduction in activity occurs in the presence of fetal calf serum [158]. Some comments on the likely mode of action of these and other Au(III) complexes are given below.

7.4.3
Mode of Action of Gold-Based Anticancer Drugs

Various gold(I and III) antitumor compounds, including auranofin, have been shown to overcome resistance to cisplatin and other anticancer drugs (e.g., References [109, 111, 124, 147, 151]) and to cause apoptotic cell death via DNA-independent processes. Mitochondria play a key role in the regulation of apoptosis and the regulation of the intracellular redox state, and almost all mechanistic studies indicate that mitochondria are the biological targets for gold antitumor compounds [31, 48]. Mitochondria contain a specific thioredoxin reductase (TrxR2) (which is different from the cytosolic form, TrxR1) and the work of Bindoli and coworkers first demonstrated that inhibition of TrxR2 was linked to mitochondrial permeability transition and the initiation of the apoptotic process [161]. The mechanism of cytotoxicity of auranofin [107–109, 112] and a diverse range of other Au(I) compounds with phosphine (both monodentate [104] and bidentate [54]), NHC [55], and phosphole [111] ligands have now been linked to inhibition of TrxR. It is assumed that inhibition is due to interaction of Au(I) with the active site selenocysteine residue. Indirect evidence for this is that the inhibition of native TrxR by Au(I) phosphole compounds is orders of magnitude stronger than that of a mutant where the active site selenocysteine is replaced by a cysteine residue [51]. Recent mass spectrometry studies of auranofin-treated TrxR1 are consistent with binding of four AuPEt$_3^+$ fragments, indicating that other binding sites apart from the Sec residue are susceptible to gold binding [48]. A crystal structure of glutathione reductase (GR, an enzyme related to TrxR but lacking the Sec residue) modified with a linear, two coordinate Au(I) phosphole complex shows Au bound to the active center thiols with S–Au–S coordination in the inactive GR product [51]. Various Au(III) compounds have also been shown to be

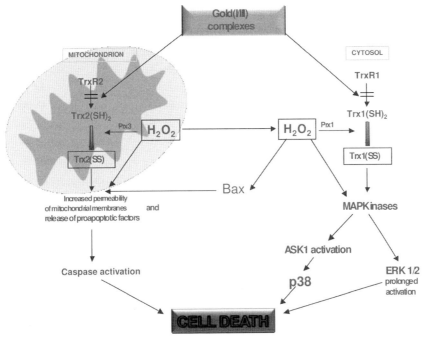

Figure 7.7 Proposed model depicting the mechanism of action of cell death induction by Au(I) and Au(III) compounds. In the mitochondrion, hydrogen peroxide produced from the mitochondrial respiratory chain oxidizes thioredoxin (Trx2), in a reaction mediated by peroxiredoxin (Prx3). Inhibition of thioredoxin reductase (TrxR2) by gold compounds prevents the reduction of the oxidized Trx2, so that H_2O_2 accumulates. These events lead to an opening of the mitochondrial permeability transition pore and/or to an increase in the permeability of the outer membrane. H_2O_2 is then released to the cytosol, causing oxidation of Trx1 that cannot be reduced back by the gold inhibited cytosolic TrxR1. Oxidized Trx1 stimulates the MAP kinase pathways leading to cell death.
Reprinted from Reference [48] with permission from Elsevier.

potent inhibitors of either mitochondrial or cytosolic TrxR [162–165] and the mechanism of cytotoxicity of three organogold(III) compounds (e.g., **13**, **14**, Figure 7.6) and also Au(III) dithiocarbamate compounds (**21**, **22**, Figure 7.6) [163] has been attributed to mitochondrial apoptotic pathways stemming from TrxR inhibition. Recent studies on the inhibition of isolated TrxR by gold(III) compounds suggest that oxidative damage to the enzyme through indiscriminate oxidation of thiol/selenol groups may be important, rather than metal coordination [48]. However, it is important to bear in mind that for almost all of the known active Au(III) compounds the active metabolites could be Au(I) species produced by Au(III) reduction *in vivo*. Recently, Bindoli and coworkers have proposed a general mechanism of action of cell death induction by Au(I) and Au(III) compounds (Figure 7.7, [48]) that involves inhibition of both mitochondrial and cytosolic TrxR and encompasses recent evidence that (i) apoptosis induction by auranofin involves peroxiredoxin 3 oxidation and is regulated by Bcl-2 family proteins [166] and (ii) both auranofin [167] and Au(III) thiocarbamates [163] stimulate mitogen-activated protein (MAP) kinase pathways that ultimately lead to cell death. While the unifying mechanism is attractive, it is important to bear in mind that various proteins involved in cell signaling and transcriptional control contain cysteine residues that are susceptible to binding by gold compounds and so multiple targets and overlapping mechanisms are likely. Detailed mechanistic studies on Au(III) thiocarbamates [163] show that they induce cell death through both apoptotic and non-apoptotic mechanisms. Apart from the involvement in the deregulation of the thioredoxin system, these compounds may also cause concomitant inhibition of the proteasome system [152, 153], another biochemical pathway that is known to trigger apoptosis. Recent results show also that auranofin reduces selenium incorporation into selenoproteins when administered to cells at a subtoxic level, suggesting that blocking selenium metabolism could be important to its therapeutic action [168]. Finally, a different mode of action is likely for the Au(III) porphyrin (**23**), which has been shown to be stable under physiological conditions and to behave essentially as an organic lipophilic cation [159]. The proposed mechanism involves induction of apoptosis through both caspase-dependent and caspase-independent mitochondrial pathways [155] involving activation of the p38 mitogen-activated protein kinase [159, 146].

7.5
Gold Complexes as Antiparasitic Agents

There is urgent need for affordable antiparasitic drugs to tackle diseases such as sleeping sickness, Chagas' disease, and malaria that are major health problems in poverty stricken areas [56]. Very recent research indicates that gold-based drugs offer enormous potential in this field due to the variety of thiol and selenol proteins that have been identified as drug targets in trypanosomes (African sleeping sickness, Chagas' disease, and leishmaniasis), malaria-causing plasmodia, and schistosomiasis [169–171]. The cysteine proteases of the trypanosomatid parasitic protozoa have been validated as drug targets for the treatment of Chagas' disease and leishmaniasis [170]. In a recent study one Au(III) compound was included amongst several metal compounds tested on the parasitic cysteine proteases, cruzain from *Trypanosoma cruzi* and cpB from *Leishmania major* [170]. While this compound was non-toxic and had no effect on *T. cruzi* growth, this study nevertheless highlighted the potential to investigate Au(I) compounds, such as auranofin, which have been shown to inhibit the cathepsins.

A recent review [169] describes drug development approaches based on paralyzing antioxidant systems in the pathogens, which protect them from attack by strong oxidants in the human host. Important differences between the redox metabolism of the host and parasite can be exploited to develop new drugs. For example, trypanosomatids have a unique redox metabolism based on the thiol-polyamine conjugate trypanothione and the flavoenzyme trypanothione reductase, which replaces glutathione reductase (GR) and probably also TrxR in these parasites. For the different TrxR enzymes in the *Plasmodium falciparum* malaria system (malaria parasite, human host, and insect vector) structural differences have been identified in the C-terminal redox center that interacts with oxidized Trx, which may be exploitable in the design of selective inhibitors of the parasitic enzyme. Reports on the potential application of gold compounds are starting to appear. The potent inhibition of TrxR by auranofin, and consequent induction of severe intracellular oxidative stress, prompted a recent investigation of its potential as an antimalarial agent. Auranofin and a few related gold complexes were shown to strongly inhibit *P. falciparum* growth *in vitro* [172].

Schistosomiasis is a tropical disease affecting more than 200 million people. It was quite recently discovered that in *Schistosoma mansoni* both TrxR and GR are absent and replaced by a unique selenium-containing enzyme, thioredoxin glutathione reductase (TGR) [173]. As expected, auranofin, aurothioglucose, and aurothiomalate are all efficient TGR inhibitors, with auranofin being the most potent (K_i = 10 nM) [171]. Furthermore, auranofin was shown to kill parasites rapidly in culture at physiological concentrations and to partially cure mice infected by *S. mansoni* (worm burden reductions of 60% [171]).

A recent review [56] has discussed the concept of "metal-drug synergism" as a further approach to the use of gold compounds in the treatment of parasitic disease, in which a conventional organic drug is complexed to a metal ion to achieve longer residence time of the drug in the organism and more efficient biological targeting. For example, the chloroquine (CQ) gold(I) complex $[Au(PPh_3)(CQ)]PF_6$ is very effective against two chloroquine resistant strains of *P. falciparum* [174].

7.6
Concluding Remarks

Gold-based medicines have been in use for thousands of years for the treatment of a diverse range of complaints and diseases and, remarkably, after 80 years, are still in clinical use as DMARDs. In the past few years an important transition has occurred from serendipity to drug design [42], and there is now an opportunity to bring together the wealth of accumulated knowledge on the biological chemistry of gold with the new ideas of molecular targeting. While gold drugs offer enormous potential as potent inhibitors of novel thiol/selenol disease targets, the drug design process must not neglect the complex biotransformation reactions of gold that are known to occur *in vivo*.

Thioredoxin reductase, a target for gold drugs, and implicated in both rheumatoid arthritis and cancer, provides a unifying link and an explanation for the long known antitumor properties of the antiarthritic drug auranofin. However, other common targets and overlapping signaling pathways are likely to exist; for example, a unifying mechanism for antitumor and anti-inflammatory effects of aurothiomalate is proposed through targeting of the PB1 domain of protein kinase C (PKC$_i$) [114].

Gold-based drugs offer great potential as anticancer drugs because they act by DNA-independent molecular mechanisms and hence have activity in tumors that are resistant to cisplatin and other anticancer drugs. The toxicological problems of $[Au(dppe)_2]^+$ that limited its clinical development were due to non-discriminate mitochondrial dysfunction and recent studies have shown that these problems can be overcome, and selective toxicity to cancer cells over normal cells achieved, by careful attention to ligand design in the modulation of the thiol/selenol reactivity of Au(I).

An attractive option in future drug design of gold-based therapeutics is the use of NHC ligands to fine-tune the Au(I) reactivity by systematic modification of the substituents on the simple imidazolium salt precursors [47, 55]. Ultimately, replacing the phosphine ligand of auranofin with an NHC ligand may offer an attractive alternative in the redesign of this orally-active DMARD, given that phosphine oxidation is likely to contribute to toxic side effects.

Finally, the next decade is likely to see the systematic development of gold-based drugs for treating parasitic diseases, by exploiting important difference in the redox metabolism of the host and parasite. New uses for old drugs is a suggested strategy for cutting down the time and costs of bringing new drugs to market, and a viable option for treating diseases of the poor [175].

Acknowledgments

I thank members of my research group and several coworkers who have made important contributions to my research in the area of gold-based therapeutic agents over many years, in particular Peter J. Sadler, Mark J. McKeage, and Aleksandra Filipovska. The Australian Research Council is acknowledged for financial support.

References

1 Higby, G.J. (1982) *Gold Bull.*, **15**, 130–140.

2 Zhao, H. and Ning, Y. (2001) *Gold Bull.*, **34**, 24–29.

3 Benedek, T.G. (2004) *J. Hist. Med. All. Sci.*, **59**, 50–89.

4 Landé, K. (1927) *Münch. Med. Wochenschr.*, **74**, 1132.

5 Kean, W.F., Forestier, F., Kassam, Y., Buchanan, W.W., and Rooney, P.J. (1985) *Semin. Arthritis Rheum.*, **14**, 180–186.

6 Forestier, J. (1935) *J. Lab. Clin. Med.*, **20**, 827–840.

7 Council, Empire Rheumatism (1960) *Ann. Rheum. Dis.*, **19**, 95–119.

8 Bau, R. (1998) *J. Am. Chem. Soc.*, **120**, 9380–9381.

9 Ruben, H., Zalkin, A., Faltens, M.O., and Templeton, D.H. (1974) *Inorg. Chem.*, **13**, 1836–1839.

10 Hill, D.T. and Sutton, B.M. (1980) *Cryst. Struct. Commun.*, **9**, 679–686.

11 Pope, J.E., Hong, P., and Koehler, B.E. (2002) *J. Rheumatol.*, **29**, 255–260.

12 Lehman, A.J., Esdaile, J.M., Klinkhoff, A. V., Grant, E., Fitzgerald, A., and Canvin, J. (2005) *Arthritis Rheum.*, **52**, 1360–1370.

13 Eisler, R. (2003) *Inflammation Res.*, **52**, 487–501.

14 Sutton, B.M., McGusty, E., Walz, D.T., and DiMartino, M.J. (1972) *J. Med. Chem.*, **15**, 1095–1098.

15 Sutton, B.M. (1986) *Gold Bull.*, **19**, 15–16.

16 Blodgett, R.C. Jr., Heuer, M.A., and Pietrusko, R.G. (1984) *Semin. Arthritis Rheum.*, **13**, 255–273.

17 Chaffman, M., Brogden, R.N., Heel, R. C., Speight, T.M., and Avery, G.S. (1984) *Drugs*, **27**, 378–424.

18 Felson, D.T., Anderson, J.J., and Meenan, R.F. (1990) *Arthritis Rheum.*, **33**, 1449–1461.

19 Sadler, P.J. (1976) *Struct. Bonding*, **29**, 171–215.

20 Brown, D.H. and Smith, W.E. (1980) *Chem. Soc. Rev.*, **9**, 217–240.

21 Razi, M.T., Otiko, G., and Sadler, P.J. (1983) *ACS Symp. Ser.*, **209**, 371–384.

22 Elder, R.C., Eidsness, M.K., Heeg, M.J., Tepperman, K.G., Shaw, C.F. III, and Schaeffer, N. (1983) *ACS Symp. Ser.*, **209**, 385–400.

23 Berners-Price, S.J. and Sadler, P.J. (1986) in *Frontiers in Bioinorganic Chemistry* (ed. A.V. Xavier), VCH Publishers, Weinheim (FDR), pp. 376–388.

24 Shaw, C.F. III (1999) in *Uses of Inorganic Chemistry in Medicine* (ed. N. P. Farrell), Royal Society of Chemistry, Cambridge, pp. 26–57.

25 Shaw, C.F. III (1999) *Chem. Rev.*, **99**, 2589–2600.

26 Mirabelli, C.K., Johnson, R.K., Sung, C. M., Faucette, L., Muirhead, K., and Crooke, S.T. (1985) *Cancer Res.*, **45**, 32–39.

27 Berners-Price, S.J., Mirabelli, C.K., Johnson, R.K., Mattern, M.R., McCabe, F.L., Faucette, L.F., Sung, C. M., Mong, S.M., Sadler, P.J., and Crooke, S.T. (1986) *Cancer Res.*, **46**, 5486–5493.

28 Messori, L. and Marcon, G. (2004) *Met. Ions Biol. Syst.*, **42**, 385–424.

29 Fricker, S.P. (1996) *Gold Bull.*, **29**, 53–59.

30 Fricker, S.P. (1996) *Transition Met. Chem.*, **21**, 377–383.

31 Barnard, P.J. and Berners-Price, S.J. (2007) *Coord. Chem. Rev.*, **251**, 1889–1902.

32 Gabbiani, C., Casini, A., and Messori, L. (2007) *Gold Bull.*, **40**, 73–81.

33 Casini, A., Hartinger, C., Gabbiani, C., Mini, E., Dyson, P.D., Keppler, B.K., and Messori, L. (2008) *J. Inorg. Biochem.*, **102**, 564–575.

34 Tiekink, E.R.T. (2008) *Inflammopharmacology*, **16**, 138–142.

35 Ott, I. (2009) *Coord. Chem. Rev.*, **253**, 1670–1681.

36 Farrell, N. (2002) *Coord. Chem. Rev.*, **232**, 1–4.

37 Hambley, T.W. (2007) *Dalton Trans.*, 4929–4937.

38 Bruijnincx, P.C.A. and Sadler, P.J. (2008) *Curr. Opin. Chem. Biol.*, **12**, 197–206.

39 Berners-Price, S.J. and Filipovska, A. (2008) *Aust. J. Chem.*, **61**, 661–668.

40 Wang, X. and Guo, Z. (2008) *Dalton Trans.*, 1521–1532.

41 Hambley, T.W. (2007) *Science*, **318**, 1392–1393.

42 Fricker, S.P. (2007) *Dalton Trans.*, 4903–4917.

43 Gunatilleke, S.S. and Barrios, A.M. (2006) *J. Med. Chem.*, **49**, 3933–3937.

44 Gunatilleke, S.S., de Oliveira, C.A.F., McCammon, J.A., and Barrios, A.M. (2008) *J. Biol. Inorg. Chem.*, **13**, 555–561.

45 Gunatilleke, S.S. and Barrios, A.M. (2008) *J. Inorg. Biochem.*, **102**, 555–563.

46 Weidauer, E., Yasuda, Y., Biswal, B.K., Cherny, M., James, M.N.G., and Bromme, D. (2007) *Biol. Chem.*, **388**, 331–336.

47 Krishnamurthy, D., Karver, M.R., Fiorillo, E., Orrú, V., Stanford, S.M., Bottini, N., and Barrios, A.M. (2008) *J. Med. Chem.*, **51**, 4790–4795.

48 Bindoli, A., Rigobello, M.P., Scutari, G., Gabbiani, C., Casini, A., and Messori, L. (2009) *Coord. Chem. Rev.*, **253**, 1692–1707.

49 Gromer, S., Urig, S., and Becker, K. (2004) *Med. Res. Rev.*, **24**, 40–89.

50 Becker, K., Gromer, S., Schirmer, R.H., and Müller, S. (2000) *Eur. J. Biochem.*, **267**, 6118–6125.

51 Urig, S., Fritz-Wolf, K., Réau, R., Herold-Mende, C., Tóth, K., Davioud-Charvet, E., and Becker, K. (2006) *Angew. Chem. Int. Ed.*, **45**, 1881–1886.

52 Gromer, S., Arscott, L.D., Williams, C. H. Jr., Schirmer, R.H., and Becker, K. (1998) *J. Biol. Chem.*, **273**, 20096–20101.

53 Chaudière, J. and Tappel, A.L. (1984) *J. Inorg. Biochem.*, **20**, 313–325.

54 Rackham, O., Nichols, S.J., Leedman, P.J., Berners-Price, S.J., and Filipovska, A. (2007) *Biochem. Pharmacol.*, **74**, 992–1002.

55 Hickey, J.L., Ruhayel, R.A., Barnard, P. J., Baker, M.V., Berners-Price, S.J., and Filipovska, A. (2008) *J. Am. Chem. Soc.*, **130**, 12570–12571.

56 Navarro, M. (2009) *Coord. Chem. Rev.*, **253**, 1619–1626.

57 Jackson-Rosario, S. and Self, W.T. (2009) *J. Bacteriol.*, **191**, 4035–4040.

58 Jackson-Rosario, S., Cowart, D., Myers, A., Tarrien, R., Levine, R., Scott, R., and Self, W.T. (2009) *J. Biol. Inorg. Chem.*, **14**, 507–519.

59 Murphy, C.J., Gole, A.M., Stone, J.W., Sisco, P.N., Alkilany, A.M., Goldsmith, E.C., and Baxter, S.C. (2008) *Acc. Chem. Res.*, **41**, 1721–1730.

60 Wang, Z. and Ma, L. (2009) *Coord. Chem. Rev.*, **253**, 1607–1618.

61 Alric, C., Serduc, R., Mandon, C., Taleb, J., Le Duc, G., Le Meur-Herland, A., Billotey, C., Perriat, P., Roux, S. and Tillement, O. (2008) *Gold Bull.*, **41**, 90–97.

62 Handel, M.L., Watts, C.K.W., DeFazio, A., Day, R.O., Sutherland, and R.L. (1995) *Proc. Natl. Acad. Sci. U.S.A.*, **92**, 4497–4501.

63 Huber, R.E. and Criddle, R.S. (1967) *Arch. Biochem. Biophys.*, **122**, 164–173.

64 Isab, A.A. and Sadler, P.J. (1982) *J. Chem. Soc., Dalton Trans.*, 135–141.

65 Berners-Price, S.J., Mazid, M.A., and Sadler, P.J. (1984) *J. Chem. Soc., Dalton Trans.*, 969–974.

66 Berners-Price, S.J. and Sadler, P.J. (1988) *Struct. Bonding (Berlin)*, **70**, 27–102.

67 Berners-Price, S.J., Jarrett, P.S., and Sadler, P.J. (1987) *Inorg. Chem.*, **26**, 3074–3077.

68 Snyder, R.M., Mirabelli, C.K., and Crooke, S.T. (1986) *Biochem. Pharmacol.*, **35**, 923–932.

69 Schmidbaur, H. (1995) *Chem. Soc. Rev.*, **24**, 391–400.

70 Pyykkö, P. (2004) *Angew. Chem. Int. Ed.*, **43**, 4412–4456.

71 Pyykkö, P. (2008) *Chem. Soc. Rev.*, **37**, 1967–1997.

72 Tiekink, E.R.T. and Kang, J-G. (2009) *Coord. Chem. Rev.*, **253**, 1627–1648.

73 Barnard, P.J., Wedlock, L.E., Baker, M. V., Berners-Price, S.J., Joyce, D.A., Skelton, B.W., and Steer, J.H. (2006) *Angew. Chem. Int. Ed.*, **45**, 5966–5970.

74 Cattalini, L., Orio, A., and Tobe, M.L. (1967) *J. Am. Chem. Soc.*, **89**, 3130–3134.

75 Best, S.L. and Sadler, P.J. (1996) *Gold Bull.*, **29**, 87–93.

76 Messori, L. and Marcon, G. (2004) *Met. Ions Biol. Syst.*, **41**, 279–304.

77 Mazid, M.A., Razi, M.T., Sadler, P.J., Greaves, G.N., Gurman, S.J., Koch, M. H.J., and Phillips, J.C. (1980) *J. Chem. Soc., Chem. Commun.*, 1261–1263.

78 Elder, R.C., Ludwig, K., Cooper, J.N., and Eidsness, M.K. (1985) *J. Am. Chem. Soc.*, **107**, 5024–5025.

79 Intoccia, A.P., Flanagan, T.L., Walz, D. T., Gutzait, L., Swagzdis, J.E., Flagiello, J., Hwang, B.Y.H., Dewey, R.H., and Noguchi, H. (1982) *J. Rheumatol.*, **9** (Suppl. 8), 90–98.

80 Coffer, M.T., Shaw, C.F. III, Hormann, A.L., Mirabelli, C.K., and Crooke, S.T. (1987) *J. Inorg. Biochem.*, **30**, 177–187.

81 Sadler, P.J. and Guo, Z. (1998) *Pure Appl. Chem.*, **70**, 863–871.

82 Shaw, C.F. III, Isab, A.A., Coffer, M.T., and Mirabelli, C.K. (1990) *Biochem. Pharmacol.*, **40**, 1227–1234.

83 Elder, R.C., Zhao, Z., Zhang, Y.F., Dorsey, J.G., Hess, E.V., and Tepperman, K. (1993) *J. Rheumatol.*, **20**, 268–272.

84 Graham, G.G. and Kettle, A.J. (1998) *Biochem. Pharmacol.*, **56**, 307–312.

85 Schuhmann, D., Kubickamuranyi, M., Mirtschewa, J., Gunther, J., Kind, P., and Gleichmann, E. (1990) *J. Immunol.*, **145**, 2132–2139.

86 Verwilghen, J., Kingsley, G.H., Gambling, L., and Panayi, G.S. (1992) *Arthritis Rheum.*, **35**, 1413–1418.

87 Takahashi, K., Griem, P., Goebel, C., Gonzalez, J., and Gleichmann, E. (1994) *Metal-Based Drugs*, **1**, 483–496.

88 Bondeson, J. (1997) *Gen. Pharmacol.*, **29**, 127–150.

89 Burmester, G.R. (2001) *Z. Rheumatol.*, **60**, 167–173.

90 Yang, J.P., Merin, J.P., Nakano, T., Kato, T., Kitade, Y., and Okamoto, T. (1995) *FEBS Lett.*, **361**, 89–96.

91 Jeon, K.I., Jeong, J.Y., and Jue, D.M. (2000) *J. Immunol.*, **164**, 5981–5989.

92 Jeon, K-I., Byun, M-S., and Jue, D-M. (2003) *Exp. Mol. Med.*, **35**, 61–66.

93 Sakurai, A., Yuasa, K., Shoji, Y., Himeno, S., Tsujimoto, M., Kunimoto, M., Imura, N., and Hara, S. (2003) *J. Cell. Physiol.*, **198**, 22–30.

94 Kataoka, K., Handa, H., and Nishizawa, M. (2001) *J. Biol. Chem.*, **276**, 34074–34081.

95 Hall, T.J., Jeker, H., Nyugen, H., and Schaeublin, M. (1996) *Inflamm. Res.*, **45**, 230–233.

96 Chircorian, A. and Barrios, A.M. (2004) *Bioorg. Med. Chem. Lett.*, **14**, 5113–5116.

97 Stoyanov, J.V. and Brown, N.L. (2003) *J. Biol. Chem.*, **278**, 1407–1410.

98 McKeage, M.J., Maharaj, L., and Berners-Price, S.J. (2002) *Coord. Chem. Rev.*, **232**, 127–135.

99 Mirabelli, C.K., Johnson, R.K., Hill, D.T., Faucette, L.F., Girard, G.R., Kuo, G.Y., Sung, C.M., and Crooke, S.T. (1986) *J. Med. Chem.*, **29**, 218–223.

100 Tiekink, E.R.T. (2002) *Crit. Rev. Oncol. Hematol.*, **42**, 225–248.

101 Barreiro, E., Casas, J.S., Couce, M.D., Sanchez, A., Sanchez-Gonzalez, A., Sordo, J., Varela, J.M., and Vazquez Lopez, E.M. (2008) *J. Inorg. Biochem.*, **102**, 184–192.

102 Casas, J.S., Castellano, E.E., Couce, M.D., Ellena, J., Sánchez, A., Sordo, J., and Taboada, C. (2006) *J. Inorg. Biochem.*, **100**, 1858–1860.

103 Casas, J.S., Castellano, E.E., Couce, M.D., Crespo, O., Ellena, J., Laguna, A., Sanchez, A., Sordo, J., and Taboada, C. (2007) *Inorg. Chem.*, **46**, 6236–6238.

104 Ott, I., Qian, X., Xu, Y., Vlecken, D.H.W., Marques, I.J., Kubutat, D., Will, J., Sheldrick, W.S., Jesse, P., Prokop, A., and Bagowski, C.P. (2009) *J. Med. Chem.*, **52**, 763–770.

105 Simon, T.M., Kunishima, D.H., Vibert, G.J., and Lorber, A. (1979) *Cancer*, **44**, 1965–1975.

106 Simon, T.M., Kunishima, D.H., Vibert, G.J., and Lorber, A. (1981) *Cancer Res.*, **41**, 94–97.

107 Rigobello, M.P., Scutari, G., Boscolo, R., and Bindoli, A. (2002) *Br. J. Pharmacol.*, **136**, 1162–1168.

108 Rigobello, M.P., Scutari, G., Folda, A., and Bindoli, A. (2004) *Biochem. Pharmacol.*, **67**, 689–696.

109 Marzano, C., Gandin, V., Folda, A., Scutari, G., Bindoli, A., and Rigobello, M.P. (2007) *Free Radical Biol. Med.*, **42**, 872–881.

110 Suresh, D., Balakrishna, M.S., Rathinasamy, K., Panda, D., and Mobin, S.M. (2008) *Dalton Trans.*, 2812–2814.

111 Viry, E., Battaglia, E., Deborde, V., Mueller, T., Réau, R., Davioud-Charvet, E., and Bagrel, D. (2008) *Chem. Med. Chem.*, **3**, 1667–1670.

112 Rigobello, M.P., Folda, A., Dani, B., Menabò, R., Scutari, G., and Bindoli, A. (2008) *Eur. J. Pharmacol.*, **582**, 26–34.

113 Stallings-Mann, M., Jamieson, L., Regala, R.P., Weems, C., Murray, N.R., and Fields, A.P. (2006) *Cancer Res.*, **66**, 1767–1774.

114 Edrogan, E., Lamark, T., Stallings-Mann, M., Jamieson, L., Pellechia, M., Thompson, E.A., Johansen, T., and Fields, A.P. (2006) *J. Biol. Chem.*, **281**, 28450–28459.

115 Mirabelli, C.K., Hill, D.T., Faucette, L. F., McCabe, F.L., Girard, G.R., Bryan, D.B., Sutton, B.M., Bartus, J.O., Crooke, S.T., and Johnson, R.K. (1987) *J. Med. Chem.*, **30**, 2181–2190.

116 Berners-Price, S.J., Girard, G.R., Hill, D.T., Sutton, B.M., Jarrett, P.S., Faucette, L.F., Johnson, R.K., Mirabelli, C.K., and Sadler, P.J. (1990) *J. Med. Chem.*, **33**, 1386–1392.

117 Rush, G.F., Alberts, D.W., Meunier, P., Leffler, K., and Smith, P.F. (1987) *Toxicologist*, **7**, 59.

118 Hoke, G.D., Rush, G.F., Bossard, G.E., McArdle, J.V., Jensen, B.D., and Mirabelli, C.K. (1988) *J. Biol. Chem.*, **263**, 11203–11210.

119 Smith, P.F., Hoke, G.D., Alberts, D.W., Bugelski, P.J., Lupo, S., and Mirabelli, C.K. (1989) *J. Pharmacol. Exp. Therap.*, **249**, 944–950.

120 Hoke, G.D., Macia, R.A., Meunier, P. C., Bugelski, P.J., Mirabelli, C.K., Rush, G.F., and Matthews, W.D. (1989) *Toxicol. Appl. Pharmacol.*, **100**, 293–306.

121 Modica-Napolitano, J.S. and Aprille, J. R. (2001) *Adv. Drug Delivery Rev.*, **49**, 63–70.

122 Chen, L.B. (1988) *Annu. Rev. Cell. Biol.*, 4, 155–181.

123 McKeage, M.J., Berners-Price, S.J., Galettis, P., Bowen, R.J., Brouwer, W., Ding, L., Zhuang, L., and Baguley, B.C. (2000) *Cancer Chemother. Pharmacol.*, **46**, 343–350.

124 Liu, J.J., Galettis, P., Farr, A., Maharaj, L., Samarasinha, H., McGechan, A.C., Baguley, B.C., Bowen, R.J., Berners-Price, S.J., and McKeage, M.J. (2008) *J. Inorg. Biochem.*, **102**, 303–310.

125 Humphreys, A.S., Filipovska, A., Berners-Price, S.J., Koutsantonis, G.A., Skelton, B.W., and White, A.H. (2007) *Dalton Trans.*, 4943–4950.

126 Caruso, F., Rossi, M., Tanski, J., Pettinari, C., and Marchetti, F. (2003) *J. Med. Chem.*, **46**, 1737–1742.

127 Caruso, F., Villa, R., Rossi, M., Pettinari, C., Paduano, F., Pennati, M., Daidone, M., and Zaffaroni, N. (2007) *Biochem. Pharmacol.*, **73**, 773–781.

128 Caruso, F., Pettinari, C., Paduano, F., Villa, R., Marchetti, F., Monti, E., and Rossi, M. (2008) *J. Med. Chem.*, **51**, 1584–1591.

129 Pillarsetty, N., Katti, K.K., Hoffman, T. J., Volkert, W.A., Katti, K.V., Kamei, H., and Koide, T. (2003) *J. Med. Chem.*, **46**, 1130–1132.

130 Higginbotham, M.L., Henry, C.J., Katti, K.V., Casteel, S.W., Dowling, P.M., and Pillarsetty, N. (2003) *Vet. Ther.*, **4**, 76–82.

131 Barreiro, E., Casas, J.S., Couce, M.D., Sánchez, A., Sánchez-González, A., Sordo, J., Varela, J.M., and Vázquez López, E.M. (2009) *J. Inorg. Biochem.*, **103**, 1023–1032.

132 Baker, M.V., Barnard, P.J., Berners-Price, S.J., Brayshaw, S.K., Hickey, J.L., Skelton, B.W., and White, A.H. (2006) *Dalton Trans.*, 3708–3715.

133 Sadler, P.J., Nasr, M., and Narayanan, V.L. (1984) in *Platinum Coordination Complexes in Cancer Chemotherapy* (eds M.P. Hacker, E.B. Douple, and I.H. Krakoff), Martinus Nijhoff Publishing, Boston, pp. 290–304.

134 Parish, R.V., Howe, B.P., Wright, J.P., Mack, J., Pritchard, R.G., Buckley, R. G., Elsome, A.M., and Fricker, S.P. (1996) *Inorg. Chem.*, **35**, 1659–1666.

135 Buckley, R.G., Elsome, A.M., Fricker, S.P., Henderson, G.R., Theobald, B.R. C., Parish, R.V., Howe, B.P., Kelland, L.R. (1996) *J. Med. Chem.*, **39**, 5208–5214.

136 Messori, L., Abbate, F., Marcon, G., Orioli, P., Fontani, M., Mini, E., Mazzei, T., Carotti, S., O'Connell, T., and Zanello, P. (2000) *J. Med. Chem.*, **43**, 3541–3548.

137 Marcon, G., Carotti, S., Coronnello, M., Messori, L., Mini, E., Orioli, P., Mazzei, T., Cinellu, M.A., and Minghetti, G. (2002) *J. Med. Chem.*, **45**, 1672–1677.

138 Messori, L., Marcon, G., Cinellu, M.A., Coronnello, M., Mini, E., Gabbiani, C., and Orioli, P. (2004) *Bioorg. Med. Chem.*, **12**, 6039–6043.

139 Casini, A., Cinellu, M.A., Minghetti, G., Gabbiani, C., Coronnello, M., Mini, E., and Messori, L. (2006) *J. Med. Chem.*, **49**, 5524–5531.

140 Gabbiani, C., Casini, A., Messori, L., Guerri, A., Cinellu, M.A., Minghetti,

G., Corsini, M., Rosani, C., Zanello, P., and Arca, M. (2008) *Inorg. Chem.*, **47**, 2368–2379.

141 Fan, M., Yang, C.T., Ranford, J.D., Lee, P.F., and Vittal, J.J. (2003) *Dalton Trans.*, 2680–2685.

142 Yang, T., Tu, C., Zhang, J., Lin, L., Zhang, X., Liu, Q., Ding, J., Xu, Q., and Guo, Z. (2003) *Dalton Trans.*, 3419–3424.

143 Shi, P., Jiang, Q., Lin, J., Zhao, Y., Lin, L., and Guo, Z. (2006) *J. Inorg. Biochem.*, **100**, 939–945.

144 Li, C.K-L., Sun, R.W-Y., Kui, S.C-F., Zhu, N., and Che, C-M. (2006) *Chem. Eur. J.*, **12**, 5253–5266.

145 Ronconi, L., Giovagnini, L., Marzano, C., Bettìo, F., Graziani, R., Pilloni, G., and Fregona, D. (2005) *Inorg. Chem.*, **44**, 1867–1881.

146 Sun, R.W-Y. and Che, C-M. (2009) *Coord. Chem. Rev.*, **253**, 1682–1691.

147 Coronnello, M., Mino, E., Caciagli, B., Cinellu, M.A., Bindoli, A., Gabbiani, C., and Messori, L. (2005) *J. Med. Chem.*, **48**, 6761–6765.

148 Casini, A., Kelter, G., Gabbiani, C., Cinellu, M.A., Minghetti, G., Fregona, D., Fiebig, H-H., and Messori, L. (2009) *J. Biol. Inorg. Chem.*, **14**, 1139–1149.

149 Fan, D.M., Yang, C.T., Ranford, J.D., Vittal, J.J., and Lee, P.F. (2003) *Dalton Trans.*, 3376–3381.

150 Shi, P., Jiang, Q., Zhao, Y., Zhang, Y., Lin, J., Lin, L., Ding, J., and Guo, Z. (2006) *J. Biol. Inorg. Chem.*, **11**, 745–752.

151 Ronconi, L., Marzano, C., Zanello, P., Corsini, M., Miolo, G., Maccà, C., Trevisan, A., and Fregona, D. (2006) *J. Med. Chem.*, **49**, 1648–1657.

152 Milacic, V., Chen, D., Ronconi, L., Landis-Piwowar, K.R., Fregona, D., and Dou, Q.P. (2006) *Cancer Res.*, **66**, 10478–10486.

153 Milacic, V. and Dou, Q.P. (2009) *Coord. Chem. Rev.*, **253**, 1649–1660.

154 Che, C-M., Sun, R.W-Y., Yu, W-Y., Ko, C-B., Zhu, N., and Sun, H. (2003) *Chem. Commun.*, 1718–1719.

155 Wang, Y., He, Q-H., Sun, R.W-Y., Che, C-M., and Chiu, J-F. (2005) *Cancer Res.*, **65**, 11553–11564.

156 Lum, C.T., Yang, Z.F., Li, H.Y., Sun, R.W-Y., Fan, S.T., Poon, R.T.P., Lin, M.C.M., Che, C-M., and Kung, H.F. (2006) *Int. J. Cancer*, **118**, 1527–1538.

157 Wang, Y, He, Q-Y., Che, C-M., and Chiu, J-F. (2006) *Proteomics*, **6**, 131–142.

158 Wang, Y., He, Q-Y., Sun, R.W-Y., Che, C-M., and Chiu, J-F. (2007) *Eur. J. Pharmacol.*, **554**, 113–122.

159 Wang, Y., He, Q-Y., Che, C-M., Tsao, S.W., Sun, R.W-Y., and Chiu, J-F. (2008) *Biochem. Pharmacol.*, **75**, 1282–1291.

160 To, Y.F., Sun, R.W-Y., Chen, Y., Chan, V.S-F., Yu, W-Y., Tam, P.K-H., Che, C-M., and Lin, C-L.S. (2009) *Int. J. Cancer.*, **124**, 1971–1979.

161 Rigobello, M.P., Callegaro, M.T., Barzon, E., Benetti, M., and Bindoli, A. (1998) *Free Radical Biol. Med.*, **24**, 370–376.

162 Rigobello, M.P., Messori, L., Marcon, G., Agostina Cinellu, M., Bragadin, M., Folda, A., Scutari, G., and Bindoli, A. (2004) *J. Inorg. Biochem.*, **98**, 1634–1641.

163 Saggioro, D., Rigobello, M.P., Paloschi, L., Folda, A., Moggach, S.A., Parsons, S., Ronconi, L., Fregona, D., and Bindoli, A. (2007) *Chem. Biol.*, **14**, 1128–1139.

164 Omata, Y., Folan, M., Shaw, M., Messer, R.L., Lockwood, P.E., Hobbs, D., Bouilaguet, S., Sano, H., Lewis, J. B., and Wataha, J.C. (2006) *Toxicol. In Vitro*, **20**, 882–890.

165 Engman, L., McNaughton, M., Gajewska, M., Kumar, S., Birmingham, A., and Powis, G. (2006) *Anti-Cancer Drugs*, **17**, 539–544.

166 Cox, A.G., Brown, K.K., Arner, E.S.J., and Hampton, M.B. (2008) *Biochem. Pharmacol.*, **76**, 1097–1109.

167 Park, S-J. and Kim, I-S. (2005) *Br. J. Pharmacol.*, **146**, 506–513.

168 Talbot, S., Nelson, R., and Self, W.T. (2008) *Br. J. Pharmacol.*, **154**, 940–948.

169 Krauth-Siegel, R.L., Bauer, H., and Schirmer, H. (2005) *Angew. Chem. Int. Ed.*, **44**, 690–715.

170 Fricker, S.P., Mosi, R.M., Cameron, B. R., Baird, I., Zhu, Y., Anastassov, V., Cox, J., Doyle, P.S., Hansell, E., Lau, G., Langille, J., Olsen, M., Qin, L., Skerlj, R., Wong, R.S.Y., Santucci, Z., and McKerrow, J.H. (2008) *J. Inorg. Biochem.*, **102**, 1839–1845.

171 Kuntz, A.N., Davioud-Charvet, E., Sayed, A.A., Califf, L.L., Dessolin, J., Arner, E.S.J., and Williams, D.L. (2007) *PLoS Med.*, **4**, 1071–1086.

172 Sannella, A.R., Casini, A., Gabbiani, C., Messori, L., Bilia, A.R., Vincieri, F.F., Majori, G., and Severini, C. (2008) *FEBS Lett.*, **582**, 844–847.

173 Alger, H.M. and Williams, D.L. (2002) *Mol. Biochem. Parasitol.*, **121**, 129–139.

174 Navarro, M., Vasquez, F., Sanchez-Delgado, R.A., Perez, H., Sinou, V., and Schrevel, J. (2004) *J. Med. Chem.*, **47**, 5204–5209.

175 Chong, C.R. and Sullivan, D.J. (2007) *Nature*, **448**, 645–646.

8
MRI Contrast Agents: State of the Art and New Trends

Daniela Delli Castelli, Eliana Gianolio, and Silvio Aime

8.1
Introduction

In 1946, Felix Bloch and Edward Purcell, both awarded with the Nobel Prize in Physics in 1952, independently discovered the magnetic resonance phenomenon. In the period between 1950 and 1970, NMR was developed and used for chemical and physical molecular studies. In 1971 Raymond Damadian showed that the nuclear magnetic relaxation times of healthy tissues and tumors differed, thus motivating scientists to consider magnetic resonance for the detection of disease. Magnetic resonance imaging (MRI) was first demonstrated on test tube samples by Paul Lauterbur in 1973. Two years later, in 1975, an important development was introduced by Richard Ernst, who proposed the use of phase and frequency encoding, and the Fourier transform. This technique is the basis of present MRI. This imaging modality is primarily used in medical settings to produce high quality images of the inside of the human body. Among the existing imaging techniques, MRI stands out thanks to the excellent spatial resolution and the outstanding capacity of differentiating soft tissues. These features have determined its widespread success in clinical diagnosis. The contrast in an MR image is the result of a complex interplay of different factors, including T_1, T_2, and proton density of the imaged tissues and instrumental parameters. When there is poor contrast between healthy and diseased regions, due to a very small difference in relaxation times or proton density, the use of contrast agents (CA) can be highly beneficial. Already in the early 1980s it was realized that some chemicals an alter markedly the relaxation times of water protons in the tissues where they distribute. Unlike contrast agents used in X-ray computed tomography and in nuclear medicine, MRI contrast agents are not directly visualized in the image. Only their effects are observed as the contrast is affected by the variation that the CA causes on water protons relaxation times, and consequently on the intensity of the NMR signal [1]. Generally, the purpose of a CA is to reduce T_1 or T_2 to obtain an hyper- or ipo-intense signal, respectively, in short times and a better signal-to-noise ratio with the acquisition of a higher number of measurements. CAs that predominantly reduce T_1 are called positive, whereas those that mainly affect T_2 are called negative. The search for

Bioinorganic Medicinal Chemistry. Edited by Enzo Alessio
Copyright © 2011 WILEY-VCH Verlag GmbH & Co. KGaA, Weinheim
ISBN: 978-3-527-32631-0

positive MRI contrast agents was oriented towards paramagnetic metal complexes because unpaired electrons display remarkable ability to reduce T_1 and T_2 of their solutions [2, 3]. Paramagnetic Mn(II) and Gd(III) chelates are the most common representatives of the T_1-positive agents, whereas iron oxide particles are the most common T_2-negative agents.

At the pre-clinical level other classes of MRI contrast agents have been proposed either based on the same mechanism of action but differing in the peculiar bio-distribution (e.g., nano-systems, dendrimers, etc.) or based on different modality to alter the contrast, such as the CEST agents (CEST = chemical exchange saturation transfer) [4], ^{19}F containing molecules, and hyperpolarized molecules. CEST agents are chemicals endowed with mobile protons in chemical exchange with the bulk water. They are able to affect the contrast in an MR image through the transfer of saturated spins to the "bulk" water magnetization, following the irradiation of the mobile protons signal with an appropriate RF pulse. Since the macroscopic effect is that of reducing the signal of the bulk water protons, they belong to the class of negative agents [5]. ^{19}F and hyperpolarized molecules differ from the other class of MRI contrast agent because they are not designed to affect the water signal, rather they provide heteronuclear images with no background signal. In this sense they are more similar to contrast agents used for X-rays. In this chapter we will not deal with these last two categories since this subject is beyond the scope of this book.

8.2
T_1 Agents

Paramagnetic substances have been under intense scrutiny as MRI contrast agents since the early days of NMR tomography. Although stable organic radicals, NO, and O_2 were also considered, it was immediately clear that paramagnetic metal complexes are the candidates of choice for this application.

On this basis, the metal ions more suitable for this application have been identified among those having the higher number of unpaired electrons, namely Mn(II) and Fe(III) (five unpaired electrons) in the transition metal series, and Gd(III) (seven unpaired electrons) among the lanthanides. A typical Gd-based contrast agent is an eight-coordinate polyaminocarboxylate complex (a "chelate") with the ninth coordination site occupied by a water molecule. Figure 8.1 gives the structures of some clinically approved Gd agents.

8.2.1
Theory of Paramagnetic Relaxation

The efficiency of a relaxation agent is commonly evaluated *in vitro* by the measure of its *relaxivity* (r_1) that, for commercial CAs as Magnevist®, Dotarem®, ProHance®, and Omniscan®, is around 3.4–3.5 mM^{-1} s^{-1} (at 20 MHz and 39 °C). It represents the relaxation enhancement of water protons in the presence of the

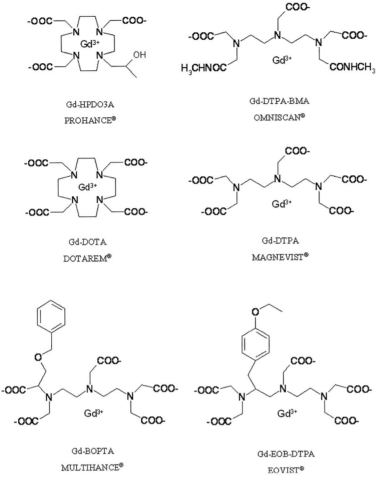

Figure 8.1 Structures of the Gd(III)-based MRI contrast agents (CA) currently used in clinical practice.

paramagnetic complex at 1 mM concentration. The observed longitudinal relaxation rate (R_1^{obs}) of the water protons in an aqueous solution containing the paramagnetic complex is the sum of three contributions [Eq. (8.1)] [6]: (i) the diamagnetic one $(R_1^°)$, whose value corresponds to proton relaxation rate measured in the presence of a diamagnetic (La, Lu, Y) complex of the same ligand; (ii) the inner sphere paramagnetic contribution, relative to the exchange of water molecules from the inner coordination sphere of the metal ion with bulk water (R_{1p}^{is}); and (iii) the outer sphere paramagnetic contribution, relative to water molecules that diffuse in the outer coordination sphere of the paramagnetic center (R_{1p}^{os}). Sometimes also a fourth paramagnetic contribution is taken into account that is due to the presence of mobile protons or water molecules (normally bound through hydrogen bonds) in the second coordination sphere of the metal ion [7]:

$$R_1^{obs} = R_1^o + R_{1p}^{is} + R_{1p}^{os} \tag{8.1}$$

The inner sphere contribution is directly proportional to the molar concentration of the paramagnetic complex, [C], to the number of water molecules coordinated to the paramagnetic center, q, and inversely proportional to the sum of the mean residence lifetime of the coordinated water protons, τ_m, and of their relaxation time, T_{1M} [Eq. (8.2)]:

$$R_{1p}^{is} = \frac{q[C]}{55.5(T_{1M} + \tau_m)} \tag{8.2}$$

The latter parameter is directly proportional to the sixth power of the distance between the metal center and the coordinated water protons (r) and depends on the overall molecular correlation time, τ_c, that, in turn, is determined by the shortest among τ_m, τ_R (reorientational correlation time of the bound water molecule, which is assumed to coincide with the reorientational correlation time of the entire metal complex), and the electronic relaxation times, T_{iE} ($i = 1, 2$), of the unpaired electrons of the metal (which depend on the applied magnetic field strength) [Eqs. (8.3) and (8.4)]:

$$\frac{1}{T_{1M}} = \frac{2}{15}\left(\frac{\mu_0}{4\pi}\right)^2 \frac{\gamma_I^2 g_e^2 \mu_B^2 S(S+1)}{r_H^6}\left[\frac{7\tau_c}{1+\omega_S^2\tau_c^2} + \frac{3\tau_c}{1+\omega_I^2\tau_c^2}\right] \tag{8.3}$$

$$\tau_c^{-1} = \tau_R^{-1} + \tau_m^{-1} + T_{iE}^{-1} \tag{8.4}$$

The outer sphere contribution depends on T_{iE}, on the distance of maximum approach between the solvent and the paramagnetic solute, on the relative diffusion coefficients and – again – on the magnetic field strength. The dependence of R_{1p}^{is} and R_{1p}^{os} on magnetic field strength is very important because it allows the determination of the principal parameters affecting the relaxivity of a paramagnetic compound. This information can be obtained through an NMR instrument in which the magnetic field is changed (field-cycling relaxometer) to obtain the measure of r_1 on a wide range of frequencies (typically 0.01–50 MHz). At the frequencies most commonly used in commercial tomographs (20–63 MHz), τ_R of the chelate is often the determinant of the observed relaxivity. A quantitative analysis of relaxivity dependence on the different structural and dynamic parameters shows that, for systems with long τ_R (e.g., a protein bound complex), the maximum attainable r_1 values can be achieved through the optimization of τ_m and T_{iE} [6]. All the available commercial Gd-based CA are monohydrated systems ($q = 1$), with a molecular weight of about 600–800 Da that corresponds to rotational correlation times τ_R of about 60–80 ps. For this class of polyaminocarboxylate complexes the exchange lifetime τ_m is typically found to be in the range of few hundreds of ns and $T_{1E} \approx$ 1 ns at 0.5 T and thus the inner sphere relaxivity, r_{1p}^{is} (i.e., the R_{1p}^{is} values normalized to the mM concentration), assumes a value of about 2.5–3.5 mM^{-1} s^{-1} at 25 °C. Therefore, as was early recognized, it is evident that at 0.5 T the overall correlation time is largely dominated by the rotational correlation time, whereas the contributions of both the exchange lifetime and the electronic relaxation play a less relevant role. Figure 8.2 summarizes the main parameters that affect the relaxivity of the T_1 contrast agents.

8.2.2
Clinically Approved T_1 Agents

Currently, several Gd-based and one Mn-based agents are available for clinical applications. Their use has led to remarkable improvements in medical diagnosis in terms of higher specificity, better tissue characterization, reduction of image artifacts, and improved functional information. Besides acting as catalyst for the relaxation of water protons, a paramagnetic MRI-contrast agent has to possess several additional properties to guarantee the safety issues required for *in vivo* applications at the administered doses, namely high thermodynamic (and possibly kinetic) stability, good solubility, and low osmolality [1].

8.2.2.1 Paramagnetic Gd(III)-Based Complexes
Extracellular Fluid Agents Extracellular fluid agents (ECF): the first CA approved for clinical use was Gd-DTPA (Figure 8.1, Magnevist, Schering AG, Germany), which, in more than 10 years of clinical experimentation, has been administered to

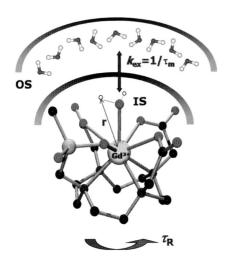

Figure 8.2 Schematic representation of all the parameters affecting the relaxivity in Gd(III) complexes; τ_m and τ_R are the mean residence lifetime of the coordinated water protons and the reorientational correlation time of the complex, respectively; OS and IS stand for outer and inner sphere water molecules, respectively, and r indicates the distance between the Gd metal ion and the water molecule protons.

many millions of patients (clinical dose 0.1 mmol kg^{-1}). Other Gd(III)-based CAs similar to Magnevist became soon available: Gd-DOTA (Dotarem, Guerbet SA, France), Gd-DTPA-BMA (Omniscan, GE Health, USA), and Gd-HPDO3A (Pro-Hance, Bracco Imaging, Italy) [8]. These CAs have very similar pharmacokinetic properties because they distribute in the extracellular fluid and are eliminated via glomerular filtration. In neurological diseases, they are particularly useful to delineate lesions as a result of the disruption of the blood–brain barrier. Two derivatives of Gd-DTPA have been successively introduced, Gd-EOB-DTPA [9] (Eovist®, Schering AG, Germany) and Gd-BOPTA [10] (MultiHance®, Bracco Imaging, Italy). They are characterized by an increased lipophilicity due to the introduction of an aromatic substituent on the carbon backbone of the DTPA ligand. This modification significantly alters the pharmacokinetics and the biodistribution of these CAs as compared to the parent Gd-DTPA, making them hepato-specific agents.

Blood Pool Agents Several systems have been studied over the last two decades for the design of macromolecular Gd(III)-complexes as MRI blood-pool contrast agents. These paramagnetic macromolecules do not diffuse across healthy vasculature and remain intravascular, thus reporting on the anatomy of the vessels bed (Figure 8.3). In addition to a higher vascular retention time, macromolecular systems are often endowed with sensibly higher relaxivities (at 0.5–1.5 T) thanks to the elongation of reorientational correlation times of slowly tumbling systems. Vasovist® is the first blood pool agent available for clinical use. It is a Gd-DTPA substituted with a diphenylcyclohexyl phosphate group. The key differentiating characteristic of Vasovist® is its reversible binding to human serum albumin (HSA): after i.v. injection, a fraction of approx. 85% of it is bound to HSA. The T_1 relaxivity rate of Vasovist® at 1.5 T is five times higher than that of standard gadolinium contrast agents.

Another approved Gd-based blood pool agent is Vistarem®, a hydrophilic high molecular weight (MW = 6473 g mol^{-1}) derivative of DOTA-Gd, whose mechanism of action relies on its large size that precludes the extravasation. The chemical

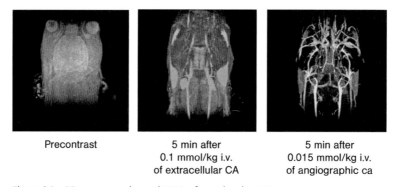

| Precontrast | 5 min after 0.1 mmol/kg i.v. of extracellular CA | 5 min after 0.015 mmol/kg i.v. of angiographic ca |

Figure 8.3 3D contrast enhanced MRA of a rat head at 2 T.

structure of Vistarem has been optimized to provide: (i) a high millimolar relaxivity in the clinical field for MRI: 29 mM^{-1} s^{-1} at 60 MHz, (ii) a high biocompatibility profile, and (iii) a high molecular volume: the apparent hydrodynamic volume is 125 times greater than that of Gd-DOTA. As a result, this molecule presents an unusual pharmacokinetic profile, as it is a rapid clearance blood pool agent (RCBPA) characterized by limited diffusion across the normal endothelium.

Biodistribution and Toxicity Studies have revealed a possible connection between a disease called nephrogenic systemic fibrosis (NSF) and the assumption of Gd (III) contrast agents [11]. NSF is a fibrosing disorder that involves predominantly the skin but also affects systemic organs such as the liver, heart, lungs, diaphragm, and skeletal muscle. It is associated with severe physical disability and possible death when multisystem disease supervenes. The cause of NSF is unknown; however, underlying kidney dysfunction is present in all cases. Gd^{3+} may act as a trigger for NSF in patients with kidney disease on the basis of its reduced clearance and possibly its chelate-binding characteristics. Gadolinium-based contrast agents are eliminated almost entirely (97%) by kidneys. Reduced renal function significantly increases the half-life of these complexes. Macrocyclic chelates bind Gd(III) more tightly than linear chelates and are more stable both *in vitro* and *in vivo*. Omniscan, the agent that most commonly is associated with NSF, is a nonionic contrast agent that uses a linear chelate endowed with a lower stability with respect to other Gd CAs. Although cause and effect correlation between Gd^{3+} release and development of NSF has not been proven, there is compelling associative evidence to recommend limiting Gd contrast agent exposure to patients with kidney disease.

8.2.2.2 Mn(II)-Based Complexes

Paramagnetic chelates of Mn(II) (five unpaired electrons) have also been considered. The main drawback appears to be related to the stability of these complexes. Manganese(II) is an essential metal, therefore evolution has selected biological structures for sequestering Mn(II) ions with high efficiency. Combined with the fact that Mn(II) forms highly labile coordination complexes, it has been difficult to design Mn(II) chelates that maintain their integrity when administered to living organisms.

The only approved Mn(II) agent is Mangafodipir trisodium (manganese-dipyridoxal diphosphate, Mn-DPDP or Teslascan®, Figure 8.4) [12]. It is a contrast agent for use in magnetic resonance imaging (MRI) of the liver. Mn-DPDP is a manganese chelate derived from vitamin B6 (pyridoxal 5-phosphate) and is specifically taken up by the hepatocytes. In particular, the agent is taken up by normal hepatocytes, resulting in increased signal on T$_1$-weighted imaging, and is excreted in the biliary system. It is the only agent that does its job by releasing metal ions to endogenous macromolecules. The huge proton relaxation enhancement brought about by the resulting Mn(II) protein adducts is responsible for the MRI visualization of hepatocytes also at low administered doses of Mn-DPDP.

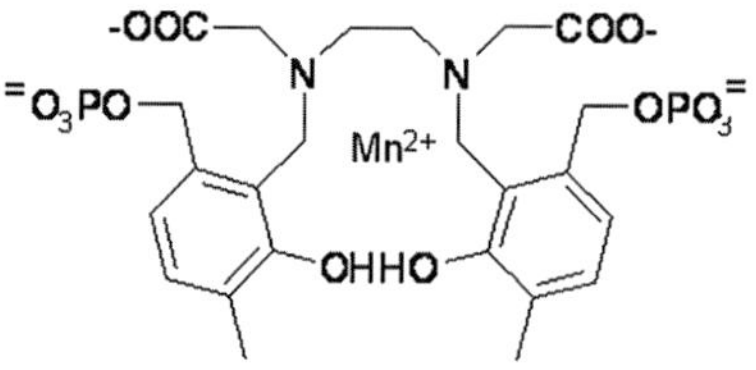

Mn-DPDP

TESLASCAN®

Figure 8.4 Structure of the contrast agent Teslascan.

Biodistribution and Toxicity The toxicity related to the use of Mn complexes resides in their low stability; in fact, chelation of Mn(II) is necessary to decrease the high acute toxicity of the free metal ion, which is a calcium-blocking agent with important effects on muscle electrophysiology and contractility. Initial evaluation of Mn-DPDP revealed it had a considerably lower acute toxicity than Mn^{2+} (LD_{50} in mice: 0.3 mmol kg^{-1} for Mn^{2+} vs. 5.5 mmol kg^{-1} for Mn-DPDP) and was highly extracted by the liver.

8.2.3
Preclinical Level

8.2.3.1 **ECF Agents**
Among the huge number of Gd complexes that have been investigated in the last 20 years, very interesting systems are represented by GdAAZTA and GdHOPO compounds. As mentioned already, the attainment of higher relaxivity can be pursued by appropriate control of the parameters that determine the inner sphere term, that is, the number (q) and the exchange lifetime (τ_m) of the water molecules directly coordinated to the paramagnetic metal center, the electronic relaxation time (T_{iE}), and the molecular reorientational time (τ_R). Whereas the lengthening of τ_R is tackled by designing slow moving macromolecular systems, the other parameters have to be optimized on the basis of the characteristic features of the chelate itself. In principle, doubling of the inner-sphere relaxivity can be obtained by going from $q = 1$ to $q = 2$ complexes. The commercially available MRI contrast agents are Gd(III) complexes with octadentate ligands, thus allowing only one water molecule to enter the inner coordination sphere ($q = 1$). The straightforward route to increase q involves a decrease of the overall denticity of the coordinating ligand. However, the simple shift from an octadentate to a heptadentate ligand may interfere with toxicological problems associated either with a decrease of the thermodynamic stability or with the replacement of the two water molecules by endogenous anions or by coordinating groups of tissue proteins [13]. Few Gd (III) chelates with heptacoordinating ligands appear to overcome these drawbacks, among them GdAAZTA and GdHOPO. The Gd(III) chelate with the heptadentate ligand AAZTA (6-amino-6-methylperhydro-1,4-diazepinetetraacetic acid) displays a relaxivity of 7.1 mM^{-1} s^{-1} at 20 MHz and 298 K. Its stability constant (log K)

has a value of 19.26, which is slightly smaller than that of $[Gd\text{-}DTPA]^{2-}$, by far the most used MRI agent, and significantly higher than that of Gd-DTPA-BMA, discussed above. When GdAAZTA was titrated with lactate or phosphate, no change in its relaxivity was detected also at concentrations of the added substrate 200 times higher than the paramagnetic chelate. Thus, one may argue that the solution structure of GdAAZTA (Figure 8.5), coupled with the occurrence of a residual negative charge, does not allow the replacement of the coordinated water molecules with other substrates [14].

Raymond and coworkers have developed HOPO (hydroxypyridinone) based ligands designed to satisfy the coordination preferences of Gd^{3+}, especially its oxophilicity (Figure 8.5). The HOPO ligands provide a hexadentate coordination for Gd^{3+}, in which all of the donor atoms are oxygens. Because Gd^{3+} favors eight or nine coordination, this design provides two to three open sites for inner-sphere water molecules. Moreover, these water molecules are in rapid exchange with bulk water. These complexes also showed a very high thermodynamic stability for Gd^{3+} binding. The contrast enhancement provided by these agents is at least twice that of commercial contrast agents, which are based on polyaminocarboxylate ligands [15].

8.2.3.2 Blood Pool Agents

The most straightforward method to increase the plasma circulation time and promote vascular confinement of a contrast agent is to covalently bind it to bio-macromolecules such as proteins, polysaccharides, and dextran.

An alternative to covalent binding has been pursued by exploiting the non-covalent reversible interaction between low molecular weight hydrophobic Gd(III)-complexes and proteins.

Human serum albumin (HSA) has been by far the most investigated protein for binding of Gd(III) chelates. Besides the attainment of high relaxivities, a high binding affinity to HSA endows the Gd(III) chelate with a long intravascular retention time, which is the property required for a good blood-pool agent for MR angiography. In blood, HSA has a concentration of about 0.6 mM and its main physiological role deals with the transport of a huge number of substrates [16].

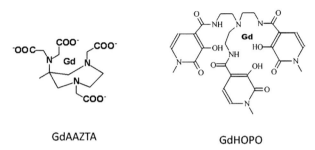

GdAAZTA   GdHOPO

Figure 8.5 Structures of GdAAZTA and GdHOPO complexes.

Caravan and coworkers [17] have tackled the problem of the residual mobility of a Gd(III) complex bound to HSA by designing a system containing two anchoring sites on the protein. Interestingly, the observed relaxivity for such an adduct, though high (60 mM^{-1} s^{-1}), is still significantly lower than that foreseen by the paramagnetic relaxation theory.

In our group a derivative of GdAAZTA containing a long aliphatic chain (GdAAZTA-C17) has been synthesized and its relaxometric properties investigated in detail [18]. The complex showed the outstanding properties of the parent complex, namely: (i) two inner sphere water molecules in fast exchange with the bulk, (ii) high thermodynamic stability in aqueous solution, and (iii) a nearly complete inertness towards the influence of bidentate endogenous anions. Already at sub-millimolar concentrations (critic micellar concentration, cmc, 0.1 mM), the presence of the hydrophobic chain induces the formation of micelles, with a re-laxivity, per Gd(III) ion, of 30 mM^{-1} s^{-1} at 20 MHz and 298 K. At concentrations lower than cmc GdAATZA-C17 displays a high binding affinity to HSA, $K_a = 2.4 \times 10^4$ M^{-1}, giving a macromolecular adduct endowed with the highest relaxivity value (84 mM^{-1} s^{-1}) reported to date for HSA-bound Gd complexes.

8.2.3.3 Responsive T$_1$ Agents

Those compounds whose contrasting properties are sensitive to a given physico-chemical variable of the micro-environment in which the probe distributes are referred to as *responsive agents*. Typical parameters of primary diagnostic relevance include pH, temperature, pO$_2$, enzymatic activity, redox potential, and concentra-tion of specific ions and low-weight metabolites.

Unfortunately, their clinical use is still uncertain mainly because an accurate measurement of one of the above-mentioned parameters with a given relaxing probe requires a precise knowledge of the local concentration of the contrast medium in the region of interest. Only if the actual concentration is known can the observed change in the relaxation rate of water protons be safely attributed to a change of the parameter to be measured.

pH-Sensitive In most of the examples so far reported the pH dependence of the relaxivity reflects changes in the hydration of the metal complex.

For instance, we found that the relaxivity of a series of macrocyclic Gd(III) complexes bearing β-arylsulfonamide groups is markedly pH-dependent (Figure 8.6), as it changes from about 8 mM^{-1} s^{-1} at pH <4 to about 2.2 mM^{-1} s^{-1} at pH > 8 [19]. It has been demonstrated that the observed decrease (about fourfold) of r_1 is the result of a switch in the number of water molecules coordinated to the Gd (III) ion from two (at acidic pH) to zero (at basic pH). This corresponds to a change in the coordination ability of the β-arylsulfonamide arm, which binds the metal ion only when it is in the deprotonated form.

In some cases the pH dependence of the relaxivity is associated with changes in the structure of the second hydration shell. Two systems of this type have been reported by Sherry and coworkers. The first case deals with a macrocyclic tetraamide derivative of DOTA (DOTA-4AmP), which displays an unusual

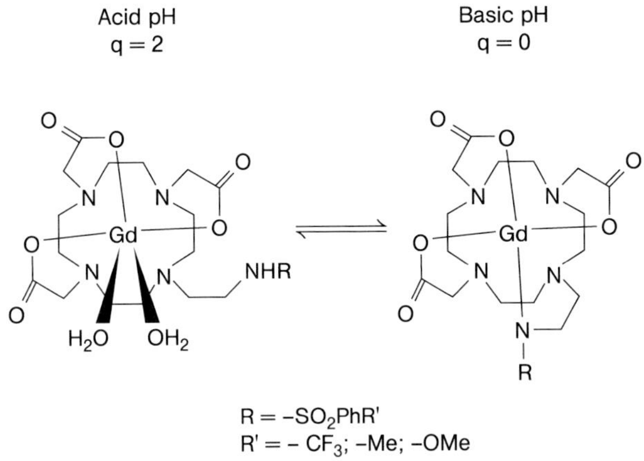

Figure 8.6 Gd(III) complex in which the chelate bears an arylsulfonamide group. The protonation/deprotonation process of the arylsulfonamide group determines a variation in the hydration of the complex.

r_1 vs. pH dependence [20]. The relaxivity of this complex increases from pH 4 to pH 6, decreases up to pH 8.5, remains constant up to pH 10.5, and then increases again. The authors suggested that this behavior is related to the formation/disruption of the hydrogen bond network between the pendant phosphonate groups and the water bound to the Gd(III) ion. The deprotonation of phosphonate occurring at pH >4 promotes the formation of the hydrogen bond network that slows down the exchange of the metal-bound water protons. Conversely, the behavior observed at pH >10.5 was accounted for in terms of a shortening of τ_m catalyzed by OH^- ions. It has been demonstrated that this complex can be successfully used *in vivo* for mapping renal and systemic pH [21].

pH-dependent probes can also be obtained when the proton concentration affects the structure of a macromolecular substrate that, in turn, results in changes of its dynamic properties. An interesting example is represented by a macromolecular Gd(III) construct formed by 30 Gd(III) units covalently linked, through a squaric acid moiety, to a polyornithine substrate (114 residues) [22]. At acidic pH the unreacted amino groups of the polymer are protonated and, therefore, tend to be localized as far apart as possible, whereas at basic pH the progressive deprotonation of the NH_3^+ groups determines an overall rigidification of the polymer structure owing to the formation of intramolecular hydrogen bonds between adjacent peptidic linkages. As expected, the reduced rotational mobility of the polymeric backbone upon increasing pH enhances the relaxivity of the system.

Redox Potential Sensitive   A diagnostic MRI probe sensitive to the *in vivo* redox potential would be very useful for detecting regions with a reduced oxygen partial pressure (P_{O_2}). This parameter is important in several pathologies, including strokes and tumors.

Very few Gd(III) chelates sensitive to the tissue oxygenation have been reported so far. Our group has investigated the potential ability of GdDOTP to act as an allosteric effector of hemoglobin [23]. In fact, it has been observed that this chelate binds specifically to the T-form of the protein, which is characterized by a lower affinity towards oxygen. The interaction is driven by electrostatic forces and leads to a significant relaxivity enhancement (about fivefold) owing to the restricted molecular tumbling of the paramagnetic complex once it is bound to the protein. Although hemoglobin can be considered as an excellent indirect target for detecting P_{O_2}, the practical applicability of the method suffers for the inability of GdDOTP to enter red blood cells.

Enzyme Responsive   One possible route to design enzyme responsive agents is to synthesize paramagnetic inhibitors whose binding to the active site of the protein can be signaled by the consequent relaxivity enhancement. An example of this approach has been provided by Anelli *et al.*, who synthesized a linear Gd(III) complex bearing an arylsulfonamide moiety, which is a well-known inhibitor of carbonic anhydrase [24].

In vitro experiments demonstrated that this complex binds quite strongly (K_A of about 1.5×10^4) to the enzyme and its relaxivity in the bound form is about fivefold higher than the free complex ($27 \ mM^{-1} \ s^{-1}$ vs. $5 \ mM^{-1} \ s^{-1}$ at 20 MHz). Unfortunately, this approach *in vivo* failed, likely owing to the small amount of enzyme circulating in the blood.

An alternative approach is to design Gd(III) complexes acting as substrate for a specific enzyme. In this direction, an example has been provided by Lauffer *et al.*, who prepared a Gd(III) chelate containing a phosphoric ester sensitive to the attack of serum alkaline phosphatase. The hydrolysis yields the exposure of an hydrophobic moiety well suitable to bind to HSA. Upon binding, there is an increase of the relaxivity as a consequence of the lengthening of the molecular reorientational time. This approach was used by the same research group to design Gd(III) complexes sensitive to TAFI (thrombin-activatable fibrinolysis inhibitor), a carboxypeptidase B involved in clot degradation [25].

8.3
T$_2$-Susceptibility Agents

The term susceptibility refers to the tendency of a certain substance to become magnetized. The intensity of magnetization, M, is related to the strength of the inducing magnetic field, B, through a constant of proportionality, k, known as the magnetic susceptibility ($M = kB$). The magnetic susceptibility is a unit-less constant that is determined by the physical properties of the magnetic material. It can take on either positive or negative values. Positive values imply that the induced magnetic field, M, is in the same direction as the inducing field, B. Negative values imply that the induced magnetic field is in the opposite direction as the inducing field. Thus T_2 susceptibility agents work upon distorting the applied magnetic field, leading to a loss of signal of the water in T_2 weighted images.

8.3.1
Iron Oxide Particles

Superparamagnetic iron oxide (SPIO) particles have the general formula Fe $(III)_2O_3M(II)O$, where M(II) is a divalent metal ion such as iron, manganese, nickel, cobalt, or magnesium. SPIO is magnetite when M(II) is Fe(II) [26].

The change in the magnetic susceptibility caused by the superparamagnetic core induces a large distortion of the externally applied magnetic field, which, in turn, leads to hypo-intensities in T_2 weighted images. Thus the areas containing particles display fast transverse relaxation rates and low signal intensity ("negative contrast"). Owing to the large magnetic susceptibility of an iron oxide particle, the signal void is much larger than the particle size, enhancing detectability at the expenses of resolution. Iron oxide particles for MRI applications are divided into two classes, namely, superparamagnetic iron oxide (SPIO) and ultrasmall superparamagnetic iron oxide (USPIO), on the basis of the overall size of the protective cover on the surface of the magnetic particle. In fact, in both types of particles the iron oxide colloid particles are encapsulated by organic material such as dextran and carboxydextran to improve their compatibility with biological systems. The difference in overall size of the particles causes marked changes in two important properties related to their use as MRI contrast agents: USPIO have longer half lifetimes in the blood vessels than SPIO because their smaller size (<50 nm) makes their uptake from macrophages more difficult. Conversely, SPIO (with diameters of the order of 100–200 nm) are very rapidly removed from the circulation by the reticuloendothelial system (RES). In terms of relaxation enhancement properties, the larger magnetic susceptibility of SPIO yields larger R_2/R_1 relaxation rates and higher T_2-shortening effects than those brought about by the smaller USPIO particles. These particles have been used to track different cell types *in vivo* [27], including T lymphocytes, macrophages, and stem cells.

8.3.2
Paramagnetic Liposomes

In the late 1980s it was reported that paramagnetic, low molecular weight, Dy(III) complexes can act as T_2-susceptibility agents in MRI images when they are unequally distributed in vessels and in the surrounding tissues. The effect can be further enhanced when the paramagnetic complexes are entrapped in vesicles such as liposomes. The observed behavior is well accounted for in terms of the field gradients created by the compartment containing the paramagnetic ions that induce the spin dephasing of the water protons diffusing in the outer region of the compartment. It is straightforward to note that any nanosized system containing paramagnetic metal ions would act as a T_2-susceptibility agent. In this context,

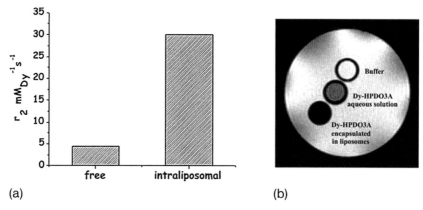

Figure 8.7 (a) Comparison between the R_2 values of 1 mM solutions of Dy-HPDO3A either free or entrapped into liposomal vesicles at 7 T and 298 K; (b) T_2 weighted image showing the difference between free and liposome-entrapped paramagnetic complex.

paramagnetic liposomes have a high potential owing to the high payload of para-magnetic complexes that can be either entrapped in their inner cavities or, upon suitable functionalization with lipophilic substituents, incorporated in their mem-brane bilayer [28]. The compartmentalization effect, which is proportional to the magnetic field strength, is clearly illustrated in Figure 8.7.

The sensitivity of such systems depends on the intrinsic paramagnetism (de-scribed by the effective magnetic moment, μ_{eff}) of the Ln ion (Dy(III) is the most effective) and on the overall concentration of the paramagnetic centers compart-mentalized in the vesicle. Hence, the incorporation of amphiphilic Dy(III) com-plexes in the liposome bilayer, in addition to the encapsulation of huge amounts of a hydrophilic Dy(III) agent in the aqueous cavity, yields a marked sensitivity en-hancement that makes these systems interesting agents for high field applications. The use of paramagnetic liposomes has been demonstrated as advantageous for developing highly-sensitive T_2-susceptibility agents as possible alternatives to the well established class of iron oxide nanoparticles [29].

Another class of liposome-based T_2-susceptibility agents is represented by mag-netoliposomes, in which iron oxide particles are encapsulated in liposomes [30].

8.4
CEST Agents

The new landscape of Molecular Imaging applications has prompted the search for new paradigms in the design of MR-imaging reporters. A possibility is the exploitation of the frequency, the key parameter of the NMR phenomenon. The roots for this class of agents rely on the well established magnetization transfer (MT) procedure, which deals with the transfer of saturated magnetization from tissue mobile protons (primarily water and labile protons from proteins) to bulk water upon radiofrequency (RF) irradiation of their semi-solid like broad NMR absorption [31]. The use of exogenous, mobile molecules, dubbed CEST (chemical exchange saturation transfer) agents, introduces the possibility of a selective RF irradiation of the sharper NMR signal of the exchangeable protons of the probe. Thus, one may design protocols in which the contrast in the MR image is generated "at will" only if the appropriate frequency corresponding to the labile protons of the exogenous agent is irradiated. Importantly, the new approach offers the possibility of detecting more than one agent in the same region.

8.4.1
Theoretical Background

The CEST contrast arises from the decrease in the intensity of the bulk water signal following saturation of the exchanging protons of the CEST agent by means of a selective RF pulse (Figure 8.8).

Hence, the basic requisite for a CEST agent is that the exchange rate between its mobile protons and the bulk water protons (k^{CEST}) has to be smaller than the frequency difference between the absorption frequencies of the exchanging spins (i.e., $k^{CEST} < \Delta\omega$).

In the CEST experiment, the RF pulse of angular frequency $\omega_2 = \gamma \times B_2$, where B_2 is the intensity of the applied RF pulse, is applied at the $\Delta\omega$ frequency

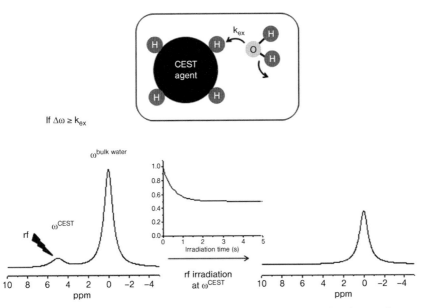

Figure 8.8 Schematic view of the saturation transfer mediated by chemical exchange. The plot reported on the arrow indicates the dependence of the bulk water magnetization on the irradiation time [Eq. (8.5), $T_1^{bw} = 1/R_1^{bw} = 1$ s; $\tau^{bw} = 1/k^{bw} = 1$ s].

offset corresponding to the resonance of the mobile protons of the contrast agent, to saturate their longitudinal magnetization (M_z^{CEST}).

Because such nuclei exchange slowly with the nuclei of the bulk water, the intensity of the longitudinal magnetization of the latter (M_z^{bw}) will decrease with a time evolution described by Eq. (8.5):

$$M_{Z(t)}^{bw} = M_0^{bw} \left[\frac{R_1^{bw}}{R_1^{bw} + k^{bw}} + \frac{k^{bw}}{R_1^{bw} + k^{bw}} e^{-(R_1^{bw} + k^{bw}) \times t} \right] \qquad (8.5)$$

where M_0^{bw} is the equilibrium longitudinal magnetization of the bulk water protons at $t = 0$ (t is the time during which the saturation pulse is switched on), R_1^{bw} is their longitudinal relaxation rate, and k^{bw} is their exchange rate.

Upon saturating the resonance of the CEST agent, the magnetization of the bulk water decreases from its equilibrium value to a steady state value ($M_{z(\infty)}^{bw}$) at $t = \infty$ [Eq. (8.6) and Figure 8.8]:

$$M_{Z(\infty)}^{bw} = M_0^{bw} \left(\frac{R_1^{bw}}{R_1^{bw} + k^{bw}} \right) \qquad (8.6)$$

Under the steady-state conditions, the extent of saturation transfer (ST) from the CEST agent to the bulk water (here termed ST_∞^*) can be expressed as in Eq. (8.7):

$$ST_{(\infty)}^* = \frac{M_0^{bw} - M_{Z(\infty)}^{bw}}{M_0^{bw}} = \frac{k^{bw}}{R_1^{bw} + k^{bw}} \qquad (8.7)$$

The ST_∞^* value at the steady state may assume values between 0 and 1 depending on whether the exchange rate k^{bw} is slower or faster than the relaxation rate R_1^{bw}.

The pseudo-first-order exchange rate constant k^{bw} can be defined as the exchange rate of the protons belonging to the CEST agent (k^{CEST}) weighted by the molar ratio fraction between the two pools of exchanging spins [Eq. (8.8)]:

$$k^{bw} = k^{CEST} \frac{[CEST \ spins]}{[BW \ spins]} \qquad (8.8)$$

ST_∞^* augments upon increasing k^{CEST} until the exchange approaches the coalescence condition ($k^{CEST} \approx \Delta\omega$). Notably, this condition depends on the magnetic field strength, which is directly related to $\Delta\omega$. In other words, the increase of the field strength improves the efficiency of the CEST experiment because it allows the exploitation of large k^{CEST} values.

In an aqueous solutions of a CEST agent, the concentration of bulk water spins is approximately $111.2 \ mol \ l^{-1}$ (i.e., $2 \times 55.6 \ mol \ l^{-1}$), whereas the concentration of the saturated protons is the product between the concentration of the CEST agent and the number of its magnetically equivalent (or pseudo-equivalent) mobile protons, n, that are saturated by the applied RF pulse. Then, k^{bw} can be expressed as in Eq. (8.9):

$$k^{bw} = k^{CEST} \frac{n \times [CEST]}{111.2} \tag{8.9}$$

This model is based on the assumption that the magnetization of the bulk water protons is not perturbed by the applied RF saturation field (i.e., no direct saturation effect on bulk water) and that the magnetization of the mobile protons of the CEST probe is completely saturated by the B_2 field. The larger is $\Delta\omega$ the higher is the probability that these assumptions hold.

However, CEST experiments are often performed using saturation pulses that balance an efficient saturation of the mobile protons of the CEST agents with a minimum spillover effect on the bulk water signal. For this reason, accurate quantification of the CEST effect needs the development of theoretical models that take into account such effects.

Usually, the extent of saturation transfer (ST) is experimentally assessed through the following expression [Eq. (8.10)]:

$$ST = \frac{(I_{off} - I_{on})}{I_{off}} \tag{8.10}$$

where I_{off} and I_{on} refer to the bulk water signal measured after two saturation experiments in which the RF irradiation field is set at the resonance frequency of the mobile protons of the CEST agent (I_{on}) and at a frequency equally spaced from the bulk water resonance (arbitrarily set to zero) but located at the opposite side (I_{off}) (Figure 8.9b). A very useful tool for characterizing the property of a CEST agent and for the identification of the resonance frequency of its mobile protons is the Z-spectrum, which reports the intensity of the bulk water protons as a function of the saturation frequency offset (Figure 8.9a).

Any Z-spectrum is characterized by a peak corresponding to the direct saturation of the bulk water signal, whose frequency is set to zero. In the presence of a CEST agent, one (or more) CEST peak appears in the spectrum, thus allowing one to determine the frequency offset of the mobile protons of the agent. In the simulation reported in Figure 8.9, the values of I_{on} and I_{off} are indicated in part (a), whereas part (b) shows the so-called ST-spectrum (or asymmetry spectrum), which plots the ST values (often expressed as percentage) as a function of the saturation offsets. The ST-spectrum provides a useful graphical access to the saturation transfer efficiency of the system under study.

8.4.1.1 The Sensitivity Issue

One of the main limiting factors of CEST agents for *in vivo* application is represented by their lower sensitivity compared to the conventional Gd- and Fe-based agents. Among the different parameters governing the CEST effect, the exchange rate of the mobile protons of the agent, k^{CEST}, has received, so far, much attention because this parameter can be quite easily modulated by changing the chemical characteristics of the exchanging group. In addition, k^{CEST} is usually

Z-spectrum

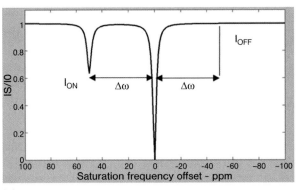

(a)

ST-spectrum

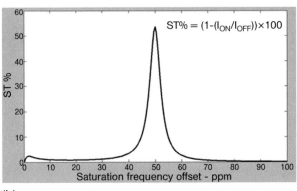

(b)

Figure 8.9 Simulated representations of the Z- (a) and ST-spectrum (b). The curves were calculated with the model described. The following parameters were used: $f = 5 \times 10^{-4}$, $k^{CEST} = 500$ s^{-1}, $R_1^{bw} = 0.33$ s^{-1}, $R_2^{bw} = 1$ s^{-1}, $R_1^{CEST} = 2$ s^{-1}, $R_2^{CEST} = 50$ s^{-1}, $B_0 = 300$ MHz, $B_2 = 6$ µT, $\Delta\omega^{CEST} = 50$ ppm.

dependent on physicochemical variables like temperature or pH, thus allowing CEST agents to be used as responsive probes (see previous section).

In principle, ST and k^{CEST} are directly correlated, but this relation is valid until k^{CEST} approaches the $\Delta\omega$ value. In fact, when coalescence occurs, the mobile protons of the CEST agent are isochronous with the bulk water signal and the CEST contrast vanishes. This is one of the reasons why attention has been focused on paramagnetic CEST agents whose large $\Delta\omega$ values allow the exploitation of higher k^{CEST} values before the coalescence between the signals occurs [28]. Figure 8.10 shows the theoretical dependence of CEST efficiency on k^{CEST} and B_2. For a given B_2, the increase of k^{CEST} causes an initial improvement of CEST efficiency, but

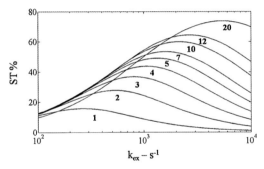

Figure 8.10 Dependence of saturation transfer (ST%) on k^{CEST} calculated at the B_2 amplitudes reported in the figure. Simulations were performed using the model described in Reference [7]. The following parameters were used: $f = 5 \times 10^{-4}$, $R_1^{bw} = 0.33$ s^{-1}, $R_2^{bw} = 1$ s^{-1}, $R_1^{CEST} = 2$ s^{-1}, $R_2^{CEST} = 50$ s^{-1}, $B_0 = 300$ MHz, $\Delta\omega^{CEST} = 50$ ppm.

beyond a certain limit, the ST effect gets worse. Furthermore, the simulated profiles indicate that, for a given $\Delta\omega$ offset, the maximum CEST contrast is achieved when $k^{CEST} = 2\pi B_2$. However, it is has to be recalled that for *in vivo* applications the B_2 values are essentially limited by SAR (specific absorption rate) constraints. This means that the CEST efficiency is controlled by the SAR requirements that define the maximum B_2 amplitude that, in turn, allows the estimation of the optimal k^{CEST} value. In addition to B_2 and k^{CEST}, the CEST efficiency is also influenced by the chemical shift offset $\Delta\omega$ of the mobile protons of the agent and by the magnetic field strength B_0 at which the CEST experiment is carried out.

CEST sensitivity can be improved considerably by increasing the number n of magnetically equivalent or pseudo-equivalent mobile protons per CEST molecule. The sensitivity of an MR agent is generally expressed as the minimum concentration necessary for detecting the MR contrast in the image; for MR-CEST experiments a ST effect of 5–10% can be considered the detection threshold. The CEST contrast is also modulated by the longitudinal relaxation time of the bulk water protons, R_1^{bw}. Low R_1 values are beneficial for the ST efficiency. Therefore, it might be expected that diamagnetic agents could be more efficient than paramagnetic ones. However, since the paramagnetic complexes used for CEST application are not particularly efficient as relaxing agents, usually this parameter does not have a relevant role in determining the ST effect. Nevertheless, high R_1 values allow a fast achievement of the steady-state condition despite a lower steady-state CEST effect.

8.4.2
Paramagnetic CEST Agents: PARACEST

CEST agents are usually classified in two main groups: diamagnetic and paramagnetic systems. The members of each class can be further divided into subgroups depending on other criteria such as their size or peculiar chemical characteristics. In this chapter we focus only on the class of paramagnetic CEST agents (PARACEST). All the PARACEST agents proposed so far are lanthanide (III) complexes endowed with two different kinds of mobile protons: (i) protons of water molecules coordinated to the Ln(III) ion and (ii) mobile protons belonging to the ligand structure.

A representative example of PARACEST agent was first reported by Sherry and coworkers in 2001 [32]. The agent was a tricationic Eu(III) macrocyclic complex, $[EuDOTA\text{-}4AmCE]^{3+}$ (Figure 8.11a), whose metal-bound water protons are in slow exchange on the NMR shift timescale (k^{CEST} of about 2600 s^{-1} and $\Delta\omega_{offset} = 50$ ppm). The Z-spectrum acquired at 22 °C and 4.7 T of a 63 mM solution of this compound showed a large CEST effect (57%) (Figure 8.11c).

Macrocyclic tetraamide derivatives of DOTA represent the primary ligand choice for PARACEST agents. In fact, this coordination cage ensures sufficiently slow water exchange rates (and also a high chemical inertness of the metal complex) as well as large shifts to fulfill the $\Delta\omega > k^{CEST}$ condition. Furthermore, the ligand provides four amide groups (typically magnetically equivalent or pseudo-equivalent) that can act as exchangeable proton pool, in addition to the coordinated water protons. The first example of the detection of a CEST contrast upon irradiation of amide protons in a PARACEST agent was reported in 2001 [5]. In this paper we reported the CEST properties of a series of Ln(III) complexes of the ligand

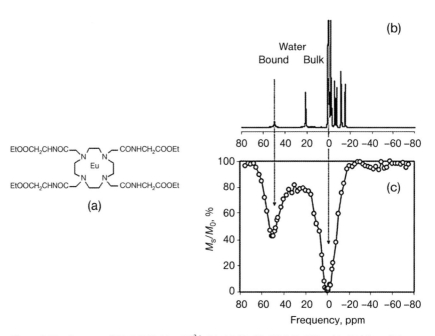

Figure 8.11 Spectra of $[EuDOTA\text{-}4AmCE]^{3+}$ (a): NMR (b) (11.7 T, 250 mM, 25 °C) and Z-spectra (c) (4.7 T, 63 mM, 22 °C) at pH 7.0.

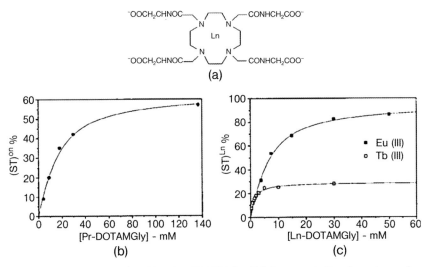

Figure 8.12 [Ln-DOTAMGly]⁻ (a) series (7 T, 312 K, pH 7.4): concentration dependence of the saturation transfer upon irradiation of the amide protons (b) (saturation scheme: $t = 4$ s single pulse, B_2 intensity 12 µT) and metal-coordinated water protons (c) (saturation scheme: 4000 shaped pulses 1 ms each, interpulse delay 10 µs, B_2 intensity 88–185 µT).

DOTAMGly (Figure 8.12, elsewhere named DOTA-4AmC), which is the product of the ester hydrolysis of the above-cited ligand DOTA-4AmCE.

The presence of four carboxylate groups confers to the Ln(III) complex an overall single negative charge that is expected to improve the *in vivo* tolerability of the agent. Among the different Ln(III) ions investigated, the Yb(III) complex was the most efficient agent. Other additional important findings of this work were: (i) the observation that the $\Delta\omega$ offset for the same proton site is strongly dependent on the magnetic properties of the Ln(III) ion, (ii) the demonstration of the remarkable pH dependence of the CEST contrast upon irradiation of the amide protons, and (iii) the exploitation of two different proton pools (amide and metal-coordinated water protons, respectively) for detecting a pH dependent and concentration independent ST upon applying the ratiometric approach described by Balaban *et al.* [4]. The CEST sensitivity displayed by [Ln-DOTAMGly]⁻ complexes is dependent on the irradiated proton site: for amide protons a CEST effect of about 5% requires a complex concentration of few millimolar, whereas for the metal-coordinated water protons the threshold lowers to 0.5–1 mM.

Bartha and coworkers have investigated the CEST properties of a series of Ln (III)-DOTA tetrasubstituted tetraamide complexes containing different dipeptides, and the tripeptide glycine-phenylalanine-lysine. The authors observed that the CEST efficiency of such systems upon saturation of the metal-bound water protons was very sensitive to the nature of the oligopeptides. In some cases, for example, for the Gly-Phe derivative, good CEST efficiencies were obtained by using safe saturation fields (e.g., specific absorption rate, SAR, under the allowed values) [33]. Moreover, Burdinski *et al.* showed that a Tm(III) complex of a tetra-imidazolyl

DOTAM ligand has a detectable CEST effect attributable to the imidazole protons likely involved in a hydrogen bonding network with the amide protons of the complex [34].

A significant sensitivity enhancement for PARACEST agents could be attained by designing polymeric systems containing a high number of PARACEST units. A couple of examples have been presented so far, involving the linkage of PARA-CEST agents, similar to those discussed in this section, to macromolecules like dendrimers [35], or to nanosized systems like perfluorocarbon nanoparticles [36]. Of particular note is the recent publication by Pagel and coworkers, who detected a CEST contrast *in vivo* after the i.v. injection of nanosized dendrimers (G2 and G5 generations) bearing Eu(III) or Yb(III)-DOTAMGly derivatives into a mouse model of MCF-7 mammary carcinoma [37]. Figure 8.13 shows the build-up of the CEST contrast in the xenografted tumor due to the accumulation of the para-magnetic nanosystem. Interestingly, both the exchanging pools, the amides pro-tons in the Yb(III) complex ($\Delta\omega = -16$ ppm) and the metal-coordinated water protons in the Eu(III) analog ($\Delta\omega = 55$ ppm), were successfully (and safely) saturated.

8.4.2.1 Responsive PARACEST Agents

The interest in designing CEST agents capable of reporting about a given physi-cochemical parameter that characterizes the local microenvironment of the probe is driven by two main factors: (i) the relative facility to correlate the parameters that control the ST efficiency to the parameters to be monitored and (ii) the possibility to make the CEST-based contrast independent of the probe concentration. This latter task can be pursued by (i) using a ratiometric approach, which requires the use of agents containing at least two sets of magnetically non-equivalent CEST-active protons (and differently responsive to the parameter of interest) or (ii) ex-ploiting peculiar properties of the Z-spectrum of the agent.

In this section we provide an overview on the responsive CEST agents in-vestigated and tested so far.

pH Reporters   The first example illustrating the potential of responsive CEST agents was reported in the pioneering work of Ward and Balaban [4], who showed the possibility of using CEST agents as concentration-independent pH reporters. They demonstrated that a diamagnetic molecule, 5,6-dihydrouracil, containing two pools of CEST-active protons whose exchange rates exhibit a different pH-dependence could be successfully used for applying the ratiometric approach to the CEST response. In 2002, we applied this concept by exploiting the peculiar property of PARACEST systems that possess the same solution structure and hydrophilic/lipophilic ratio (hence the same *in vivo* biodistribution), but exhibit very different NMR properties depending on the coordinated Ln(III) ion [38]. The two isostructural [Yb-DOTAMGly]⁻ and [Eu-DOTAMGly]⁻ complexes were se-lected for their ability to promote efficient CEST contrast upon saturation of the amide protons ($\Delta\omega = -15$ ppm) of the former and metal-coordinated water protons ($\Delta\omega = 50$ ppm) of the latter. Interestingly, the amide protons CEST

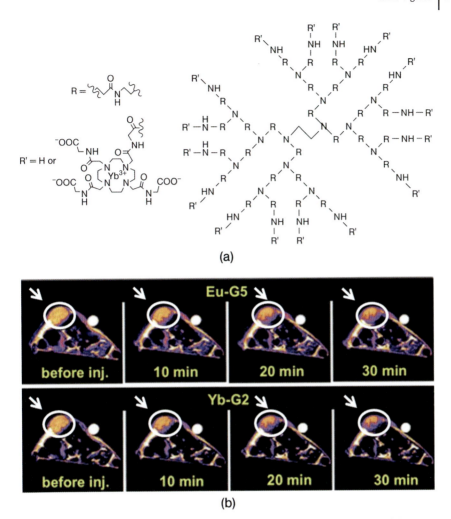

(a)

(b)

Figure 8.13 (a) Schematic representation of the structure of Yb-G2. (b) Variation of the signal intensity in the MRI-CEST image of mice bearing MCF-7 mammary carcinoma (the carcinoma is enclosed in the white circle) after injecting Eu-G5 or Yb-G2 probes. The water signal intensity decreases after injection of the probes due to the saturation transfer.

response is pH-dependent, whereas that associated to the water exchange is pH-invariant (at least in the pH interval 5–8), thus improving the pH-responsiveness of the system.

Subsequently, we showed that PARACEST pH probes can be designed by using a single Ln-complex containing two CEST-active proton sites on the same molecule [39]. Among the investigated [Ln-DOTAMGly]$^-$ complexes (Ln = Pr, Nd, Eu), the Pr(III)-chelate displayed the most sensitive ratiometric plot, which was nicely tuned within the physiopathological pH interval (Figure 8.14).

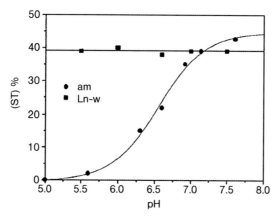

Figure 8.14 pH dependence of the saturation transfer for a 30 mM solution of [Pr-DOTAMGly]$^-$ (7 T, 312 K). Filled squares: saturation of the exchanging metal-coordinated water; filled circles: saturation of the mobile amide protons.

Within this research field, Sherry and coworkers have highlighted that the two magnetically non-equivalent amide protons of the primary amide groups in the [Yb-DOTAM]$^{3+}$ complex have a slightly different pH dependence and, consequently, a different pH dependence of their ST effects, when selectively saturated [40].

Temperature Reporters   The interest in designing MRI thermometers is basically driven by the necessity to develop low invasive methodologies for measuring and controlling temperature *in vivo*. The availability of methods for *in vivo* assessment of temperature can be very helpful for monitoring heat-based anticancer therapies like hyperthermia and thermal ablation. The first example of a concentration-independent temperature reporter CEST agent was described by us in 2004 [41]. Unlike the pH dependence, the temperature responsiveness of the PARACEST agent [Pr-DOTAMGly]$^-$ was mainly dominated by the variation of the CEST effect arising from the irradiation of the metal-coordinated water protons.

Enzyme Reporters   Most work in this field has been carried out so far by Pagel and coworkers, who designed PARACEST probes specifically engineered to act as substrate for proteases, such as caspases that are enzymatic biomarkers of apoptosis [42]. Such Ln-complexes contained a CEST-active amide proton ($\Delta\omega = -51$ ppm) that, upon cleavage of the peptidic bond by the proteolytic enzyme, turns into a primary amine group ($\Delta\omega = 8$ ppm). As a consequence, the CEST effect measured at -51 ppm becomes sensitive to the presence of the caspase. Though the sensitivity displayed by this agent is rather low (about 5 mM, 14 T, B_2 intensity 31 μT) mainly due to the presence of only one CEST-active proton per molecule, it was claimed that it could detect amounts of caspase-3 of the order of 5–50 nM.

Tóth and coworkers have reported a versatile and general approach for the preparation of PARACEST agents responsive to enzymes. The approach was based on the coupling between an enzyme-specific substrate and a lanthanide-chelating unit through a self-immolative spacer. After the enzymatic action, the substrate is removed and the spacer is spontaneously eliminated, thus resulting in a change in the PARACEST properties of the probe. Upon changing the substrate linked to the Ln-chelating unit it is possible to make these systems specifically responsive to a wide variety of enzymes [43].

Metabolite Reporters   The examples illustrating the *in vitro* potential of CEST agents to act as sensitive reporters of concentration changes of metabolites and small ions of diagnostic relevance are quite numerous.

The first one was reported by our group in 2002, and dealt with the design of a coordinatively unsaturated PARACEST agent, $[Yb\text{-}MBDO3AM]^{3+}$ (Figure 8.15a), able to report about lactate concentration, an important biomarker of anaerobic glycolysis [44]. The interaction between the probe, containing six CEST-active amide protons, and lactate was quite strong (K_A of 8000) and, interestingly, the exchange rate between free and bound lactate was slow on the NMR frequency timescale. Thus it was possible to detect separately in the NMR spectrum (and consequently also in the Z-spectrum) both the free and lactate-bound forms of the

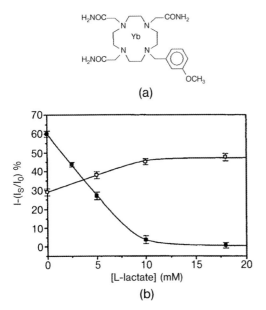

(a)

(b)

Figure 8.15   Saturation transfer for a 9.3 mM solution of $[Yb\text{-}MBDO3AM]^{3+}$ (a) as a function of L-lactate concentration (7 T, 312 K). (b) Open circles: saturation of the amide protons of the complex bound to lactate (resonating at −15.5 ppm from bulk water); filled circles: saturation of the amide protons of the free complex (resonating at −29.1 ppm from bulk water).

CEST probe. As expected, the CEST effect observed upon irradiating the amide protons of the free probe ($\Delta\omega = 29.1$ ppm) decreased upon increasing the lactate concentration, whereas the reverse occurred when the saturation offset was placed on the corresponding resonances of the probe bound to lactate ($\Delta\omega = 15.5$ ppm) (Figure 8.15b). Despite some limitations associated with the broadness of the amide resonance of the free probe, it is worth noting that this system is sensitive to changes in lactate concentration up to 10 mM, that is, just in the range of diagnostic relevance.

In 2003, Sherry and coworkers reported a PARACEST agent whose CEST effect was made sensitive to the glucose concentration through the presence of two phenylboronic moieties on a macrocyclic Eu(III) complex [45]. Upon interaction with the sugar ($K_A = 380$, which is very strong compared with other phenylboronic-based probes), the exchange rate of the metal-bound water protons is slowed down, thus affecting the bandwidth of the corresponding CEST peak in the Z-spectrum.

8.4.2.2 Cell Labeling

The unique peculiarity of CEST agents, that is, the possibility to visualize more than one agent in the same image, is extremely advantageous for designing MRI cell-tracking experiments in which two (or more) cell lines, each one labeled with a specific CEST probe, are simultaneously injected and visualized over time by MRI. As proof of principle, we demonstrated the potential of CEST agents by labeling rat HTC cells (HTC = hepatoma tumor cells) with two PARACEST probes sharing the same ligand (the already cited DOTAMGly) but differing in the coordinated Ln(III) ion [Eu(III) or Tb(III)] [46]. As pointed out in previous sections, the $\Delta\omega$ values of the water protons coordinated to the metal in the two complexes are different ($\Delta\omega = 50$ ppm for [Eu-DOTAMGly]$^-$ and $\Delta\omega = -600$ ppm for [Tb-DOTAMGly]$^-$) and allow the selective saturation of the two CEST-active sites. Cells were labeled by exploiting the pinocytosis route, which allowed a

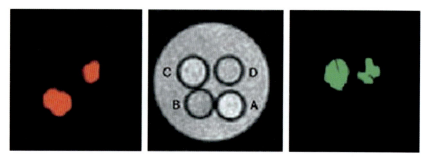

Figure 8.16 (a) MR-CEST difference image, in false colors, after saturation of the Tb-bound water protons. (b) MR proton density image of four capillaries containing: (A) unlabeled HTC cells, (B) HTC cells labeled with [Tb-DOTAMGly]$^-$, (C) HTC cells labeled with [Eu-DOTAMGly]$^-$, and (D) cellular pellet obtained by mixing pellets B and C. (c) MR-CEST difference image, in false colors, after saturation of the Eu-bound water protons.

sufficient number of probe molecules to be internalized and detected in an MR-CEST experiment. Figure 8.16 illustrates the results obtained by visualizing a phantom made of four capillaries containing unlabeled cells as control, cells labeled with each of the two probes, and a pellet obtained by mixing the two differently labeled cells. The CEST response, displayed in false colors, is a clear demonstration of the potential of CEST agents in this field and it represents the first experimental evidence of a simultaneous visualization of different CEST agents in the same image region.

8.5
Concluding Remarks

In the last two decades, the chemistry of paramagnetic metal chelates has been the subject of an extraordinary interest for the potential use of these complexes as contrast agents for MRI investigations. In particular, this effort has led to the synthesis of several hepta- and octa-coordinating ligands able to wrap around the Ln(III) ions to yield chelates characterized by very high thermodynamic stabilities. Moreover, much attention has been devoted to relate the structure of Ln(III) chelates to the exchange rate of the coordinated water molecule(s). These studies have dramatically extended the range of available exchange rates, which now goes from a few nanoseconds to several microseconds. The acquisition of an in-depth knowledge of the structural, electronic, and dynamic determinants of the observed relaxivity has prompted interesting activities aimed at making it responsive to specific characteristics of the biological environment of the contrast agent. This approach may find unique applications in medical diagnosis. Finally, the huge work on Gd(III) chelates has led to the exploitation of these ligands with lanthanide(III) ions other than Gd, favoring the development of the novel class of contrast agents called CEST. Their potential can be very high as it allows us to exploit more contrast agents in the same MR-image as each of them will be activated "at will" by selecting the characteristic irradiation frequency. These peculiarities will lead to the development of new diagnostic protocols precluded with conventional agents (i.e., multiple detection of different cellular epitopes or the simultaneous tracking of different cells subpopulations).

References

1 Brücher, E. and Sherry, A.D. (2001) in *The Chemistry of Contrast Agents in Medical Magnetic Resonance Imaging* (eds A.E. Merbach and E. Tóth), John Wiley & Sons, Ltd., Chichester, UK.

2 Engelstadt, B.L. and Wolf, G.L. (1988) in *Magnetic Resonance Imaging* (eds D.D. Stark and W.G. Bradley Jr), The V.C. Mosby Company, St. Louis.

3 Rinck, P.A. (1993) *Magnetic Resonance in Medicine*, Blackwell Scientific Publications, London.

4 Ward, K.M. and Balaban, R.S. (2000) *Magn. Reson. Med.*, **44**, 799–802.

5 Terreno, E., Aime, S., Barge, A., Delli Castelli D., and Nielsen F.U. (2001) *J. Inorg. Biochem.*, **86**, 452–452.

6 Banci, L., Bertini, I., and Luchinat, C. (1991) *Nuclear and Electronic Relaxation*, VCH, Weinheim,.

7 Botta, M. (2000) *Eur. J. Inorg. Chem.*, **3**, 399–407.

8 Weinmann, H.J., Mühler, A., and Radüchel, B. (2000) in *Biomedical Magnetic Resonance Imaging and Spectroscopy* (ed. I.R. Young), John Wiley & Sons, Ltd, Chichester, UK.

9 Schmitt-Willich, H., Brehm, M., Evers, C.L.J., Michl, G., Muller-Fahrnow, A., Petrov, O., Platzek, J., Raduchel, B., and Sulzle, D. (1999) *Inorg. Chem.*, **38**, 1134–1144.

10 Uggeri, F., Aime, S., Anelli, P.L., Botta, M., Brocchetta, M., Dehan, C., Ermondi, G., Grandi, M., and Paoli, P. (1995) *Inorg. Chem.*, **34**, 633–642.

11 Perazella, M.A. (1995) *Curr. Opin. Nephrol. Hypertens.*, **18**, 519–525.

12 Aime, S., Botta, M., Fasano, M., and Terreno, E. (1998) *Chem. Soc. Rev.*, **27**, 19–29.

13 Padovani, B., Lecesne, R., and Raffaelli, C. (1996) *Eur. J. Radiol.*, **23**, 205–211.

14 Aime, S., Calabi, L., Cavallotti, C., Gianolio, E., Giovenzana, G.B., Losi, P., Maiocchi, A., Palmisano, G., and Sisti, M. (2004) *Inorg. Chem.*, **43**, 7588–7590.

15 Datta, A. and Raymond, K.N. (2009) *Acc. Chem. Res.*, **42**, 938–947.

16 Carter, D. and Ho, J.X. (1994) *Adv. Prot. Chem.*, **45**, 153–203.

17 Zhang, Z.D., Greenfield, M., Spiller, T. J., McMurry, R.B., Lauffer, R.B., and Caravan, P. (2005) *Angew. Chem. Int. Ed.*, **44**, 6766–6769.

18 Gianolio, E., Giovenzana, G.B., Longo, D., Longo, I., Menegotto, I., and Aime, S. (2007) *Chemistry*, **13**, 5785–5797.

19 Lowe, M.P., Parker, D., Reany, O., Aime, S., Botta, M., Castellano, G., Gianolio, E., and Pagliarin, R. (2001) *J. Am. Chem. Soc.*, **123**, 7601–7609.

20 Zhang, S.R., Wu, K.C., and Sherry, A.D. (1999) *Angew. Chem Int. Ed.*, **38**, 3192–3194.

21 Raghunand, N., Howison, C., Sherry, A. D., Zhang, S.R., and Gillies, R.J. (2003) *Magn. Reson. Med.*, **49**, 249–257.

22 Aime, S., Botta, M., Geninatti Crich, S., Giovenzana, G., Palmisano, G., and Sisti, M. (1999) *Bioconjugate Chem.*, **10**, 192–199.

23 Aime, S., Ascenzi, P., Comoglio, E., Fasano, M., and Paoletti, S. (1995) *J. Am. Chem. Soc.*, **117**, 9365–9366.

24 Anelli, P.L., Bertini, I., Fragai, M., Lattuada, L., Luchinat, C., and Parigi, G. (2000) *Eur. J. Inorg. Chem.*, **4**, 625–630.

25 Nivorozhkin, A.L., Kolodziej, A.F., Caravan, P., Greenfield, M.T., Lauffer, R.B., and McMurry, T.J. (2001) *Angew. Chem. Int. Ed.*, **40**, 2903–2906.

26 Muller, R.N., Vander Elst, L., Roch, A., Peters, J.A., Csajbok, E., Gillis, P., and Gossuin, Y. (2005) *Adv. Inorg. Chem.*, **57**, 239–292.

27 Cunningham, C.H., Arai, T., Yang, P.C., McConnell, M.V., Pauly, J.M., and Conolly, S.M. (2005) *Magn. Reson. Med.*, **53**, 999–1005.

28 Terreno, E., Cabella, C., Carrera, C., Delli Castelli, D., Mazzon, R., Rollet, S., Stancanello, J., Visigalli, M., and Aime, S. (2007) *Angew. Chem. Int. Ed.*, **46**, 966–968.

29 Delli Castelli, D., Terreno, E., Cabella, C., Chaabane, L., Lanzardo, S., Tei, L., Visigalli, M., and Aime, S. (2009) *NMR Biomed.*, **22**, 1084–1092.

30 Mehta, R.C., Pike, G.B., and Enzmann, D.R. (1996) *Top. Magn. Reson. Imag.*, **8**, 214–230.

31 Woods, M., Woessner, D.E., and Sherry, A.D. (2006) *Chem. Soc. Rev.*, **35**, 500–511.

32 Zhang, S., Winter, P., Wu, K., and Sherry, A.D. (2001) *J. Am. Chem. Soc.*, **123**, 1517–1518.

33 Suchy', M., Li, A.X., Bartha, R., and Hudsona, R. (2008) *Bioorg. Med. Chem.*, **16**, 6156–6166.

34 Burdinski, D., Lub, J., Pikkemaat, J.A., Langereis, S., Gruell, H., and ten Hoeve, W. (2008) *Chem. Biodiv.*, **5**, 1505–1512.

35 Pikkemaat, J.A., Wegh, R.T., Lamerichs, R., van de Molengraaf, R.A., Langereis, S., Burdinski, D., Raymond, A.Y., Janssen, H.M., de Waal, B.F., Willard, N.P., Meijer, E.W., and Grull, H. (2007) *Contrast Media Mol. Imag.*, **2**, 229–239.

36 Winter, P.M., Cai, K., Chen, J., Adair, C. R., Kiefer, G.E., Athey, P.S., Gaffney, P.J., Buff, C.E., Robertson, J.D.,

Caruthers, S.D., Wickline, S.A., and Lanza, G.M. (2007) *Magn. Reson. Med.*, **56**, 1384–1388.

37 Meser Ali, M., Yoo, B., and Pagel, M.D. (2009) *Mol. Pharm.*, **6**, 1409–1416.

38 Aime, S., Barge, A., Delli Castelli, D., Fedeli, F., Mortillaro, A., Nielsen, F.U., and Terreno, E. (2002) *Magn. Reson. Med.*, **47**, 639–648.

39 Aime, S., Delli Castelli, D., and Terreno, E. (2002) *Angew. Chem. Int. Ed.*, **41**, 4334–4336.

40 Michaudet, L., Burgess, S., and Sherry, A.D. (2002) *Angew. Chem. Int. Ed.*, **41**, 1919–1921.

41 Terreno, E., Delli Castelli, D., Cravotto, G., and Aime, S. (2004) *Invest. Radiol.*, **39**, 235–243.

42 Yoo, B. and Pagel, M.D. (2006) *J. Am. Chem. Soc.*, **128**, 14032–14033.

43 Chauvin, T., Durand, P., Bernier, M., Meudal, H., Doan, B.T., Noury, F., Badet, B., Beloeil, J.C., and Tóth, E. (2008) *Angew. Chem. Int. Ed.*, **47**, 4370–4372.

44 Aime, S., Delli Castelli, D., Fedeli, F., and Terreno, E. (2002) *J. Am. Chem. Soc.*, **124**, 9364–9365.

45 Zhang, S., Trokowski, R., and Sherry, A. D. (2003) *J. Am. Chem. Soc.*, **125**, 15288–15289.

46 Aime, S., Carrera, C., Delli Castelli, D., Geninatti Crich, S., and Terreno, E. (2005) *Angew. Chem. Int. Ed.*, **44**, 1813–1815.

9
Metal-Based Radiopharmaceuticals

Roger Alberto

9.1
Introduction

The non-invasive diagnosis and therapy of socially relevant diseases such as cancer or neurodegeneration is one of the major objectives of nuclear medicine and radiopharmacy [1, 2]. As important as diagnosis or therapy is prognosis and radiopharmacy provides modalities to early assess success or failure of, for example, common chemotherapy. The basic tools of radiopharmacy are radionuclides, radioactive isotopes of particular elements, produced in cyclotrons or in nuclear reactors. Some of these radionuclides for radiopharmaceutical purposes are isotopes of main group elements – [123]I, ^{125}I, and ^{131}I are probably the best known amongst them. With the advent of small, on-site cyclotrons, clinics also apply more and more ^{11}C and ^{18}F in nuclear medicine. The latter two nuclides are positron (β^+) emitters and are used in positron emission tomography (PET) on a routine clinical basis. Since this section attempts to scope the presence and future of metal-based radiopharmaceuticals, albeit of high importance, these nuclides will not be further considered herein and the reader is referred to published reviews and book chapters [3].

According to their weight in the periodic table, most radionuclides origin from metallic or semi-metallic elements. In contrast to carbon, iodine, or fluorine, it is clear to every inorganic chemist that the combination of a metal with a ligand, a molecule, or a biologically active biomolecule is not routine but requires several building blocks and steps to be achieved successfully. This is the challenge of metal-based radiopharmaceutical chemistry but, at the same time, also the incentive to find innovative and novel approaches towards ligands, complexes, and radio-bioconjugates to benefit from in health care. Some metallic radionuclides are β^+-emitters whereas others, for example, the so-called single photon emission computer tomography (SPECT), are γ-emitters. Among them ^{99m}Tc is one of the most important. There is an ongoing controversy between the PET and the SPECT communities about advantages and disadvantages or, rather, about superiority. This issue will also not be discussed here, but can be found in more specialized literature [2, 4, 5]. Detailed explanation of the biological and physical background

Bioinorganic Medicinal Chemistry. Edited by Enzo Alessio
Copyright © 2011 WILEY-VCH Verlag GmbH & Co. KGaA, Weinheim
ISBN: 978-3-527-32631-0

of the two imaging techniques SPECT and PET or combined modalities such as SPET/CT and PET/CT would exceed the space available here and the reader is also referred to excellent recent reviews from praxis and theory [6–9].

Diagnosis–therapy aspects are further categories that have to be considered. Typically, β^+ and γ-emitters are used in diagnosis, whereas β^- and α-emitters are applied in radionuclide-based therapy. Details will be outlined in the following sections, which are organized according to therapeutic and diagnostic topics. For some elements, copper in particular, radionuclides exist for both diagnosis and therapy. In some rare cases (^{131}I being a good example) the same radionuclide is – or rather was – used for both purposes the same time.

A comprehensive state-of-the-art analysis in metal-based radiopharmaceuticals would greatly exceed the space available for this chapter. It is rather the intention to provide the reader with basic information about chemical and biological requirements to critically evaluate the pros and cons of metal-based radiopharmaceuticals. These criteria shall be underscored with a few selected examples from literature and for the different categories of metallic radionuclides arranged according to the therapy–diagnosis concept. This chapter also aims to provoke the scientific curiosity of (especially young) researchers to contribute to this field by new approaches that could ultimately lead to novel, urgently needed radiopharmaceuticals.

9.1.1
Selection of the Radionuclide

Among the approximately 1700 known radionuclides, very few meet the requirements for diagnostic or therapeutic application in humans. From a physical point of view, the radionuclide must have (i) an optimal half-life, (ii) an emission of a particle or a photon of reasonable energy for the particular purpose of application, and (iii) a cost-efficient synthesis to keep the price for the radiopharmaceutical low (see below). Whereas the first two points are given by nature, the last point is a commercial issue.

The optimal half-life of the radiopharmaceutical depends on the application and is strongly affected by the pharmacokinetics of the compound or the biomolecule to which the radionuclide is attached to. Half-lives should be sufficiently short to release the patient from the hospital as fast as possible but must be long enough to allow good images with high target/background ratio at the site of interest. Thus, for diagnostic imaging, half-lives from minutes to a few hours are preferable. Of course, the *in vivo* behavior of the radiopharmaceutical is inherently coupled to the selection of the radionuclide. However, there is not always a choice between different radionuclides, but the selection is dictated by the particular application. If the radiopharmaceutical irreversibly accumulates at the target site, absolute amounts of administered radioactivity can be decreased, which leads to a lower dose burden for the patient.

The situation is somewhat different for therapeutic radionuclides. These are particle emitters but, preferentially, are accompanied by the emission of a photon of appropriate energy to allow detection by γ-cameras. The percentage of γ-emission should be low to reduce the whole body dose for the patient. If the α- or β-decay is accompanied by no or very little γ-emission (as in case of ^{90}Y), it is not easily possible to assess the accumulation in, for example, tumor tissue. Optimal half-lives for therapeutic radiopharmaceuticals can be on the order of hours to days, again depending on the pharmacokinetics of the targeting biomolecule and its *in vivo* behavior. If rapid and irreversible accumulation in tumor tissue is observed, accompanied by rapid clearance from the rest of the body, radionuclides with short to long half-life times are appropriate. If accumulation is slow, the radionuclides decay before a reasonable dose can be deposited at the target site. It should be emphasized at this point that there is no "recipe" for the radionuclide to be selected, neither for diagnosis nor for therapy, since a limited number are immediately available for the labeling of a particular biomolecule. Table 9.1 gives an overview of the physical properties of the most important radionuclides for therapeutic and diagnostic applications.

9.1.2
Production of Radionuclides

Radionuclides have, ideally, half-lifetimes of some hours to days. This brings up the issue that a radionuclide or a ready-to-inject radiopharmaceutical can generally not be "stored" until usage. With exceptions, radiopharmaceuticals of more complex structures are generally not produced on one site and then shipped to another for application. It is more convenient to prepare them on site immediately prior to administration to the patient. Some compounds actually are shipped after preparation, ^{18}FDG (^{18}F-2-deoxyglucose) being a notable example. However, this increases the price and is, ultimately, not an ideal situation. Normally, the radionuclides are prepared in nuclear reactors or cyclotrons and then shipped as solutions to the hospitals where the final radiopharmaceutical is synthesized and quality controlled.

There are commonly two ways of preparing radionuclides. The cheapest one is by irradiating a target of the corresponding element with neutrons. The (n, γ) reaction yields then the neighboring isotope with mass number increased by one. This is unstable and decays under emission of an electron to the next higher element or isobar. An example for the activation of natural ^{185}Re to ^{186}Re is given in Scheme 9.1. If the thermal neutron cross section and the neutron flux in the reactor are high, very high specific activities can be achieved. The specific activity $[(A_{spec})$ (Bq g^{-1} or Bq mol^{-1})] is a measure for the ratio between radioactive (hot) and inactive (cold) isotopes. Further relevant examples of radionuclides prepared along this route are ^{153}Sm and ^{177}Lu (Scheme 9.1), both important in systemic cancer therapy. Many radionuclides are prepared by this route though the aspect of specific activity limits qualitatively the preparation of receptor targeting biomolecules [11].

A favorable situation for the production of radionuclides is encountered when the product of the (n, γ) reaction is a radioisotope that decays not to a stable but

Table 9.1 Decay properties of some radionuclides for diagnostic and therapeutic purposes.

Isotope	Half-life	Decay mode[a]	E_β (keV) (%)	E_γ (keV) (%)	Mean range (μm) [10]
For diagnostic purposes					
^{67}Ga	78.3 h	EC, Auger		93 (38), 185 (21)	
^{99m}Tc	6.0 h	IT		140 (90)	
^{111}In	67.2 h	Auger, γ		171 (90), 247 (94)	
^{61}Cu	3.4 h	β^+, EC	527 (50)	656 (10), 283 (13)	
^{62}Cu	0.16 h	β^+, EC	1315 (98)	1173 (0.3)	
^{64}Cu	12.9 h	β^-, β^+, EC	190 (37) (β^-), 278 (18) (β^+)	1346 (0.5)	
^{68}Ga	1.1 h	β^+, EC	836 (88)	1077 (3)	
^{86}Y	14.7 h	β^+, EC	550 (13)	1077 (83), 777 (22)	
^{89}Zr	78.5 h	β^+, EC	395 (23)	909 (99.9)	
For therapeutic purposes					
^{47}Sc	3.35 d	β^-, γ	0.14 (68%), 0.20 (32%)	159 (68%)	810
^{67}Cu	2.58 d	β^-, γ	0.19 (20%), 0.12 (57%)	93 (16%), 185 (49%)	710
^{90}Y	2.67 d	β^-	0.93 (100%)		3900
^{111}Ag	7.45 d	β^-, γ	0.36 (93%)	342 (6.7%)	1800
^{198}Au	2.7 d	β^-, γ	0.31 (99%)	412 (95%)	1600
^{213}Bi	46 min	α, γ	5.87 (2%) (α), 0.49 (65%) (α), 0.32 (32%) (α)	440 (27%)	
^{225}Ac	10 d	α	5.83 (51%) (α)	100 (3.5%)	
^{149}Pm	2.21 d	β^-, γ	0.37 (97%)	286 (2.9%)	1800
^{153}Sm	1.95 d	β^-, γ	0.23 (43%), 0.20 (35%)	103 (28%)	1200
^{177}Lu	6.71 d	β^-, γ	0.15 (79%)	208 (11%)	
^{186}Re	3.78 d	β^-, γ	1075 (71%)	137 (9%)	1800
^{188}Re	17 h	β^-, γ	2100 (100%)	155 (15%)	2000

[a] EC = electron capture; IT = internal transition

again to a radioactive daughter. Since the radioactive daughter belongs to a different (the next higher) element, the daughter and the parent can be chemically separated. Chemical separation of parent and daughter can be performed by means of an adsorption chromatography column and gives the radionuclide of

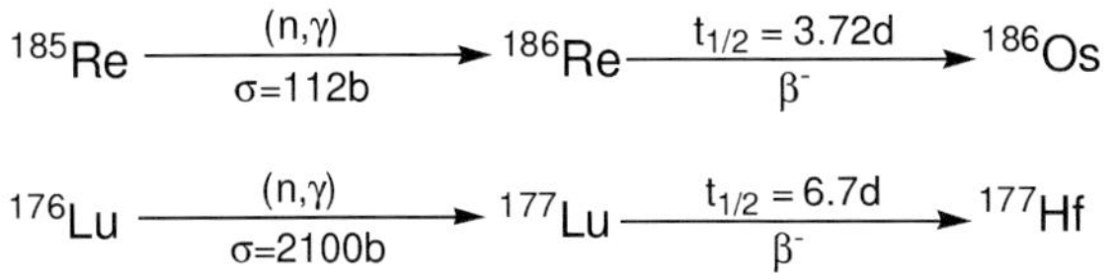

$$^{185}\text{Re} \xrightarrow[\sigma=112b]{(n,\gamma)} {}^{186}\text{Re} \xrightarrow[\beta^-]{t_{1/2} = 3.72d} {}^{186}\text{Os}$$

$$^{176}\text{Lu} \xrightarrow[\sigma=2100b]{(n,\gamma)} {}^{177}\text{Lu} \xrightarrow[\beta^-]{t_{1/2} = 6.7d} {}^{177}\text{Hf}$$

Scheme 9.1 Nuclear reactor produced therapeutic radionuclides.

interest – the so-called no-carrier-added (NCA) radionuclide – in the highest A_{spec}. For the preparation of radiopharmaceuticals, targeting biomolecules can then be labeled at – or purified to – extremely low concentrations; hence, target binding of the radio-bioconjugate is not in competition with cold biomolecules, which are useless for imaging purposes. To give an example, 37 MBq of ^{99m}Tc ($t_{1/2} = 6$ h) corresponds to about 4 pmol of ^{99m}Tc. After labeling, ideally, the same amount of radiopharmaceutical is prepared and administered. If the parent has a sufficiently long half-life, the parent– daughter system can be turned into a radionuclide generator. In such a generator, the parent is loaded on the column by adsorption to the stationary phase. The daughter, ideally with lower affinity for the stationary phase, can be eluted at regular time intervals. The prototypical generator is the ^{99}Mo/^{99m}Tc system, which is worldwide in extensive use and represents one of the decisive bases for ^{99m}Tc being the working horse in nuclear medicine. Other parent–daughter based radionuclides are ^{90}Y from a ^{90}Sr/^{90}Y generator and ^{188}Re from a ^{188}W/^{188}Re generator system; both are important radionuclides for therapeutic purposes [12].

The preparation of radionuclides from generators introduces "on line" availability in clinics, a very convenient situation for routine investigations. It impacts, however, on the chemical research towards novel radiopharmaceuticals since all labeling procedures have to be brought to a level where the synthesis can be performed in laboratories not specially equipped for syntheses. Preparations must be performed in "kits" according to the "shake and bake" principle. Yields must be >98% and, after the reaction, the compound must be ready for administration, without purification or separation. Obviously (as will be outlined later in the selected examples) these constraints compel high requirements to radiopharmaceutical chemistry and account for the essential endeavors of developing a radiopharmaceutical.

Related to the (n, γ) production is the radionuclide preparation by neutron induced fission from highly enriched ^{235}U. If the radionuclide of interest is a fission product, it can again be obtained in the NCA form by the usual separation prodcesses. The parent ^{99}Mo for the ^{99m}Tc generator or ^{131}I are examples of such radionuclides from fission products.

A second way of preparing radionuclides is target irradiation with positively charged baryons such as protons (p$^+$) or deuterons (d$^+$). The irradiated nucleus undergoes different nuclear reactions, the (p,n) and (p, α) being the most frequent ones. The PET radionuclide ^{18}F, for instance, is prepared by a (p,n) reaction with ^{18}O as the target. Radionuclides prepared along this route are often β$^+$-emitters, but some others decay by electron capture (EC), thereby emitting a photon which is used in SPECT. The radionuclide ^{111}In, for instance, is obtained from ^{112}Cd by a (p,2n) reaction in NCA quality and is widely used in diagnostic radio-pharmaceuticals. However, cyclotron production is generally expensive, with drawbacks in terms of the availability of such radionuclides. Although the ^{68}Ge/^{68}Ga system is has long been known, commercially available and reliable generators have been introduced only recently. As in the case of the ^{99}Mo/^{99m}Tc system, the PET radionuclide ^{68}Ga becomes thereby immediately available at hospitals, reducing the price and rendering application much more attractive. Nowadays, small cyclotrons of limited energy are installed and important early radionuclides can be produced on site. However, this does not apply to metallic elements in the focus of our interest, which still need to be prepared at larger cyclotrons [13].

PET and SPECT are two modalities in molecular imaging. Biomolecules labeled with ^{18}F or ^{11}C are very attractive since they do not affect the nature of the molecule to the same extent as metal-based radiopharmaceuticals do. PET is nowadays clearly favored, owing to the (still) higher resolution and less influence on the biological properties of the radiopharmaceutical. Metallic PET radionuclides, however, do not have the latter advantage and, as common SPECT radionuclides such as ^{111}In or ^{99m}Tc, require chelators to be stabilized under physiological conditions, thereby affecting the biological properties (e.g., pharmacokinetic or receptor affinity) of the molecule to which they are attached. Furthermore, the resolution of PET is physically limited since the positron has a limited spatial range before it is annihilated. SPECT radionuclides emit the γ-photons from where they actually are and the resolution is limited by the physical distribution of the radiopharmaceutical. These arguments should be kept in mind when following the controversial discussions of SPECT vs. PET and the selection of the corresponding radionuclides.

9.1.3
Concepts and Chemistry of Labeling

The labeling of a targeting biomolecule with metallic radionuclides requires chelators that stabilize the metal-ions against release from the carrier. Decomplexation can occur for thermodynamic or kinetic reasons if equilibrium is established. A radiopharmaceutical, once administered, is under non-equilibrium conditions and so, therefore, high thermodynamic stability together with kinetic stability is a basic requirement. Release from the chelator goes along with – or is paralleled by – trans-metalation, a process in which the radionuclide is bound by other potential coordinating sites *in vivo*. Competing sites are numerous in biological systems: functionalities in the side chains of amino acids in proteins, small molecules in blood plasma or in cells such as glutathione or carbohydrates. Even the negatively charged surface of cells is a trap for cations. Trans-metalation of radionuclides will lead inevitably to high imaging background, slow excretion, and accumulation at the target site. Accordingly, to receive a clear picture of the *in vivo* behavior of a radiopharmaceutical, the labeled metal complex must be of very high

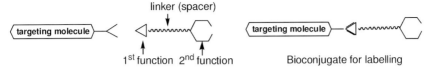

Scheme 9.2   Principle of bifunctional chelator (BFC) and bioconjugate ready to be labeled in a kit formulation.

thermodynamic stability and/or kinetically robust. As will be seen later, examples for both concepts exist in research and in applications.

The concept of labeling biomolecules with metallic radionuclides involves a so-called bifunctional chelator (BFC). A BFC consists of as the first function a group that can be bound covalently to a functional group on the biomolecule. This conjugating function is connected via a linker (or spacer) to the second function, which represents the chelator designed for the particular radionuclide (Scheme 9.2). The linker, though not directly involved in conjugation or complexation, is important since it influences lipo- or hydrophilicity of the radiopharmaceutical. It also separates the bioactive part from the metal-complex, thereby minimizing steric effects imposed to the receptor binding part of the radiopharmaceutical. Structure–activity relationships that have been carried out with, for example, labeled carbohydrates corroborate the relevance of an appropriate selection of spacer type and length.

Major endeavors are undertaken for synthesizing tailor-made chelators for particular biomolecules and radionuclides fulfilling the requirements outlined above. For high thermodynamic stability, multidentate macrocyclic or acyclic chelators such DTPA or DOTA have been used (see below). Since labeling is generally performed at high dilution, the concept of 1:1 metal to chelator complex is usually followed since the rate of formation for 1:2 complexes would be slow or require high concentrations of the ligand to accelerate product formation.

Besides the pure physicochemical aspects, further limitations in labeling chemistry emerge from attempted routine application in clinics. Labeling with a metallic radionuclide should be a one-step process and consist, for example, of mixing the radionuclide solution with the biomolecule conjugated to the BFC (Scheme 9.3). This concept of labeling chemistry does not cause severe problems if the radionuclide is present as an aqua-ion in one single defined oxidation state. Most of the radionuclide families discussed later are of this nature, which makes their application comparably facile. However, if redox chemistry is involved, as for ^{99m}Tc or ^{188}Re, the situation becomes more difficult. In this case the labeling chemistry not only consists in complexation but requires also reduction and stabilization of precursor complexes by auxiliary ligands (L) and finally trans-metalation to the biomolecule (Scheme 9.3). To combine all these issues in one or two synthetic steps is a major challenge for radiopharmaceutical chemistry and there are elegant examples for the realization of this concept.

Finally, the procedure of labeling chemistry should produce a well-defined complex that can be characterized by chemical methods. Since the concentration

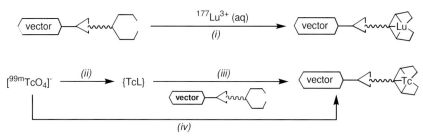

Scheme 9.3 Different pathways to metal-based radiopharmaceuticals. (i) Simple substitution to the bioconjugate for $^{177}Lu^{3+}$. The case of $[^{99m}TcO_4]^-$: (ii) reduction to a lower oxidation state stabilized by weak ligands L; (iii) substitution to the bioconjugate; and (iv) all-in-one as ideally found in kit formulation for routine application.

of the radiopharmaceutical is extremely low, characterization is often performed with non-radioactive analogs on the macroscopic level. Such a "cold" congener of a radiopharmaceutical is compared by chromatographic methods with the radio-pharmaceutical. If retention time of the cold, fully characterized compound matches that of the radiolabeled analog, identity and therefore characterization is confirmed. Early labeling methods did not follow this concept but still resulted in useful – though not characterized – radiopharmaceuticals. For instance, the re-duction of disulfide bonds in antibodies and subsequent labeling reaction with simple pertechnetate $[^{99m}TcO_4]^-$ gave unspecific but stably labeled proteins. Nowadays, such radiopharmaceuticals are not approved by drug administrations, but they were at that time.

The problem of characterization is a minor issue with aquo-ions, but may be-come a major one with radioelements such as Tc for which multiple oxidation states and, hence, coordination environments exist. Therefore, current and novel labeling chemistry must aim at a well-defined design of compounds rather than just achieving stable conjugation such as the one mentioned above. It stands to reason that rational drug design must rely on well-defined compounds, otherwise the search for new lead compounds will remain a trial and error approach.

9.1.4
Different Generations of Metal-Based Radiopharmaceuticals

The original concept of radiopharmaceuticals was based on the synthesis of radio-metal complexes with adequate properties for, for example, accumulating in a par-ticular target-organ or for visualizing organs such as the kidneys for assessing their proper function by following rate of excretion. These radiopharmaceuticals belong to the so-called first generation – also called metal-essential since only the entire compound will give the radiopharmaceutical whereas the single components are physiologically inactive [5]. The term first generation should not imply that they are out-dated since research is still going on towards compounds with improved

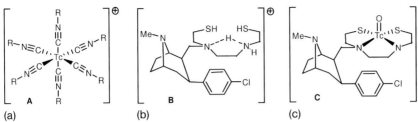

(a)　　　　　　　　　　(b)　　　　　　　　　　(c)

Figure 9.1 (a) Metal-essential radiopharmaceutical Cardiolite® [R = $CH_2C(CH_3)_2(OCH_3)$]; (b) a tropane derivative that does not penetrate the BBB (blood–brain barrier); and (c) neutral radiopharmaceutical TRODAT-1, capable of binding to the dopamine transporter.

biological properties. Metal-essential first-generation radiopharmaceuticals are found in particular for ^{99m}Tc (Figure 9.1) and they make up a vast portion of the nuclear medicinal diagnostics carried out worldwide.

The second generation of radiopharmaceuticals are metal-tagged rather than metal-essential. A biomolecule is derivatized with a BFC as outlined before and then labeled with the metal. The vector (a peptide for instance) guides the radio-nuclide to its target but the radionuclide is generally not required for the bio-molecule to exert its action. Still, the metal bound to the BFC influences the physiological behavior of the vector but its presence is not absolutely required [14]. Ideally, accumulation at the target site is positively influenced by metal-binding but usually the complex plays a negative rather than a positive role. There are some exceptions to the rule that second-generation radiopharmaceuticals are metal-nonessential. Tropane analogs as radiopharmaceuticals for the assessment of some neurodegenerative diseases are important tools in molecular imaging [15]. They bind the dopamine transporter with high affinity. Metal-labeled tropane analogs have to penetrate the blood–brain barrier (BBB) and should therefore be of low molecular weight and high lipophilicity. The tropane derivative shown in Figure 9.1b has a positive charge, and therefore does not penetrate the BBB and is inactive. Once labeled with ^{99m}Tc, the receptor targeting radio-conjugate (TRO-DAT-1) is neutral and highly lipophilic, does penetrate the BBB, and is a very good radiopharmaceutical for imaging the dopamine transporter [16, 17].

As will be seen in Section 9.2, first- and second-generation radiopharmaceuticals are both under extensive research for molecular imaging purposes. The two generations do not follow each other but are developed in parallel.

A less extensively investigated class (not generation) of radiopharmaceuticals are mimics of existing vectors. Such compounds are both metal-essential and target-ing. The concept behind them is to assemble by metal coordination a molecular structure that mimics a natural, receptor binding substrate. Pioneering work in this direction were complexes of Tc^{V} that assembled to a structure of some si-milarity to estradiol. Instability and synthetic difficulties prevented further devel-opment but the concept is convincing and merits attention. To replace, for example, phenyl rings in pharmacophores with cyclopentadienyls or arene complexes goes towards the mimic concept but has not yet been exploited, although the success of some cold bioorganometallic compounds encourages the following of this concept [18].

9.2
Selected Examples: Therapeutic Radiopharmaceuticals

The following sections exemplify the general concepts outlined above for metal-based radiopharmaceuticals with results from current studies. The examples are primarily selected as representative for the current status of labeling and radio-pharmaceutical chemistry.

Radionuclides that are hard β^-- or α-emitters are applied for therapeutic or palliative purposes. They deposit the decay energy of the particle within a short range, which will lead to very high local doses accompanied by corresponding biological damage, necrosis, or apoptosis of the targeted cells. A selection of radionuclides is given in Table 9.1.

9.2.1
The "3+ Family": Yttrium and Some Lanthanides

Surprisingly, most metallic radionuclides applied in radiopharmacy belong to the groups 3, 13, or to the lanthanide elements. Thus all these radionuclides are present in their triply charged form M^{3+}, generally as binary aquo-ions or stabilized by some weak ligands in aqueous solution. Examples of these radionuclides for therapeutic purposes are ^{90}Y, ^{153}Sm, ^{177}Lu, and ^{166}Ho (all reactor produced). Owing to their high charge and their position in the periodic table, they are hard acids that require appropriate chelators for stabilization under biological conditions. The important difference between corresponding members of the 3+ family is the ionic radius, which differs widely from Y^{3+} (90 pm) to Sm^{3+} (110 pm), and the coordination number (CN), which ranges from 6 to 8. Size and CN are the factors determining the selection of the appropriate chelators for each of these 3+ cations [19].

According to the chemical properties of the 3+ family, most of the BFCs are of the polyamino-polycarboxylate type. These octadentate ligands, for which EDTA is the prototype, may be acyclic or macrocyclic. Classically, diethylenetriaminopentaacetic acid (DTPA) is the most widely used for 3+ cations. One of the binding sites normally bears the first function of the BFC and is conjugated to the targeting molecule, typically a peptide or an antibody. This design leaves seven groups for coordination to the metal. For ^{111}In^{3+}, DTPA is an almost ideal BFC, its stability is, however, not perfect for ^{90}Y^{3+}, for instance. To keep all eight coordination sites untouched and available for coordination, syntheses of backbone-modified DTPA systems have been published (Figure 9.2) that provide a much better stability to the complex [20, 21].

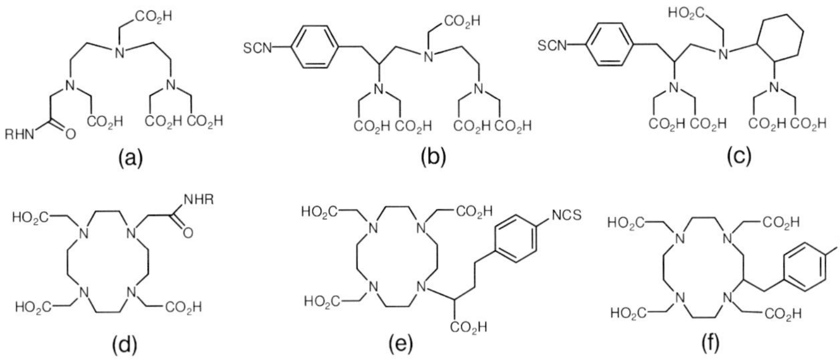

Figure 9.2 (a)–(c) DTPA-based BFCs and (d)–(f) DOTA-based BFCs with the first function at different positions in the basic ligand framework.

For systemic cancer radiotherapy, ^{90}Y is of particular interest due to its high energy pure β^-decay. The metal center Y^{3+} is preferentially stabilized by DOTA, attached to peptides and antibodies (Figure 9.2). DOTA stabilizes Y^{3+} to a sufficient extent to avoid trans-metalation *in vivo*. However, the exchange of one carboxylate group in the coordination sphere for an amide reduces the stability constant of the complex from log $K = 23.5$ to 21.9 [22]. Nevertheless, stability is sufficient for *in vivo* application. The importance of kinetic stability becomes obvious. Comparing the stabilities for Y^{3+} with DOTA and DTPA, one finds that the log K for Y-DTPA is 22 and for an amide derivative of DOTA is around 20. These constants are still very high and one would expect that Y-DTPA is sufficiently stable, which is not the case. Y-DTPA dissociates under physiological conditions rather rapidly. This disadvantage is compensated by a very rapid complex formation that is about four orders of magnitude faster for DTPA than for DOTA [23]. Conditions for rapid labeling at high dilution are critical: factors such as pH, temperature, and the presence of metal contaminants influence this rate substantially [24]. It has also to be considered that heating temperature and reaction time are limited if proteins are labeled with ^{90}Y due to their tendency to denaturate. Some novel radiopharmaceuticals with ^{90}Y for therapeutic purposes and based on DOTA are in clinical trials and an antibody-based radioimmunotherapeutical agent (Zevalin® or ibritumomab-tiuxetan) has been approved by the FDA for application in the treatment of some types of non Hodgkin lymphoma. Labeling with Y^{3+} is complete after 5 min at room temperature due to the presence of a relatively large amount of chelator (or antibody). Peptides were frequently labeled with ^{90}Y; a good example for DOTA-labeling is the cyclic dimeric peptide SU015, a DOTA-conjugated vitronectin receptor antagonist. At low peptide concentration, elevated temperature was required to achieve quantitative labeling within 5–10 min [25].

One of the drawbacks of ^{90}Y is the complete absence of detectable photon emission accompanying its β^--decay. To follow localization of ^{90}Y-radiopharmaceuticals, the ^{111}In surrogates are often prepared and co-administered. ^{111}In has an EC decay of reasonable energy to be seen from outside. However, there are substantial differences between ^{111}In- and ^{90}Y complexes and corresponding radiopharmaceuticals. These differences are of structural and physicochemical impact. Although both are trivalent, In^{3+} is smaller than Y^{3+}. The Y^{3+} cation fits perfectly into the DOTA ring but In^{3+} is too small. In contrast, DTPA is tailor-made for In^{3+} but not for Y^{3+}. The co-ordination number of In^{3+} is six or seven, whereas Y^{3+} tries to maintain eight [26]. For In^{3+}, only a few complexes with CN higher than seven are known [27, 28]. An X-ray structure analysis of In-DOTA-monoamide shows CN = 8 in the solid state, but it is likely that the amide dissociates in solution. The different coordination chemistry of In- and Y-DOTA complexes is then mirrored by different hydrophilicity and solubility. The In^{3+} complex is usually more hydrophilic than the Y^{3+} analog and *in vivo* behaviors are different.

A very interesting radionuclide from the 3+ family is the reactor-produced ^{153}Sm. This radionuclide is mainly used for the treatment of bone cancer lesions or for palliative treatment of bone cancer [29, 30]. The ligands for encapsulating ^{153}Sm^{3+} are hexadentate phosphonates, the most prominent one being ethylenediaminetetramethylphosphonate (EDTMP) together with some other octadentate ligands designed on the same principle (Figure 9,3). These ^{153}Sm based radiopharmaceuticals are metal-essential and are guided into the hydroxyapatite matrix of the bones by the phosphonate groups [29, 31, 32].

A last example from the 3+ family of therapeutic radionuclides is ^{177}Lu [33]. The coordination chemistry required for the preparation of metal-based radiopharmaceuticals is very comparable to that for Y^{3+} or Sm^{3+}. DOTA proved also here to be the chelator of choice and has been coupled to peptides, especially to octreotide and derivatives, and proteins. Some targeting radiopharmaceuticals for therapeutic purposes are currently in clinical trials [34–37].

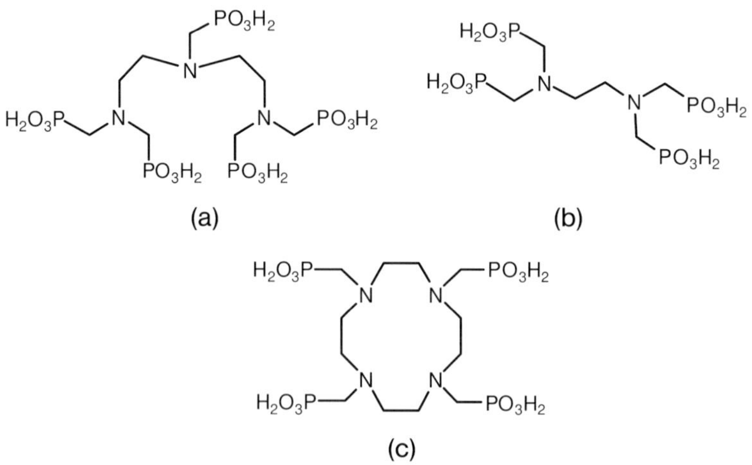

Figure 9.3 Some phosphonate-based multidentate chelators for stabilizing ^{153}Sm^{3+}: (a) DTPMP; (b) EDTMP; and (c) DOTMP.

All radionuclides from the 3+ family previously discussed show great promise as metal-based radiopharmaceuticals, as corroborated by the state of clinical trials. It has to be kept in mind, however, that the application with respect to vector is limited to hydrophilic biomolecules. The chelators, which cannot substantially be changed for thermodynamic and kinetic stability reasons, govern the physico-chemical properties of the tagged vector. The pendent complexes are relatively large, which limits their use for the labeling of small molecules and makes the labeling of pharmacophores for penetrating the BBB essentially impossible. Nonetheless, a major target of molecular imaging and radiotherapy is cancer and for targeted therapy with amino acid based macromolecules these radionuclides are certainly very well suited.

9.2.2
Rhenium

There are two rhenium radionuclides with appropriate properties for therapy, ^{186}Re and ^{188}Re. Whereas the former has superior physical properties, it cannot be produced in NCA quality and therefore its A_{spec} is relatively low. The latter is received from the ^{188}W/^{188}Re generator and therefore in no-carrier-added quality [38–40]. The half-life (17 h) is at the lower limit for, for example, antibody delivery. Still, its A_{spec} makes it very attractive and most of the studies have been performed with generator-produced ^{188}Re (Table 9.1)

Rhenium for therapeutic application is attractive because it is the higher homolog of technetium, and their chemical properties are therefore assumed to be similar or even equal. This point of view leads to the paradigm of a *matched pair*, which means that ^{99m}Tc is used for diagnosis and the very same kit with ^{188}Re for therapy. Such a matched pair would be very attractive since it allows us to perform dosimetry in advance, an important point for clinical decision making. As outlined before, ^{111}In and ^{90}Y are also considered as a matched pair but, although the same vector-tagged radiopharmaceuticals were applied, the coordination chemistry is too different to suggest an "equal" *in vivo* behavior. The structural features of Rhenium and Tc compounds are similar indeed but there are substantial differences in their chemistry, which are responsible for labeling with rhenium being more difficult than with technetium. Their chemistry will be given in more detail in the technetium section. Rhenium is harder to reduce and, correspondingly, easier to be reoxidized; hence, it displays increased instability towards air. The tetradentate MAG$_3$ ligand is frequently used to stabilize the $[^{188}\text{Re}^V=O]^{3+}$ core in the form of the $[^{188}\text{ReO(MAG}_3)]^-$ complex (Scheme 9.4). Even this stable complex is rather sensitive towards reoxidation to $[^{188}\text{ReO}_4]^-$ whereas the Tc analog is stable. Rhenium is less reactive than technetium. The standard reduction of $[MO_4]^-$ with Sn^{2+} is essentially immediate for $[^{99m}\text{TcO}_4]^-$ but requires harsher conditions or longer reaction times for $[^{188}\text{ReO}_4]^-$. Going to the low valent cores $[M^I(CO)_3]^+$, the Tc complex can be prepared in a one-step kit (Section 9.3.2) whereas the Re analog again requires two steps and stronger conditions or longer reaction times [41, 42]. The question of whether ^{99m}Tc chemistry has a role for

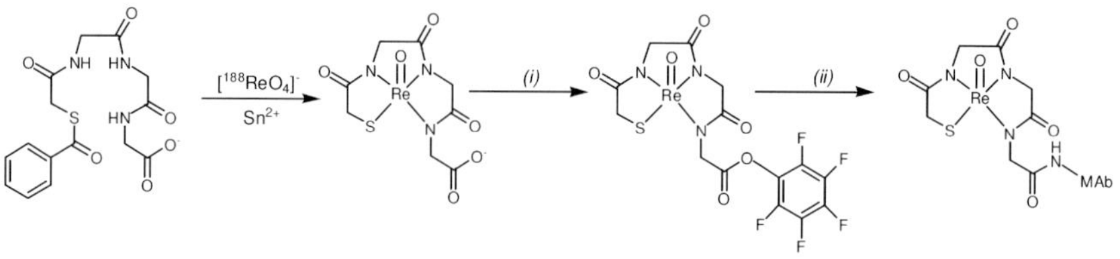

Scheme 9.4 Prelabeling method with ^{188}Re, complex formation with MAG$_3$ at elevated temperature: (i) activation of the first function and (ii) covalent coupling to a monoclonal antibody (MAb).

predicting developments for rhenium has been addressed recently [43]. As a consequence of the more difficult reduction of $[^{188}ReO_4]^-$ vs. $[^{99m}TcO_4]^-$, the so-called prelabeling method is used for preparing vector-tagged radiopharmaceuticals with Re. In this method, the complex with the radio-metal is prepared first and only coupled in the last step to the antibody. The final radiopharmaceutical is purified and quality assessed before administration. Since the half-life time is longer, instant kit preparation is not immediately required and the procedures are approved by authorities (Scheme 9.4) [44, 45].

One rhenium-essential radiopharmaceutical is in clinical use: the meso-2,3-dimercaptosuccinic acid, DMSA, complex $[ReO(DMSA)_2]^-$ is a versatile agent for medullary thyroid carcinoma and other tumors [46–48].

Antibodies have been labeled as mentioned before either by the prelabeling method or, the majority, by direct labeling methods. Direct labeling methods are less feasible for peptides since the reduction of disulfide bonds will lead to severe structural changes in the receptor binding motive with loss of affinity. An elegant way to bypass this problem has been performed recently with the direct labeling of salmon calcitonin, a disulfide-containing cyclic octapeptide. Reduction of the peptide with phosphines and reaction with $[^{188}ReO_4]^-$ or $[^{99m}TcO_4]^-$ gave recyclized products. Closer inspection of the nature of products shows a strong difference between Re and Tc. Whereas each complex induces ring-closure of the peptide, the oxidation state was +V for Re but +III for Tc, mirroring the greater oxidation power of the latter (Scheme 9.5) [49–51]. As shown in Scheme 9.5, not only monocyclic but also bicyclic products were found by mass spectrometry.

Newer developments in rhenium-based radiopharmaceuticals are low-valent organometallic carbonyl complexes. The labeling precursor $[^{188}Re(OH_2)_3(CO)_3]^+$ has been introduced, following a method similar to that used for preparing the ^{99m}Tc analog, whose chemistry will be discussed below [52–54]. Preparation of the $[^{188}Re(OH_2)_3(CO)_3]^+$ precursor is more time consuming and reaction with biomolecule tagged ligands slower than for ^{99m}Tc. The complexes are very inert and physiologically stable. Radiopharmaceuticals prepared with these M^I precursors are essentially identical and the matched-pair idea can probably be applied. The chemistry of the two organometallic aquo-ions has been compared recently [55].

As evident from this brief state of the art description and a recent review [56], the chemistry of Re-based radiopharmaceuticals is somehow still at its infancy as

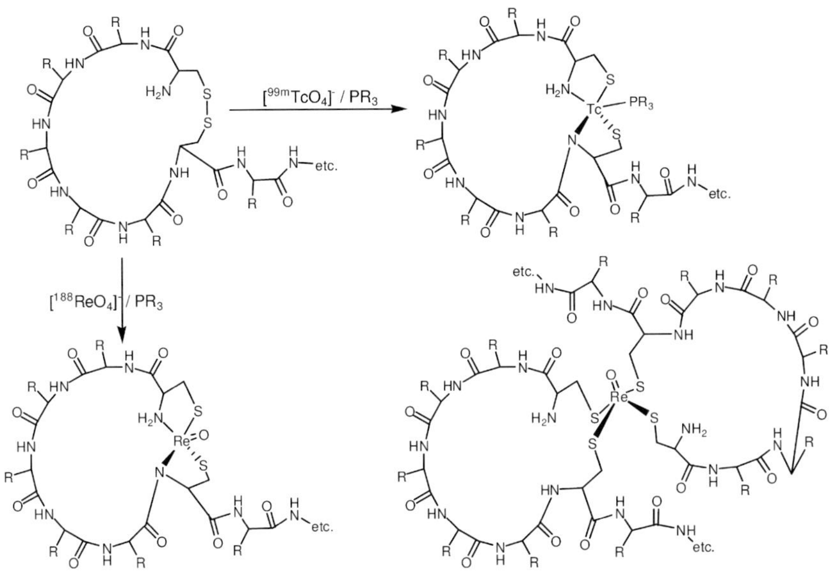

Scheme 9.5 Reduction of salmon calcitonin with phosphines and subsequent direct labeling yielded +III for Tc ($^{99m}Tc^{III}$) but +V for Re ($^{188}Re^V$) complexes, as evident from mass spectrometric analyses.

compared to Tc. This may be related to the higher attraction of radionuclides of the 3+ family for therapy, but also to the greater difficulty of preparing well-defined ^{188}Re radiopharmaceuticals in quantitative yield. Once a novel ^{99m}Tc radiopharmaceutical has been found, it is likely that more endeavors will be undertaken to realize the corresponding ^{188}Re analog.

9.2.3
α-Emitters: Bismuth and Actinium

The radionuclides ^{213}Bi and ^{225}Ac are attractive α-emitters for therapy due to their very high linear energy transfer (LET), which results in a substantial local dose. For example, α-particles from ^{225}Ac have a LET of about 89 keV μm^{-1} (energy deposition per path length), which is much higher than the 0.22 keV mm^{-1} for the high energy β-emitter ^{90}Y. The mean range of α-particles is only a few cell diameters, offering the prospect of matching the cell-specific nature of targeted molecular carriers with radiation having a similar range of action [57]. ^{225}Ac has a 10 d half-life, ^{213}Bi, a daughter product of ^{225}Ac, 46 min. Both radionuclides belong to the 3+ family since actinium is present as Ac^{3+} and bismuth as Bi^{3+}. An interesting approach has been proposed for the use of ^{225}Ac, a so-called *targeted atomic nanogenerator* consisting of trapping the daughter products of ^{225}Ac such as ^{221}Fr inside the cell, thereby strongly increasing the dose and cell damage [58].

Excellent biological results have been obtained *in vitro* and *in vivo*, but side effects elsewhere in the body need better control [59]. Scheme 9.6 shows the decay series of ^{225}Ac as a base for the targeted atomic nanogenerator.

A nice systematic study on appropriate chelators for Ac^{3+} has been performed. It turned out that DTPA, TETA, DOTPA, and others were too unstable to find application *in vivo* [60]. It was found that DOTA is again the best ligand for ^{225}Ac^{3+} with <1% released from the chelator after 10 days. Accordingly, most labeling studies are performed with the prelabeling concept with DOTA. Prelabeling is applied since incorporation of Ac^{3+} into the macrocyclic chelate is slow and requires heating for completion. At a 10 mM concentration of DOTA-NCS, 30 min at 60° C were sufficient to achieve quantitative complexation, followed by coupling to the antibody and subsequent column purification [59]. More details about the biology of targeted α-particle therapy have been published recently [61].

The application of α-emitters for targeted therapy would be a very convenient and efficient concept. One has, however, to consider the general availability of the radionuclides, their half-lifetimes, and the way of handling these α-emitters, which is not routine, requiring additional precautions to be taken to protect the personnel involved in syntheses and application from exposure to these nuclides.

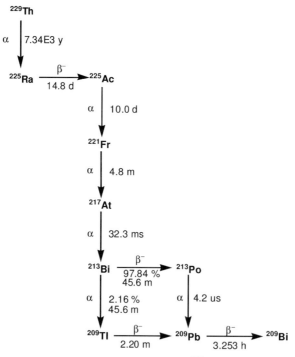

Scheme 9.6 Full decay scheme leading to ^{225}Ac. The subsequent cascades of α- and β-decays are the basis for the targeted atomic nanogenerator [58].

9.3
Diagnostic Metal-Based Radiopharmaceuticals

9.3.1
The "3+ Family Continued": Indium and Gallium

Two of the more important radionuclides for diagnostic purposes, ^{111}In and ^{68}Ga, belong also to the 3+ family. ^{111}In is cyclotron produced and decays by electron capture (EC) under photon emission used in SPECT. ^{68}Ga is the daughter of cyclotron produced ^{68}Ge and is a positron emitter applied in PET. The latter radionuclide became important since it is available in a ^{68}Ge/^{68}Ga generator system. The processing of generator produced ^{68}Ga for medical application and the generator itself have been described in detail recently [62]. ^{111}In is received in batches from the producer; no generator system is available.

As expected for group 13, the chemistry of both elements is governed by the hard acid cations Ga^{3+} and In^{3+}. As was the case with $^{90}Y^{3+}$, typical middle to hard multidentate chelators of the polyamino-polycarboxylate type (Figure 9.2) are employed to stabilize these cations *in vivo*. One of the earliest FDA approved ^{111}In radiopharmaceuticals is ^{111}In-Oxine. This neutral In^{3+} complex with three deprotonated hydroxyquinoline ligands is widely used for the labeling of blood products such as white blood cells, platelets granulocytes, and others. It is believed that its lipophilic nature allows the complex to diffuse across lipid membranes; the metal is then irreversibly trapped by ligand exchange with intracellular macromolecules [63]. The solid state structure (Figure 9.4) shows a symmetric coordination environment; it is, however, likely that the coordination is highly fluxional in solution and does not correspond to the one fixed in the solid state [64].

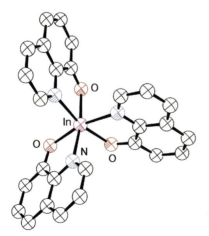

Figure 9.4 X-ray structure of In-Oxine, a radiopharmaceutical applied in the labeling of white platelets.

The vast majority of ^{111}In-based radiopharmaceuticals employ DTPA or DOTA-type ligands, which are both tailor-made for the size and charge of In^{3+} (see also Section 9.2.1). One of the carboxylate arms in DTPA or DOTA is covalently bound to peptides. Octreoscan® (^{111}In-pentetreotide) is the commonly used brand name of a somatostatin receptor targeting peptide derivatized with ^{111}In-DTPA, which is frequently used in clinical routine and in research [65, 66]. The acyclic nature of DTPA makes labeling a fast process since the cation does not have to be incorporated into a macrocycle. The structural flexibility of In^{3+} is mirrored by the different existing solid-state structures. Whereas In^{3+} prefers a CN of six or seven, structures are also known with a CN of eight. If one of the carboxylate arms is conjugated to an amine, the resulting amide does coordinate in the solid state but is believed to be released in solution, leaving In^{3+} with a CN of seven. The same features are also encountered for DOTA and derivatized DOTA complexes in the solid state and in solution. Figure 9.5 shows several solid state structures for different DOTA, NOTA (Figure 9.6), and DTPA complexes of In^{3+} [67–69].

The chemistry for Ga^{3+} is different with respect to stability and kinetics as compared to that of In^{3+}. Principally, many radiopharmaceuticals have been prepared using the DOTA ligand, which provides sufficient stability for *in vivo* applications. Recently, DOTA has been derivatized with two tyrosines bound to

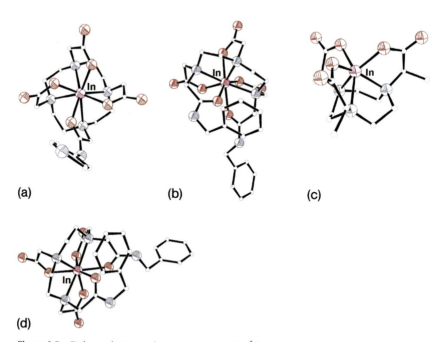

(a)   (b)   (c)

(d)

Figure 9.5 Eight- and six-coordinate complexes of In^{3+} with DOTA, NOTA, and DTPA-based ligands; (a) with one amide group; (b) with two amide groups; (c) with CH$_3$-NOTA; and (d) with DTPA (O = red, N = blue) [67–69].

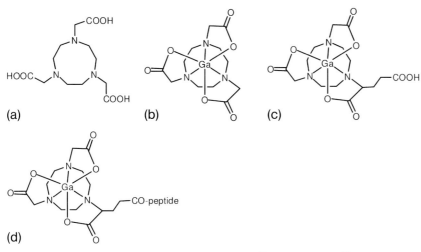

Figure 9.6 (a) Basic NOTA ligand; (b) corresponding Ga^{3+} complex; (c) Ga^{3+} complex with the NODASA BFC having one function left to be bound to a peptide, as shown in (d).

two of the amines, leaving the four amines and two carboxylate groups for coordination to Ga^{3+}. The tyrosine should target amino acid transporters thereby opening an approach to new metabolic tracers. The compound is promising since uptake in some cancer cells has been assessed [70].

It is also often found that both the SPECT nuclide $^{111}In^{3+}$ and the PET nuclide $^{68}Ga^{3+}$ are applied with the same biomolecule. Octreoscan is thereby often taken as a standard [71]. In contrast to In^{3+} or Y^{3+}, it has been found that NOTA and derivatives provide complexes with Ga^{3+} of exceptionally high stability, with log K being 31.0, whereas the corresponding DOTA complex has a log K of "only" 21.3 [72, 73]. However, NOTA is a hexadentate ligand and derivatization of one of its side chains would affect the stability of the complex. Recently, Mäcke *et al.* have introduced an elegant derivative of NOTA that has two acetic acid and one succinic acid as side chains. In this ligand, NODASA (Figure 9.6), one of the acid groups in the succinic acid coordinates to Ga^{3+}, completing the hexadentate coordination, whereas the second carboxylate is conjugated to the peptide (Figure 9.7). Labeling of these bioconjugates with $^{68}Ga^{3+}$ is very fast and quantitative, which is convenient for routine clinical application [74]. Alternatively, NOTA-type ligands have also been synthesized with phosphonate groups bound to the secondary amines (NOTP, Figure 9.7) [75].

Interest in ^{68}Ga is strongly increasing. The introduction of ^{68}Ga-based kits for the production of radiopharmaceuticals would not immediately require an on-site cyclotron, which renders its application very convenient for routine use. The labeling procedures are very simple, fast, and efficient, characteristics that lead one to expect a rapid expansion of this radionuclide in research, development, and application.

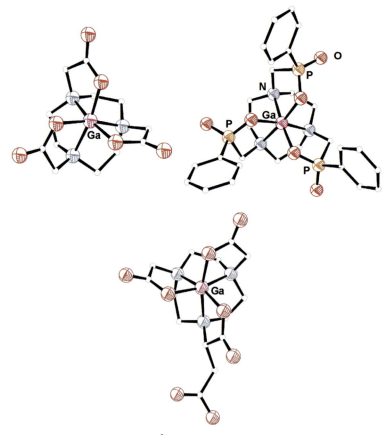

Figure 9.7 X-ray structures of Ga^{3+} complexes with NOTA-based ligands.

9.3.2
Technetium

With respect to routine clinical application, ^{99m}Tc is still the most widely used radionuclide in imaging. This is certainly due to its immediate availability in hospitals and its comparably low price. The vast majority of ^{99m}Tc-based radiopharmaceuticals are of the first generation type and used for, for example, myocardial, renal, and bone imaging [76]. The ease of availability is contrasted by the difficulty of synthesizing compounds with the requirements outlined in the first section. Whereas the radionuclides of the 3+ family exist under physiological conditions in one oxidation state only, technetium has at least eight different oxidation states, of which probably three or four are accessible in water and afford air-stable complexes. The fundamental chemistry and that related to the synthesis of radiopharmaceutical compounds has been comprehensively reviewed [77]. Furthermore, ^{99m}Tc is exclusively available in its most stable form $[^{99m}TcO_4]^-$.

All chemistry has to start ultimately from $[^{99m}TcO_4]^-$, which entails reduction processes in (almost) any preparation of ^{99m}Tc radiopharmaceuticals. Thus, labeling procedures are different compared to those for the 3+ centers (Scheme 9.3).

As with other relevant metallic radionuclides, ^{99m}Tc has to be stabilized by multidentate ligands, tagged to a vector for second-generation radiopharmaceuticals. For Tc-essential first-generation radiopharmaceuticals, the ^{99m}Tc center and the ligands must form a physiologically stable complex. To fulfill these conditions, ligands are required that match the electronic properties of the metal.

The oxidation state +V is the one best developed in technetium chemistry. Reduction of $[^{99m}TcO_4]^-$ with Sn^{2+} in the presence of stabilizing glucoheptonate ligands leads to complexes containing the $[Tc=O]^{3+}$ or the $[O=Tc=O]^+$ core. The remaining coordination sites are occupied by tetradentate ligands with a mixture of N, S, or O-donor atoms. Figure 9.8 depicts some of the most frequently applied ligands in connection with ^{99m}Tc^V. The chemistry with the $[^{99m}Tc=O]^{3+}$ or $[O=^{99m}Tc=O]^+$ core has been the subject of different reviews summarizing the state of the art. Since there have not been many changes over the last few years in Tc^V chemistry, the reader is referred to those publications [1, 2, 14, 15, 66, 79–83]. This chapter will rather focus on Tc chemistry in oxidation states different from +V and in which interesting progress, relevant for chemistry and radiopharmacy, has been made.

In the recent past, three novel building blocks have emerged from studies in basic technetium chemistry. These are the nitrido core $[Tc\equiv N]^{2+}$ (Tc^V) [81, 84, 85], an umbrella type complex [TcNS3] (Tc^{III}) [86, 87], and the carbonyl $[Tc(CO)_3]^+$ (Tc^I) core [53, 54, 88]. For these three cores, a plethora of basic complexes have been synthesized and their relevance for molecular imaging shown in many examples. Figure 9.9 depicts some representative complexes.

The related fundamental and applied chemistry underscores the fact that even relatively complex Tc-compounds can be synthesized, including redox reactions, coordination, and labeling, often in one step and in quantitative yield as required for routine application. These achievements underline the importance of inorganic chemistry for life sciences in general and radiopharmaceutical chemistry in particular. Since this chapter cannot be comprehensive, the following section will highlight some selected topics for the labeling of small molecules with one of the new cores described in the previous section.

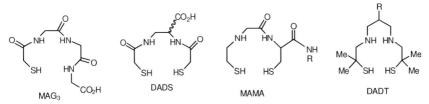

Figure 9.8 Selection of tetradentate ligands used to stabilize the $[Tc=O]^{3+}$ core. MAG$_3$ and DADS lead to mono-negatively charged complexes, MAMA and DADT ligands give neutral complexes [78].

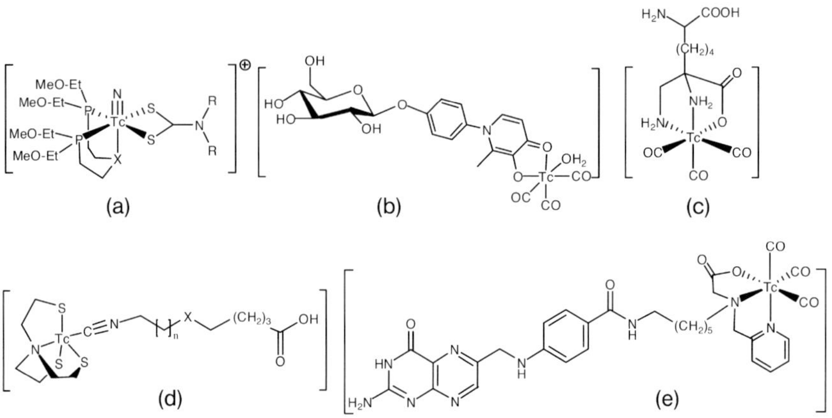

Figure 9.9 Examples of small targeting molecules labeled with different ^{99m}Tc cores from recent studies: (a) myocardial imaging agent based on the [Tc≡N]$^{2+}$ core [89, 90]; (b) labeled carbohydrate for cancer cell targeting based on the [^{99m}Tc(CO)$_3$]$^+$ core [91]; (c) labeled amino acid as metabolic tracer [92]; (d) labeled fatty acid based on the [4+1] approach [93]; and (e) labeled folate [94].

9.3.3
Targeted Radiopharmaceuticals: Technetium-Labeled Carbohydrates

Progress towards the labeling of carbohydrates is of particular importance since, if successful, it would account for complementing ^{18}F-deoxyglucose (^{18}FDG) in PET with ^{99m}Tc for SPECT. Evidently, however, it is a tremendous challenge to design a glucose derivative with a pendent metal complex that is still recognized by the GLUT-1 transporter. After internalization into cells by active transportation, a ^{99m}TcDG must be a substrate for hexokinase, the first enzyme in the glucose metabolism, and to be phosphorylated at the 6-OH position. Phosphorylation, but no subsequent conversion, leads to trapping of the negatively charged product together with ^{99m}Tc in the cell. Therefore, the radionuclide accumulates in rapidly proliferating sites like cancerous cells. Despite many attempts, no functionally active ^{99m}Tc substitute of ^{18}FDG has been obtained so far. The principle feasibility has been doubted due to the high selectivity of GLUT-1 for intact glucose. Orvig and Bowne have provided recently an excellent review of the biological and chemical backgrounds for labeled glucose [95].

Most of the examples followed the pure pendent approach. It seems to be very difficult to mimic the structures and the functionalities of a carbohydrate with a ^{99m}Tc complex while keeping its molecular weight and size close to that of the parent. Figure 9.10 depicts some of the glucose derivatives labeled according to the pendent approach.

When two glucose molecules (glucosamines) were combined by an ECD (ethylenedicysteine) ligand, labeling with the [^{99m}Tc = O]$^{3+}$ moiety yielded an overall neutral complex. Since this radiopharmaceutical consists of two glucose

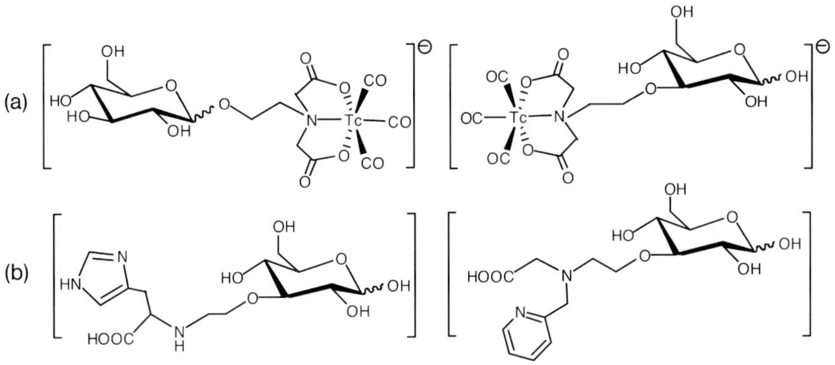

Figure 9.10 Labeled carbohydrates (a) and glucose derivatives for labeling with the $[^{99m}Tc(CO)_3]^+$ core (b) [96–98].

molecules, it was much larger than in the natural substrate. Still, tumor uptake was observed, probably via GLUT-1, but the tumor/blood ratio was generally <1 [99]. In a comparable approach, but using only mono-functionalized glucosamine, Liu *et al.* have conjugated to glucose different tetradentate ligands designed for the $[^{99m}Tc=O]^{3+}$ core. The compound with MAG$_3$ (Figure 9.8) showed promising results [100]. Similarly, DTPA was coupled to glucose and labeled with ^{99m}Tc. This compound allowed visualization of the tumor and the accumulated activity remained constant over a longer period of time. This finding indicated phosphorylation; however, the mechanism of trapping via this pathway has not been proven unambiguously [101]. Despite a promise beginning, these approaches have not lead yet to a breakthrough, probably due to the impossibility of improving the conjugates by specially designed ligands for the $[^{99m}Tc=O]^{3+}$ core.

The higher flexibility with respect to ligands has induced studies with the $[^{99m}Tc(CO)_3]^+$ core and glucose or glucosamine. Early attempts with tridentate N,O ligands conjugated at positions C1 and C3 gave some insight into the complexity of the problem but provided no solution [102]. Whereas derivatization, characterization, and labeling were performed, hexokinase inhibition or GLUT-1 uptake gave ambiguous results. For the future design of glucose derivatives according to the pendent approach, it seems important to use a linker of appropriate length between glucose and the ligand, as shown by Schibli *et al.* [98, 103]. Orvig *et al.* have developed several strategies towards glucose conjugation with bidentate ligands. These approaches followed the pendent concept but they also introduced novel complexes that consisted of intramolecular coordination to the ^{99m}Tc core. In initial attempts, 3-hydroxypyridinone was attached at C1 or C2 in glucose [91]. The rhenium-based model complexes were synthesized by reaction with [Re(OH$_2$)$_3$(CO)$_3$]$^+$ and labeled in quantitative yields. The compounds were tested for reactivity towards hexokinase but no inhibition was found, neither for the free glucose derivative nor for the corresponding Re(I) complexes. Presumably, the spacer between glucose and the ligand was not appropriate since Schibli *et al.*

reported some glucose derivatives that showed very slow but measurable activity with hexokinase [98].

A less common approach has been presented by Gottschaldt *et al.* They bound two glucose molecules to a central bipyridine ligand and studied the model complexes as well as the labeling. Conceptually, the approach resembles the one of Liu *et al.* mentioned before. In addition, these compounds also display inherent fluorescence, which makes detection by fluorescence microscopy attractive [104, 105].

The current state in ^{99m}Tc-labeled carbohydrate research seems to confirm the supposed unattainability of the objective, that is, to prepare ^{99m}Tc-glucose. It seems likely that the pendent approach will not be successful and that different concepts such as mimics of carbohydrates have more promise. Although no direct success for metal-based carbohydrates has been achieved, the fundamental results produced in the course of these studies are instrumental for the next steps.

9.3.4
Technetium-Essential Radiopharmaceuticals for Heart Imaging

The development of agents for heart imaging based on cationic complexes has been a focus of radiopharmaceutical chemistry since the early 1980s. These efforts led to several cationic complexes that are in clinical routine. Among them, Sestamibi (Cardiolite$^{®}$) is certainly the most important one and strongly contributed to public health care and life saving [79, 106, 107]. It should be emphasized here that Sestamibi is an organometallic compound that is synthesized in a very elegant approach. Its quantitative preparation could be reduced to a formulation that satisfies all regulations from the authorities and the producer. Recently, the patent expired and generic compounds are entering the market. Cardiolite$^{®}$ is widely applied in clinical routine although its pharmacological properties, liver accumulation in particular, are not ideal [108]. These drawbacks have encouraged research groups to find further complexes with improved *in vivo* behavior. These competitors are mainly based on the $[Tc{\equiv}N]^{2+}$ and the $[Tc(CO)_3]^{+}$ core, while maintaining the mono-cationic charge of Sestamibi. A few selected examples of these novel compounds are presented below.

Mono-cationic nitrido compounds have been investigated intensely for heart imaging. The mixed ligand approach by Duatti *et al.*, based on the $[Tc{\equiv}N]^{2+}$ core, gave this strategy a new impetus since the ligand sphere can conveniently be tuned by one ligand without the requirements of synthesizing new complexes from scratch. This strategy exceeds what was possible before with the promising TcN-NOEt complex (Figure 9.11). It is experimentally proven that ether groups in the ligand's periphery improve the biodistribution. For example, the complex ^{99m}TcN-DBODC5 (Figure 9.11b) was extracted by the rat heart very efficiently and liver activity was washed out after about 60 min, allowing for images with substantially improved target/non-target ratios [89, 90, 109, 113].

The nitrido approach has been extended by introducing crown ethers on the thiocarbamate ligand bound to the $[^{99m}TcN(PNP)]^{2+}$ moiety [109, 114]. The key

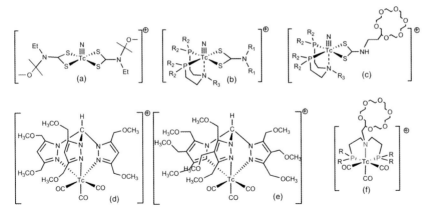

Figure 9.11 Lead compounds for myocardial imaging based on the $[^{99m}Tc{\equiv}N]^{2+}$ and the $[^{99m}Tc(CO)_3]^+$ core: (a) TcN-NOEt; (b) TcN-DBODC5 (R_1 = -(CH$_2$)$_2$O-Et, R_2 = -(CH$_2$)$_3$O-CH$_3$, R_3 = -(CH$_2$)$_2$O-Et) [90]; (c) a nitrido complex with additional crown ethers [109]; (d) and (e) tris-pyrazolylmethane based complexes [110, 111]; and (f) a Tc(I) complex with crown ethers attached to the ligand [112].

finding was that crown ether groups are useful for modifying the biological properties of cationic compounds. In rats, some of these compounds displayed heart/liver ratios four to five times higher than Cardiolite®. The crown ether strategy was extended with the same ligand but to the $[^{99m}Tc(CO)_3]^+$ core. Liu *et al.* could again show that the corresponding complexes had biological properties superior to Cardiolite® and were about identical to those of the respective complex with the $[^{99m}TcN(PNP)]^{2+}$ core [112].

A further significant improvement was achieved by Santos *et al.*: tris(pyrazolyl) methane is an excellent ligand for the $[^{99m}Tc(CO)_3]^+$ core. By introducing ether groups to the pyrazole rings they received within a very short time a stable heart uptake and a liver clearance rate not achieved by any of the other ^{99m}Tc complexes for myocardial imaging [110, 111]. These results provide an impetus for pursuing the search for novel radiopharmaceuticals to yield faster and more accurate diagnosis of coronary artery disease in humans. All these systems have in common that ether groups greatly enhance the biological behavior of the technetium-essential complexes. The reason for this finding is not completely understood but ongoing attempts towards novel myocardial imaging agents always include this structural feature.

Despite these progresses, it remains a major challenge to develop novel myocardial imaging agents with improved pharmacokinetics and biodistribution. The examples shown above are encouraging and the chemistry behind them inspires future fundamental and applied research. As a striking fact, they underline that the very same ligand can be selected for different metal cores. Thus, improving the biological behavior can not only be done by variations of the ligand – an often tedious procedure – but also by altering the metal core bound to the same chelator.

These two examples from technetium-based small molecule radiopharmaceuticals for diagnostic purposes are representative of the complexity of chemical structures to be achieved under difficult synthetic conditions as imposed by requirements from routine application. Many more complexes and compounds have been reported but synthesizing them directly from saline and in kit formulations makes biological studies or attempted applications very unlikely [78]. Still, the incentive for achieving quantitative formation of complexes with sophisticated structures remains a driving force for fundamental and applied chemistry in metal-based radiopharmaceuticals. The question of how to prepare a particular complex is not only a matter of development but demands profound insight into the mechanisms of formation. This concerns not only technetium but also the chemistry of the other elements described in this chapter.

9.4
Perspectives and Conclusion

Metal-based radiopharmaceuticals play an important role in diagnosis and imaging of various diseases. For example, myocardial imaging agents are essential for diagnosis and prognosis of heart diseases, thereby contributing to the enhancement of quality and duration of life, especially in the developed world.

If labeled compounds interact with single molecule targets, the method is also referred to as *molecular imaging*. Selected compounds are in clinical use, and some are in advanced stages of clinical trials, especially with radionuclides from the 3+ family. Still, radiopharmaceuticals, metal-based or not, are experiencing more and more competition from non-radioactive methods for molecular imaging such as MRI or fluorescence spectroscopy in its widest sense. Despite the instrumental and methodological progress of these methods, it should not be forgotten that the contributions of MRI (also in combination with other methods such as CT) are essentially diagnostic and the aspect of therapy is lacking. The therapeutic aspect, however, is of utmost importance and only metal-based radiopharmaceuticals are able to combine both, often even in the very same compound and just with two different radioisotopes. Radiopharmaceuticals are often designed for functional imaging and not for receiving static pictures. As long as the sensitivity of contrast agents as used in MRI cannot be dramatically increased, imaging on the molecular level is not possible. The extremely low concentrations of radiopharmaceuticals together with their sensitivity are therefore the only modality so far to achieve noninvasive imaging on the true molecular level. All these characteristics of radionuclide-based pharmaceuticals encourage and trigger research in the metal-based radiopharmaceutical chemistry although, admittedly, the success for public health care has not been overwhelming thus far. The perspective of combining diagnosis and therapy in the same compound is an incentive that encourages further endeavors in fundamental and in applied research.

References

1 Fichna, J. and Janecka, A. (2003) *Bioconjugate Chem.*, **14**, 3–17.

2 Yang, D.J., Kim, E.E., and Inoue, T. (2006) *Ann. Nucl. Med.*, **20**, 1–11.

3 Eisenhut, M. and Mier, W. (2003) in *Handbook of Nuclear Chemistry*, Vol. **4** (eds A. Vertes, S. Nagy, Z. Klencsar, and F. Roesch), Kluwer Academic Press, Dordrecht, pp. 257–277.

4 Bedford, M. and Maisey, M.N. (2004) *Eur. J. Nucl. Med. Mol. I*, **31**, 208–221.

5 Blower, P. (2006) *Dalton Trans.*, 1705–1711.

6 Dobrucki, L.W. and Sinusas, A.J. (2010) *Nat. Rev. Cardiol.*, **7**, 38–47.

7 Cherry, S.R. (2009) *Semin. Nucl. Med.*, **39**, 348–353.

8 Delbeke, D., Schoder, H., Martin, W. H., and Wahl, R.L. (2009) *Semin. Nucl. Med.*, **39**, 308–340.

9 Josephs, D., Spicer, J., and O'Doherty, M. (2009) *Target. Oncol.*, **4**, 151–168.

10 Jennings, W.A. (1994) *Nucl. Instrum. Methods A*, **346**, 548–549.

11 Mirzadeh, S., Mausner, L.F., and Garland, M.A. (2003) in *Handbook of Nuclear Chemistry*, Vol. **4** (eds A. Vertes, S. Nagy, Z. Klencsar, and F. Roesch), Kluwer Academic Press, Dordrecht, pp. 1–44.

12 Roesch, F. and Knapp, F.F. (2003) in *Handbook of Nuclear Chemistry*, Vol. **4** (eds A. Vertes, S. Nagy, Z. Klencsar, and F. Roesch) Kluwer Academic Press, Dordrecht, pp. 81–108.

13 Qaim, S.M. (2003) in *Handbook of Nuclear Chemistry*, Vol. **4** (eds A. Vertes, S. Nagy, Z. Klencsar, and F. Roesch), Kluwer Academic Press, Dordrecht, pp. 47–76.

14 Liu, S. (2004) *Chem. Soc. Rev.*, **33**, 445–461.

15 Johannsen, B. and Pietzsch, H.J. (2002) *Eur. J. Nucl. Med. Mol. I*, **29**, 263–275.

16 Kung, M.-P., Stevenson, D.A., Plössl, K., Meegalla, S.K., Beckwith, A., Essman, W.D., Mu, M., Lucki, I., and Kung, H.F. (1997) *Eur. J. Nucl. Med.*, **24**, 372–380.

17 Kung, H.F., Kung, M.P., Wey, S.P., Lin, K.J., and Yen, T.C. (2007) *Nucl. Med. Biol.*, **34**, 787–789.

18 Chi, D.Y. and Katzenellenbogen, J.A. (1993) *J. Am. Chem. Soc.*, **115**, 7045–7046.

19 Mäcke, H.R. and Good, S. (2003) in *Handbook of Nuclear Chemistry*, Vol. **4** (eds A. Vertes, S. Nagy, Z. Klencsar, and F. Roesch), Kluwer Academic Press, Dordrecht, pp. 279–310.

20 Safavy, A., Smith, D.C., Bazooband, A., and Buchsbaum, D.J. (2002) *Bioconjugate Chem.*, **13**, 317–326.

21 Safavy, A., Smith, D.C., Bazooband, A., and Buchsbaum, D.J. (2002) *Bioconjugate Chem.*, **13**, 327–332.

22 Liu, S. and Edwards, D.S. (2001) *Bioconjugate Chem.*, **12**, 7–34.

23 Kodama, M., Koike, T., Mahatma, A.B., and Kimura, E. (1991) *Inorg. Chem.*, **30**, 1270–1273.

24 Kukis, D.L., Denardo, S.J., Denardo, G. L., O'Donnell, R.T., and Meares, C.F. (1998) *J. Nucl. Med.*, **39**, 2105–2110.

25 Liu, S., Cheung, E., Ziegler, M.C., Rajopadhye, M., and Edwards, D.S. (2001) *Bioconjugate Chem.*, **12**, 559–568.

26 Yang, L.W., Liu, S., Wong, E., Rettig, S.J., and Orvig, C. (1995) *Inorg. Chem.*, **34**, 2164–2178.

27 Maecke, H.R., Kaden, T.A., Riesen, A., Ritter, W., and Studer, M. (1989) *Br. J. Cancer*, **59**, 307–307.

28 Riesen, A., Kaden, T.A., Ritter, W., and Maecke, H.R. (1989) *J. Chem. Soc., Chem. Commun.*, 460–462.

29 Boni, G., Ricci, S., Pastina, I., Grosso, M., Cresti, N., Chiacchio, S., Galli, L., Alsharif, A., and Mariani, B. (2003) *J. Nucl. Med.*, **44**, 412–416.

30 Modoni, S., Urbano, N., Cocco, G., and Fusco, V. (2006) *Ann. Oncol.*, **17** (VII), 160.

31 Sinzinger, H., Chehne, F., Palumbo, B., Pirich, C., Kratzik, C., Weiss, K., and Palumbo, R. (2001) *Eur. J. Nucl. Med.*, **28**, 1055.

32 Nestle, U., Lehmann, J., Bock, S., Farmakis, G., Stoeckle, M., and Kirsch,

C.M. (2006) *Eur. J. Nucl. Med. Mol. I*, **33**, S335.

33 Schmitt, A., Bernhardt, P., Nilsson, O., Ahlman, H., Kolby, L., Maecke, H.R., and Forssell-Aronsson, E. (2004) *J. Nucl. Med.*, **45**, 1542–1548.

34 Ocak, M., Kabasakal, L., Uslu, E., Decristoforo, C., Mut, S., and Uslu, I. (2006) *Eur. J. Nucl. Med. Mol. I*, **33**, S307.

35 Buchsbaum, D.J., Rogers, B.E., Khazaeli, M.B., Mayo, M.S., Milenic, D. E., Kashmiri, S.V.S., Anderson, C.J., Chappell, L.L., Brechbiel, M.W., and Curiel, D.T. (1999) *Clin. Cancer Res.*, **5**, 3048–3055.

36 Nayak, T.K., Atcher, R.W., Prossnitz, E. R., and Norenberg, J.P. (2008) *Nucl. Med. Biol.*, **35**, 673–678.

37 Tijink, B.M., Neri, D., Leemans, C.R., Budde, M., Dinkelborg, L.M., Stigter-Van Walsum, M., Zardi, L., and Van Dongen, G.A.M.S. (2006) *J. Nucl. Med.*, **47**, 1127–1135.

38 Knapp, F.F., Beets, A.L., Guhlke, S., Zamora, P.O., Bender, H., Palmedo, H., and Biersack, H.J. (1997) *Anticancer Res.*, **17**, 1783–1795.

39 Knapp, F.F., Mirzadeh, S., Beets, A.L., O'Doherty, M.J., Blower, P.J., Verdera, E.S., Gaudiano, J.S., Kropp, J., Guhlke, S., Palmedo, H., and Biersack, H.J. (1998) *Appl. Radiat. Isot.*, **49**, 309–315.

40 Knapp, F.F. (1994) *Eur. J. Nucl. Med.*, **21**, 1151–1165.

41 Park, S.H., Seifert, S., and Pietzsch, H. J. (2006) *Bioconjugate Chem.*, **17**, 223–225.

42 Schibli, R., Schwarzbach, R., Alberto, R., Ortner, K., Schmalle, H., Dumas, C., Egli, A., and Schubiger, P.A. (2002) *Bioconjugate Chem.*, **13**, 750–756.

43 Blower, P.J. and Rosales, R.T. (2008) *Technetium-99m Radiopharmaceuticals: Status and Prospective*, International Atomic Energy Agency, Vienna, pp. 271–287.

44 Zamora, P.O., Marek, M.J., and Knapp, F.F. (1997) *Appl. Radiat. Isot.*, **48**, 305–309 (in Russian).

45 Guhlke, S., Schaffland, A., Zamora, P. O., Sartor, J., Diekmann, D., Bender,

H., Knapp, F.F., and Biersack, H.J. (1998) *Nucl. Med. Biol.*, **25**, 621–631.

46 Singh, J., Powell, A.K., Clarke, S.E.M., and Blower, P.J. (1991) *J. Chem. Soc., Chem. Commun.*, 1115–1117.

47 Bolzati, C., Boschi, A., Uccelli, L., Duatti, A., Franceschini, R., and Piffanelli, A. (1999) *Eur. J. Nucl. Med. I*, **26**, S645.

48 Choudhry, U., Greenland, W.E.P., Goddard, W.A., Maclennan, T.A.J., Teat, S.J., and Blower, P.J. (2003) *Dalton Trans.*, 311–317.

49 Greenland, W.E.P. and Blower, P.J. (2005) *Bioconjugate Chem.*, **16**, 939–948.

50 Pervez, S., Mushtaq, A., and Arif, M. (2001) *Appl. Radiat. Isot.*, **55**, 647–651.

51 Laznicek, M., Laznickova, A., Trejtnar, F., Lorenc, P., Varvarigou, A., Bouziotis, P., and Archimandritis, S. (2002) *Anticancer Res.*, **22**, 2125–2130.

52 Alberto, R., Egli, A., Schibli, R., Abram, U., Waibel, R., Schwarzbach, R., and Schubiger, P.A. (1999) *Technetium and Rhenium in Nuclear Medicine and Chemistry* (eds U. Mazzi and M. Nicolini), SG Editorali, Padua, Italy, pp. 27–34.

53 Alberto, R., Ortner, K., Wheatley, N., Schibli, R., and Schubiger, A.P. (2001) *J. Am. Chem. Soc.*, **123**, 3135–3136.

54 Alberto, R., Schibli, R., Schubiger, A.P., Abram, U., Pietzsch, H.J., and Johannsen, B. (1999) *J. Am. Chem. Soc.*, **121**, 6076–6077.

55 Alberto, R. (2009) *Eur. J. Inorg. Chem.*, 21–31.

56 Dilworth, J.R. (2005) in *Metallotherapeutic Drugs and Metal-Based Diagnostic Agents* (eds M. Gielen and E.R.T. Tiekink), John Wiley & Sons, Ltd., Chichester, UK, pp. 463–486.

57 Zalutsky, M.R. and Vaidyanathan, G. (2000) *Curr. Pharm. Design*, **6**, 1433–1455.

58 Mcdevitt, M.R., Ma, D.S., Lai, L.T., Simon, J., Borchardt, P., Frank, R.K., Wu, K., Pellegrini, V., Curcio, M.J., Miederer, M., Bander, N.H., and Scheinberg, D.A. (2001) *Science*, **294**, 1537–1540.

59 Miederer, M., Mcdevitt, M.R., Sgouros, G., Kramer, K., Cheung, N.K.V., and Scheinberg, D.A. (2004) *J. Nucl. Med.*, **45**, 129–137.

60 Mcdevitt, M.R., Ma, D.S., Simon, J., Frank, R.K., and Scheinberg, D.A. (2002) *Appl. Radiat. Isotopes*, **57**, 841–847.

61 Miederer, M., Scheinberg, D.A., and McDevitt, M.R. (2008) *Adv. Drug. Deliver. Rev.*, **60**, 1371–1382.

62 Zhernosekov, K.P., Filosofov, D.V., Baum, R.P., Aschoff, P., Bihl, H., Razbash, A.A., Jahn, M., Jennewein, M., and Rosch, F. (2007) *J. Nucl. Med.*, **48**, 1741–1748.

63 Choi, H.O. and Hwang, K.J. (1987) *J. Nucl. Med.*, **28**, 91–96.

64 Green, M.A. and Huffman, J.C. (1988) *J. Nucl. Med.*, **29**, 417–420.

65 Goldsmith, S.J. (2009) *Future Oncol.*, **5**, 75–84.

66 Khan, M.U. and Coleman, R.E. (2008) *Nucl. Med. Biol.*, **35**, S77–S91.

67 Hsieh, W.Y. and Liu, S. (2004) *Inorg. Chem.*, **43**, 6006–6014.

68 Matthews, R.C., Parker, D., Ferguson, G., Kaitner, B., Harrison, A., and Royle, L. (1991) *Polyhedron*, **10**, 1951–1953.

69 Liu, S., He, Z.J., Hsieh, W.Y., and Fanwick, P.E. (2003) *Inorg. Chem.*, **42**, 8831–8837.

70 Burchardt, C., Riss, P.J., Zoller, F., Maschauer, S., Prante, O., Kuwert, T., and Roesch, F. (2009) *Bioorg. Med. Chem. Lett.*, **19**, 3498–3501.

71 Buchmann, I., Henze, M., Engelbrecht, S., Eisenhut, M., Runz, A., Schaefer, M., Schilling, T., Haufe, S., Herrmann, T., and Haberkorn, U. (2007) *Eur. J. Nucl. Med. Mol. I*, **34**, 1617–1626.

72 Clarke, E.T. and Martell, A.E. (1991) *Inorg. Chim. Acta*, **181**, 273–280.

73 Clarke, E.T. and Martell, A.E. (1991) *Inorg. Chim. Acta*, **190**, 37–46.

74 Velikyan, I., Maecke, H., and Langstrom, B. (2008) *Bioconjugate Chem.*, **19**, 569–573.

75 Prata, M.I.M., Santos, A.C., Geraldes, C.F.G.C., and De Lima, J.J.P. (1999) *Nucl. Med. Biol.*, **26**, 707–710.

76 Alberto, R. (2005) in *Contrast Agents III: Radiopharmaceuticals – from Diagnostics to Therapeutics*, Vol. **252** (ed. W. Krause), Springer, Heidelberg, pp. 1–44.

77 Alberto, R. (2006) in *Bioorganometallics* (ed. G. Jaouen), Wiley-VCH Verlag GmbH, Weinheim, pp. 97–124.

78 Banerjee, S., Pillai, M.R.A., and Ramamoorthy, N. (2001) *Semin. Nucl. Med.*, **31**, 260–277.

79 Banerjee, S.R., Maresca, K.P., Francesconi, L., Valliant, J., Babich, J.W., and Zubieta, J. (2005) *Nucl. Med. Biol.*, **32**, 1–20.

80 Boschi, A., Duatti, A., and Uccelli, L. (2005) *Top. Curr. Chem.*, **252**, 85–115.

81 Schwochau, K. (1994) *Angew. Chem., Int. Ed.*, **33**, 2258–2267.

82 Warner, R.R.P. and O'Dorisio, T.M. (2002) *Semin. Nucl. Med.*, **32**, 79–83.

83 Alberto, R. (2003) in *Comprehensive Coordination Chemistry II*, Vol. **5** (eds J. A. McCleverty and T.S. Meer), Elsevier Science, Amsterdam, pp. 127–271.

84 Bolzati, C., Uccelli, L., Boschi, A., Malago, E., Duatti, A., Tisato, F., Refosco, F., Pasquali, R., and Piffanelli, A. (2000) *Nucl. Med. Biol.*, **27**, 369–374.

85 Duatti, A., Boschi, A., Bolzati, C., Uccelli, L., Benini, E., Sabba, N., Moretti, E., Piffanelli, A., Refosco, F., and Tisato, F. (2002) *J. Nucl. Med.*, **43**, 137–138.

86 Pietzsch, H.J., Tisato, F., Refosco, F., Leibnitz, P., Drews, A., Seifert, S., and Spies, H. (2001) *Inorg. Chem.*, **40**, 59–64.

87 Drews, A., Pietzsch, H.J., Syhre, R., Seifert, S., Varnas, K., Hall, H., Halldin, C., Kraus, W., Karlsson, P., Johnsson, C., Spies, H., and Johannsen, B. (2002) *Nucl. Med. Biol.*, **29**, 389–398.

88 Alberto, R. and Abram, U. (2005) in *Nuclear Chemistry (Radiochemistry and Radiopharmaceutical Chemistry in Life Science)*, Vol. **4** (eds A. Vertes, S. Nagy, and Z. Klencsar), Kluwer Academic Publishers, Dordrecht, pp. 211–256.

89 Boschi, A., Bolzati, C., Uccelli, L., Duatti, A., Benini, E., Refosco, F., Tisato, F., and Piffanelli, A. (2002) *Nucl. Med. Commun.*, **23**, 689–693.

90 Boschi, A., Uccelli, L., Bolzati, C., Duatti, A., Sabba, N., Moretti, E., Di Domenico, G., Zavattini, G., Refosco, F., and Giganti, M. (2003) *J. Nucl. Med.*, **44**, 806–814.

91 Ferreira, C.L., Bayly, S.R., Green, D.E., Storr, T., Barta, C.A., Steele, J., Adam, M.J., and Orvig, C. (2006) *Bioconjugate Chem.*, **17**, 1321–1329.

92 Liu, Y., Pak, J.-K., Schmutz, P., Bauwens, M., Mertens, J., Knight, H., and Alberto, R. (2006) *J. Am. Chem. Soc.*, **128**, 15996–15997.

93 Walther, M., Jung, C.M., Bergmann, R., Pietzsch, J., Rode, K., Fahmy, K., Mirtschink, P., Stehr, S., Heintz, A., Wunderlich, G., Kraus, W., Pietzsch, H.J., Kropp, J., Deussen, A., and Spies, H. (2007) *Bioconjugate Chem.*, **18**, 216–230.

94 Muller, C., Hohn, A., Schubiger, P.A., and Schibli, R. (2006) *Eur. J. Nucl. Med. Mol. I*, **33**, 1007–1016.

95 Bowen, M.L. and Orvig, C. (2008) *Chem. Commun.*, 5077–5091.

96 Storr, T., Fisher, C.L., Mikata, Y., Yano, S., Adam, M.J., and Orvig, C. (2005) *Dalton Trans.*, 654–655.

97 Dumas, C., Petrig, J., Frei, L., Spingler, B., and Schibli, R. (2005) *Bioconjugate Chem.*, **16**, 421–428.

98 Schibli, R., Dumas, C., Petrig, J., Spadola, L., Scapozza, L., Garcia-Garayoa, E., and Schubiger, P.A. (2005) *Bioconjugate Chem.*, **16**, 105–112.

99 Yang, D.J., Kim, C.G., Schechter, N.R., Azhdarinia, A., Yu, D.F., Oh, C.S., Bryant, J.L., Won, J.J., Kim, E.E., and Podoloff, D.(2003) *Radiology*, **226**, 465–473.

100 Chen, Y., Li, L., Liu, F., and Liu, B. (2006) *Bioorg. Med. Chem. Lett.*, **16**, 5503–5506.

101 Chen, Y., Xiong, Q., Huang, Z., and He, L. (2007) *Eur. J. Nucl. Med. Mol. I*, **34**, S363.

102 Petrig, J., Schibli, R., Dumas, C., Alberto, R., and Schubiger, P.A. (2001) *Chem. Eur. J.*, **7**, 1868–1873.

103 Dumas, C., Schibli, R., and Schubiger, P.A. (2003) *J. Org. Chem.*, **68**, 512–518.

104 Gottschaldt, M., Koth, D., Muller, D., Klette, I., Rau, S., Gorls, H., Schafer, B., Baum, R.P., and Yano, S. (2007) *Chem. Eur. J.*, **13**, 10273–10280.

105 Gottschaldt, M. and Schubert, U.S. (2009) *Chem. Eur. J.*, **15**, 1548–1557.

106 Nunn, A.D. (1990) *Semin. Nucl. Med.*, **20**, 111–118.

107 Jain, D. (1999) *Semin. Nucl. Med.*, **29**, 221–236.

108 Llaurado, J.G. (2001) *J. Nucl. Med.*, **42**, 282–284.

109 Liu, S. (2007) *Dalton Trans.*, 1183–1193.

110 Maria, L., Cunha, S., Videira, M., Gano, L., Paulo, A., Santos, I.C., and Santos, I. (2007) *Dalton Trans.*, 3010–3019.

111 Maria, L., Fernandes, C., Garcia, R., Gano, L., Paulo, A., Santos, I.C., and Santos, I. (2009) *Dalton Trans.*, 603–606.

112 He, Z., Hsieh, W.-Y., Kim, Y.-S., and Liu, S. (2006) *Nucl. Med. Biol.*, **33**, 1045–1053.

113 Hatada, K., Riou, L.M., Ruiz, M., Yamamichi, Y., Duatti, A., Lima, R.L., Goode, A.R., Watson, D.D., Beller, G. A., and Glover, D.K. (2004) *J. Nucl. Med.*, **45**, 2095–2101.

114 Liu, S., He, Z.J., Hsieh, W.Y., and Kim, Y.S. (2006) *Nucl. Med. Biol.*, **33**, 419–432.

10
Boron and Gadolinium in the Neutron Capture Therapy of Cancer

Ellen L. Crossley, H.Y. Vincent Ching, Joseph A. Ioppolo, and Louis M. Rendina

10.1
Introduction

Tumors of the central nervous system (CNS), particularly those of the brain, are an important public health problem, particularly in light of the increased life expectancy of the general population in Western countries. Although the median age for the diagnosis of brain tumors is typically over 55, they are the most common solid tumor in children [1–4]. In the United States, brain cancer accounts for approximately 1.4% of all cancer cases, with the estimated 5-year relative survival rate being 30% [3]. However, the latter does not provide a complete picture of this particular disease as there exists a dramatic variation in survival depending upon the histological sub-type. For example, meningiomas are the most common brain tumor and account for more than 32% of all brain cancers, but 96% of cases are classified as non-malignant and are thus amenable to surgery and/or conventional radiotherapy [3]. Glioblastoma multiforme (GBM) is the most common brain malignancy and, in contrast to meningiomas, it is an intractable, aggressive disease that to date has proven resistant to all forms of treatment. In the United States, the median survival time for the disease is less than 12 months, and the 5-year survival rate is approximately 3.4% [5–8].

Surgery is the primary means of removing malignant brain tumors such as GBM, but due to their highly infiltrative growth pattern it is virtually impossible to remove them completely. Extensive surgical resection and/or high-dose conventional radiotherapy leads to some degree of neurological deterioration, and the complete eradication of every remaining tumor cell is a great, perhaps insurmountable, challenge [9]. Indeed, approximately 80–90% of GBM cases are known to re-occur locally. Surgery and radiotherapy are considered to be only palliative measures, thereby signifying the need for innovative methods such as binary therapies for the treatment of this devastating disease [9, 10].

A binary therapy employs two components that in combination result in a lethal effect. It is based upon the principle that the two components become highly toxic

Bioinorganic Medicinal Chemistry. Edited by Enzo Alessio
Copyright © 2011 WILEY-VCH Verlag GmbH & Co. KGaA, Weinheim
ISBN: 978-3-527-32631-0

only when combined whereas on their own they are normally quite benign [11]. Since each component of a binary system can be manipulated independently, their combination can be localized at the target site, for example, a solid tumor, thus sparing healthy cells from being destroyed. At least one of the two components must be confined to a target site for the technique to be successful [11]. Examples of binary systems in medicine include photodynamic therapy (PDT) [12, 13], photon activation therapy (PAT) [14, 15], and neutron capture therapy (NCT). This chapter is primarily concerned with NCT.

10.2
Boron Neutron Capture Therapy

Boron neutron capture therapy (BNCT) was first proposed as a binary therapy over 70 years ago [16] and it is used for the treatment of certain solid cancers, predominately GBM. Japan is considered to be the leading country in this area of medical research and many GBM patients have been treated by this therapy over the past four decades [9, 17–19]. Clinical trials have also been conducted in the United States, Finland, Sweden, Italy, Argentina, the Czech Republic, and The Netherlands [9, 20–23]. Besides GBM, some limited clinical trials have also been initiated for head and neck cancer, metastatic liver cancer, and cutaneous and intra-cerebral malignant melanoma [22, 24, 25].

The two key components of BNCT are ^{10}B nuclei and neutrons possessing kinetic energies of about 0.025 eV, which are commonly referred to as "thermal" neutrons [20, 26]. ^{10}B is an excellent nuclide for NCT as it is non-radioactive and it also possesses a very large neutron capture cross-section for thermal neutrons (3838 barn) that is several orders of magnitude higher than that of other nuclides found in the human body such as ^{1}H, ^{12}C, ^{14}N, and ^{16}O (Table 10.1). This is an important feature as the undesired nuclear reactions involving neutron capture by

Table 10.1 Thermal neutron capture cross-sections for selected nuclides [11].

Nuclide	Neutron capture cross-section (barn)	Natural abundance (%)
^{1}H	3.3×10^{-1}	99.8
^{6}Li	9.4×10^{2}	7.5
^{10}B	3.8×10^{3}	19.8
^{12}C	3.4×10^{-3}	98.9
^{14}N	1.8	99.6
^{16}O	1.8×10^{-4}	99.8
^{126}I	6.0×10^{3}	–[a]
^{135}Xe	2.6×10^{6}	–[a]
^{157}Gd	2.6×10^{5}	15.7
^{235}U	5.8×10^{2}	7.2×10^{-1a}

[a]Radioactive.

endogenous nuclei (which are present in far greater concentrations in tissue than that of ^{10}B) are minimized but necessarily contribute to the overall radiation dose of the patient. The thermal neutrons possess kinetic energies that are too low to cause any significant damage to human tissue but their depth of penetration is very limited. Epithermal neutrons (0.5 eV–1.0 keV) are normally used in the clinic to treat deep-seated tumors in a non-invasive manner. The numerous collisions of epithermal neutrons with H-atoms in tissue results in their conversion into thermal neutrons due to a loss of kinetic energy [27]. Indeed, the conversion of epithermal into thermal neutrons is essential for the neutron capture reaction to proceed effectively.

Upon capturing a thermal neutron, ^{10}B forms the excited ^{11}B* nucleus, which is very short-lived and decays into two high-energy particles (the ^{7}Li nucleus and an alpha particle) that are capable of destroying cells due to their very high linear energy transfer (LET) characteristics; the nuclear fission reaction is accompanied by the release of a tremendous amount of energy (about 2.4 MeV) [11]. The major (94%) reaction for the neutron capture process is presented in Eq. (10.1). Importantly, the high LET particles possess very short path lengths (<10 μm) and so their cytotoxic effect is largely confined to the cell in which the ^{10}B agent is located. BNCT is thus an excellent example of an internal hadron therapy [28].

$$^{10}B + {}^1n \rightarrow [{}^{11}B^*] \rightarrow {}^4He^{2+} + {}^7Li^{3+} + \gamma + 2.31 MeV \qquad (10.1)$$

The use of ^{10}B is also advantageous from a chemical perspective, as boron can form hydrolytically-stable bonds with carbon, oxygen, and nitrogen, and so it can readily be incorporated into suitable BNCT agents [11]. These agents are classified as global, tumor targeted, or tumor specific [29]. Global agents readily distribute throughout the whole body in a uniform manner, possess little or no tumor specificity, and are essentially non-toxic (even when administered in large doses). In this case, accurate positioning of the neutron beam at the tumor site is essential. Tumor-targeted agents do not necessarily possess any inherent tumor specificity but can bind avidly to specific components of tumor cells such as their DNA, thereby increasing the cell kill by placing the ^{10}B in close proximity to critical cellular components prior to neutron irradiation. The tumor-targeting by these agents is usually based upon the much greater rate of cell division associated with many tumors compared to that of normal tissue. Finally, tumor-specific agents can selectively target cancer cells owing to the overexpression of some critical protein, for example [29]. An excellent example of a tumor-specific agent is a class of boronated thymidines that specifically target kinases that are overexpressed by tumor cells [30].

10.2.1
Key Criteria for BNCT Agents

BNCT agents need to satisfy several criteria to both maximize damage to cancer cells and minimize damage to healthy cells. First, the agent must achieve a sufficiently high boron concentration within the tumor (about 10–30 μg of ^{10}B per gram of tumor for global agents) to maximize the probability of a neutron capture reaction and also minimize the background radiation dose resulting from the capture of thermal neutrons by endogenous nuclei (Table 10.1); lower cellular concentrations of ^{10}B are feasible when it is localized near key cellular components such as DNA [11, 31–33]. Second, the agent must persist within the cell for a sufficient period of time that allows the tumor to be irradiated with thermal neutrons [11]. Third, the agent should be selective enough to achieve a tumor-to-healthy tissue ratio greater than 3:1, and it should also clear from the blood rapidly so that the tumor-to-blood ratio is greater than 1:1 [29]. The latter ensures that damage to the surrounding network of blood vessels is minimized [11]. Fourth, for tumors of the CNS, some lipophilicity is required for them to traverse the blood–brain barrier (BBB) although, in some cases, the BBB is compromised in GBM patients [34]. Moreover, there must exist some balance between lipophilicity and hydrophilicity in ^{10}B agents. If the ^{10}B agent is too hydrophilic then it will not cross the BBB. If it is too lipophilic then the compound will bind nonspecifically to proteins or remain trapped inside cell membranes [11]. Finally, the general toxicity of the agent, in particular a global agent, must also be considered as it should display sufficiently low toxicity towards healthy cells and also ensure that sufficiently large quantities of boron can be incorporated into the tumor prior to neutron irradiation.

10.2.2
Clinical BNCT Agents

The boronated L-phenylalanine analog L-4-dihydroxyborylphenylalanine (BPA, **1**) and the polyhedral borane sodium mercaptoundecahydrododecaborate(2–) (BSH, sodium borocaptate, **2**) are the only two agents being used in clinical trials for BNCT [18, 35, 36]. A third agent, $Na_2[B_{10}H_{10}]$ (GB-10, **3**), has also been approved for use in humans [37, 38]. Although BPA and BSH have been used for many years as BNCT agents, the mechanisms of their selective accumulation in tumors are not clearly established. BPA most likely enters tumor cells by means of a normal amino-acid transport process owing to its structural similarity to the amino acid L-phenylalanine [29, 33]. It is taken up faster in GBM tumors than in healthy tissue *in vivo* [39], and it has a higher tumor-to-blood ratio than BSH [33]. The selectivity of BSH is attributed to the formation of disulfide bonds with thiol groups in proteins [29]. BSH does not penetrate the normal BBB or cellular membranes but rather appears to target tumors by crossing the disrupted BBB [7, 33, 40].

BSH was the first BNCT agent to be administered in patients with high-grade gliomas in the 1960s [41] while BPA was first trialed in the 1980s to treat malignant melanoma and, in 1994, it was first administered to GBM patients [36]. BPA is usually administered to the patient as its water-soluble D-fructose or, more recently, its D-mannitol derivative [36, 42]. Despite its reasonable tumor selectivity and very low toxicity, dose escalation trials with BPA as the single boron delivery agent have reached a limiting threshold both in terms of the amount of BPA administered to the patient (typically 250–900 mg per kg body weight) and in the escalation of the beam dose as normal brain tolerance becomes limiting and leads to somnolence syndrome, a phenomenon that is attributed to the significant amounts of boron distributed in normal brain tissue [43]. At the present time, BSH and BPA are used both separately and in combination to treat GBM, malignant melanoma, and several other types of cancers [9, 17, 22]. The combination of BPA and BSH was originally utilized to minimize the heterogeneous ^{10}B distribution resulting from the different mechanisms by which BPA and BSH accumulate within the tumor [9]. The current drug dosage regimen for BNCT typically involves the administration of BSH to the patient approximately 12 h prior to neutron irradiation followed by the administration of BPA approximately 1 h before irradiation [17, 19, 44]. Clinical trials employing a combination of both BPA and BSH have shown an improved effectiveness for the treatment of GBM [17, 44–46] and head and neck tumors when compared to the single agent [47, 48]. Very recent studies have shown that the uptake of BSH and BPA was improved when administered along with D-mannitol, as studies have shown that this polyol appears to disrupt the BBB [36, 42].

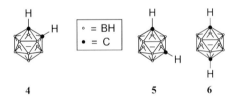

4   5   6

10.3
Role of Medicinal Inorganic Chemistry in BNCT

In the context of this chapter, some key areas of medicinal inorganic chemistry are examined in terms of their potential application to BNCT and important advances are summarized in the following section. Many of the agents described herein make use of boron-rich moieties known as dicarba-*closo*-dodecaboranes(12) (*closo*-carboranes) [11, 32, 49, 50]. *closo*-Carboranes possess a pseudo-aromatic polyhedral structure containing carbon and boron that occupies a volume similar to the three-dimensional sweep of a phenyl group. These boron clusters have the general formula $C_2B_{10}H_{12}$ and can exist as one of three isomers: *closo*-1,2-, 1,7-, or 1,12-carborane (**4–6**, respectively), which differ in the relative positions of the two carbon atoms. *closo*-Carboranes are air and moisture stable, and can be incorporated readily into organic structures. A single boron atom can be removed selectively from the hydrophobic *closo*-carborane cluster to generate an anionic *nido*-carborane species that is known to exhibit excellent water solubility. The chemistry of carboranes is beyond the scope of this chapter but the reader is directed to some excellent reviews on this subject [32, 51, 52].

10.3.1
Platinum–Boron Agents

Chromosomal DNA is an important cellular target for BNCT agents and numerous strategies have been investigated for the delivery of large numbers of boron atoms to this critical macromolecule [29, 31, 53]. Platinum-based DNA-binding agents, such as cisplatin, have been used in the clinic for the treatment of cancer for almost four decades [54–56] and, in recent years, boronated platinum complexes have been studied for their potential application in BNCT. Pt-B agents may potentially display additive or even synergistic biological effects associated with the DNA-binding reactions of the Pt-B agent coupled with the neutron capture reactions associated with the ^{10}B nuclei. Furthermore, an additional advantage is that such agents can be radiolabeled by the use of ^{195m}Pt ($t_{1/2} = 4$ d), which would allow their *in vivo* tumor uptake and biodistribution to be monitored using standard gamma imaging techniques [57].

The first examples of mono- and multi-nuclear platinum(II)–amine complexes containing carboranes have been reported [58–60]. These compounds were shown to bind covalently to DNA in a similar manner to that of cisplatin and related platinum(II)-amine complexes. Carborane-containing mono- and di-nuclear platinum(II)-2,2′;6′,2″-terpyridine (terpy) based DNA-metallointercalators have also been reported and these are known to bind DNA in a non-covalent manner [61–63]. The hydrophilic analog **7**, bearing a pendant glycerol group, has also been synthesized and it represents the first example of a highly water-soluble metal-carborane complex [64]. A preliminary biological assessment of selected carborane-containing platinum(II)-terpy complexes has shown excellent *in vitro* antitumor activity, particularly against cisplatin-resistant cell lines, but the platinum(II)-amine complexes were found to be quite toxic, thus precluding the delivery of sufficient quantities of boron to chromosomal DNA. Platinum-(II) and -(IV) complexes containing bis(1,2-carboranylpropoxycarbonyl)-2,2′-bipyridine ligands (e.g., **8**) have also been described but biological data are yet to be reported [65, 66].

7

8

10.3.2
Boronated Porphyrins

Often used in conjunction with PDT [67], boronated porphyrins and related macrocycles have been heavily studied for their potential use in BNCT. This area of research has already been extensively reviewed [68–71] and therefore only a couple of historically-important compounds and some recent developments, particularly those involving metals, will be discussed here. BOPP [2,4-bis(α,β-dihydroxyethyl)-deuteroporphyrin IX, **9**] is a porphyrin derivative that is known to accumulate in animal model brain tumors [50, 72]. More recent studies of this agent using convection-enhanced delivery (CED) [73] revealed a significant increase in boron uptake by the tumor [74]. The metalated porphyrin CuTCPH (**10**) possesses an inherently low photosensitivity, low toxicity, and potential for ^{64}Cu and ^{67}Cu radiolabeling for application in positron emission tomography (PET) imaging and has been shown to accumulate in tumors in several animal models [75–77]. Tumor ablation by thermal neutron irradiation has also been demonstrated for **10** [75]. Recently, a methoxy derivative of **10** showing a higher tumor uptake than the parent species has also been reported [78]. A series of anionic porphyrin cobaltacarborane conjugates [79] containing up to 16 carborane clusters have been synthesized and have been shown to accumulate in lysosomes of HEp2 cells. Charged trimethylamine or phosphonate-substituted carboranylporphyrins have also been reported to accumulate in lysosomes of human carcinoma HEp2 cells as well as human GBM T98G cells [80]. A porphyrin compound that contains four *nido*-carborane cages has recently been shown to induce an appreciable response in an animal melanotic melanoma model upon BNCT [81, 82].

9

10

10.3.3
Boronated Phosphonium Salts

A new strategy for potentially targeting tumors in BNCT is to employ delocalized lipophilic cations (DLCs). It is well known that DLCs including Rhodamine 123, tetraphenylphosphonium (TPP) chloride and triphenylmethylphosphonium (TPMP) iodide accumulate selectively into cancer cells [83–89]. The mitochondrial membrane potential of a tumor cell is known to be about 60 mV higher than that of a healthy cell and this factor accounts for the significant differences in DLC uptake [90].

The few boron-containing DLCs reported to date are *closo-* and *nido-*carborane analogs of TPMP iodide [91], dequalinium chloride (DEQ-B) [92], TPP chloride [93], and Nile Blue [93]. The first boronated analogs of TPMP for potential use in BNCT have also been described [91]. In particular, the salt **11** was found to have a favorably low toxicity towards the SF268 glioblastoma cell line in the absence of neutrons. The DEQ-B salt was found to have similar tumor uptake and retention properties to Rhodamine 123 and TPP chloride, and it was also found to accumulate selectively in human epidermoid carcinoma and rat glioma *in vitro* [92]. Very recently, 1,2- and 1,7-carborane derivatives of TPP were reported by Tsibouklis and coworkers [93]. In the case of the TPP phosphonium salt **12**, the anionic *nido-*7,8-carborane is not covalently attached to the cation and so the exact mechanism of boron delivery and tumor selectivity for tumor cells is unclear. *In vitro* boron uptake studies with the human prostate epithelial carcinoma demonstrated that selected boronated DLC derivatives such as **12** achieved a reasonable (up to 4.2:1) cancer to healthy tissue selectivity, despite the absence of any covalent link between the lipophilic phosphonium cation and the boron entity. Nile Blue DLC derivatives, in which the carborane moiety is covalently linked to the structure, demonstrated a similar selectivity to agents such as **12** [93]. These results suggest that a covalent link between the phosphonium center and the boron moiety may not be a necessary criterion for the selective delivery of boron to tumor sites.

10.3.4
Radiolabeling of Boronated Agents for *In Vivo* Imaging

The *in vivo* evaluation of tumor uptake and biodistribution pertaining to boronated agents is a difficult problem. There exists only a few useful methods for determining boron concentrations in humans, an important parameter that is critical in improving patient treatment outcomes. The optimal dosage of BNCT agent and time and duration of neutron irradiation each relies upon an accurate assessment of boron levels *in vivo*. Current protocols in BNCT involve a determination of boron levels in the blood by means of ICP-MS and ICP-AES but this is not a direct measure of tumor boron concentration [94, 95]. Although boron nuclei (both ^{10}B and ^{11}B) can be detected non-invasively by means of magnetic resonance imaging (MRI) [96–99], this technique is quite limited for numerous reasons, including the low sensitivity and short relaxation times of these quadrupolar nuclei [97]. There has been some progress using other NMR-sensitive nuclides attached to boronated agents, such as ^{19}F [100, 101] and ^{157}Gd [102, 103]. However, the large majority of work pertaining to the imaging of boronated agents *in vivo* concerns radiolabeling techniques. This includes the incorporation of radionuclides into new and existing BNCT agents for single-photon emission computed tomography (SPECT) and PET imaging studies.

All radionuclides in medicine can be divided into those used either for therapy or imaging. Of those used for imaging, these radionuclides can be further divided into those used for SPECT (e.g., ^{99m}Tc) and PET (e.g., ^{18}F) [104–106]. There have been *in vitro* studies using tritium (^{3}H)-labeled carboranes [107] but this nuclide is not used clinically as it is a β-emitter with a very long half-life (12.3 years) [108]. Most studies to date, however, concern boron compounds that have been radiolabeled by radiohalogenation using isotopes such as ^{18}F, or by complexation with a radioactive metal such as ^{57}Co or ^{99m}Tc [40].

10.3.5
Radiofluorination of BPA

The PET nuclide ^{18}F is the most widely used radioisotope in medical imaging due to its low-energy positron emission and short half-life (110 min) [109]. The most widely-used PET agent is 2-[^{18}F]-fluorodeoxy-D-glucose (FDG), which accumulates in most malignant tumors due to their increased glucose transport and glycolysis [110]. However, the high glucose metabolism in normal brain tissue precludes the use of FDG as a tracer for the detection of gliomas [111]. Other ^{18}F imaging agents include ^{18}F-DOPA, which has been used for probing cerebral dopamine metabolism [112, 113] and neuroendocrine tumors in humans [114].

^{18}F-BPA has been used extensively to determine the tumor uptake and biodistribution of the parent clinical BNCT agent BPA. The synthesis of racemic ^{18}F-BPA (4-borono-2-[^{18}F]-fluoro-D,L-phenylalanine) was first reported in 1991 [115]. Early studies in mice showed an effective accumulation of the agent in FM3A mammary carcinoma and in B16 melanoma [115, 116]. More recently, the distribution of enantiomerically-pure ^{18}F-L-BPA has been studied in rats bearing F98 gliomas [117–119]. For example, the L-isomer was shown to accumulate selectively in the tumors, achieving high tumor to normal brain ratios (about 3:1) a few hours after injection [117]. ^{18}F-BPA is known to accumulate selectively in GBM *in vivo*. The PET scans were found to be consistent with the MR images, and the calculated uptake was found to parallel those results determined by the direct boron analyses of tissue biopsies [120]. It was also found that the accumulation of ^{18}F-BPA correlated with the grade of tumor malignancy, and it was confirmed that L-BPA was taken up more efficiently than its racemic form [121]. The use of ^{18}F-BPA and PET imaging has since been used to estimate the concentration of ^{10}B in tumors [122, 123]. ^{18}F-BPA has also been used in the clinic to image and treat other human cancers [124]. For example, it has been used in the study and treatment of metastatic malignant melanoma in the thorax and brain [125, 126]. In one study, it was even possible to clearly detect a brain lesion that was difficult to image using traditional MR and CT techniques [126]. Head and neck malignancies have also been treated and imaged [25, 127]. The first clinical use of ^{18}F-BPA in BNCT for head and neck malignancies was reported in 2006 [128]. The patient receiving treatment by this method has since experienced a continued regression of the tumor. Furthermore, low-grade brain tumors such as schwannoma and meningioma [129] and recurrent cancer in the oral cavity and cervical lymph node metastases [130] have been treated in this way. A recent promising computational study has highlighted the possibility of using ^{18}F-BPA for the treatment of malignant spinal tumors by BNCT, for which conventional therapies are rarely effective [131].

10.3.6
Carborane Radiohalogenation

Although there are no known radiofluorinations of carboranes, they are readily labeled using other radiohalogens. Following on from the work of Hawthorne and coworkers [132–134], *nido*-carboranes linked to targeting vectors have been treated with radio-iodide (e.g., $^{125}I^-$ or $^{131}I^-$) in the presence of an oxidizing agent such as IodoGen® (1,3,4,6-tetrachloro-3α,6α-diphenylglucoluril) or chloramine-T (sodium *N*-chloro-*p*-toluenesulfonamide) [52]. Similarly, radiobromide ion (e.g., $^{76}Br^-$) in the presence of chloramine-T has been used to label monoclonal antibodies (MAb) (e.g., **13**) [104]. The radiohalogenation of *closo*-carboranes using ^{125}I and ^{76}Br has also been explored [135].

13

There have been similar radiohalogenations of other types of boron clusters such as $[B_{12}H_{12}]^{2-}$ [136–140], *closo*-$[CB_{11}H_{12}]^-$, and *closo*-$[B_{10}H_{10}]^{2-}$ derivatives [141]. In addition, the incorporation of the α-emitting ^{211}At nuclide into $[B_{12}H_{12}]^{2-}$ derivatives [142], *nido*-carborane [143], and the "venus flytrap cluster" (VFC) [144] (*vide infra*) for tumor treatment have been reported.

10.3.7
Radiometallacarboranes

Radiometallacarboranes are *nido*-carboranes complexed to radioactive nuclides (such as ^{99m}Tc) and can – in principle – be used to evaluate the uptake and biodistribution of carborane-containing agents for BNCT. More broadly, existing tumor-targeting agents can be modified to incorporate metal chelating groups such as *nido*-carborane [145] for both the imaging and treatment of tumors. It is essential that these complexes are robust *in vivo* as many types of radiometal complexes are readily dissociated in blood plasma by transferrin, an iron transport glycoprotein [40, 146]. Two excellent examples of stable, biologically-compatible radiometallacarboranes are ^{57}Co complexes of the VFC ligand first reported by Hawthorne and coworkers [147] and the ^{99m}Tc complexes of *nido*-carborane first reported by Valliant and coworkers [148].

The VFC ligand contains two dicarbollide fragments connected by a bifunctional pyrazole molecule. It possesses a highly pre-organized structure and it forms very stable complexes with cobalt(III) [149, 150]. The ^{57}Co-VFC complex (**14**) conjugated to an antibody has been used for both *in vitro* and *in vivo* animal studies [147, 149]. For example, **14** has been conjugated to the anti-carcinoembryonic antigen monoclonal antibody known as T84.66. It has shown excellent localization in carcinoembryonic antigen-producing tumors [147] such as the human colon carcinoma LS174T [149]. Although the long half-life of ^{57}Co (271 d) [151] precludes its use in humans, the shorter half-life (17 h) nuclide ^{55}Co could be used instead [134].

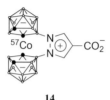

14

The ^{99m}Tc radionuclide is widely used in medicine for imaging and diagnosis due to its favorable low-energy γ-radiation (140 keV), short half-life (6 h), and ease of production (see Chapter 9) [52, 152–154]. Traditional syntheses of ^{99m}Tc radiometallacarboranes normally require long reaction times (about 24 h) [148]. More recently, however, microwave-assisted methodologies have allowed these complexes to be prepared in a timeframe of only a few minutes [32, 155–157]. These complexes are stable *in vivo* and are readily linked to suitable tumor-targeting agents. Rhenium(I) and ^{99m}Tc radiometallacarboranes have been prepared (e.g., **15**) that show a high affinity for the estrogen receptors ER$_\alpha$ and ER$_\beta$. These targeting vectors are known to be very useful as the upregulation of estrogen is observed in both breast and ovarian cancers [158].

15

10.4
Gadolinium Neutron Capture Therapy

^{10}B is not the only nuclide that possesses a large effective neutron capture cross-section. Two other nuclides that possess even larger neutron capture cross-sections are ^{135}Xe (2.6 × 10^6 barn) and ^{157}Gd (2.6 × 10^5 barn) (Table 10.1) [11, 159]. ^{135}Xe is a radioactive noble gas ($t_{1/2} = 9.1$ h) that cannot be incorporated into tumor-selective agents and so it has no potential as a NCT agent. In contrast, ^{157}Gd is non-radioactive and it possesses the largest effective neutron capture cross-section of all the naturally-occurring elements [160–163]. Indeed, on the basis of the neutron capture cross-sections alone there would exist an approximate 66-fold enhancement in the number of NC-events per cell when ^{157}Gd is used instead of ^{10}B. In contrast to ^{10}B, however, where the nuclear capture process yield fission end products, ^{157}Gd neutron capture instead involves the formation of an excited ^{158}Gd* nucleus that expels a high-energy gamma photon (about 7.9 MeV) to afford the (stable) ^{158}Gd isotope [Eq. (10.2)]:

$$^{157}\text{Gd} + {}^1\text{n} \rightarrow [^{158}\text{Gd}^*] \rightarrow {}^{158}\text{Gd} + \gamma + 7.94\,\text{MeV} \tag{10.2}$$

To a much lesser extent, a parallel reaction also occurs in which the excited ^{158}Gd* nucleus interacts with inner-core electron(s), resulting in the production of high-energy internal conversion electrons, and this process is accompanied by secondary Auger and Coster-Kronig (ACK) electron emission [164–166]. There are approximately five ACK electrons produced per neutron capture event and their path-lengths have a range of only a few nanometers [159, 167], in contrast to the high LET particles produced by the capture of thermal neutrons by the ^{10}B nucleus, which can travel up to 9 μm [168, 169]. ^{157}Gd must therefore be localized in very close proximity to critical cellular components such as DNA if the neutron capture reaction is to be exploited effectively. Whereas at a first glance this may appear to be a significant limitation in the use of ^{157}Gd for the treatment of solid tumors, any damage resulting from the neutron capture reactions involving this isotope would be highly localized and thus if the Gd agent can be selectively delivered to tumor cells then no damage would result to nearby healthy cells irrespective of the nature of the sub-cellular components that are targeted. In the case of BNCT, the cell nucleus is the principal target of choice for the ^{10}B agent owing to the significantly enhanced cell kill upon neutron irradiation and this criterion would also apply to a ^{157}Gd agent.

The ^{157}Gd neutron capture reactions are much more complex than those of ^{10}B but the biologically relevant products appear to be the ACK electrons [165, 169, 170]. Studies by Martin *et al.* [171, 172] have shown that upon thermal neutron capture Gd^{3+}-bound DNA induces a breaking of double-stranded DNA. Free Gd^{3+} ions bind to the negatively charged phosphodiester backbone of DNA and it is proposed that DNA double-stranded cleavage can be attributed to the ACK electrons [160, 161, 173, 174]. When an ACK electron is emitted it interacts with a water molecule to produce a hydroxyl radical, which in turn locally propagates the oxidative damage leading to double-strand DNA breaks [163, 175].

10.4.1
Archetypal Gd NCT Agents

In sharp contrast to the development of boronated agents for potential use in BNCT, the development of new GdNCT agents is still only in its infancy. Free Gd^{3+} ions are highly toxic and must therefore be incorporated into chelate complexes [161, 176]. It has been known for some time that paramagnetic Gd(III) complexes, for example, Gd-DTPA (gadopentetate dimeglumine, Magnevist®), are highly-efficient water relaxation agents that can improve image contrast in MRI [176–180]. Gd(III) complexes are also used in therapy and these agents can enhance tumor response to chemotherapeutic agents such as cisplatin [181] or, more commonly, act as a radiosensitizer in the treatment of cancer [182–185]. Gadolinium may also play an important role in therapeutic techniques such as synchrotron stereotactic radiotherapy (SSR) [175]. This therapy shares some parallels with GdNCT as it relies upon the selective delivery of Gd to the cell nucleus to significantly enhance the efficacy of the treatment. De Stasio and coworkers have demonstrated that motexafin-Gd (**16**), a Gd(III) complex of the pentadentate texaphyrin ligand [183, 186], was accumulated by approximately 90% of glioblastoma cell nuclei *in vitro* [175]. These results show the potential of this complex to act as a GdSSR or GdNCT agent. The use of other Gd(III) complexes as potential GdNCT delivery agents to brain tumors has also been described [167, 187–189]; however, the major concern for using such complexes in a clinical context for GdNCT is the limited number of tumor cell nuclei that have been shown to incorporate Gd [163]. In fact, the number of Gd compounds reported to date that have a capacity to aggregate selectively in tumor cell nuclei is very limited. Thus, the search for new types of Gd(III) complexes with high nuclear affinity has been proposed [163]. A dinuclear Pt(II) derivative of DTPA (17) can selectively deliver Gd to tumor cells and, most importantly, deliver the lanthanide ion to the cell nuclei most likely by virtue of the intercalating Pt-terpy centers [190]. Clearly, there is still a great deal of research that needs to be completed before GdNCT clinical trials are even considered feasible but the preliminary studies in this area, including those involving animal models [191], are encouraging.

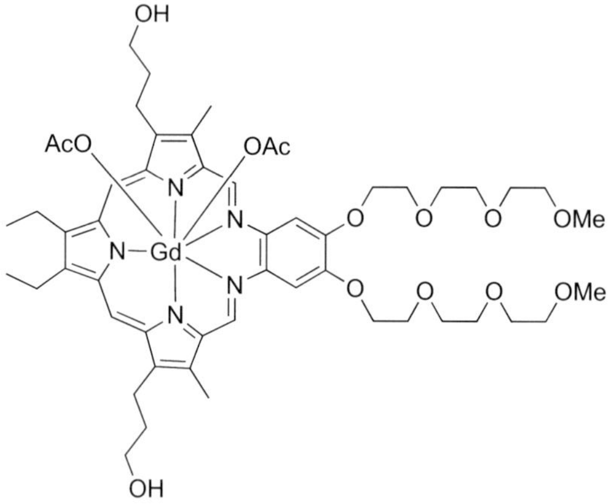

16

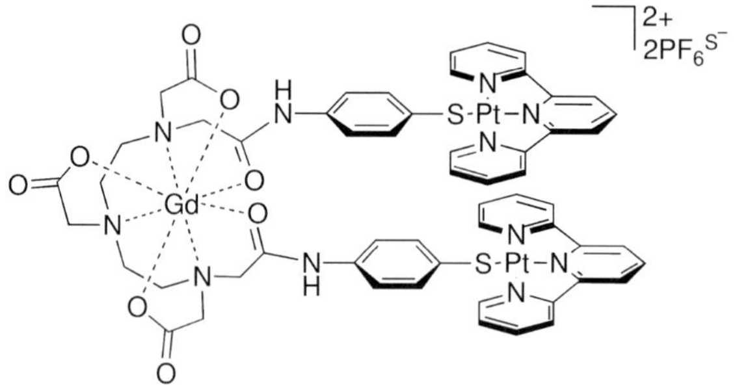

17

10.4.2
Hybrid Boron–Gadolinium Agents

Gadolinium(III) has been incorporated into selected boronated agents to monitor the uptake and distribution of boron concentrations in tissue by exploiting the MRI-contrasting characteristics of the lanthanide ion. One example is a carborane-gadolinium-DTPA species [192, 193]. More recently, a BPA derivative of Gd-DTPA [194] and a carborane containing Gd-DOTA agent [195] have been prepared and evaluated. The syntheses of other B-Gd hybrid compounds have also been investigated for NCT [196–198]. The possible additive or synergistic effects resulting from the incorporation of both boron and gadolinium within the same agent for potential application in NCT are yet to be elucidated and further research must be completed to determine the efficacy of a hybrid B-Gd strategy.

10.5
Conclusions and Future Outlook

Despite the number of remarkable results achieved in the clinic to date, BNCT as a front-line treatment for GBM appears to lie at the cross-roads at the present time. Most worldwide clinical trials have now ceased despite clearly demonstrating that the therapy involving BPA and/or BSH is safe. In countries such as Japan, a great deal of research endeavor and investment over the past four decades have made BNCT one of the front-line therapies in the treatment of GBM. There are certainly several limitations associated with clinical BNCT, for example, the use of a nuclear reactor as the neutron source, but the viability of using new accelerator-based technologies for the generation of epithermal neutrons has been demonstrated and is an important breakthrough [199–203].

One author has recently questioned whether, after five decades of worldwide research, BNCT is worth pursuing for the treatment of GBM [204]. Perhaps an even more valid question to ask at the present time is whether BNCT *utilizing BPA and/or BSH* is worth pursuing for the treatment of this devastating disease. This question has yet to be fully answered but, in terms of new chemistry, two factors associated with the boronated agents themselves require critical attention if BNCT is to ever become a front-line therapy or, at the very least, an adjunct therapy in combination with conventional radiotherapy and chemotherapy for the treatment of refractory tumors such as GBM: (i) a dramatic enhancement of tumor selectivity is required as BPA and BSH display only marginal to moderate selectivities for GBM and other solid cancers and (ii) the inherent difficulty in delivering large quantities of ^{10}B to each tumor cell to achieve a sufficiently high therapeutic index must be addressed. Each of these factors is non-trivial and undoubtedly there will exist major challenges in addressing them in the future but it does not mean that the task is impossible. New classes of boronated agents and tumor delivery systems must be developed and assessed for their *in vivo* efficacy and it is here that medicinal inorganic chemists can play an important role, not only in terms of the synthesis and biological assessment of new agents for BNCT but also in the real-time monitoring of tumor uptake and biodistribution of such agents *in vivo*. Furthermore, within tumor cells certain exogenous and endogenous metal ions may exhibit important additive or synergistic effects when irradiated with thermal neutrons in the presence of ^{10}B (or ^{157}Gd) nuclei, an area of research that has not been investigated to date. Indeed, the cutting-edge field of metalloneurochemistry is still in its infancy and one would expect it to apply not only to healthy brain tissue but also to brain tumors such as GBM [205]. Finally, perhaps it is time for a paradigm shift in BNCT where elements other than boron are explored in terms of their capacity to undergo neutron capture reactions inside tumor cells. Clearly, ^{157}Gd with its enormous neutron capture cross-section offers great promise although no suitable Gd agents have yet entered clinical trials for GdNCT. Some recently-developed Gd agents have shown great promise, at least *in vitro*, and the considerable use of Gd in medicine as a MRI contrast agent has certainly paved the way for the use of this metal in the clinic for NCT.

Whereas the treatment planning, dosimetry, neutron beam characteristics and sources, pharmacology, and patient management protocols each play a critical role in the future success of NCT, the reality is that the nature of the NCT agent itself is the keystone to any significant clinical advances if this binary therapy is to continue being a viable option for the treatment of intractable tumors such as GBM. No matter what the future holds, medicinal inorganic chemists are clearly greatly under-represented in this exciting field and they will certainly play an increasingly important role in the future development of NCT.

References

1 Chang, D. (2003) *Statistics on Incidence, Survival Rates and Mortality Associated with Brain Tumors in Australia*, North Shore Private Hospital, Australia.

2 Reardon, D.A., Rich, J.N., Friedman, H.S., and Bigner, D.D.J. (2006) *Clin. Oncol.*, **24**, 1253–1265.

3 CBTRUS, Statistical Report: Primary Brain Tumors in the United States. Hinsdale, IL, Central Brain Tumor Registry of the United States, 2007–2008.

4 Horner, M.J., Ries, L.A.G., Krapcho, M., Neyman, N., Aminou, R., Howlader, N., Altekruse, S.F., Feuer, E. J., Huang, L., Mariotto, A., Miller, B.A., Lewis, D.R., Eisner, M.P., Stinchcomb, D.G., and Edwards, B.K. (eds) (2009) *SEER Cancer Statistics Review, 1975–2006*, National Cancer Institute, Bethesda, MD, http://seer.cancer.gov/

csr/1975_2006/, based on November 2008 SEER data submission, posted to the SEER Web site.

5 Yamanaka, R. (2008) *Trends Mol. Med.*, **14**, 228–235.

6 Stewart, L.A. (2002) *Lancet*, **359**, 1011–1018.

7 van Rij, C.M., Wilhelm, A.J., Sauerwein, W.A.G., and van Loenen, A.C. (2005) *Pharm. World Sci.*, **27**, 92–95.

8 Henson, J.W. (2006) *Arch. Neurol.*, **63**, 337–341.

9 Yamamoto, T., Nakai, K., and Matsumura, A. (2008) *Cancer Lett.*, **262**, 143–152.

10 Soloway, A.H., Barth, R.F., Gahbauer, R.A., Blue, T.E., and Goodman, J.H.J. (1997) *Neuro-Oncol.*, **33**, 9–18.

11 Soloway, A.H., Tjarks, W., Barnum, B.A., Rong, F.-G., Barth, R.F., Codogni, I.M.,

and Wilson, J.G. (1998) *Chem. Rev.*, **98**, 1515–1562.

12 Allison, R.R., Bagnato, V.S., Cuenca, R., Downie, G.H., and Sibata, C.H. (2006) *Future Oncol.*, **2**, 53–71.

13 Garcia-Zuazaga, J., Cooper, K.D., and Baron, E.D. (2005) *Expert Rev. Anticancer Ther.*, **5**, 791–800.

14 Suortti, P. and Thomlinson, W. (2003) *Phys. Med. Biol.*, **48**, R1–R35.

15 Laster, B.H., Thomlinson, W.C., and Fairchild, R.G. (1993) *Radiat. Res.*, **133**, 219–224.

16 Locher, G.L. (1936) *Am. J. Roentgenol. Radi.*, **36**, 1–13.

17 Yamamoto, T., Nakai, K., Tsurubuchi, T., Matsuda, M., Shirakawa, M., Zaboronok, A., Endo, K., and Matsumura, A. (2009) *Appl. Radiat. Isot.*, **67**, S25–S26.

18 Nakagawa, Y. and Hatanaka, H.J. (1997) *Neuro-Oncol.*, **33**, 105–115.

19 Matsumura, A., Yamamoto, T., Tsurubuchi, T., Matsuda, M., Shirakawa, M., Nakai, K., Endo, K., Tokuue, K., and Tsuboi, K. (2009) *Appl. Radiat. Isot.*, **67**, S12–S14.

20 Barth, R.F., Coderre, J.A., Vicente, M. G.H., and Blue, T.E. (2005) *Clin. Cancer Res.*, **11**, 3987–4002.

21 Capala, J., Stenstam Britta, H., Skold, K., Munck af Rosenschold, P., Giusti, V., Persson, C., Wallin, E., Brun, A., Franzen, L., Carlsson, J., Salford, L., Ceberg, C., Persson, B., Pellettieri, L., and Henriksson, R.J. (2003) *Neuro-Oncol.*, **62**, 135–144.

22 Menendez, P.R., Roth, B.M.C., Pereira, M.D., Casal, M.R., Gonzalez, S.J., Feld, D.B., Santa Cruz, G.A., Kessler, J., Longhino, J., Blaumann, H. Jimenez Rebagliati, R., Calzetta Larrieu, O.A., Fernandez, C., Nievas, S.I., and Liberman, S. (2009) *J. Appl. Radiat. Isot.*, **67**, S50–S53.

23 Skold, K., Stenstam, B.H., Hopewell, J., Diaz, A.J., Giusti, V., and Pellettieri, L. (2008) Efficacy of BNCT for GBM: assessment of clinical results from Studsvik, Sweden. Presented at 13th International Congress on Neutron Capture Therapy, November 2–7, 2008, Florence.

24 Barth, R.F. and Joensuu, H. (2007) *Radiother. Oncol.*, **82**, 119–122.

25 Kankaanranta, L., Seppala, T., Koivunoro, H., Saarilahti, K., Atula, T., Collan, J., Salli, E., Kortesniemi, M., Uusi-Simola, J., Makitie, A., Seppanen, M., Minn, H., Kotiluoto, P., Auterinen, I., Savolainen, S., Kouri, M., and Joensuu, H. (2007) *Int. J. Radiat. Oncol. Biol. Phys.*, **69**, 475–482.

26 Hatanaka, H. (1986) *Neutron Capture Therapy*, Niushimura, Japan.

27 Fairchild, R.G. and Bond, V.P. (1985) *Int. J. Radiat. Oncol. Biol. Phys.*, **11**, 831–840.

28 Braccini, S. (2007) *Nucl. Phys. B, Proc. Suppl.*, **172**, 8–12.

29 Hawthorne, M.F. (1998) *Mol. Med. Today*, **4**, 174–181.

30 Tjarks, W., Tiwari, R., Byun, Y., Narayanasamy, S., and Barth, R.F. (2007) *Chem. Commun.*, 4978–4991.

31 Crossley, E.L., Ziolkowski, E.J., Coderre, J.A., and Rendina, L.M. (2007) *Mini Rev. Med. Chem.*, **7**, 303–313.

32 Valliant, J.F., Guenther, K.J., King, A. S., Morel, P., Schaffer, P., Sogbein, O. O., and Stephenson, K.A. (2002) *Coord. Chem. Rev.*, **232**, 173230.

33 Yokoyama, K., Miyatake, S.-I., Kajimoto, Y., Kawabata, S., Doi, A., Yoshida, T., Asano, T., Kirihata, M., Ono, K., and Kuroiwa, T.J. (2006) *Neuro-Oncol.*, **78**, 227–232.

34 Steichen, J.D., Weiss, M.J., Elmaleh, D. R., and Martuza, R.L. (1991) *J. Neurosurg.*, **74**, 116–122.

35 Coderre, J.A., Elowitz, E.H., Chadha, M., Bergland, R., Capala, J., Joel, D.D., Liu, H.B., Slatkin, D.N., and Chanana, A.D.J. (1997) *Neuro-Oncol.*, **33**, 141–152.

36 Barth, R.F. (2009) *Appl. Radiat. Isot.*, **67**, S3–S6.

37 Hawthorne, M.F. and Lee Mark, W.J. (2003) *Neuro-Oncol.*, **62**, 33–45.

38 Diaz, A., Stelzer, K., Laramore, G., and Wiersema, R. (2002) in *Research and Development in Neutron Capture Therapy* (eds M.W. Sauerwein, R. Moss, and A. Witting), Monduzzi Editore, International Proceedings Division, Bologna, p. 993.

39 Bergenheim, A.T., Capala, J., Roslin, M., and Henriksson, R.J. (2005) *Neuro-Oncol.*, **71**, 287–293.

40 Sivaev, I.B. and Bregadze, V.V. (2009) *Eur. J. Inorg. Chem.*, 1433–1450.

41 Hatanaka, H. (1975) *J. Neurol.*, **209**, 81–94.

42 Cruickshank, G.S., Ngoga, D., Detta, A., Green, S., James, N.D., Wojnecki, C., Doran, J., Hardie, J., Chester, M., Graham, N., Ghani, Z., Halbert, G., Elliot, M., Ford, S., Braithwaite, R., Sheehan, T.M.T., Vickerman, J., Lockyer, N., Steinfeldt, H., Croswell, G., Chopra, A., Sugar, R., and Boddy, A. (2009) *Appl. Radiat. Isot.*, **67**, S31–S33.

43 Coderre, J.A., Hopewell, J.W., Turcotte, J.C., Riley, K.J., Binns, P.J., Kiger, W. S., and Harling, O.K. (2004) *Appl. Radiat. Isot.*, **61**, 1083–1087.

44 Matsuda, M., Yamamoto, T., Kumada, H., Nakai, K., Shirakawa, M., Tsurubuchi, T., and Matsumura, A. (2009) *Appl. Radiat. Isot.*, **67**, S19–S21.

45 Kawabata, S., Miyatake, S.-I., Kajimoto, Y., Kuroda, Y., Kuroiwa, T., Imahori, Y., Kirihata, M., Sakurai, Y., Kobayashi, T., and Ono, K. (2003) *J. Neurooncol.*, **65** (2), 159–165.

46 Miyatake, S.-I., Kawabata, S., Kajimoto, Y., Aoki, A., Yokoyama, K., Yamada, M., Kuroiwa, T., Tsuji, M., Imahori, Y., Kirihata, M., Sakurai, Y., Masunaga, S.-I., Nagata, K., Maruhashi, A., and Ono, K. (2005) *J. Neurosurg.*, **103**, 1000–1009.

47 Kato, I., Ono, K., Sakurai, Y., Ohmae, M., Maruhashi, A., Imahori, Y., Kirihata, M., Nakazawa, M., and Yura, Y. (2004) *Appl. Radiat. Isot.*, **61**, 1069–1073.

48 Kato, I., Fujita, Y., Maruhashi, A., Kumada, H., Ohmae, M., Kirihata, M., Imahori, Y., Suzuki, M., Sakrai, Y., Sumi, T., Iwai, S., Nakazawa, M., Murata, I., Miyamaru, H., and Ono, K. (2009) *Appl. Radiat. Isot.*, **67**, S37–S42.

49 Srivastava, R.R., Singhaus, R.R., and Kabalka, G.W. (1997) *J. Org. Chem.*, **62**, 4476–4478.

50 Kahl, S.B. and Koo, M.S. (1990) *J. Chem. Soc., Chem. Commun.*, 1769–1771.

51 Bregadze, V.I. (1992) *Chem. Rev.*, **92**, 209–223.

52 Armstrong, A.F. and Valliant, J.F. (2007) *Dalton Trans.*, 4240–4251.

53 Hartman, T. and Carlsson, J. (1994) *Radiother. Oncol.*, **31**, 61–75.

54 Lippard, S.J. (1978) *Acc. Chem. Res.*, **11**, 211–217.

55 Rosenberg, B., VanCamp, L., Trosko, J. E., and Mansour, V.H. (1969) *Nature*, **222**, 385–386.

56 Galanski, M., Jakupec, M.A., and Keppler, B.K. (2005) *Curr. Med. Chem.*, **12**, 2075–2094.

57 Dowell, J.A., Sancho, A.R., Anand, D., and Wolf, W. (2000) *Adv. Drug Delivery Rev.*, **41**, 111–126.

58 Woodhouse, S.L. and Rendina, L.M. (2001) *Chem. Commun.*, 2464–2465.

59 Woodhouse, S.L. and Rendina, L.M. (2004) *Dalton Trans.*, 3669–3677.

60 Todd, J.A. and Rendina, L.M. (2004) *Inorg. Chem. Commun.*, **7**, 289–291.

61 Todd, J.A. and Rendina, L.M. (2002) *Inorg. Chem.*, **41**, 3331–3333.

62 Todd, J.A., Turner, P., Ziolkowski, E.J., and Rendina, L.M. (2005) *Inorg. Chem.*, **44**, 6401–6408.

63 Woodhouse, S.L., Ziolkowski, E.J., and Rendina, L.M. (2005) *Dalton Trans.*, 2827–2829.

64 Crossley, E.L., Caiazza, D., and Rendina, L.M. (2005) *Dalton Trans.*, 2825–2826.

65 Yoo, J. and Do, Y. (2005) *Bull. Korean Chem. Soc.*, **26**, 231–232.

66 Yoo, J. and Do, Y. (2009) *Dalton Trans.*, 4978–4986.

67 MacDonald, I.J. and Dougherty, T.J. (2001) *J. Porphyrins Phthalocyanines*, **5**, 105–129.

68 Bregadze, V.I., Sivaev, I.B., Gabel, D., and Wohrle, D. (2001) *J. Porphyrins Phthalocyanines*, **5**, 767–781.

69 Evstigneeva, R.P., Zaitsev, A.V., Luzgina, V.N., Ol'shevskaya, V.A., and Shtil, A.A. (2003) *Curr. Med. Chem. Anti-Cancer Agents*, **3**, 383–392.

70 Renner, M.W., Miura, M., Easson, M. W., and Vicente, M.G.H. (2006) *Anticancer Agents Med. Chem.*, **6**, 145–157.

71 Ratajski, M., Osterloh, J., and Gabel, D. (2006) *Anticancer Agents Med. Chem.*, **6**, 159–166.

72 Hill, J.S., Kahl, S.B., Kaye, A.H., Styll, S.S., Koo, M.S., Gonzales, M.F., Vardaxis, N.J., and Johnson, C. (1992) *Proc. Natl. Acad. Sci. USA*, **89**, 1785–1789.

73 Bobo, R.H., Laske, D.W., Akbasak, A., Morrison, P.F., Dedrick, R.L., and Oldfield, E.H. (1994) *Proc. Natl. Acad. Sci. USA*, **91**, 2076–2080.

74 Ozawa, T., Afzal, J., Lamborn, K.R., Bollen, A.W., Bauer, W.F., Koo, M.-S., Kahl, S.B., and Deen, D.F. (2005) *Int. J. Radiat. Oncol. Biol. Phys.*, **63**, 247–252.

75 Miura, M., Morris, G.M., Micca, P.L., Lombardo, D.T., Youngs, K.M., Kalef-Ezra, J.A., Hoch, D.A., Slatkin, D.N., Ma, R., and Coderre, J.A. (2001) *Radiat. Res.*, **155**, 603–610.

76 Miura, M., Joel, D.D., Smilowitz, H.M., Nawrocky, M.M., Micca, P.L., Hoch, D.A., Coderre, J.A., and Slatkin, D.N.J. (2001) *Neuro-Oncol.*, **52**, 111–117.

77 Kreimann, E.L., Miura, M., Itoiz, M.E., Heber, E., Garavaglia, R.N., Batistoni, D., Rebagliati, R.J., Roberti, M.J., Micca, P.L., Coderre, J.A., and Schwint, A.E. (2003) *Arch. Oral Biol.*, **48**, 223–232.

78 Wu, H., Micca, P.L., Makar, M.S., and Miura, M. (2006) *Bioorg. Med. Chem.*, **14**, 5083–5092.

79 Hao, E., Sibrian-Vazquez, M., Serem, W., Garno, J.C., Fronczek, F.R., and Vicente, M.G.H. (2007) *Chem. Eur. J.*, **13**, 9035–9042.

80 Easson, M.W., Fronczek, F.R., Jensen, T.J., and Vicente, M.G.H. (2008) *Bioorg. Med. Chem.*, **16**, 3191–3208.

81 Soncin, M., Friso, E., Jori, G., Hao, E., Vicente, M.G.H., Miotto, G., Colautti, P., Moro, D., Esposito, J., Rosi, G., and Fabris, C. (2008) *J. Porphyrins Phthalocyanines*, **12**, 866–873.

82 Jori, G., Soncin, M., Friso, E., Vicente, M.G.H., Hao, E., Miotto, G., Colautti, P., Moro, D., Esposito, J., Rosi, G., Nava, E., Sotti, G., and Fabris, C. (2009) *Appl. Radiat. Isot.*, **67**, S321–S324.

83 Fantin, V.R. and Leder, P. (2006) *Oncogene*, **25**, 4787–4797.

84 Galeano, E., Nieto, E., Garcia-Perez, A.I., Delgado, M.D., Pinilla, M., and Sancho, P. (2005) *Leukemia Res.*, **29**, 1201–1211.

85 McKeage, M.J., Maharaj, L., and Berners-Price, S.J. (2002) *Coord. Chem. Rev.*, **232**, 127–135.

86 Modica-Napolitano, J.S. and Aprille, J.R. (2001) *Adv. Drug Delivery Rev.*, **49**, 63–70.

87 Modica-Napolitano, J.S., Nalbandian, R., Kidd, M.E., Nalbandian, A., and Nguyen, C.C. (2003) *Cancer Lett.*, **198**, 59–68.

88 Rowe, T.C., Weissig, V., and Lawrence, J.W. (2001) *Adv. Drug Delivery Rev.*, **49**, 175–187.

89 Takasu, K., Shimogama, T., Saiin, C., Kim, H.-S., Wataya, Y., Brun, R., and Ihara, M. (2005) *Chem. Pharm. Bull.*, **53**, 653–661.

90 Modica-Napolitano, J.S. and Aprille, J.R. (1987) *Cancer Res.*, **47**, 4361–4365.

91 Ioppolo, J.A., Clegg, J.K., and Rendina, L.M. (2007) *Dalton Trans.*, 1982–1985.

92 Adams, D.M., Ji, W., Barth, R.F., and Tjarks, W. (2000) *Anticancer Res.*, **20**, 3395–3402.

93 Calabrese, G., Gomes, A.C.N.M., Barbu, E., Nevell, T.G., and Tsibouklis, J. (2008) *J. Mater. Chem.*, **18**, 4864–4871.

94 Linko, S., Revitzer, H., Zilliacus, R., Kortesniemi, M., Kouri, M., and Savolainen, S. (2008) *Scand. J. Clin. Lab. Invest.*, **68**, 696–702.

95 Laakso, J., Kulvik, M., Ruokonen, I., Vahatalo, J., Zilliacus, R., Farkkila, M., and Kallio, M. (2001) *Clin. Chem.*, **47**, 1796–1803.

96 Kabalka, G.W., Davis, M., and Bendel, P. (1988) *Magn. Reson. Med.*, **8**, 231–237.

97 Kabalka, G.W., Tang, C., and Bendel, P.J. (1997) *Neuro-Oncol.*, **33**, 153–161.

98 Glover, G.H., Pauly, J.M., and Bradshaw, K.M. (1992) *J. Magn. Reson. Imag.*, **2**, 47–52.

99 Bradshaw, K.M., Schweizer, M.P., Glover, G.H., Hadley, J.R., Tippets, R., Tang, P.P., Davis, W.L., Heilbrun, M.

P., Johnson, S., and Ghanem, T. (1995) *Magn. Reson. Med.*, **34**, 48–56.

100 Porcari, P., Capuani, S., D'Amore, E., Lecce, M., La Bella, A., Fasano, F., Campanella, R., Migneco, L.M., Pastore, F.S., and Maraviglia, B. (2008) *Phys. Med. Biol.*, **53**, 6979–6989.

101 Porcari, P., Capuani, S., D'Amore, E., Lecce, M., La Bella, A., Fasano, F., Migneco, L.M., Campanella, R., Maraviglia, B., and Pastore, F.S. (2009) *Appl. Radiat. Isot.*, **67**, S365–S368.

102 Tatham, A.T., Nakamura, H., Wiener, E.C., and Yamamoto, Y. (1999) *Magn. Reson. Med.*, **42**, 32–36.

103 Rozijn, T.H., Van der Sanden, B.P.J., Heerschap, A., Creyghton, J.H.N., and Bovee, W.M.M. (1999) *J. Magn. Reson. Mater. Phys., Biol. Med.*, **9**, 65–71.

104 Adam, M.J. and Wilbur, D.S. (2005) *Chem. Soc. Rev.*, **34**, 153–163.

105 Levin, C.S. (2005) *Eur. J. Nucl. Med. Mol. Imaging*, **32**, S325–S345.

106 Tolmachev, V. and Stone-Elander, S. (2010) *Biochim. Biophys. Acta*, **1800**, 487–510.

107 Mizusawa, E., Dahlman, H.L., Bennett, S.J., Goldenberg, D.M., and Hawthorne, M.F. (1982) *Proc. Natl. Acad. Sci. USA*, **79**, 3011–3014.

108 Lucas, L.L. and Unterweger, M.P. (2000) *J. Res. Natl. Inst. Stand. Technol.*, **105**, 541–549.

109 Ametamey, S.M., Honer, M., and Schubiger, P.A. (2008) *Chem. Rev.*, **108**, 1501–1516.

110 Liu, Y. (2009) *Ann. Nucl. Med.*, **23**, 17–23.

111 Jacobs, A.H., Thomas, A., Kracht, L.W., Li, H., Dittmar, C., Garlip, G., Galldiks, N., Klein, J.C., Sobesky, J., Hilker, R., Vollmar, S., Herholz, K., Wienhard, K., and Heiss, W.-D. (2005) *J. Nucl. Med.*, **46**, 1948–1958.

112 Garnett, E.S., Firnau, G., and Nahmias, C. (1983) *Nature*, **305**, 137–138.

113 Volkow, N.D., Fowler, J.S., Gatley, S.J., Logan, J., Wang, G.-J., Ding, Y.-S., and Dewey, S. (1996) *J. Nucl. Med.*, **37**, 1242–1256.

114 Becherer, A., Szabo, M., Karanikas, G., Wunderbaldinger, P., Angelberger, P.,

Raderer, M., Kurtaran, A., Dudczak, R., and Kletter, K. (2004) *J. Nucl. Med.*, **45**, 1161–1167.

115 Ishiwata, K., Ido, T., Mejia, A.A., Ichihashi, M., and Mishima, Y. (1991) *Appl. Radiat. Isot.*, **42**, 325–328.

116 Ishiwata, K., Ido, T., Kawamura, M., Kubota, K., Ichihashi, M., and Mishima, Y. (1991) *Nucl. Med. Biol.*, **18**, 745–751.

117 Chen, J.C., Chang, S.M., Hsu, F.Y., Wang, H.E., and Liu, R.S. (2004) *Appl. Radiat. Isot.*, **61**, 887–891.

118 Wang, H.-E., Liao, A.-H., Deng, W.-P., Chang, P.-F., Chen, J.-C., Chen, F.-D., Liu, R.-S., Lee, J.-S., and Hwang, J.-J. (2004) *J. Nucl. Med.*, **45**, 302–308.

119 Hsieh, C.-H., Chen, Y.-F., Chen, F.-D., Hwang, J.-J., Chen, J.-C., Liu, R.-S., Kai, J.-J., Chang, C.-W., and Wang, H.-E. (2005) *J. Nucl. Med.*, **46**, 1858–1865.

120 Kabalka, G.W., Smith, G.T., Dyke, J.P., Reid, W.S., Longford, C.P., Roberts, T.G., Reddy, N.K., Buonocore, E., and Hubner, K.F. (1997) *J. Nucl. Med.*, **38**, 1762–1767.

121 Imahori, Y., Ueda, S., Ohmori, Y., Kusuki, T., Ono, K., Fujii, R., and Ido, T. (1998) *J. Nucl. Med.*, **39**, 325–333.

122 Imahori, Y., Ueda, S., Ohmori, Y., Sakae, K., Kusuki, T., Kobayashi, T., Takagaki, M., Ono, K., Ido, T., and Fujii, R. (1998) *Clin. Cancer Res.*, **4**, 1833–1841.

123 Nichols, T.L., Kabalka, G.W., Miller, L.F., Khan, M.K., and Smith, G.T. (2002) *Med. Phys.*, **29**, 2351–2358.

124 Menichetti, L., Cionini, L., Sauerwein, W.A., Altieri, S., Solin, O., Minn, H., and Salvadori, P.A. (2009) *Appl. Radiat. Isot.*, **67**, S351–S354.

125 Busse, P.M., Harling, O.K., Palmer, M.R., Kiger, W.S. 3rd, Kaplan, J., Kaplan, I., Chuang, C.F., Goorley, J.T., Riley, K.J., Newton, T.H., Santa Cruz, G.A., Lu, X.-Q., and Zamenhof, R.G.J. (2003) *Neuro-Oncol.*, **62**, 111–121.

126 Kabalka, G.W., Nichols, T.L., Smith, G.T., Miller, L.F., Khan, M.K., and Busse, P.M. (2003) *J. Neurooncol.*, **62**, 187–195.

127 Fuwa, N., Suzuki, M., Sakurai, Y., Nagata, K., Kinashi, Y., Masunaga, S., Maruhashi, A., Imahori, Y., Kodaira, T.,

Tachibana, H., Nakamura, T., and Ono, K. (2008) *Br. J. Radiol.*, **81**, 749–752.

128 Aihara, T., Hiratsuka, J., Morita, N., Uno, M., Sakurai, Y., Maruhashi, A., Ono, K., and Harada, T. (2006) *Head Neck*, **28**, 850–855.

129 Havu-Auren, K., Kiiski, J., Lehtio, K., Eskola, O., Kulvik, M., Vuorinen, V., Oikonen, V., Vahatalo, J., Jaaskelainen, J., and Minn, H. (2007) *Eur. J. Nucl. Med. Mol. Imag.*, **34**, 87–94.

130 Ariyoshi, Y., Miyatake, S.-I., Kimura, Y., Shimahara, T., Kawabata, S., Nagata, K., Suzuki, M., Maruhashi, A., Ono, K., and Shimahara, M. (2007) *Oncol. Rep.*, **18**, 861–866.

131 Nakai, K., Kumada, H., Yamamoto, T., Tsurubuchi, T., Zaboronok, A., and Matsumura, A. (2009) *Appl. Radiat. Isot.*, **67**, S43–S46.

132 Mizusawa, E.A., Thompson, M.R., and Hawthorne, M.F. (1985) *Inorg. Chem.*, **24**, 1911–1916.

133 Varadarajan, A., Sharkey, R.M., Goldenberg, D.M., and Hawthorne, M.F. (1991) *Bioconjugate Chem.*, **2**, 102–110.

134 Hawthorne, M.F. and Maderna, A. (1999) *Chem. Rev.*, **99**, 3421–3434.

135 Eriksson, L., Tolmachev, V., and Sjoeberg, S. (2003) *J. Labelled Comp. Radiopharm.*, **46**, 623–631.

136 Tolmachev, V., Koziorowski, J., Sivaev, I., Lundqvist, H., Carlsson, J., Orlova, A., Gedda, L., Olsson, P., Sjoeberg, S., and Sundin, A. (1999) *Bioconjugate Chem.*, **10**, 338–345.

137 Koryakin, S.N., Yadrovskaya, V.A., Savina, E.P., Ul'yanenko, S.E., and Brattsev, V.A. (2002) *Pharm. Chem. J.*, **36**, 459–461.

138 Yadrovskaya, V.A., Ul'yanenko, S.E., Savina, E.P., Koryakin, S.N., and Brattsev, V.A. (2001) *Pharm. Chem. J.*, **35**, 408–410.

139 Ul'yanenko, S.E., Yadrovskaya, V.A., Savina, E.P., Bozadzhiev, L.L., and Brattsev, V.A. (2000) *Pharm. Chem. J.*, **34**, 73–75.

140 Ul'yanenko, S.E., Yadrovskaya, V.A., Savina, E.P., Brattsev, V.A., and Borisov, G.I. (2000) *Pharm. Chem. J.*, **34**, 232–233.

141 Wilbur, D.S., Hamlin, D.K., Srivastava, R.R., and Chyan, M.-K. (2004) *Nucl. Med. Biol.*, **31**, 523–530.

142 Wilbur, D.S., Chyan, M.-K., Hamlin, D. K., and Perry, M.A. (2009) *Bioconjugate Chem.*, **20**, 591–602.

143 Sjoestroem, A., Tolmachev, V., Lebeda, O., Koziorowski, J., Carlsson, J., and Lundqvist, H. (2003) *J. Radioanal. Nucl. Chem.*, **256**, 191–197.

144 Wilbur, D.S., Chyan, M.-K., Hamlin, D. K., Kegley, B.B., Risler, R., Pathare, P. M., Quinn, J., Vessella, R.L., Foulon, C., Zalutsky, M., Wedge, T.J., and Hawthorne, M.F. (2004) *Bioconjugate Chem.*, **15**, 203–223.

145 Bregadze, V.I., Sivaev, I.B., and Glazun, S.A. (2006) *Anticancer Agents Med. Chem.*, **6**, 75–109.

146 Moerlein, S.M. and Welch, M.J. (1981) *Int. J. Nucl. Med. Biol.*, **8**, 277–287.

147 Paxton, R.J., Beatty, B.G., Hawthorne, M.F., Varadarajan, A., Williams, L.E., Curtis, F.L., Knobler, C.B., Beatty, J.D., and Shively, J.E. (1991) *Proc. Natl. Acad. Sci. USA*, **88**, 3387–3391.

148 Valliant, J.F., Morel, P., Schaffer, P., and Kaldis, J.H. (2002) *Inorg. Chem.*, **41**, 628–630.

149 Beatty, B.G., Paxton, R.J., Hawthorne, M.F., Williams, L.E., Rickard-Dickson, K.J., Do, T., Shively, J.E., and Beatty, J. D. (1993) *J. Nucl. Med.*, **34**, 1294–1302.

150 Hawthorne, M.F., Varadarajan, A., Knobler, C.B., Chakrabarti, S., Paxton, R.J., Beatty, B.G., and Curtis, F.L. (1990) *J. Am. Chem. Soc.*, **112**, 5365–5366.

151 Martin, R.H., Burns, K.I. W., and Taylor, J.G.V. (1997) *Nucl. Instrum. Methods Phys. Res. Sect. A*, **390**, 267–273.

152 Dilworth, J.R. and Parrott, S.J. (1998) *Chem. Soc. Rev.*, **27**, 43–55.

153 Jurisson, S.S. and Lydon, J.D. (1999) *Chem. Rev.*, **99**, 2205–2218.

154 Liu, S. and Edwards, D.S. (1999) *Chem. Rev.*, **99**, 2235–2268.

155 Sogbein, O.O., Merdy, P., Morel, P., and Valliant, J.F. (2004) *Inorg. Chem.*, **43**, 3032–3034.

156 Sogbein, O.O., Green, A.E.C., and Valliant, J.F. (2005) *Inorg. Chem.*, **44**, 9585–9591.

157 Green, A.E.C., Causey, P.W., Louie, A. S., Armstrong, A.F., Harrington, L.E., and Valliant, J.F. (2006) *Inorg. Chem.*, **45**, 5727–5729.

158 Causey, P.W., Besanger, T.R., and Valliant, J.F. (2008) *J. Med. Chem.*, **51**, 2833–2844.

159 Salt, C., Lennox, A.J., Takagaki, M., Maguire, J.A., and Hosmane, N.S. (2004) *Russ. Chem. Bull.*, **53**, 1871–1888.

160 Brugger, R.M. and Shih, J.A. (1989) *Strahlenther. Onkol.*, **165**, 153–156.

161 Shih, J.A. and Brugger, R.M. (1992), in *Progress in Neutron Capture Therapy for Cancer. Proceedings of the 4th International Symposium* (eds B.J. Allen, B.V. Harrington, and D.E. Moore), Springer, pp. 183–186.

162 Miller, G.A. Jr, Hertel, N.E., Wehring, B.W., and Horton, J.L. (1993) *Nucl. Technol.*, **103**, 320–331.

163 De Stasio, G., Rajesh, D., Casalbore, P., Daniels, M.J., Erhardt, R.J., Frazer, B. H., Wiese, L.M., Richter, K.L., Sonderegger, B.R., Gilbert, B., Schaub, S., Cannara, R.J., Crawford, J.F., Gilles, M.K., Tyliszczak, T., Fowler, J.F., Larocca, L.M., Howard, S.P., Mercanti, D., Mehta, M.P., and Pallini, R. (2005) *Neurol. Res.*, **27**, 387–398.

164 Goorley, T. and Nikjoo, H. (2000) *Radiat. Res.*, **154**, 556–563.

165 Goorley, T., Zamenhof, R., and Nikjoo, H. (2004) *Int. J. Radiat. Biol.*, **80**, 933–940.

166 Greenwood, R.C., Reich, C.W., Baader, H.A., Koch, H.R., Breitig, D., Schult, O.W.B., Fogelberg, B., Backlin, A., and Mampe, W. (1978) *Nucl. Phys. A*, **304**, 327–428.

167 De Stasio, G., Casalbore, P., Pallini, R., Gilbert, B., Sanita, F., Ciotti, M.T., Rosi, G., Festinesi, A., Larocca, L.M., Rinelli, A., Perret, D., Mogk, D.W., Perfetti, P., Mehta, M.P., and Mercanti, D. (2001) *Cancer Res.*, **61**, 4272–4277.

168 Tokuuye, K., Tokita, N., Akine, Y., Nakayama, H., Sakurai, Y., Kobayashi, T., and Kanda, K. (2000) *Strahlenther. Onkol.*, **176**, 81–83.

169 Laster, B.H., Shani, G., Kahl, S.B., and Warkentien, L. (1996) *Acta Oncol.*, **35**, 917–923.

170 Stepanek, J., Larsson, B., and Weinreich, R. (1996) *Acta Oncol.*, **35**, 863–868.

171 Martin, R.F., D'Cunha, G., Pardee, M., and Allen, B.J. (1988) *Int. J. Radiat. Biol.*, **54**, 205–208.

172 Martin, R.F., D'Cunha, G., Pardee, M., and Allen, B.J. (1989) *Pigment Cell Res.*, **2**, 330–332.

173 Adelstein, S.J. and Kassis, A.I. (1996) *Acta Oncol.*, **35**, 797–801.

174 Lobachevsky, P.N. and Martin, R.F. (2004) *Int. J. Radiat. Biol.*, **80**, 861–866.

175 De Stasio, G., Rajesh, D., Ford, J.M., Daniels, M.J., Erhardt, R.J., Frazer, B. H., Tyliszczak, T., Gilles, M.K., Conhaim, R.L., Howard, S.P., Fowler, J.F., Esteve, F., and Mehta, M.P. (2006) *Clin. Cancer Res.*, **12**, 206–213.

176 Hermann, P., Kotek, J., Kubicek, V., and Lukes, I. (2008) *Dalton Trans.*, 3027 –3047.

177 Runge, V.M., Clanton, J.A., Lukehart, C.M., Partain, C.L., and James, A.E. Jr (1983) *Am. J. Roentgenol.*, **141**, 1209– 1215.

178 Choppin, G.R. and Schaab, K.M. (1996) *Inorg. Chim. Acta*, **252**, 299–310.

179 Felix, R., Schorner, W., Laniado, M., Niendorf, H.P., Claussen, C., Fiegler, W., and Speck, U. (1985) *Radiology*, **156**, 681–688.

180 Stack, J.P., Antoun, N.M., Jenkins, J.P., Metcalfe, R., and Isherwood, I. (1988) *Neuroradiology*, **30**, 145–154.

181 William W.N. Jr, Zinner R.G., Karp D. D., Oh Y.W., Glisson B.S., Phan, S.-C., and Stewart D.J. (2007) *J. Thorac. Oncol.*, **2**, 745–750.

182 Magda, D., Lepp, C., Gerasimchuk, N., Lee, I., Sessler, J.L., Lin, A., Biaglow, J. E., and Miller, R.A. (2001) *Int. J. Radiat. Oncol. Biol. Phys.*, **51**, 1025– 1036.

183 Magda, D. and Miller, R.A. (2006) *Semin. Cancer Biol.*, **16**, 466–476.

184 Mehta, M.P., Shapiro, W.R., Phan, S.C., Gervais, R., Carrie, C., Chabot, P., Patchell, R.A., Glantz, M.J., Recht, L.,

Langer, C., Sur, R.K., Roa, W.H., Mahe, M.A., Fortin, A., Nieder, C., Meyers, C. A., Smith, J.A., Miller, R.A., and Renschler, M.F. (2009) *Int. J. Radiat. Oncol. Biol. Phys.*, **73**, 1069–1076.

185 Renschler, M.F. (2004) *Eur. J. Cancer*, **40**, 1934–1940.

186 Bernhard, E.J., Mitchell, J.B., Deen, D., Cardell, M., Rosenthal, D.I., and Brown, J.M. (2000) *Cancer Res.*, **60**, 86–91.

187 Takagaki, W., Oda, Y., Miyatake, S., Kikuchi, H., Kobayashi, T., Kanda, K., and Ujeno, Y. (1992) in *Progress in Neutron Capture Therapy for Cancer. Proceedings of the 4th International Symposium* (eds B.J. Allen, B.V. Harrington, and D.E. Moore), Springer, pp. 407–410.

188 Yasui, L.S., Andorf, C., Schneider, L., Kroc, T., Lennox, A., and Saroja, K.R. (2008) *Int. J. Radiat. Biol.*, **84**, 1130–1139.

189 Cerullo, N., Bufalino, D., and Daquino, G. (2009) *Appl. Radiat. Isot.*, **67**, S157–S160.

190 Crossley, E.L., Aitken, J.B., Vogt, S., Harris, H.H., and Rendina, L.M. (2010) *Angew. Chem. Int. Ed.*, **49**, 1231–1233.

191 Hofmann, B., Fischer, C.O., Lawaczeck, R., Platzek, J., and Semmler, W. (1999) *Invest. Radiol.*, **34**, 126–133.

192 Nemoto, H., Cai, J., Nakamura, H., Fujiwara, M., and Yamamoto, Y.J. (1999) *Organomet. Chem.*, **581**, 170–175.

193 Nakamura, H., Fukuda, H., Girald, F., Kobayashi, T., Hiratsuka, J.i., Akaizawa, T., Nemoto, H., Cai, J., Yoshida, K., and Yamamoto, Y. (2000) *Chem. Pharm. Bull.*, **48**, 1034–1038.

194 Takahashi, K., Nakamura, H., Furumoto, S., Yamamoto, K., Fukuda, H., Matsumura, A., and Yamamoto, Y. (2005) *Bioorg. Med. Chem.*, **13**, 735–743.

195 Aime, S., Barge, A., Crivello, A., Deagostino, A., Gobetto, R., Nervi, C., Prandi, C., Toppino, A., and Venturello, P. (2008) *Org. Biomol. Chem.*, **6**, 4460–4466.

196 Bandyopadhyaya, A.K., Narayanasamy, S., Barth, R.F., and Tjarks, W. (2007) *Tetrahedron Lett.*, **48**, 4467–4469.

197 Zhang, T., Matsumura, A., Yamamoto, T., Yoshida, F., Sakurai, Y., Kumada, H., Yamamoto, K., and Nose, T. (2002) Proceedings of the 10th International Congress on Neutron Capture Therapy, pp. 819–824.

198 Kulakov, V.N., Bregadze, V.I., Sivaev, I. B., Nikitin, S.M., Gol'tyapin, Y.V., Khokhlov, V.F. (2001) *Frontiers in Neutron Capture Therapy: Proceedings of the 8th International Symposium on Neutron Capture Therapy for Cancer*, Vol. **2** (eds M.F. Hawthorne, K. Shelly, and R.J. Wiersema), Springer, pp. 843–846.

199 Halfon, S., Paul, M., Steinberg, D., Nagler, A., Arenshtam, A., Kijel, D., Polacheck, I., and Srebnik, M. (2009) *Appl. Radiat. Isot.*, **67**, S278–S281.

200 Alfuraih, A., Chin, M.P.W., and Spyrou, N.M. (2008) *J. Radioanal. Nucl. Chem.*, **278**, 681–684.

201 Mameli, A., Greco, F., Fidanzio, A., Fusco, V., Cilla, S., D'Onofrio, G., Grimaldi, L., Augelli, B.G., Giannini, G., Bevilacqua, R., Totaro, P., Tommasino, L., Azario, L., and Piermattei, A. (2008) *Nucl. Instrum. Methods Phys. Res. Sect. B*, **266**, 3656–3660.

202 Tanaka, H., Sakurai, Y., Suzuki, M., Takata, T., Masunaga, S., Kinashi, Y., Kashino, G., Liu, Y., Mitsumoto, T., Yajima, S., Tsutsui, H., Takada, M., Maruhashi, A., and Ono, K. (2009) *Appl. Radiat. Isot.*, **67**, S258–S261.

203 Tanaka, H., Sakurai, Y., Suzuki, M., Masunaga, S., Kinashi, Y., Kashino, G., Liu, Y., Mitsumoto, T., Yajima, S., Tsutsui, H., Maruhashi, A., and Ono, K. (2009) *Nucl. Instrum. Methods Phys. Res. Sect. B*, **267**, 1970–1977.

204 Linz, U. (2008) *Technol. Cancer Res. Treat.*, **7**, 83–88.

205 Burdette, S.C. and Lippard, S.J. (2003) *Proc. Natl. Acad. Sci. USA*, **100**, 3605–3610.

11
Essential Metal Related Metabolic Disorders

Yasmin Mawani and Chris Orvig

11.1
Introduction: What is Essentiality?

The physiological importance of metals in humans, especially in blood, is well known. At low concentrations, essential metals play an important role in metabolism, enzymatic processes, and as functional components of proteins. At high concentrations, these metals can lead to serious health problems and even death [1]. The ability of our bodies to maintain a constant internal state with varying external conditions is essential for survival. This is called homeostasis, a state in which the nutrient flow within an organism is at controlled equilibrium. The importance of this equilibrium can be seen in Figure 11.1, where extreme deficiency or overload of the essential metal, if untreated, can lead to death.

For an element to be considered essential, it must have a specific role, where deficiency of that element results in adverse affects that are reversed upon re-supply. Thus, it is important to distinguish nutritional effects from pharmacological effects, identifying an essential biochemical function for these metals [2]. In this chapter we overview some of the metabolic disorders that can lead to, or derive from, deficiency or overload of metal ions, and the effects that the perturbation of homeostasis of these metal ions can have on our body.

11.2
Iron Metabolic Diseases: Acquired and Genetic

11.2.1
Iron Homeostasis

Iron is an essential metal, necessary for cytochromes, hemoglobin, myoglobin, and for the function of many non-heme enzymes as well. Iron can be found in its ferric (Fe^{3+}) and ferrous (Fe^{2+}) states and thus is involved in many redox reactions. Excessive amounts can be toxic, with free iron leading, like copper, to Fenton chemistry, toxicity to the liver, and death. Too little iron can lead to cognitive

Bioinorganic Medicinal Chemistry. Edited by Enzo Alessio
Copyright © 2011 WILEY-VCH Verlag GmbH & Co. KGaA, Weinheim
ISBN: 978-3-527-32631-0

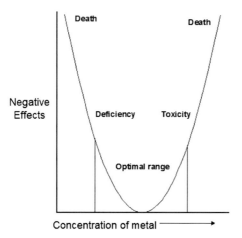

Figure 11.1 Dose–effect curve demonstrating the biological effect of the concentration of an essential metal.

decline, weakness, and death. Thus homeostasis of iron is important, especially since there is no active physiological pathway for excretion [3].

In a normal healthy adult, 1–2 mg is obtained from the diet, and 1–2 mg leaves the body each day (Table 11.1). Iron is absorbed in the duodenum of the small intestine, circulating in the plasma bound to transferrin. Premenopausal women have lower iron stores as a result of blood loss through menstruation [3].

Non-heme iron binds to mucosal membrane sites, is internalized, and then is either retained by the mucosal cell or is transported to the basolateral membrane where it is bound to transferrin (Tf) in the plasma pool. Acidity in the stomach, along with ferrireductase, reduces iron from its ferric to its ferrous state, increasing iron's solubility, making it more bioavailable. Divalent metal transporter DMT1 is a non-specific metal transporter that transfers iron across the apical membrane and into the cell through a proton-coupled process. Heme iron, on the other hand, does not require stomach acid to be solubilized. It is taken up by the enterocyte where it

Table 11.1 Distribution of iron in the body.

Dietary iron	1–2 mg day^{-1}
Muscle (myoglobin)	300
Bone marrow	300
Plasma transferrin (transport)	3
Circulating erythrocytes (hemoglobin)	1800
Liver parenchyma	1000
Reticuloendothelial macrophages	600
Menstruation, other blood loss	1–2

is either stored as ferritin or transferred across the basolateral membrane into the plasma. This receptor/transporter has not been identified [3, 4].

Diseases resulting from defects in iron metabolism are amongst the most common diseases in humans; herein we briefly discuss diseases of primary iron overload, secondary iron overload, and iron deficiency [3]. Primary iron overload disorders, also known as primary hemochromatosis, are caused by genetic defects leading to iron accumulation in tissues. Secondary or acquired iron overload, on the other hand, is iron accumulation caused by non-genetic disorders.

11.2.2
Diseases of Primary Iron Overload: Hemochromatosis

There are many causes of genetic iron overload disorders known as hemochromatosis. The most common is type 1 hereditary hemochromatosis (HH), which is caused by an inborn error of iron metabolism, leading to an increase in intestinal absorption of iron. Iron overload disorders lead to accumulation of the metal in the body, causing irreversible tissue and organ damage and fibrosis [5]. There are four types of hereditary hemochromatosis (HH) described below. Table 11.2 gives an overview of the four forms of HH.

11.2.2.1 Type 1 Hereditary Hemochromatosis
Type 1 hereditary hemochromatosis is the most common autosomal recessive disorder amongst Caucasians, presenting in 1 in 200–400 individuals. This hereditary disorder is caused by a mutation of the HFE (hemochromatosis) gene located on chromosome 6, resulting in an increase of iron absorption from the intestine, leading to liver cirrhosis, diabetes mellitus, and bronze skin pigmentation. It is caused by two mutations in the gene: a substitution of a tyrosine for a cysteine at position 282 (C282Y) and histidine for an aspartic acid at position 63 (H63D) [5].

Pathogenesis of HFE-related hemochromatosis is difficult to describe as the function of the HFE gene has not been clearly established. The HFE protein is found in the intestinal crypt cell of the duodenum where it complexes to transferrin receptor 1 (TfR1), which is the receptor by which cells acquire holo-transferrin. Under normal conditions, the HFE and TfR1 help to regulate uptake of iron by crypt cells. In type 1 HH, the mutated HFE protein is believed to impair the TfR1 uptake of iron, causing a deficiency of iron in duodenal crypt cells. As a result of the low levels of iron in the crypt cells, an overexpression of DMT1 occurs, increasing iron absorption. Most patients suffering from type 1 HH absorb two to three times the amount iron, compared to that of a healthy individual, from dietary sources [3, 5].

11.2.2.2 Type 2 Hereditary Hemochromatosis: Juvenile Hemochromatosis
Juvenile hemochromatosis is a rare, autosomal recessive disorder caused by a mutation of the HJV gene (type 2A juvenile hemochromatosis gene) or of the HAMP (hepcidin antimicrobial peptide) gene [5]. It manifests as hypogonadotropic

Table 11.2 Genetic diseases of iron overload: genetic defects, pathology, and common symptoms.

Disease	Genetic defect	Pathology	Symptoms
Type 1 HH[a]	Mutation of the HFE (hemochromatosis) gene located on chromosome [5]	HFE gene is believed to facilitate uptake of transferrin iron into crypt cells, causing iron overload	Liver cirrhosis, diabetes mellitus, and bronze skin pigmentation (same as type 2 HH)
Type 2 HH[a]	Autosomal recessive disease caused by mutation of the HJV gene on chromosome 1 or in the HAMP gene on chromosome [5]	Mutations of the HAMP gene that encodes for hepcidin results in iron overloading. Function of HJV not fully understood	Hypogonadrotropic hypogonadism, cardiac disease, liver cirrhosis, diabetes, and skin pigmentation (same as type 1 HH)
Type 3 HH[a]	Autosomal recessive disorder caused by mutation of the TfR2 gene on chromosome 7 [4]	TfR2 gene is implicated in the uptake of iron by hepatocytes through a receptor-mediated endocytosis	Symptoms are the same as seen in type 1 HH
Type 4 HH[a]: African Iron Overload	Mutation in the SLC40A1 gene on chromosome 2q32, which encodes for the protein ferroportin [5]	Mutation of the SLC40A1 gene which encodes for ferroportin 1	Cirrhosis, cardiomyopathy, impaired immune function
Neonatal hemochromatosis	Unknown [3]	Unknown	Iron accumulation in the liver and fetal organs, leads to death
Aceruloplasminemia	Autosomal recessive disorder caused by a mutation in the CP gene [6]	It is believed that CP plays are role in ferric iron uptake by transferrin	Progressive neurodegeneration of the retina and basal ganglia
Friedrich's ataxia	Abnormal expansion of a GAA repeat in the FRDA gene on chromosome 9, encodes for the protein frataxin [7]	Unknown	Progressive gait and limb ataxia, lack of tendon reflexes, dysarthria and weakness of the limbs
Hallervorden-Spatz syndrome	Thought to be autosomal recessive [8]	Unknown. Causes iron deposition in the brain	Cognitive decline and extrapyramidal dysfunction

[a]HH denotes hereditary hemochromatosis.

hypogonadism, cardiac disease, liver cirrhosis, diabetes, and skin pigmentation. Iron overload occurs at an early age, leading to severe organ impairment before the age of 30, manifesting with increased severity to that of type 1 HH. Cardiac failure generally leads to death in individuals affected with juvenile hemochromatosis. Juvenile hemochromatosis, caused by mutations of the HAMP gene that encodes for hepcidin, results in more severe iron overloading. The function of HJV protein is unknown; however, patients with either type of juvenile HH present with low urinary hepcidin levels. It is thus believed that both the HJV and HAMP genes have the same pathophysiological effect [3, 5].

11.2.2.3 Type 3 Hereditary Hemochromatosis
Type 3 HH is an autosomal recessive disorder caused by mutations in the transferrin receptor 2 (TfR2) gene. While the role of TfR2 is not fully elucidated, there is evidence that it is highly expressed in the liver, and thus involved in iron uptake by hepatocytes through a receptor-mediated mechanism. Symptoms are the same as seen in type 1 HH [5].

11.2.2.4 Type 4 Hereditary Hemochromatosis: African Iron Overload
African iron overload is a hemochromatosis that occurs predominantly in those of African descent, affecting up to 10% of some rural populations is sub-Saharan Africa. Formerly known as "bantu siderosis," it is a predisposition to iron overload that manifests because of excessive intake of dietary iron [3]. Unlike primary HH, it is not caused by a mutation in the HFE gene, but rather by mutations in the SLC40A1 gene on chromosome 2q32, which encodes for the protein ferroportin 1 [5]. Ferroportin is an export protein for iron, and mutations lead to an autosomal dominant hereditary condition characterized by high serum ferritin concentration, normal transferrin saturation, and iron accumulation [9]. It manifests itself in Africans who drink beer that is made in nongalvanized steel drums, because of high levels of iron in the beer [3].

11.2.2.5 Neonatal Hemochromatosis
Neonatal hemochromatosis (NH) is a rare condition that occurs during pregnancy, in which iron accumulates in the liver and extrahepatic sites of the fetus, causing extensive liver damage. It has similar pathology to HFE-associated hemochromatosis (type 1 HH). Without vigorous therapy it is fatal to the fetus, leading to the death within hours to days of birth [10]. The pathophysiology is unknown, but there is no genetic linkage to the HLA complexes. Though often unsuccessful, liver transplantation is the only primary treatment [3].

11.2.3
Diseases of Iron Overload: Accumulation of Iron in the Brain

Pathological brain iron accumulation is seen in common disorders, including Parkinson's disease, Alzheimer's disease, and Huntington disease. In disorders of systematic iron overload such as hemochromatosis, there is no accumulation of brain iron. This suggests that there is a fundamental difference that exists between brain and systematic iron metabolism and transport [8]. Three iron-loading disorders of iron metabolism that result in accumulation in the brain are described below. A summary of these diseases can be found in Table 11.2.

11.2.3.1 Aceruloplasminemia

Aceruloplasminemia is an autosomal recessive iron metabolism disorder characterized by progressive neurodegeneration of the retina and basal ganglia. It is associated with inherited mutations in the ceruloplasmin gene leading to iron overload [6].

Ceruloplasmin is a blue copper oxidase that is synthesized in hepatocytes and secreted as a holoprotein binding six copper atoms. Copper does not affect the rate of synthesis or secretion of apoceruloplasmin, but failure to incorporate copper results in an unstable protein lacking oxidase activity. Though ceruloplasmin is a copper protein, the role of ceruloplasmin in copper uptake has not been elucidated; however, there is some evidence that demonstrates ceruloplasmin ferroxidase activity, suggesting a role for ceruloplasmin in ferric iron uptake by transferrin. This is consistent with evidence from animal studies that anemia that develops in copper-deficient animals is unresponsive to iron, but not to ceruloplasmin administration [6, 11]. The presence of neurological symptoms in aceruloplasminemia is unique among the known inherited and acquired disorders or iron metabolism [6].

11.2.3.2 Hallervorden–Spatz Syndrome (HSS)

This is an iron metabolic disorder that results in excessive iron storage in the brain [12]. Iron accumulation in the brain in an individual suffering from Hallervorden–Spatz syndrome (HSS) is so excessive that post-mortem the basal ganglia are rust colored. The pathophysiology of HSS is unknown; however, it is known that it is an autosomal recessive disorder manifesting as massive iron deposition in the globus pallidus and substantia nigra. It results in cognitive decline and extrapyramidal dysfunction [8].

11.2.3.3 Friedreich's Ataxia

Friedreich's ataxia (FRDA) is an autosomal recessive neurodegenerative disease that affects 1 in 50 000, and is caused by a mutation in the FRDA gene. It is believed that Friedrich ataxia is the result of accumulation of iron in mitochondria leading to excess production of free radicals, which results in cellular damage and death [7, 13]. The disease is characterized by progressive gait and limb ataxia, with lack of tendon reflexes in the legs, dysarthria, and weakness of the limbs. The gene associated with the disease has been mapped to chromosome 2q13 and encodes for the protein frataxin. The function of the protein is unknown, but a deficiency in the activity of iron-sulfur (Fe-S) cluster-containing subunits of mitochondrial respirator complexes and increased iron content in the heart of patients suffering from FRDA have been reported [13].

11.2.4
Acquired Iron Overload Disorders

Acquired iron overload disorders are common because there is no physiological pathway for excretion of excessive iron. Secondary iron overload adversely affects the function of the heart, the liver, and other organs. As with other acquired iron overload disorders, it is generally treated by chelation therapy (Section 11.2.5) [14].

The main causes of iron overload in chronic hepatic diseases are alcohol-induced hepatocyte damage, chronic liver failure, and chronic iron transfusion therapy [14, 15]. Alcohol intake can lead to chronic liver failure, which induces increased iron absorption, chronic hemolysis, ineffective erythropoiesis, and increased ability of transferrin to deliver iron to the liver. Hepatocyte damage leads to an increase in iron and ferritin release to the extracellular fluid and plasma, as well as an increase in cytokine-mediated hepatocellular iron uptake [15].

11.2.5
Treatment of Iron Overload Disorders: Chelation Therapy

As there is no mechanism for iron excretion, iron loss is almost exclusively by blood loss. As a result, when absorption exceeds excretion, iron overload is inevitable; thus the use of chelators to remove excess iron is necessary to prevent oxidative stress and eventual organ failure [16]. These chelators must bind strongly to non-transferrin bound iron, as this iron is available for Fenton chemistry, while having limited access to both the brain and fetus, and must prevent the iron from participating in redox chemistry [17].

The three most commonly used iron (III) chelators (Figure 11.2) are: desferrioxamine B, deferiprone, and most recently deferasirox. Desferrioxamine B is a

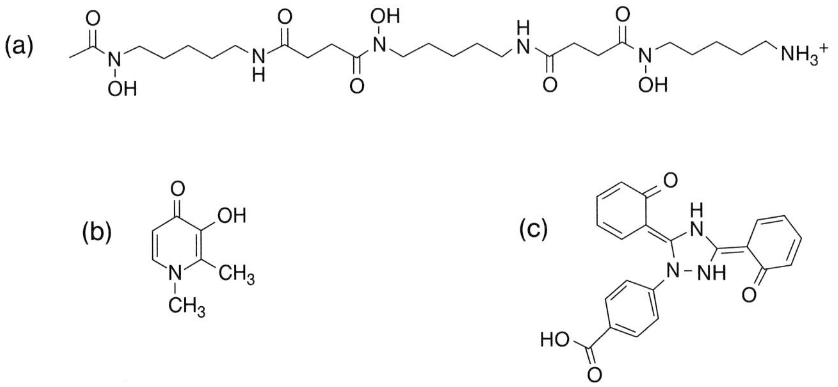

Figure 11.2 (a) Desferrioxamine B; (b) deferiprone; and (c) deferasirox.

siderophore, one of many strong chelators produced by microorganisms, containing either catecholate, hydroxamate, or hydroxy acid functionalities to bind to metals. Desferrioxamine B, which binds strongly to iron by its hydroxamates in a 1:1 ratio forming a charged octahedral complex (because of the NH_3^+ group), has poor oral availability, and a short retention time in the body, meaning it has to be administered intravenously. Deferiprone is an oral bidentate hydroxypyridonate ligand, binding to iron in a 3:1 ligand to metal ratio, forming a neutral complex with Fe(III) [16, 18]. Lastly, deferasirox belongs to a new class of oral tridentate chelators, containing an N-substituted bis-hydroxyphenyltriazole, forming a strong Fe(III) complex, binding in a 2:1 ligand to metal ratio [18, 19].

11.2.6
Iron Deficiency

Iron deficiency anemia (IDA) is caused by low iron levels and low hemoglobin, or abnormal levels of two out of the following three iron status tests: erythrocyte protoporphyrin, transferrin saturation, or serum ferritin [20]. Iron deficiency anemia overwhelmingly occurs in toddlers and women of a reproductive age [21]. According to the WHO, 35–75% of child-bearing women in developing countries and 18% in industrialized countries are anemic, while 43% of women in developing nations and 12% in industrialized nations suffer, or have suffered, from anemia [22]. In comparison IDA occurs in only 1–2% of men and 2% of women over the age of 50 years [21].

Transfer of iron from mother to fetus results in an increase in maternal iron absorption during pregnancy, which is regulated by the placenta. Significant decreases in serum ferritin is observed between 12 and 25 weeks of pregnancy. If maternal iron levels decrease, transferrin receptors to the fetus increase so as to increase the uptake of iron by the placenta. A lack of synthesis of placental ferritin often prevents excessive iron transport to the fetus. Evidence is accumulating that the capacity of this system may be inadequate to maintain iron transfer to the fetus when the mother is deficient, leading to detrimental effects to the cognitive function of the fetus [22].

Iron deficiency may be caused by prolonged low dietary intake, increased iron requirement due to pregnancy, loss of blood through gastrointestinal (GI) bleeding or menstruation, or gastrointestinal malabsorption of iron. In adults over the age of 50, GI blood loss is an important cause of iron deficiency. Amongst patients over the age of 50 suffering from IDA, 11% of these cases was a result of GI cancer [21].

Iron deficiency can have many negative effects on an individual's health, including changes in immune function, cognitive development, temperature regulation, energy metabolism, and work performance [20]. A decrease in cognitive function due to iron deficiency is not well understood; however, it is proposed that a decrease in iron-dependent dopamine D2 receptors in the cortex is observed, altering dopamine neurotransmission, causing a decrease in cognitive function [23].

11.3
Copper Metabolic Diseases

11.3.1
Copper Homeostasis

Copper is an essential metal that is required for cellular respiration, iron oxidation, pigment formation, neurotransmitter biosynthesis, antioxidant defense, peptide amidation, central nervous system development, and connective tissue formation [24]. Physiologically, copper exists in two redox states, cuprous (Cu^+) and cupric (Cu^{2+}), with many known enzymes requiring it. Copper is found complexed to proteins in its ionic form; free Cu ions, like free Fe ions, catalyze the formation of free radicals, resulting in Fenton chemistry [25].

Our ability to tightly regulate copper is cardinal to keep these processes in check. Imbalances in copper homeostasis can lead to neurodegeneration, growth retardation, and mortality [24, 26]. Table 11.3 describes the location and function of some important copper-dependent proteins [24–28].

Copper homeostasis is maintained by a balance between intestinal absorption and excretion. Copper is absorbed from the gut and transported to the liver, the main storage area for copper, where it is subsequently redistributed to all tissues and organs. Copper is then returned to the liver to be excreted by the bile, the principal route for copper elimination (Figure 11.3) [29]. While copper is also excreted in sweat and urine, this excretion is not significant enough to contribute to homeostasis.[26]

Ceruloplasmin accounts for approximately 90% of the copper content found in plasma, but is not believed to be involved as a specific copper transport vehicle. Those suffering from aceruloplasminemia (see Section 11.2.3.1), an autosomal recessive disorder, lack a functional form of the protein, but do not exhibit any signs of copper deficiency. In contrast, only 5% of the total copper in serum is

Table 11.3 Important copper-dependent proteins involved in copper transport and homeostasis.

Protein/enzyme	Location	Function
Cu/Zn superoxide dismutase	Cytosol	Antioxidant defense (superoxide dismutation)
Cytochrome *c* oxidase	Mitochondria	Mitochondrial respiration
Ceruloplasmin	Plasma	Iron and copper transport (ferrioxidase)
Lysyl oxidase	Elastin and collagen	Connective tissue formation
Dopamine-β-hydroxylase	Storage vesicle	Catecholamine production
Tyrosinase	Storage vesicle	Melanin formation
Peptidylglycine α-amidating mono-oxygenase	Storage vesicle	Peptide amidation (activation of peptides)
Metallothionein	Liver and kidneys	Storage and chaperon

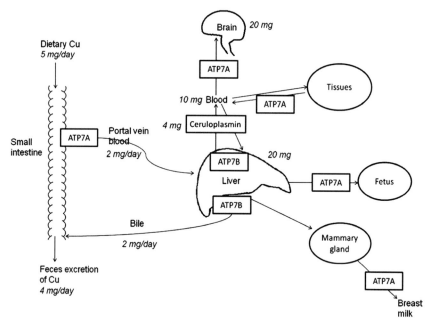

Figure 11.3 Transport and distribution of copper, and functions of ATP7A and APT7B.

bound to albumin, but this protein is considered to be very important for copper uptake [30]. Cellular copper uptake is mediated through energy-dependent transporters localized in the plasma membrane. Copper can enter cells by the high affinity carrier Ctr1, a membrane transport protein found in most tissues [31].

The mechanism of copper transport and homeostasis in the brain has not been completely elucidated; however, it has been well established that copper plays an important role in brain development. Copper transporting P-type ATPases possess six metal-binding sites (MBS) in the N-terminal part of the molecule to pump copper ions through physiological barriers, including the blood–brain barrier. Identification of ATP7A, a copper transporting ATPase that has a loss-of-function in Menkes disease (Section 11.3.2), has helped to gain insight into copper transport mechanisms in the brain [30, 31].

Once copper is transported into the cytosol by the Ctr1 transporter protein, the chaperone Atx1 shuttles copper to the ATPases (ATP7A/ATP7B), the copper chaperone for superoxide dismutase, CCS, delivers copper to superoxide dismutase (SOD) and Cox17, Sco1, and Sco3 are the chaperones involved in transporting copper to mitochondria and cytochrome oxidase (Cox). Metallothionein chelates most of the excess copper in the cell once Ctr1 transports it into the cell. Metallothionein plays an important role in scavenging free copper, along with other heavy metals, but it also may play a role in copper storage. ATP7A works to pump out excess copper in nonhepatic cells, while in hepatic cells this role is carried out by the P-type ATPase ATP7B, the protein that undergoes a loss-of-function in Wilson's disease (Section 11.3.3). The role of the ATPases ATP7A and

Table 11.4 Comparison of the hereditary disorders of copper metabolism.

	Wilson's disease	Menkes disease
Genetics	Autosomal recessive ATP7B	X-linked ATP7A
Onset	Late childhood: liver 20s–30s: neuropsychiatric problems	Early infancy
Pathogenesis	Copper overload caused by defected biliary copper excretion	Copper deficiency caused by defected copper transport across the brain, placenta, and GI tract
Presentation	Cirrhosis, liver disease, neuropsychiatric symptoms	Hypopigmentation, abnormal hair growth, failure to thrive, seizures, mental retardation

ATP7B can been seen in Figure 11.3. These two copper transporting proteins are the only ones that are presently known to be associated with specific copper metabolic diseases [24, 30]. Table 11.4 gives a comparison of the pathogenesis of Wilson's disease and Menkes disease [24].

11.3.2
Copper Deficiency: Menkes Disease

Menkes disease (MD) is an X-linked disorder that is diagnosed by stunted growth, hypopigmentation, brittle hair, arterial tortuosity (twisting of the arteries), and neurodegeneration. These characteristic disorders are caused by a mutation in the encoding of the copper transporting gene ATP7A, resulting in impaired activity of the cuproenzymes [24]. The CNS central nervous system pathology is less affected in patients suffering from MD, but that possess some ATP7A activity. This suggests that there is a hierarchic order of copper distribution, and that under copper deficiency copper will distribute preferentially in the brain.

Menkes is a rare, but serious, disease the affects 1 in 250 000 [32]. The defective gene, ATP7A, belongs to the large family of cation transporter P-type ATPases, and is found in muscle, kidney, lung, and brain. The functional role of ATP7A, originally described by Llanos and Mercer, can be seen in Figure 11.3 [33]. Only a trace amount is found in the liver, and is thus responsible for copper transport in non-hepatic cells. All P-type ATPases have similar amino acid sequences, where there is a conserved phosphorylation motif that contains an aspartic acid, an ATP-binding site, and a CPC (cysteine-proline-cysteine) domain that acts to bind copper when transferred from the metal binding site (MBS) [30]. The N-terminus contains six MBSs, each with the sequence GMTXCXXC, where X denotes a non-conserved amino acid, and copper binds to the CXXC motif [32].

When ATP7A is inactivated, copper becomes trapped in the endothelial cells of the blood–brain barrier (BBB) and the brain becomes severely deficient, leading to the profound neurological symptoms manifested in Menkes disease [30].

Deficiency in copper results in impaired function in some of the cuproenzymes and copper proteins described in Table 11.2. For example, a lack of function of lysyl oxidase results in connective tissue and skeletal defects [34]. Cytochrome *c* oxidase impairment results in deficient energy production (ATP production), causing altered nerve conduction, seizures and myopathy. Loss of function of tyrosinase results in hypopigmentation of the skin. Altered function of Cu/Zn SOD leads to oxidative stress to cells, which results in degeneration of the central nervous system and mitochondrial defects [26].

Treatment of MD with copper is not effective because copper transport into the brain is dependent on the function of ATP7A; however, administered copxper-histidine is taken up by the brain more efficiently, though the mechanism is unknown [24]. Copper-histidine therapy also results in normalization of serum copper, ceruloplasmin, dopamine, and norepinephrine levels in patients who have undergone this treatment course. Connective tissues disorders, however, still persist for these patients, indicating that copper-histidine does not bind effectively to lysyl oxidase [32].

11.3.3
Copper Overload: Wilson's Disease

Wilson's disease (WD) is an autosomal recessive disorder that causes cirrhosis, liver disease, progressive neurological disorders, or psychiatric illness, affecting 1 in 30 000 individuals [35]. There is an impairment of biliary copper excretion, leading to hepatocyte copper accumulation and copper-mediated liver damage. Leakage of copper can occur in the plasma, and eventually overload is seen in all tissues. ATP7B, the affected copper-transporter gene, is a P-type ATPase [24]. WD and MD P-type ATPases are functionally homologous, sharing 67% protein identity [33]. Their pathophysiologies are quite different, WD ATPase being found mainly in the liver and kidney, whereas ATP7A is in muscle, kidney, lung, and the brain [32]. Like Menkes, Wilson's disease ATPase possesses six Cys-X-X-Cys (CXXC) metal binding sites in the N terminus. There are some amino acid differences, but this does not affect the copper binding [32].

Copper can act as a prooxidant as it physiologically exists in two different valence states. Thus an excess of copper in the liver leads to organ damage caused by oxidative stress. Free copper ions participate in Fenton or Haber–Weiss chemistry, generating reactive oxygen species, which have been shown to form in HepG2 cells. Apoptosis commonly causes liver damage in Wilson's disease [35]. Although ATP7B is expressed in some regions of the brain, in WD, copper overload seen in extrahepatic tissues is due to accumulation in the plasma following liver injury. A complete reversal of non-hepatic tissue accumulation is seen after liver transplantation [24].

11.3.4
Treatment of Wilson's Disease: Chelation Therapy

Wilson's disease was once an untreatable disorder, inevitably leading to death; now, if caught at an early enough stage, the disease is treatable with copper chelators

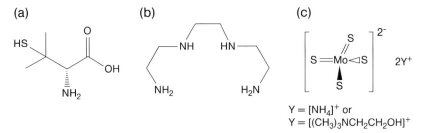

Figure 11.4 (a) D-Penicillamine; (b) trientine; and (c) tetrathimolybdate.

and zinc salt therapy [36, 37]. The three most common chelating agents are D-penicillamine, trientine, and tetrathiomolybdate. Penicillamine was the first oral agent for the treatment of WD. It is believed that penicillamine binds in a bidentate fashion, and can either bind in a 1:1 or a 1:2 ligand to metal ratio, forming a Cu(I, II) mixed-valence chelate that is unusually strong, binding the first copper through the amino nitrogen and thiol sulfur, and the second copper through a deprotonated carboxyl group, both acting as bidentate donors (Figure 11.4) [36, 38, 39].

Some patients have a hypersensitivity to penicillamine, a metabolite of the antibiotic penicillin. In these cases trientine is prescribed. Trientine is a tetradentate ligand, coordinating to copper in a 1:1 ratio by its four amines [36]. Tetrathiomolybdate is believed to form a polymetallic clusters with copper, binding up the three coppers per ligand with Mo(IV) in a tetrahedral arrangement and an overall 2– charge on the complex [40]. Tetrathiomolybdate is especially successful in treating patients with neurological manifestations from Wilson's disease [36].

More recently, zinc salts have been used in the treatment of copper overload. Zinc (Section 11.4) stimulates metallothionein production, helping to sequester copper. Since copper has a higher affinity for metallothionein than zinc, excess copper binds to it, helping to excrete excess copper [41]. Treatment is lifelong with either oral chelating agents or zinc salts. If unresponsive, liver transplantation is necessary for patients suffering from WD [35].

11.4
Zinc Metabolic Diseases

11.4.1
Zinc Homeostasis

Zinc is a co-factor in over 200 biologically important enzymes (e.g., alcohol dehydrogenase, carbonic anhydrase, carboxypeptidase), particularly enzymes involved in protein synthesis [41]. Between 3% and 10% of all proteins in mammals bind to zinc. The uniqueness of zinc is that unlike other abundant transition metals such as iron and copper it lacks redox activity [42]. It is one of the most abundant trace elements in the body, where it is present in all tissues and fluids. The average amount of zinc in healthy adults is 1.4–2.3 g [41]. Muscle and bone

Table 11.5 Distribution of zinc in the body.

Tissue	% Of total body zinc
Skeletal muscle	57
Bone	29
Skin	6
Liver	5
Brain	1.5
Kidneys	0.7
Heart	0.4
Hair	∼0.1
Blood plasma	∼0.1

contain 57% and 29%, respectively, accounting for greater than 85% of whole body zinc (Table 11.5) [2]. The prostate contains high concentrations of zinc because of the zinc-containing enzyme acid phosphatase. In the plasma, two-thirds of zinc is bound to albumin, with the rest bound to α2-globulins. Albumin-bound zinc in plasma is the metabolically active form of zinc in the body. Zinc in the bone is relatively inert, except during periods of calcium mobilization [2, 41].

Changes in zinc absorption and excretion in the GI tract are the primary mechanisms for maintaining zinc homeostasis [41]. Fecal excretion responds almost immediately to increases and decreases in zinc intake, but only by a relatively small amount. In contrast, zinc absorption responds at a slower rate, but has a greater capacity to deal with large fluctuations in dietary zinc [2]. Zinc is primarily excreted through the feces (70–80%), and the rest from sweat and urine [41]. Renal loss of zinc tends to be low, and remains constant over a large range of intake, while fecal losses increase. Intake can vary tenfold, in humans normal intake is from 107 to 231 μmol day^{-1}, but can be as little as 22 μmol day^{-1} or as much as 306 μmol day^{-1}, and still achieve a healthy balance [2].

Zinc is a critical structural component of thousands of zinc-finger proteins with diverse functions. In the absence of zinc, zinc-containing proteins can no longer be synthesized. When exposed to toxic levels of heavy metals, such as mercury or zinc, most cells can induce metallothionein synthesis. This occurs by the metal-regulatory transcription factor MTF-1, but the mechanism by which it senses elevated metal levels activating the transcription of metallothionein genes is not established [43]. Unregulated free zinc has also been implicated in the formation of beta-amyloid plaques associated with Alzheimer's disease [42]. Several proteins are responsible for cell transport and buffering of zinc. They include the zinc transporter family (ZnT) proteins, zinc-regulated metal transporters, iron regulated metal transporter-like proteins (ZIP), and three distinct forms of metallothionein. While ZnT proteins have not been fully elucidated, the rapid changes occurring in extracellular Zn^{2+}, and the number of pathways that exist for the permeation of this ion, suggest an important role for the ZnT protein in the regulation of zinc [42].

11.4.2
Zinc Overload: Zinc Toxicity

Zinc is essential for many processes; however, excess zinc can be toxic [44]. The absorption of zinc from the gastrointestinal tract occurs via the duodenum, and the amount of zinc absorbed depends on a multitude of factors. Isotope studies suggest that 20–40% of dietary zinc is absorbed; however, the presence of other trace metals such as mercury, copper, and cadmium reduce zinc absorption. Zinc, as mentioned earlier, stimulates metallothionein production. Copper has a higher affinity for metallothionein than zinc, thus more copper is bound to metallothionein, causing it to be excreted. Thus, Zinc toxicity generally manifests as copper deficiency. Excretion of zinc-metallothionein helps prevent the absorption of zinc, and thus its toxic side effects [41]. Some examples of common cases of zinc toxicity are described below.

Meta-fume fever or zinc shakes are a manifestation of acute zinc toxicity, caused by the inhalation of zinc oxide, commonly seen in steel welding and brass industries. It manifests as fever, headache, myalgia, fatigue, dyspnea, and can have adverse affects on pulmonary function [41, 44]. In the presence of high levels of zinc oxide, initially workers will exhibit some or all of the aforementioned symptoms, but as exposure continues, tolerance increases. It has been postulated that this is a result of an increase in metallothionein synthesis, which can bind heavy metals, preventing toxic side effects [45].

Toxicity can also occur if excessive levels of zinc are ingested. Zinc is considered to be nontoxic if consumed orally; however, toxicity can occur if excessive levels of zinc are consumed, the symptoms are nausea, vomiting, epigastric pain and fatigue [44]. At high intakes, 100–300 mg day^{-1} vs. RDA of 15 mg day^{-1}, induced copper deficiency is observed, with symptoms of neutropenia and anemia, impaired immune function, and adverse effects on the ratio of low-density-lipoprotein to high density-lipoprotein (LDL/HDL). These symptoms have been observed even at amounts closer to the RDA of 15 mg day^{-1}. Typically, in acute cases of zinc toxicity, chelation therapy is sufficient to remove excess zinc [41, 44].

Zinc is often used to treat sickle cell anemia and nonresponsive celiac disease; patients receiving this treatment can develop zinc toxicity. Since elimination of excess zinc is slow, often decreasing zinc intake and increasing copper intake is not enough to restore copper homeostasis. Until excretion of the zinc occurs, intestinal absorption of copper is blocked [44].

11.4.3
Treatment of Zinc Toxicity: Chelation Therapy

The chelation of excess zinc can inhibit the toxic zinc effects and subsequent neuronal death. The three main chelators used are TPEN, pyrithione, and EDTA (Figure 11.5). EDTA (dicalcium ethylenediaminetetraacetic acid) used to be the chelator of choice to sequester zinc; however, it cannot cross the BBB, leading to the development of both *N,N,N',N'*-tetrakis(2-pyridylmethyl)ethylenediamine (TPEN) and the lower affinity 1-hydroxypryidine-2-thione (pyrithione) [46]. TPEN binds with high

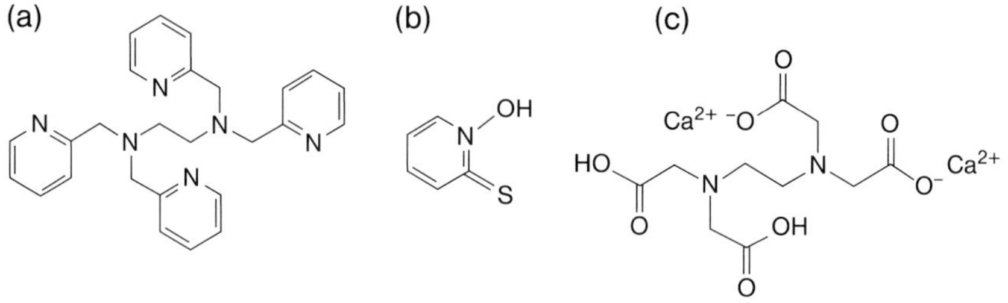

Figure 11.5 (a) TPEN; (b) pyrithione; and (c) EDTA, dicalcium salt.

affinity to divalent transition metals, but with low affinity to Ca^{2+}, protecting the heart tissue from ischemic damage [47]. TPEN coordinates to zinc in a hexadentate fashion, forming an octahedral complex [48]. Crystallographic studies indicate that the low-affinity pyrithione coordinates zinc with four ligands, binding in a bidentate fashion to the metal, to form a complex in a 2:2 ligand to metal ratio [49].

11.4.4
Acquired Zinc Deficiency

Hypogonadal dwarf syndrome has been associated with zinc deficiency characterized by hypogonadism and growth retardation along with anemia, dysomia, altered taste, and poor wound healing. An abnormally low zinc concentration in breast milk has been associated with the development of zinc deficiency in breast-fed newborns, characterized by acrodermatitis (see below), irritability, and delayed growth [41].

Zinc is also lost through sweat, seminal emissions, menstruation, hair, and nails. These whole body surface losses increase and decrease proportionally with intake and can be a significant source of zinc loss. Seminal emissions can be a significant source of zinc loss with frequent ejaculations as semen is rich in zinc. As with other integumental sources of zinc loss, the amount of zinc lost due to semen ejaculation decreases with zinc depletion [2].

Zinc deficiency symptoms manifest in the gastrointestinal tract, the central nervous system, the immune system, skeletal, and reproductive systems. These symptoms are generally attributed to an impairment of zinc-dependent metabolic functions in all tissues [50]. Symptoms of zinc deficiency include growth failure, altered taste, impaired wound healing, and lowered immunity [51, 52].

11.4.5
Genetic Zinc Deficiency: Acrodermatitis Enterophathica

Acrodermatitis enterophathica (AE) is an autosomal recessive disorder, resulting in malabsorption of zinc, and severe zinc deficiency [53, 54]. This deficiency is caused by defective uptake of zinc from the duodenum and jejunum [53]. The disease usually manifests at infancy, and is characterized by diarrhea, dermatitis, alopecia, loss of appetite, and failure to thrive [54, 55]. It was first discovered by

Barnes and Moynahan in 1973, when it was observed that this inability to absorb zinc sufficiently could be cured by zinc supplementation [56]. Clinical diagnosis is difficult because of a variety of symptoms in different people [54]. Typically, AE is diagnosed by low plasma and serum zinc concentrations.

There is evidence that the AE mutation affects zinc metabolism in human fibroblasts. The activity of the zinc-dependent enzyme 5′ nucleotidase, and the zinc content in fibroblast cells, are significantly reduced in someone suffering from AE. Homozygosity mapping has been used to identify the AE gene to a telomeric region localized at the chromosomal region 8q24.3 [54]. This chromosome encodes for zinc-uptake proteins, amongst others [57]. Sequence analysis of DNA from patients suffering from AE has identified the gene SLC39A4 located in chromosome 8q24.3, containing a mutation in both alleles of the gene [53, 57]. The gene encodes for hZIP4, a protein that is part of a family of zinc/iron-regulated transporter-like-proteins. In contrast to other hZIP proteins, hZIP4 has a large histidine-rich N-terminus, which is thought to be responsible for binding dietary zinc in the intestinal lumen. It is thus speculated that AE patients retain some mechanism for intestinal zinc uptake, and thus supplementation of 2–20-fold excess of the RDA of zinc is effective in treating AE affected patients [57].

11.5
Diseases Related to Imbalances in Electrolytic Metabolism: P, the Alkali Metals, and the Alkaline Earths

> The preservation of electrolyte homeostasis and thus water balance is vital to our functioning. No idea could be thought, no muscle moved without proper balance of salts in our body. It is the responsibility of mainly the kidneys to maintain this vital "milie interiur".
> Kleta and Bockenhauer [58].

Maintenance of fluid and electrolyte homeostasis intra- and extracellularly is critical for normal muscular function, nerve function, hydration, and pH balance. The kidneys play a key role in maintaining homeostasis of electrolyte and fluid balance in a wide variety of environments. Renal disease thus almost always leads to several electrolytic disorders [59].

The role of alkali and alkaline earth metals in clinical medicine, nutrition, and physiology is of interest; for example, calcium and phosphorus make up the structural component of our bones, potassium has a potent protective effect against cardiovascular disease, and magnesium deficiency causes cardiac arrhythmia and irritation of the nervous system [1]. Maintenance of fluid and electrolyte balance is especially important as the homeostasis of these metals is inter-related. Magnesium deficiency causes potassium deficiency; vitamin D affects the serum concentrations of both phosphorus and calcium. Many diseases cause a multitude of electrolyte disorders and are all related to the renal and kidney function (Figure 11.6). Table 11.6 gives a summary of diseases that affect the electrolyte balance [60–63].

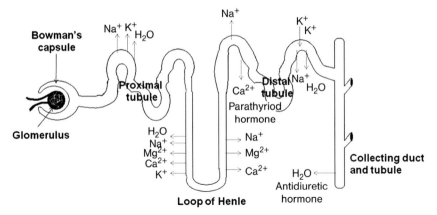

Figure 11.6 Renal physiology and electrolyte reabsorption.

In this section we discuss the effects of disturbances of the following electrolytes: sodium, magnesium, potassium, phosphates, and calcium.

11.5.1
Sodium

Sodium is the most abundant cation in the body and is generally found in the extracellular compartment, with the exception of blood cells, which have a high intracellular sodium concentration. Sodium is critical for osmoregulation, nerve function, and muscle function. Absorption of sodium occurs through the jejunum and ileum of the small intestine, and is excreted through sweat and urine [64].

Despite many foods being high in sodium, and dietary variations in salt and water intake, plasma sodium concentration is finely tuned within a small range of 135–145 mM [65]. Healthy kidneys maintain a constant sodium level by adjusting the amount of sodium excreted in urine, thus disorders of sodium metabolism are generally only seen in patients whose kidney function is impaired. The kidneys help maintain sodium balance by stimulating adrenal glands to secrete aldesterone, which causes the kidneys to retain sodium and excrete potassium. Urinary dilution is determined by arginine vasopressin, also known as antidiuretic hormone (ADH), which is produced by the pituitary gland. ADH causes the kidneys to conserve fluid, increase blood volume, helping to regulate the body's retention of water [66]. When ADH is produced, concentrated urine is produced by water reabsorption across the renal collecting ducts (Figure 11.6).

Disorders of sodium balance are a common clinical manifestation, seen in hospitalized patients and the elderly. In mild cases sodium disorders generally are symptomless, but in extreme cases can lead to morbidity. Most cases of sodium balance disorders, are iatrogenic – meaning they are unavoidable in certain medical treatments. Disorders of sodium balance – hyponatremia and hypernatremia – are the most common electrolytic disturbances seen in hospitalized patients and clinical medicine, and are discussed below [65].

Hypocalcemia	Hypomagnesemia	Hypophosphatemia	Hypokalemia
Factitious: Sample contamination with EDTA *Normal plasma ionized with reduced total calcium:* Hypoalbuminaemia Reduced plasma ionized with normal total calcium Respiratory alkalosis *Reduced plasma ionized with normal total calcium:* Hypoparathyroidism Malabsorption of vitamin D Vitamin D resistant rickets Rhabdomyolysis Pacreatitis Hyperphosphatemia	*Redistribution:* Hungry bone syndrome Catecholamine excess Massive blood transfusion *Gastrointestinal disorders (GI):* Malabsorption Short-bowel syndrome GI tract fistulas, diarrhea Pancreatitis *Renal losses:* Reduced sodium reabsorption Bartter's and Gitelman's syndromes *Endocrine disorders:* Hypercalcemia Hyperparathyroidism Hyperthyroidism Hyperaldosteronism *Alcoholism, diuretics*	*Redistribution:* Malnutrition Diabetic ketoacidosis Respiratory alkalosis Hormones Rapid cell proliferation (hungry bone syndrome, acute leukemia, Burkitt's lymphoma) *Increased urinary excretion:* Hyperparathyroidism Vitamin D and PTH deficiency Fanconi syndrome Alcoholism Drugs – diuretics Glucocorticoids Kidney transplant *Decreased intestinal absorption:* Phosphate binders Diarrhea, vomiting	*Inadequate intake* *Increased excretion:* Diarrhea, laxative abuse *Renal losses:* Loop, thiazide diuretics Metabolic alkalosis Osmotic dieresis Bartter's and Gitelman's syndromes Glucocorticoids *Magnesium depletion* *Renal tubular acidosis*

Hypercalcemia	Hypermagnesemia	Hyperphosphatemia	Hyperkalemia
Factitious *Carcinomas:* Primary (lymphoma) Secondary (breast, renal) Paget's disease Hyperparathyroidism Renal failure Vitamin A and D toxicity Drugs – lithium, diuretics Endocrine Addison's disease Familial hypocalcuric hypercalcaemia	Acidosis Hemolysis Excessive administration of Mg^{2+} salts (enemas, cathartics) in the presence of renal failure	*Factitious* *Redistribution:* Trauma Rhabdomyolysis Acidosis (keto and lactic) Tumor lysis Bisphosphonate therapy Vitamin D toxicity *Renal retention:* Renal failure Hypoparathyroidism Pseudohypoparathyrodism	*Factitious:* Thrombocytosis Leukocytosis Hemolysis *Impaired potassium excretion:* Renal insufficiency or failure Mineralocorticoid deficiency *Pseudohypoaldosteronism* *Drugs – diuretics, NSAIDS*

11.5.1.1 Hyponatremia

Hyponatremia is an electrolytic disorder that is defined as a decrease in serum sodium concentration below 135 mM. Hyponatremia can be associated with low, normal, or high osmotic pressure. The three types of hyponatremia are: hypotonic, hypertonic, and isotonic. Hypertonic hyponatremia is caused by osmotically active particles in the extracellular fluid, which results in a shift of water from intracellular to extracellular fluid (this occurs with glucose and is associated with hyperglycemia), causing the dehydration of cells. Isotonic hyponatremia is called pseudohyponatremia and is caused by high lipid or protein levels in the serum [67]. Hypotonic hyponatremia can be classified according to the volume status of a patient as hypovolemic, hypervolemic, and euvolemic hyponatremias. Figure 11.7 shows the causes of hypotonic hyponatremia [65, 68, 69].

An increase in tonicity stimulates the thirst center and ADH secretion from the pituitary glands, which act on receptors in the renal tubules resulting in increased water reabsorption. When there is a decrease in tonicity, thirst is inhibited, which in turn inhibits ADH secretion, causing diuresis. Most cases of hyponatremia are a result of a decrease in urine output from the kidneys [69].

Hypervolemic hyponatremia indicates an increased total body sodium and water content, where the increase in water is greater than that of sodium, resulting in edema. Hypervolemia is generally caused by congestive heart failure, liver cirrhosis, and renal disease and is quite easy to diagnose [69].

Hypovolemia is a state of decreased blood volume, and is associated with both extrarenal loss and renal loss (Figure 11.7). Hypovolemic hyponatremia is caused by a deficit in total body sodium and water, with sodium loss being significantly lower. Extrarenal loss occurs through skin losses, such as burns, and perspiration. Renal losses are associated with diuretics, salt wasting, and Addison's disease. Addison's disease is a rare endocrine disorder that causes adrenal insufficiency. This results in hyponatremia and hyperkalemia (see Section 11.5.3 potassium) due to a loss of production of the hormone aldosterone [70].

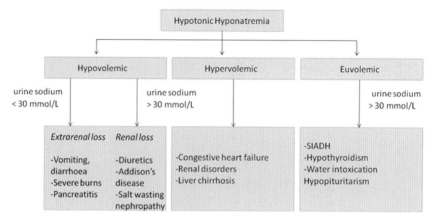

Figure 11.7 Causes of hypotonic hyponatremia.

Lastly, euvolemic hyponatremia is caused by low total body water levels, but normal total body sodium. It is often difficult to distinguish between euvolemic and hypovolemic hyponatremias [65]. Euvolemic hyponatremia is caused by hypothyroidism (see Section 11.5.5 calcium), water intoxication, and syndrome of inappropriate ADH secretion (SIADH) [69].

A decrease in plasma sodium concentration creates an osmotic gradient between intracellular and extracellular fluid in brain cells, causing movement of water into cells, increasing intracellular volume, resulting in neurological symptoms. Chronic, symptomless hyponatremia is generally treated with fluid restriction to no more than 1 l of fluids a day. If this fails then Demeclocycline, a drug that inhibits ADH action in the kidneys, is administered for treatment of asymptomatic hyponatremia due to SIADH [65].

Symptoms of hyponatremia begin to show at sodium concentrations between 125 and 130 mM, which include nausea and malaise. As sodium levels drop below 120 mM, severe hyponatremia shows as headache, lethargy, restlessness, and disorientation [65, 68, 71]. If sodium levels fall rapidly, seizure, coma, brain damage, and respiratory arrest can occur. If the drop is more gradual, the brain can self-regulate to prevent swelling by active transport of sodium chloride, potassium, and, as a last resort, other organic solutes from intracellular to the extracellular compartments. As a result, gradual changes in sodium concentration tend to be asymptomatic, resulting in water loss due to the active transport of solute, helping to prevent brain swelling. Owing to the many causes of hyponatremia, the source of the hyponatremia must be established to properly manage the low sodium levels [69].

11.5.1.2 Hypernatremia

Hypernatremia, which is much less common than hyponatremia, is an electrolyte disturbance of sodium that is defined as a rise in serum sodium concentration above 145 mM. It reflects a state of hypertonicity – a state of sodium gain or water loss. Generally, a net water loss accounts for most causes of hypernatremia, commonly caused by administration of IV fluids or sodium loading to increase muscle size. Hypernatremia is commonly observed in patients whose sensation of thirst is impaired by an altered mental status, or those suffering from adipsia. Severe symptoms are generally only seen at plasma concentrations above 158 mM [65, 72].

In healthy patients, water homeostasis is maintained by two osmoregulated mechanisms: thirst and ADH release. An increase in plasma osmolality stimulates thirst and ADH release. Thirst stimulates increased water intake while ADH, by its action on the distal renal tubules, increases tubular reabsorption of water and reduces urine output. Increased intake of water and decreased urine output results in a net gain of water, returning plasma osmolality to normal values [73]. A fall in plasma osmolality, on the other hand, suppresses ADH and thirst sensations. In an individual suffering from adipsia, the hypothalamus contains lesions, which prevents the release of ADH and thus impairs osmoregulated thirst [65].

11.5.2
Magnesium

Magnesium is the fourth most abundant cation in the body and the second most prevalent intracellular cation [74]. Magnesium is an essential trace metal that is primarily found in bone and intracellularly bound as a metallo-cofactor; the total distribution of magnesium in the body is reported in Table 11.7 [75, 76]. Magnesium is essential for over 300 phosphate transfer reactions inherent to energy transfer as it plays a vital role in the production of adenosine triphosphate (ATP), and is involved in protein and nucleic acid synthesis. Unlike serum calcium and phosphate levels, parathryorid hormone (PTH) and vitamin D do not regulate magnesium metabolism [74].

Normal serum levels of magnesium are 0.8–1.2 mM (plasma levels of 0.7–0.96 mM), where the ionic concentration of magnesium is approximately the same in the extra- and intracellular compartments [75, 77]. Intracellular magnesium homeostasis is well managed and thus changes in magnesium homeostasis are caused by extracellular disturbances [76]. Average daily magnesium intake is approximately 8–20 mmol (300–350 mg), and total body magnesium levels depend on kidney function and gastrointestinal absorption [74]. Of this, 40% is absorbed in the jejunum and ileum of the small intestine by passive absorption. Urinary loss accounts for 2.5–8 mmol day^{-1}, filtration by the glomerus (Figure 11.6) is 100 mmol day^{-1} in the kidneys. Of this only 10–15% is absorbed into the proximal tubule, 60–70% is reabsorbed in the cortical thick ascending limb of the loop of Henle, 10–15% in the distal tube and 5% is excreted [75, 76]. Owing to its abundance in our diet, there is no indication that magnesium homeostasis is under specific tight hormonal control, despite its importance to many physiological processes in the body [76].

11.5.2.1 Hypomagnesemia
Hypomagnesemia is an electrolytic disturbance caused by low serum magnesium levels of less than 0.7 mM [60]. It is often found as a clinical manifestation, occurring in up to 65% of patients in critical care and 12% of general hospitalized patients, and by various medications that affect renal activity. There are, however, a few genetic disorders such as Fanconi syndrome, Bartter syndrome, and

Table 11.7 Distribution of magnesium in the body in a 70 kg man.

	Amount (mmol)	% Total
Bone	600	60–65
Intracellular (bound)	365	37
(free)	25	3
Extracellular fluid (ECF)	10	1

Gitleman's syndrome that result in hypomagnesemia [74]. Hypomagnesemia is characterized by neurological, cardiovascular, and metabolic disorders. Common symptoms, include weakness, confusion, delirium, convulsions, tachyarrhytmias, and renal tubular acidosis [60].

Hypomagnesemia is associated with magnesium depletion due to decreased intake or redistribution, but is usually associated with alcoholism and with gastrointestinal absorption or irregular kidney function; a summary of the causes of hypomagnemia can be seen in Table 11.6 [74]. Hypomagnesemia may also cause hypocalcemia due to impaired synthesis of the parathyroid hormone (see Section 11.5.5 calcium) [60]. Impaired gastrointestinal (GI) absorption is one of the most common causes of hypomagnesemia. Short-bowel syndrome is a malabsorption disease caused by extensive resection of the small intestine [78]. Diarrhea, malabsorption, and – rarely – inadequate intake can result in a reduced amount of GI uptake [61].

Renal loss by the kidneys accounts for the most common cause of magnesium deficiency. Reduced sodium reabsorption results in hypomagnesemia because proximal tubular reabsorption of magnesium is proportional to sodium reabsorption [61]. Renal tubular defects such as renal tubular acidosis or Fanconi syndrome results in reduced reabsorption of magnesium. Renal tubular acidosis is a metabolic acidosis that is caused by impaired acid excretion by the kidneys [79]. Fanconi syndrome is a disease that is characterized by a general disorder of proximal renal tubule function, resulting in impaired absorption of sodium, phosphate (see Section 11.5.1 sodium and Section 11.5.4 phosphorus), and magnesium. The causes of Fanconi syndrome are multifaceted, occurring by both genetic and toxic factors [80].

Total body Mg depletion may exist in the presence of normal total plasma magnesium level. Hypomagnesemia can also cause hypocalcemia (see Section 11.5.5 calcium) and hypokalemia (see Section 11.5.3 potassium). If the underlying hypomagnesemia is not corrected, the associated hypocalcemia and hypokalemia are often resistant to therapy [60]. Classical Bartter's syndrome, Gitleman's syndrome (see Section 11.5.3 potassium), and Familial hypomagnesemia are all inherited disorders of renal magnesium wasting [59].

Treatment of hypomagnesemia occurs through the administration of Mg^{2+} salts, whose anions include sulfate and chloride – the anion does not modulate the biological effects. Acute magnesium deficiency is treated with intravenous magnesium sulfate.

11.5.2.2 **Hypermagnesemia**

Hypermagnesemia, a rare electrolytic disorder, is defined as plasma level greater than 1.0 mM. An ingestion of a large amount of magnesium with normal renal function is generally not toxic, as up to 200 mmol per day of Mg can be excreted [60]. High magnesium content from foods or from supplements are not connected with adverse health effects [76]. Generally, someone who does not suffer from renal or bowel disorders will not suffer from hypermagnesemia. Elevated magnesium serum levels are also observed when magnesium-containing cathartics, antacids, and enemas are given [60, 74].

Hypermagnesemia can cause drowsiness, hyporeflexia, coma, reparatory paralysis, hypotension, conduction defects or sinatrial and atrioventricular nodal block and asystole. Excess Mg suppresses the release of acetylcholine and blocks transmission at the neuromuscular junction. Treatment requires excretion of magnesium ion, which is usually achieved by dialysis. Calcium chloride ($CaCl_2$) is often given to rapidly treat the cardiac conduction defects [60].

11.5.3
Potassium

Potassium is the most abundant intracellular cation – with only 2% found in the extracellular fluid. Potassium plays a vital role in maintaining intracellular osmotic pressure, with the ratio between intracellular and extracellular potassium concentrations in polarizing membranes. These polarized membranes help conduct nerve impulses and muscle cell contraction. β-Adrenergic and insulin levels affect potassium's concentration in and out of cells. Insulin and β-Adrenergic move potassium into cells; the mechanism by which β-Adrenergic moves K^+ into cells has not been identified. Small changes in serum potassium concentration can have detrimental clinical manifestations [63].

In a healthy individual, serum potassium levels are between 3.5 and 5 mM [77]. Potassium intake is unregulated, with the average adult consuming between 40 and 100 mmol day^{-1}. Excretion increases with intake, so homeostasis of potassium is usually easily maintained. Once potassium is absorbed, insulin is released, stimulating N^+/K^+ ATPase activity, which helps to facilitate its entrance into cells. A rise in potassium also stimulates aldosterone secretion, inducing its excretion within 30 min of ingestion; 90% of potassium is excreted by the kidneys and the other 10% by the stool [63].

Hyperkalemia and hypokalemia can result from an impairment in renal activity or in a perturbation of transcellular homeostasis. A summary of causes of disturbances in potassium homeostasis can be seen in Table 11.6.

11.5.3.1 Hypokalemia
Hypokalemia is a disorder of potassium deficiency that is defined as serum potassium concentration <3.6 mM. Hypokalemia reflects a disruption in normal homeostasis except in the case of factitious or pseudohypokalemia. The causes of hypokalemia are summarized in Table 11.6. In factitious hypokalemia, potassium levels are falsely low due to an elevated white blood cell count. When a blood sample is taken to be tested, the abnormal leukocytes may uptake some of the potassium, resulting in falsely low potassium levels [63].

For the most part, hypokalemia is asymptomatic, and is only identified by low serum potassium levels. In some cases it is associated with increases of life-threatening cardiac arrhythmias in patients with ischemia or scarred cardiac muscle. Severe hypokalemia, <2.5 mM, can result in rhabdomyolysis, respiratory arrest, and paralysis [63].

The most common causes of hypokalemia are either abnormal loss or inadequate dietary intake. Abnormal losses can occur from diarrhea or excessive use of laxatives, metabolic acidosis caused by Bartter's syndrome, diuretics, and magnesium depletion. Bartter's syndrome is an autosomal recessive disorder characterized by diverse abnormalities of electrolytic homeostasis, including hypokalemia and metabolic alkalosis [81]. Bartter's syndrome is caused by a defect in the thick ascending limb of the loop of Henle, affecting reabsorption of potassium (see Figure 11.6) [59, 81].

Both loop and thiazide diuretics increase sodium and chloride delivery to the collecting duct, stimulating potassium secretion, and causing chloride depletion [82]. Both classes of diuretics also increase magnesium excretion, causing depletion of this cation. Magnesium depletion (see Section 11.5.2 magnesium) promotes renal potassium losses, causing more potassium depletion [60].

A wide variety of drugs are β-adrenergic agonist drugs that shift potassium from the intracellular fluid, resulting in hypokalemia; these drugs include decongestants and bronchodilators. Theophylline (a bronchodilator) and caffeine stimulate cell membrane Na/K ATPases and promote potassium entry into cells. A few cups of coffee can decrease serum K levels by 0.4 mM. Accidental ingestion of barium salts can block potassium exit from cells, resulting in hypokalemia [67].

Owing to the risk of hyperkalemia, intravenous treatment is rarely used to treat hypokalemia [63]. As long as the patient is not suffering from cardiac arrhythmia, generally oral replacement therapy is adequate. Potassium replacement is usually given as KCl, because most causes of hypokalemia require repletion of chloride and potassium. When magnesium depletion is the cause of hypokalemia the magnesium depletion must be treated [60].

11.5.3.2 Hyperkalemia

Hyperkalemia is defined as a serum potassium concentration greater than 5.0 mM and is far less common than the aforementioned hypokalemia [63]. Severe hyperkalemia occurs at a serum potassium concentration greater than 6.5 mM. Elevated potassium serum and plasma levels usually indicate a disorder of potassium homeostasis; however, factitious hyperkalemia can release potassium from cells into the blood sample [63].

The word factitious indicates an artificially high or low level of electrolyte and is usually an *in vitro* manifestation. Factitious or pseudohyperkalemia results from an artificially high serum potassium concentration when plasma levels are normal. It is a phenomenon that occurs after phlebotomy and is attributed to potassium release from cells after collection of a blood sample [83]. Factitious hyperkalemia is usually caused by the release of potassium from white blood cells or platelets, which is the case for myeloproliferative disorders such as leukemia or thrombocytosis. It can also result if homolysis occurs [63, 84].

Familial pseudohyperkalemia (FP) is a dominantly inherited, asymptomatic, red blood cell trait that presents as high serum potassium concentrations. It is caused by a condition called "leaky red blood cell," that shows a temperature dependent release of potassium from red blood cells when the blood sample is stored at room temperature. It is attributed to a passive leakage of potassium across red blood cell membranes. If the sample is analyzed soon after being collected, high plasma potassium concentration is not observed. It is only observed if the blood sample is left at room temperature for a few hours [85, 86].

True hyperkalemia is usually caused by renal failure, adrenal insufficiency, massive tissue breakdown during trauma, rhabdomyolysis, and hemolysis during tumor lysis syndrome (see Section 11.5.4 phosphorus). Hyperkalemia can result in alternations in cardiac excitability, resulting in electrocardiograms that show heart arrhythmias, which can lead to death [63].

The Na-K pump is a major determinant of extrarenal potassium disposal. It produces an influx of potassium and efflux of sodium such that inhibition of the Na-K pump results in a decrease of intracellular potassium manifesting as severe hyperkalemia [87].

The treatment of acute severe hyperkalemia ultimately involves the removal of potassium from the body – this is usually accomplished by hemodialysis; however, because of the cardiac risks, especially during a rapid increase in potassium levels, calcium gluconate is administered immediately. While calcium has no effect on potassium levels, it has a positive effect on the heart, helping to stabilize cardiac arrhythmias [87].

11.5.4
Phosphorus

Phosphorus is ubiquitous throughout the body; in a healthy 70 kg adult the total body phosphorus content is 700 g, and serum phosphate levels are between 0.89 and 1.44 mM [74]. Phosphorus is an essential element that is usually found bound to oxygen as a phosphate. Most of this phosphorus, approximately 80%, is found in hydroxyapatite, the main mineral component of bone and teeth, $[Ca_{10}(PO_4)_6(OH)_2]_n$, 9% in skeletal muscle, $\sim 11\%$ in other tissues, and approximately 0.1% in the extracellular fluid [62, 74].

Phosphate is a vital component of nucleic acids forming the backbone of DNA and RNA and constitutes an important component of the lipid bilayer. Most intracellular phosphate is bound to adenosine triphosphate (ATP) and phosphocreatine. Phosphocreatine is an important energy store for skeletal muscle and the brain. Most of the inorganic phosphorus found in the body is free, existing as $H_2PO_4^-$, HPO_4^{2-}, PO_4^{3-}. Approximately 10% is found complexed to proteins, and 5% to calcium, magnesium, and sodium [62].

Phosphorus, as mentioned above, is an essential component of energy stores in the form of ATP and phosphocreatine. ATPases are a class of enzymes that catalyze the decomposition of adenosine triphosphate (ATP) to adenosine diphosphate (ADP), releasing a free phosphate ion and energy. The Na^+/K^+-ATPase, or sodium–potassium pump, is responsible for active transport of potassium into cells and sodium out of cells against respective concentration gradients. This is extremely important in maintaining resting potential and the regulation of cellular volume.

Phosphorus is found abundantly in protein-rich foods such as meat, dairy, and eggs, but is also high in cereal grains; the average dietary intake is 800–1400 mg day^{-1}. It is absorbed in GI tract in the duodenum and jejunum mostly through passive transport as phosphates. There is, however, a sodium-dependent active transport stimulated by the active form of vitamin D, 1,25-dihydroxyvitamin D3 (calcitriol) [74]. Hypophosphatemia, hypophosphatasia, and hyperphosphatemia are all disorders of phosphate metabolism and are discussed below.

11.5.4.1 Hypophosphatemia

Hypophosphatemia is an electrolytic disorder defined as a phosphate plasma level below 0.65 mM, with normal phosphate levels ranging from 0.89 to 1.44 mM. It is most commonly observed in patients who have been hospitalized, occurring in 0.24–3.1% of all hospitalization cases [74, 88]. Hypophosphatemia does not necessarily mean phosphate depletion; it is a disease that can occur in the presence of low, normal, or high total body phosphate. In the case of normal or high total body phosphate there is a shift from the extracellular pool to the intracellular compartment, resulting in hypophosphatemia. The term phosphate depletion thus refers to a reduction of the total body phosphate. Moderate and severe cases of hypophosphatemia are characterized by plasma phosphate concentrations of 0.32–0.65 mM and <0.32 mM, respectively [88].

There are three mechanisms by which hypophosphatemia can occur: internal redistribution, loss of phosphate from the body through increased urinary excretion, and decreased intestinal absorption [62, 74, 88]. The three causes are discussed below, and summarized in Table 11.6.

Decreased intestinal absorption is rarely caused by malnutrition, as phosphates are ubiquitously found in foods. It is more likely to occur because of vitamin D deficiency, prolonged use of phosphate binders, or alcoholism [62, 89]. Decreased vitamin-D synthesis results in an increase in PTH production, which causes a decrease in intestinal absorption of phosphorus, resulting in hypophosphatemia. Familial disorders of vitamin D metabolism such as vitamin-D resistance rickets and X-linked vitamin-D-resistant rickets are associated with hypophosphatemia [74, 90].

Increased redistribution is the cause of most cases of hypophosphatemia. An acute shift in phosphate from the extracellular to the intracellular compartment is primarily responsible for the lowering of the serum phosphate. Any treatment that stimulates glycolysis leads to the formation of phosphorylated glucose compounds, resulting in an intracellular shift of phosphorus [74]. This can include intravenous glucose intake, causing the release of insulin, increasing the cellular uptake of glucose and phosphate. The refeeding of malnourished patients suffering from anorexia or chronic alcoholism results in the refeeding syndrome. During the anabolic period following refeeding there is an influx of phosphate into the cells A patient suffering from respiratory alkalosis (hyperventilation), causes a fall in carbon dioxide, which results in a fall of CO_2 in the cell. This stimulates glycolysis, leading to accelerated production of phosphorylate metabolites and a rapid shift of phosphate into the cells [88, 89].

Hypophosphatemia caused by increased urinary excretion of phosphorus is commonly seen in patients suffering from hyperparathyroidism. This is because there is an increased loss of phosphate due to the inhibition of the co-transporter by PTH. Patients usually present with hypercalcemia, moderate hypophosphatemia, and decreased renal tubular reabsorption. The Fanconi syndrome causes an impairment of the proximal tubule leading to a urinary loss of compounds normally reabsorbed by the proximal tube; this can cause hypophosphatemia, amongst other electrolyte disorders [74]. Patients using diuretics will often manifest with hypophosphatemia [88, 89].

The major mechanisms responsible for symptoms of hypophosphatemia are a decrease in intracellular ATP and 2,3-DPG (diphosphoglycerate). Under certain clinical situations, such as diabetic acidosis and in patients with prolonged hyperventilation, symptoms are non-existent because phosphate depletion is not prolonged. Clinical manifestations are usually not observed unless plasma phosphorus levels fall below 0.32 mM [74]. In the skeletal muscle and bone, phosphate deficiency may result in myopathy, weakness, and bone pain caused by mobilization of the bone mineral in an attempt to maintain a normal serum phosphate concentration resulting in osteopenia and osteomalacia (see Section 11.5.5 calcium). Usually, these symptoms only occur in conjunction with vitamin D deficiency [88].

Cardiovascular, neurological, and hematological effects are generally only seen in severe cases of phosphate depletion, and can lead to respiratory failure, coma, seizures, parethesias, hemolysis, leukocyte, and platelet dysfunction [62, 88].

To treat hypophosphatemia, one must determine the cause of the phosphate depletion. In most cases increasing dietary intake of phosphate-rich foods (such as cow's milk) is often enough [62]. However, in extreme cases, oral or intravenous phosphate repletion must be used, but can lead to hypocalcemia, metastatic calcification, hyperkalemia, hypernatremia, metabolic acidosis, and hyperphosphatemia [62, 74].

11.5.4.2 Hypophosphatasia

Hypophosphatasia is a rare, but sometimes fatal, hereditary form of rickets or osteomalacia characterized by defective bone mineralization and a deficiency of the tissue-nonspecific isoenzyme of alkaline phosphatase (TNSALP) [91]. The disease is characterized by a below normal circulating alkaline phosphatase activity, increased urinary excretion of phosphoethanolamine, and inorganic pyrophosphates [92]. Pyrophosphates are necessary to inhibit the formation of calcium-phosphate crystals, leading to the defect in bone mineralization seen in patients suffering from hypophosphatasia [93].

Hypophosphatasia can manifest itself at various ages, and is currently recognized in five different clinical forms: perinatal (which is usually lethal), infantile, childhood, adult, and odontohypophosphatasia [91]. While perinatal and infantile forms are considered to be autosomal recessive traits, childhood and adult forms of hypophosphatasia are usually found to be autosomal recessive, but can occasionally be dominant [92]. TNSALP is vital for proper skeletal mineralization and hypophosphatasia is the result of a mutation in the TNSALP gene, manifesting itself as rickets in children and osteomalacia in adults [91, 93]. The disease manifests itself in various ways, with symptoms including anything from stillbirths to loss of teeth at an adult age. Shorter limbs and deformity, along with hypomineralization are all results of the mutation in the TNSALP gene [91].

11.5.4.3 Hyperphosphatemia

Hyperphosphatemia is a disease of high serum phosphate concentration that can occur because of increased intake or increased absorption in the gastrointestinal tract, increased production of phosphates or decreased loss such as a decrease in urinary excretion, and, lastly, there is pseudohyperphosphatemia [74].

An increase in dietary phosphate intake, overwhelming renal excretory capacity, can result from administration of too much phosphate or by overdose of vitamin D. Premature babies who are fed cows milk can develop neonatal or hypocalcemic tetany due to hyperphosphatemia. This can result in muscle twitching, jitters, tremors, and convulsions [74].

An increase in plasma phosphate from endogenous sources is observed in tumour-lysis syndrome (TLS), rhabdomyolysis, bowel infarction, malignant hyperthermia (see Section 11.5.5 calcium), and severe hemolysis. TLS is a group of metabolic complications usually caused by chemotherapy treatments for lymphomas and leukemias. It is characterized by electrolyte abnormalities such as hyperkalemia, hyperphosphatemia, hyperuricemia, and hypercalcemia leading to acute renal failure; hyperphosphatemia is the most common of the aforementioned electrolytic disturbances caused by TLS [94]. Rhabdomyolysis is a disease that, due to injury to muscle tissue, causes the rapid breakdown of skeletal muscle. This damage can be caused by physical injury or chemical or biological factors, resulting in hyperkalemia, hyperphosphatemia, hyperuricemia, and hypermagnesemia [95]. Phosphorus, and other electrolytes, are released from damaged muscle and accumulate, causing hyperphosphatemia [96]. Malignant hyperthermia can be a fatal genetic metabolic disorder of skeletal muscle, which can trigger rhabdomyolysis because of sustained muscle rigidity, leading to hyperphosphatemia [97]. Metabolic acidosis or other acid–base disorders (such as lactic acidosis, diabetic ketoacidosis, and respirator acidosis) release phosphorus from endogenous stores, resulting in hyperphosphatemia [74].

Reduced loss of phosphates can be a result of renal failure, hypoparathyroidism, acromegaly, tumoral calcinosis, bisphosphonate therapy, magnesium deficiency, or multiple myeloma [74]. Renal failure results in phosphate retention, which leads to the development of hyperphosphatemia, hypocalcemia, and increased PTH levels [98]. Hypoparathyroidism is caused by decreased activity of the parathyroid hormone, which leads to hypocalcemia (see Section 11.5.5 calcium) and hyperphosphatemia due to increased levels of serum phosphorus. Tumoral calcinosis is a rare condition that results in abnormal calcifications around joints, hyperphosphatemia, but normal PTH levels. It is thought that hyperphosphatemia in tumoral calcinosis is the result of increased tubular reabsorption of phosphate, despite normal PTH levels. The elevated serum calcium-phosphate solubility product can lead to the deposition of calcium-phosphate salts at the joints [99]. Bisphosphonates, in high dosages, can cause hyperphosphatemia by enhancing renal tubular reabsorption of phosphorus [100].

Untreated hyperphosphatemia can lead to secondary hyperparathyroidism, renal osteodystrophy, increased deposition of calcium-phosphate complexes in soft tissues, inhibition of 1-α-hydroxylase, which forms the active form of vitamin D, calcitriol, and mortality [95, 101]. Treatment is usually carried out by decreased protein intake or, especially in the case of kidney disease, by phosphate binders such as aluminum, magnesium, or calcium [101, 102]. More recently the salt lanthanum carbonate has been used to bind excess serum phosphate, with decreased side effects [101].

11.5.5
Calcium

Calcium is vital to many biological functions, having a key structural component in skeleton and teeth. It plays a vital role in skeletal and myocardial muscle function, and in neurotransmission. Approximately 90% of calcium is found in hydroxyapatite in bones and teeth. The remaining 10% is found in the extracellular fluid and the cytoplasm, where half of it is free, ionized calcium, 40% is bound to albumin and globulin, and the other 10% is bound to bicarbonate, phosphate, and citrate [103].

To fully understand calcium disorders, it is important to understand both the role of the parathyroid and vitamin D. They are discussed below, followed by disorders of calcium metabolism: hypocalcemia, hypercalcemia, disorders of the parathyroid hormone, disorders of vitamin D synthesis, calcifications, osteomalacia, and milk-alkali syndrome.

11.5.5.1 Vitamin D

Vitamin D has two main forms, vitamin D_2 (ergocalciferol) and vitamin D_3 (cholecalciferol). Vitamin D_2 is synthesized by plants, while D_3 is synthesized in humans. Vitamin D_3 is synthesized in the skin by direct exposure to UV B radiation from the sun or is obtained directly from the diet. The use of sunscreen, darker skin, or lack of exposure to sunlight can lead to vitamin D deficiency [104].

Vitamin D is metabolized by the liver to 25-hydroxyvitamin D_3 [25(OH)D], which is then metabolized by the kidneys to 1,25-dihydroxyvitamin D_3 [1,25 $(OH)_2D$], also known as calcitriol, the active form of vitamin D. The production of calcitriol is stimulated by the PTH, and is regulated by the serum concentrations of calcitriol, parathyroid hormone, calcium, and phosphate [104].

Vitamin D helps the intestinal absorption of calcium and phosphate. Calcium enters the cell through membrane proteins. In the intestinal cell, calcitriol enters, binds to the vitamin D receptor, and the calcium binding protein is synthesized, regulating the active transport of calcium through the cell [104]. There is negative feedback through calcium that decreases PTH and direct negative feedback from calcitriol to PTH.

11.5.5.2 Parathyroid Hormone

The expression of the parathyroid hormone (PTH) gene that synthesizes the parathyroid hormone is regulated by several factors, most importantly the concentration of serum calcium, but vitamin D and serum phosphate levels also regulate PTH synthesis. The active form of vitamin D_3, calcitriol, decreases the transcription of the PTH gene [105]. When there is a decrease in ionized calcium levels in the blood, there is an increase in PTH. The PTH receptor is expressed by osteoblasts (which build bone), osteocytes, and bone-lining cells, but not osteoclasts (which destroy bone). A decrease in extracellular calcium leads to a rapid increase in PTH secretion, whereas an increase in calcium concentration inhibits the secretion of PTH. The surface of the calcium-sensing receptor of the PT cell contains a large extracellular domain for the binding of cations [106].

The primary function of PTH is maintenance of plasma calcium levels. This occurs by withdrawing the calcium from bone tissue, reabsorbing it from the glomerular filtrate, and indirectly increasing its intestinal absorption by stimulating calcitriol production. Additionally, PTH promotes an increase in urinary excretion of phosphorus and bicarbonate, seeking a larger quantity of free calcium available in circulation [107].

11.5.5.3 Hypocalcemia

Hypocalcemia is a metabolic disorder occurring when plasma calcium concentration is <2.10 mM. It can be a result of a multitude of diseases, summarized in Table 11.6. It is almost always caused by an impairment of the parathyroid hormone or impairment of vitamin D action or synthesis; however, it can be caused by an increase in tissue sequestration of calcium, which is observed in pancreatitis, burns, or toxic shock. Osteomalacia or rickets may also be a cause of hypocalcemia.

Factitious hypocalcemia is usually caused by blood sample contamination with EDTA. Decreased parathyroid activity resulting in hypoparathyroidism and pseudohypoparathyroidism are discussed later in this section. Malabsorption of vitamin D or impairment of vitamin D synthesis are discussed below. Hypocalcemia is characterized by tetany, laryngospa, asphyxia, cramps, altered mental status, seizures, muscle spasms, hypotension, alopecia, and coarse dry skin [60].

11.5.5.4 Hypercalcemia

Hypercalcemia is defined as a total plasma calcium level greater than 2.55 mM [60]. Usually, high levels of serum calcium indicate primary hyperparathyroidism or malignancy. Other causes of hypercalcemia include vitamin D intoxication, tuberculosis, some fungal infections, Addison's disease, milk-alkali syndrome, vitamin A intoxication, thiazide diuretics, familial hypocalciuric hypercalcemia, prolonged immobilization in patients with high skeletal turnover, and recovery from acute renal failure (especially associated with rhabdomyolysis). They all amount to less than 10% of all causes of hypercalcemia [108].

Factitious hypercalcemia and hypocalcemia (or pseudohypercalcemia and pseudohypocalcemia, respectively) are quite common. Calcium in serum is either bound to proteins, principally albumin, or is free (ionized) calcium. When measuring the total plasma or serum concentration of calcium, it is the ionized calcium levels that are important. With patients suffering from chronic illness or malnutrition, serum albumin may be low (hypoalbuminemia), causing a reduction in the total, but not in the ionized, serum calcium. This is referred to as "factitious" hypocalcemia. In contrast, hyperalbuminemia, caused by volume depletion or multiple myeloma, leads to increased protein binding of calcium, which results in an elevated serum total calcium concentration without any rise in the serum ionized calcium concentration. This is referred to as "factitious" hypercalcemia [60].

Symptoms of hypercalcemia are usually not observed until >3 mM plasma concentration. Symptoms include nausea, vomiting, constipation, muscular weakness, lethargy, myalgia, hypotonia, and ectopic calcification. Mental disturbances such as psychosis, apathy, confusion, depression, and rarely coma are also observed. Hypercalcemia may also predispose the patient to peptic ulcer and pancreatitis [60].

11.5.5.5 Diseases Related to the Parathyroid Hormone

Hypoparathyroidism Hypoparathyroidism is caused by a decrease in the function of the parathyroid glands, which leads to a decrease in the level of parathyroid hormone, which can lead to hypocalcemia. Two mechanisms may alter the function of the parathyroid hormone, limiting its control on calcium: (i) insufficient PTH production by the parathyroids, which is called hypoparathyroidism, and (ii) a resistance against its action in target tissues, which is called pseudohypoparathyroidism, which is discussed below. Both cases result in significantly reduced levels of plasma calcium associated with hyperphosphatemia [107].

Causes of low PTH include destruction of the parathyroid by surgery, autoimmune disorders, metastatic disease, reduction in parathyroid function caused by hypomagnesemia, PTH gene defects and calcium sensing receptor mutations or parathyroid agenesis caused by DiGeorge syndrome, isolated x-linked hypoparathyroidism, and mitochondrial neuropathies [107]. Symptoms of hypoparathyroidism are the same as those of hypocalcemia. Laboratory measurements present as hypocalcemia, hyperphosphatemia, and low PTH levels. Generally in hypoparathyroidism calcitriol levels are low as well. Hypomagnesemia may cause hypocalcemia because it promotes PTH secretion levels to drop, in addition to causing renal and bone resistance to PTH action.

Pseudohypoparathyroidism Pseudohypoparathyroidism (PHP) is a group of abnormalities characterized by clinical indications of hypoparathyroidism (hypocalcemia, hyperphosphatemia), but with the presence of high levels of PTH in the absence of chronic renal failure or magnesium deficiency [107]. In 1942, Albright *et al.* investigated a patient with seizures, hypocalcemia, and hyperphosphatemia, but that did not respond to parathyroid extract [109]. It was concluded that the patient was resistant to PTH and the disease was called pseudohypoparathyroidism or PHP. Pseudohypoparathyroidism manifests as short stocky build, round face, and cutaneous ossification. This physical appearance is known as Albright's hereditary osteodystrophy (AHO). A decade later, Albright *et al.* described a patient with typical features of AHO, but manifested with normal calcium and phosphate serum levels and the absence of PTH resistance. This disease was called pseudopseudohypoparathyroidism (PPHP) [109].

PTH exerts itself by stimulating the intracellular formation of cAMP. The signal-transducing protein known as G_s (a GTP binding protein), consisting of α, β, and γ subunits, helps to stimulate cAMP formation by coupling to receptors for hormones, such as PTH. The $G_{s\alpha}$ subunit activity in patients with PHP has been shown to be reduced significantly ($\sim 50\%$). Thus, patients suffering from AHO show resistance to other hormones, such as TSH (thyroid-stimulating hormone). Surprisingly, individuals suffering from PPHP do not lack the activity of the $G_{s\alpha}$ protein, despite the lack of PTH resistance [109, 110]. The gene presents with parental imprinting phenomenon, which explains the phenotypic variations of the disease, depending on whether the mutations origins are maternal or paternal [107]. The gene is present as a single copy per haploid, and heterozygous loss of function is present in both PHP and PPHP patients [109]. Patients who manifest with PHP were found to result exclusively from maternal transmission and patients with PPHP from paternal [107, 109].

11.5.5.6 Hyperparathyroidism (HPT)

Primary Hyperparathyroidism Primary hyperparathyroidism is a disease caused by excessive secretion of the parathyroid hormone, manifesting as hypercalcemia. Adenoma, hyperplasia, and carcinoma have all been attributed to primary HPT [111]. When extracellular calcium binds to the calcium receptor in the parathyroid cell, PTH secretion is inhibited. At the kidneys this interaction between calcium and the calcium receptor inhibits the 1-hydroxy-vitamin D [112].

In primary hyperparathyroidism mutations of the calcium receptor gene have not been identified. However, significant reductions in calcium receptor mRNA levels have been detected in parathyroid adenomas, a benign glandular tumor. The pathophysiological significance of this observation is unclear because the reductions could be secondary to the chronic hypercalcemia and not the primary case [112].

Primary HPT is associated with a reduction in bone mineral density, generally a silent symptom [112]. Symptoms are generally those of hypercalcemia. Symptomatic patients with primary HPT presented with nephrolithiasis, osteitis fibrosa cystical, muscle atrophy, hyperreflexia, gait abnormalities, and other neuromuscular signs and present with hypercalcemia and its associated symptoms. When primary HPT occurs in the absence of hypercalcemia, patients tend to be asymptomatic. Nephrolithiasis (formation of kidney stones, calcification) is the major clinical manifestation of primary HPT [111].

For patients whose serum calcium levels are >3 mM, the recommended course of treatment is a parathyroidectomy. Hormone replacement therapy has seen some success. Bisphosphonates have also seen some use, as they decrease serum calcium levels and decrease bone loss, but do not affect PTH levels [111].

Secondary Hyperparathyroidism   Secondary hyperparathyroidism is caused by an increase in PTH because of hypocalcemia, without disturbances in calcitriol, calcium, or phosphorus metabolism. Secondary hyperparathyroidism is usually caused by chronic kidney disease (CKD). PTH levels begin to rise in the course of CKD when normal serum levels of phosphorus, calcitriol, and calcium are observed. A decrease in calcitriol leads to a decrease in calcium absorption, which leads to a decrease in ionized calcium. This leads to an increase in PTH (because of decreased ionized Ca levels), which normalizes calcitriol and restores serum calcium values. As loss of kidney function progresses, phosphorus is retained and calcitriol production decreases [111, 113].

Symptoms include many of those seen in primary HPT; arthritis, bone pain, myopathy, tendon rupture, extraskeletal calcifications, and low bone turnover. Secondary HPT is easier to prevent than to treat. Vitamin D replacement and phosphorus reduction, from phosphate binders, are the mainstays of prevention [111].

11.5.5.7   Osteomalacia

Osteomalacia is a clinical syndrome of vitamin D deficiency, characterized by major deficits in bone mineralization and by relatively minor changes in calcium and phosphate homeostasis in the blood, resulting in proximal muscle weakness. In children this is called rickets, in which structural changes related to defective bone mineralization are added to the features of osteomalacia that are seen in the adult manifestation of this disease [114].

Osteomalacia due to vitamin D deficiency is rare in North America because of the routine fortification of milk and other dairy products with vitamin D. Vitamin D depletion in adults leads to osteomalacia, the characteristic histologic feature of which is defective mineralization of osteoid, leading to its accumulation on bone surfaces. Clinically, osteomalacia is manifested by progressive generalized bone pain, muscle weakness, hypercalcemia, pseudofractures, and, in its late stages, by a waddling gait. Patients with various GI disorders are at risk [115]. Treatment involves calcium supplements and vitamin D supplements.

11.5.5.8   Milk-Alkali Syndrome

First identified in 1923, this is a disease that historically was caused by the use of dairy products and alkaline powders, such as milk and bicarbonate used to treat peptic ulcers [116]. In 1985, with the introduction of proton pump inhibitors and antibiotic therapies of *Helicobacter pylori* for the treatment of these ulcers, milk-alkali syndrome became rare. There has been an increase in incidence of milk-alkali syndrome as a result of malignancy, primary hyperparathyroidism, and over-the-counter calcium supplements used to combat osteoporosis [117]. Milk-alkali syndrome is caused by a triad of hypercalcemia, metabolic alkalosis, and renal insufficiency.

11.5.5.9   Calcifying Disorders

Calcification   Serum calcium is tightly regulated by parathyroid hormone and calcitriol. Despite this careful regulation, calcification and ossification of cutaneous and subcutaneous tissues may occur. Cutaneous calcification can be divided into four major categories: dystrophic, metastatic, idiopathic, and iatrogenic. Dystrophic calcification is the result of local tissue injuries and abnormalities. Metastatic calcification is caused by abnormal calcium and phosphate metabolism. Calcification generally results in secondary ossification of tissues, but primary ossification is rare. Idiopathic calcification occurs without identifiable underlying tissue abnormalities or abnormal calcium or phosphate metabolism. Iatrogenic calcification occurs as a result of intravenous infusion of calcium chloride or calcium gluconate [103, 118].

Calcinosis Calciphylaxis causes vascular calcification and skin necrosis that occurs as a result of end stage renal disease (ENRD). It also presents itself in individuals with chronic renal failure, hypercalcemia, hyperphosphatemia, and secondary hyperparathyroidism. It is associated with poor phosphate and calcium control [7, 119].

11.6
Metabolism of Other Trace Elements

11.6.1
Questionable Essentiality of Nickel, Molybdenum, and Vanadium

If an element is to be considered essential, then it must have a specific role, a specific enzyme or cofactor, and being deficient of this element must result in a disease that can be reversed by its supply. There has been much debate on the essentiality of nickel, molybdenum, and vanadium for human nutrition. All these elements have demonstrated roles in animal models, and there is strong evidence that they have physiological effects in humans, but are they necessary for wellbeing? Can we live without these metals? Briefly, the necessity of these metals is discussed in this section. For a more detailed overview, the reader is asked to refer to the detailed articles referenced within this section.

11.6.1.1   Nickel
The essentiality of nickel in humans is questionable, because its metabolism is not well understood. Like other trace metals, deficiency is not observed, and it is impossible to achieve induced deficiency due to the abundance of trace amounts of nickel in the diet. No enzymes or cofactors that require nickel are known in humans or other higher organisms. The most consistent result obtained from studies on nickel-depleted animals is that the concentration of other metals such as iron, zinc, and copper are decreased, indicating some physiological function of nickel [120, 121]. Nickel has been shown to be essential for several different animals, including chickens, rats, and cows, with depletion affecting these animals' gestation periods, number of offspring, activity of some enzymes and causing anemia. The essentiality of nickel for bacteria and plant species has been demonstrated with the identification of almost a dozen nickel-dependent enzymes [122].

Exposure to high levels of nickel, occurring by ingestion, inhalation, or, less commonly, by contact dermatitis, can have adverse side effects on human health. Accumulation in the body from chronic nickel exposure results in cardiovascular, lung, and kidney diseases. Nickel is also considered to be a carcinogen, with industry workers having a higher occurrence of nasal and lung cancers from inhalation of nickel [120]. The mutagenic effects caused by nickel exposure are likely due to nickel-induced oxidative damage, depletion of glutathione, and the activation of other transcription factors that are sensitive to oxidative stress.

11.6.1.2 Molybdenum

It has been well established that molybdenum is essential for plant growth but its essentiality in humans has not been well established. There has been no documented molybdenum deficiency reported, and concrete results from animal studies fed molybdenum-deficient diets have been unsuccessful because it is impossible to eliminate low levels of molybdenum [123].

Molybdenum enzymes play an important role in catalyzing redox reactions that characteristically involve oxygen transfer [124]. In mammals sulfite oxidase catalyzes the terminal step in metabolism of sulfur-containing amino acids. Aldehyde oxidases, whose function has not been fully elucidated, are also found in mammals, as are xanthine oxidoreductases, which are involved in purine ring metabolism [123–125]. Molybdenum deficiency is not reported in healthy humans but deficiency has been observed in patients undergoing prolonged total parental nutrition (TPN) [125].

With the exception of nitrogenase, the molybdenum cofactor (MoCo) is the active site in all molybdenum-containing enzymes [124]. Nitrogenase, on the other hand, contains an Fe-Mo cofactor cluster as the active site [126]. The MoCo active site contains a molybdenum covalently bound to one or two dithiolates on a tricyclic pterin. Defects in the synthesis of the molybdenum cofactor results in a loss of the activity of all molybdenum enzymes. Human molybdenum cofactor deficiency, not to be confused with molybdenum deficiency, is a hereditary metabolic disorder characterized by neurodegeneration, resulting in death during early childhood [124].

Individuals suffering from molybdenum cofactor deficiency only exhibit sulfite oxidase deficiency. This is attributed to the xanthine oxidase substrate remaining intact, and thus being reused. Aldehyde oxidase deficiency has not been reported in these individuals. It is believed that the neurodegenerative effects seen in molybdenum cofactor deficiency are a result of a decrease of sulfate compounds (formed from sulfite) in the brain by the lack of sulfite oxidase activity [125].

Molybdenum is relatively nontoxic: amounts greater than 100 mg kg^{-1} of the metal need to be ingested for toxic effects to be observed. Signs include failure of red blood cells to mature, and high levels of uric acid in the blood. An increase in molybdenum ingestion affects copper levels as it is an antagonist to copper [125].

11.6.1.3 Vanadium

In the 1970s it was believed that vanadium was an essential nutrient; however, more recent research indicates that this is simply due to its pharmacological effect in the body. Vanadium deficiency has not been successfully investigated in humans; however, there are suggestions from animal deprived diets that vanadium plays a role in phosphoryl transfer enzymes, regulation of Na/K-ATPase, adenylate cyclase, and protein kinases [127]. Pharmacologically, vanadium compounds have demonstrated insulin-mimetic activity and anticarcinogenic effects in humans [128]. There has also been evidence that vanadate (V^{5+}) compounds can affect proteins, lipid structures, and cellular functions, demonstrating *in vivo* effects on oxidative stress and other biological processes [129].

11.6.2
Chromium

Chromium in biological systems has been studied since the nineteenth century when carcinogenic effects of hexavalent chromium were first discovered [130]. There is still some debate over whether chromium, in its trivalent form, is an essential trace metal in humans, as evidenced in the many reviews [130–133]. As the American Society of Clinical Nutrition recommends 20 μg-Cr day^{-1}, we will consider it an essential metal and leave it up to the reader to follow the debate [131].

Chromium can exist in all oxidation states from -2 to $+6$, but is generally found in biological systems in its $+3$ and $+6$ oxidation states. Divalent chromium is readily oxidized to trivalent chromium under aerobic conditions, and thus is not available to biological systems. Hexavalent chromium is a strong oxidizing agent, generally found as chromate (CrO_4^{2-}) or dichromate ($Cr_2O_7^{2-}$). This form is extremely bioavailable, crossing cell membranes, reacting with protein and nucleic acids, and exhibiting carcinogenic effects. Trivalent chromium is the most stable form found in living organisms, but is not as bioavailable as its hexavalent counterpart, with less than 3% typically crossing cell membranes [132].

Chromium has been deemed an essential nutrient by experimental observation on induced deficiency in laboratory animals and clinical observations in humans [130]. It was first observed that chromium was necessary to maintain normal glucose tolerance in 1959 in rats [134]. Induced chromium deficiency has been demonstrated in patients, with a reversal in this deficiency in patients supplemented with TPN (total parental nutrition), a diet enhanced with chromium. Improved glucose tolerance has been observed in those who are supplemented with chromium, as further evidence of chromium's essentiality [135].

The most commonly observed signs of deficiency are glucose tolerance impairment, glycosuria, and elevations in insulin serum. In animal models, induced chromium deficiency has resulted in a decrease in longevity, impaired growth and immunity, and decrease in reproductive function, amongst other observations. Chromium has an effect on glucose metabolism in humans; however, it is no longer believed to be because of a substance called the glucose tolerance factor (GTF), as GTF activity has now been found to be independent of chromium content. It is now somewhat accepted that chromodulin or LMWCr (low-molecular-weight chromium binding substance) in animals is essential for insulin signalling; however, chromodulin has only been isolated in mammals, but not in humans, and is poorly characterized [131, 132].

11.6.3
Cobalt and Vitamin B$_{12}$

Cobalt is an essential metal for many proteins and is the central component of vitamin B$_{12}$ (cobalamin). Most cobalt proteins contain a corrin ring, but recent studies have shown that there are a handful of noncorrin cobalt-containing enzymes, including methionine aminopeptidase, prolidase, and nitrile hydratase. Since little is known about their function, only cobalamin will be discussed in this chapter [136].

The corrin ring of cobalamin is almost identical to the heme of hemoglobin, differing only by one deleted alpha methane bridge and by the metal (cobalt instead of iron). Cyanocobalamin, a common inactive precursor form of the vitamin, contains cyanide as an axial ligand, which stabilizes the vitamin. It is the common form used in supplementation, with the cyanide removed in the body, yielding an active form of the vitamin, such as methylcobalamin. There have been rare reported cases of vitamin B$_{12}$ deficiency caused by an inability to metabolize cyanocobalamin into its active form [137].

Since humans cannot produce vitamin B$_{12}$ it must be obtained from the diet. The average American ingests 5–15 µg of vitamin B$_{12}$ per day; less than 1 µg is necessary for normal function [137]. Vitamin B$_{12}$ deficiency can result from insufficient dietary intake, defects in metabolism of vitamin B$_{12}$, or pernicious anemia. Vitamin B$_{12}$ deficiency is of particular interest to those who have a vegetarian diet, because B$_{12}$ is not found in plant-based foods. Cobalamin deficiency results in megoblastic anemia, whose symptoms are indistinguishable from a folic acid deficiency. The symptoms include anemia, decreased white blood cell count, hypercellular bone marrow, and abnormal maturation. Cobalamin deficiency can also result in neuropsychiatric disturbances due to demyelinization of nerves and the spinal cord [138].

Errors in cobalamin metabolism, including malabsorption of cobalamin, intrinsic factor deficiency, and transcobalamin deficiency, result in cobalamin deficiency. Absorption of vitamin B$_{12}$ occurs by the binding of intrinsic factor-B$_{12}$ complex to the ileum receptors, releasing vitamin B$_{12}$, which then attaches to transcobalamins. A deficiency in the intrinsic factor of transcobalamin can thus manifest as vitamin B$_{12}$ deficiency [139].

Pernicious anemia is the most common cause of vitamin B$_{12}$ deficiency, affecting an estimated 1.9% of the population over the age of 60. The term pernicious refers to a condition associated with chronic atrophic gastritis, which is caused by an autoantibody of the intrinsic factor. This results in a decrease in the absorption of vitamin B$_{12}$, resulting in megoblastic anemia. Although the disease is asymptomless until end stage, the gastric lesions that cause it can be predicted before the onset of anemia [140].

11.6.4
Manganese

The function of manganese is not well understood, despite it being essential; no case of manganese deficiency in humans has ever been observed. Animal studies indicate that manganese is an essential cofactor in enzymes such as hexokinase, superoxide dismutase, and xanthine oxidase. No cases of manganese deficiency have been reported; however, those fed a manganese-deficient diet developed fine, scaly skin, and had an increase in serum calcium, phosphorus, and alkaline phosphatase levels. Deficiency of manganese in mammals manifests as severe skeletal and reproductive abnormalities [141]. Manganese is involved in metabolism of proteins, lipid, and carbohydrates; however, the action of manganese in enzymes is not manganese-specific, because Mn^{2+} resembles Mg^{2+} and can substitute in enzymes *in vivo* [141]. Absorption of manganese occurs in its divalent and tetravalent forms, with an intake of 2–9 mg day^{-1}. Some manganese can be absorbed by the gastrointestinal tract and the lungs, but this accounts for a very minute amount, and the source is usually not identifiable. Elimination of manganese occurs by excretion into the bile [142].

Absorption of manganese from the gastrointestinal tract varies inversely with the amount of calcium present. It is also affected by serum iron levels, because both iron and manganese have the same transport system. The β-globulin protein, transmanganin, transports part of the manganese in its trivalent form to tissue stores, mainly in the liver and muscle. Additionally, some transport of the divalent forms occurs through α-macroglobulin. Experiments in animals have indicated that excretion from the brain is slower than from the rest of the body, which helps to explain the neurological effects observed in manganese overload [141].

Manganese toxicity or manganism is a central nervous system disease first observed in the 1800s due to exposure to high concentrations of manganese oxides, manifesting as Parkinson's-like symptoms. The effects of low levels of manganese from industry and environmental sources are not well known as there is no available biological indicator for manganese exposure, but are currently under investigation. Characteristic symptoms of manganese toxicity are neurological and include aberrant behavior and hallucination, later manifesting as Parkinson's-like disease [143].

11.6.5
Selenium

Selenium is essential for cell homeostasis and apoptosis, and in higher doses has been shown to exhibit anticarcinogenic effects [144–146]. If ingested in toxic amounts, selenium can lead to "blind staggers" and "alkali disease" [147]. The daily requirement for selenium in adults, 55 μg day^{-1}, is met by most Americans; however, inadequate intake is seen European, Asian, and some African populations. Selenium deficiency is associated in China with Keshan disease, a type of juvenile cardiomyopathy, in populations with intake of less than 25 μg day^{-1} [144].

Selenium is in the same group of the periodic table as sulfur and can exchange for sulfur, forming the selenium analogs selenomethionine (Se-meth) and selenocysteine (Se-cyst). In foods, organic forms of selenium are the predominant form found, containing selenium in its reduced form, selenide (Se^{2-}). Selenite (SeO_3^{2-}) and selenate (SeO_4^{2-}) are the inorganic forms of selenium predominately found in dietary supplements. When inorganic selenium is ingested it is reduced to selenide by glutathione and NADPH. Selenomethionine is generally metabolized to selenocysteine, and incorporated into selenoproteins; however, selenomethionine is indistinguishable from methionine and thus can be incorporated in proteins [145]. Some of the selenoproteins such as glutathione peroxidases and thioredoxin reductases are important for detoxification and antioxidant activities. Selenium also plays a key role in cell cycle apoptosis, which is essential to maintain tissue homeostasis, but the mechanism is not well understood in humans [144].

11.7
Conclusions

As demonstrated in this chapter, disorders of the metabolism of essential metal ions can have detrimental effects. Thus, it is of the utmost importance that we fully elucidate the mechanisms by which we uptake and excrete these metals, and the effects, both biological and medical, of these essential metals. This is especially true for the trace metal ions whose essentially is still unconfirmed, as diseases associated with these metals can be left undiagnosed, leading to adverse health effects.

References

1 Irnius, A., Speiciene, D., Tautkus, S., and Kareiva, A. (2007) *Mendeleev Commun.*, **17**, 216–217.

2 King, J.C., Shames, D.M., and Woodhouse, L.R. (2000) *J. Nutr.*, **130**, 1360S–1366S.

3 Andrews, N.C. (1999) *New Engl. J. Med.*, **341**, 1986–1995.

4 Beard, J.L., Dawson, H., and Piñero, D. J. (1996) *Nutr. Rev.*, **54**, 295–317.

5 Franchini, M. (2006) *Am. J. Hematol.*, **81**, 202–209.

6 Harris, Z.L., Klomp, L.W.J., and Gitlin, J.D. (1998) *Am. J. Clin. Nutr.*, **67**, 972S–977S.

7 Delatycki, M.B., Williamson, R., and Forrest, S.M. (2000) *J. Med. Genet.*, **37**, 1–8.

8 Swaiman, K.F. (1991) *Arch. Neurol.*, **48**, 1285–1293.

9 Gordeuk, V.R., Caleffi, A., Corradini, E., Ferrara, F., Jones, R.A., Castro, O., Onyekwere, O., Kittles, R., Pignatti, E., Montosi, G., Garuti, C., Gangaidzo, I.

T., Gomo, Z.A.R., Moyo, V.M., Rouault, T.A., MacPhail, P., and Pietrangelo, A. (2003) *Blood Cells Mol. Dis.*, **31**, 299–304.

10 Knisely, A.S., Mieli-Vergani, G., and Whitington, P.F. (2003) *Gastroenterol. Clin. North Am.*, **32**, 877–889.

11 Harris, Z.L., Takahshi, Y., Miyajima, H., Serizawa, M., MacGillivray, R.T.A., and Gitlin, J.D. (1995) *Proc. Natl. Acad. Sci. USA*, **92**, 2539–2543.

12 Taylor, T.D., Litt, M., Kramer, P., Pandolfo, M., Angelini, L., Nardocci, N., Davis, S., Pindea, M., Hattori, H., Flett, P.J., Cilio, M.R., Bertini, E., and Hayflick, S.J. (1996) *Nat. Genet.*, **14**, 479–481.

13 Rötig, A., Lonlay, P.D., Chretien, D., Foury, F., Koenig, M., Sidi, D., Munnich, A., and Rustin, P. (1997) *Nat. Genet.*, **17**, 215–217.

14 Kushner, J.P., Porter, J.P., and Olivieri, N.F. (2001) *Hematol. Am. Soc. Hematol. Educ. Program*, **1**, 47–61.

15 Piperno, A. (1998) *Haematologica*, **83**, 447–455.

16 Turcot, I., Stintzi, A., Xu, J., and Raymond, K.N. (2000) *J. Biol. Inorg. Chem.*, **5**, 634–641.

17 Hider, R.C. and Zhou, T. (2005) *Ann. N. Y. Acad. Sci.*, **1054**, 141–154 (Cooley's Anemia: Eighth Symposium).

18 Neufeld, E.J. (2006) *Blood*, **107**, 3436–3441.

19 Hasinoff, B.B., Patel, D., and Wu, X. (2003) *Free Radic. Biol. Med.*, **35**, 1469–1479.

20 Looker, A.C., Dallman, P.R., Carroll, M.D., Gunter, E.W., and Johnson, C.L. (1997) *JAMA*, **277**, 973–976.

21 Brady, P.G. (2007) *South. Med. J.*, **100**, 966–967.

22 Allen, L.H. (2000) *Am. J. Clin. Nutr.*, **71** (Suppl.), 1280S–1284S.

23 Pollitt, E. (1993) *Annu. Rev. Nutr.*, **13**, 521–537.

24 Madsen, E. and Gitlin, J.D. (2007) *Annu. Rev. Neurosci.*, **30**, 317–337.

25 deBie, P., Muller, P., Wijmenga, C., and Klomp, L.W.J. (2009) *J. Med. Genet.*, **44**, 673–688.

26 Cater, M.A. and Mercer, J.F.B. (2006) in *Topics in Current Genetics* (eds M. Tamás and E. Martinoia), Springer, Berlin/Heidelberg, pp. 101–129.

27 Shim, H. and Harris, Z.L. (2003) *J. Nutr.*, **133**, 1527S–1531S.

28 Prohaska, J.R. (2008) *Am. J. Clin. Nutr.*, **88**, 826S–829S.

29 Horn, N. and Tumer, Z. (1999) *J. Trace Elem. Exp. Med.*, **12**, 297–313.

30 Zatta, P. and Frank, A. (2007) *Brain Res. Rev.*, **54**, 19–33.

31 Madsen, E. and Gitlin, J.D. (2007) *Curr. Opin. Gastroenterol.*, **23**, 187–192.

32 Strausak, D., Mercer, J.F.B., Dieter, H. H., Stremmel, W., and Multhaup, G. (2001) *Brain Res. Bull.*, **55**, 175–185.

33 Llanos, R.M. and Mercer, J.F.B. (2002) *DNA Cell Biol.*, **21**, 259–270.

34 Smith-Mungo, L.I. and Kagan, H.M. (1998) *Matrix Biol.*, **16**, 387–398.

35 Roberts, E.A. and Sarkar, B. (2008) *Am. J. Clin. Nutr.*, **88**, 851S–854S.

36 Schilsky, M.L. (2001) *Curr. Gastroenterol. Rep.*, **3**, 54–59.

37 Hoogenraad, T.U. (2006) *Brain Dev.*, **28**, 141–146.

38 Hosny, N.M. and Sherif, Y.E. (2009) *Phosphorus Sulfur Silicon Relat. Elem.*, **184**, 2786–2798.

39 Sugiura, Y. and Tanaka, H. (1972) *Mol. Pharmacol.*, **8**, 249–255.

40 George, G.N., Pickering, I.J., Harris, H. H., Gailer, J., Klein, D., Lichtmannegger, J., and Summer, K.-H. (2003) *J. Am. Chem. Soc.*, **125**, 1704–1705.

41 Barceloux, D.G. (1999) *Clin. Toxicol.*, **37**, 279–292.

42 Sekler, I., Sensi, S.L., Hershfinkel, M., and Silverman, W.F. (2007) *Mol. Med.*, **13**, 337–343.

43 Palmiter, R.D. (2004) *Proc. Natl. Acad. Sci. USA*, **101**, 4918–4923.

44 Fosmire, G.J. (1990) *Am. J. Clin. Nutr.*, **51**, 225–227.

45 Kaye, P., Young, H., and O'Sullivan, I. (2002) *Emerg. Med.*, **19**, 268–269.

46 Cai, L., Li, X.-K., Song, Y., and Cherian, M.G. (2005) *Curr. Med. Chem.*, **12**, 2753–2763.

47 Shmist, Y.A., Kamburg, R., Ophir, G., Kozak, A., Shneyvays, V., Appelbaum, Y.J., and Shainberg, A. (2005) *J. Pharmacol. Exp. Ther.*, **313**, 1046–1057.

48 Barnett, B.L., Kretschmar, H.C., and Hartman, F.A. (1977) *Inorg. Chem.*, **16**, 1834–1838.

49 Blindauer, C.A., Razi, M.T., Parsons, S., and Sadler, P.J. (2006) *Polyhedron*, **25**, 513–520.

50 Hambidge, M. (2000) *J. Nutr.*, **130**, 1344S–1349S.

51 Hendricks, K.M. and Walker, W.A. (1988) *Nutr. Rev.*, **46**, 401–408.

52 Rink, L. and Haase, H. (2006) *Trends Immunol.*, **28**, 1–4.

53 Kury, S., Dreno, B., Bezieau, S., Giraudet, S., Kharfi, M., Kamoun, R., and Moisan, J.-P. (2002) *Nat. Genet.*, **31**, 239–240.

54 Wang, K., Pugh, E.W., Griffen, S., Doheny, K.F., Mostafa, W.Z., al-Abossi, M.M., el-Shanti, H., and Gitschier, J. (2001) *Am. J. Hum. Genet.*, **68**, 1055–1060.

55 Michalczyk, A., Varigos, G., Catto-Smith, A., Blomeley, R.C., and Ackland, M.L. (2003) *Hum. Genet.*, **113**, 202–210.

56 Moynahan, E.J. (1974) *Lancet*, **2**, 399–400.

57 Wang, K., Zhou, B., Kuo, Y.-M., Zemansky, J., and Gitshier, J. (2002) *Am. J. Hum. Genet.*, **71**, 66–73.

58 Kleta, R. and Bockenhauer, D. (2006) *Nephron Physiol.*, **104**, 73–80.

59 Simon, D.B., Nelson-Williams, C., Bia, M.J., Ellison, D., Karet, F.E., Molina, A.M., Vaara, I., Iwata, F., Cushner, H.M., Koolen, M., Gainza, F.J., Gitelman, H.J., and Lifton, R.P. (1996) *Nat. Genet.*, **12**, 24.

60 Baker, S.B. and Worthley, L.I.G. (2002) *Crit. Care Resusc.*, **4**, 307–315.

61 Swaminathan, R. (2003) *Clin. Biochem. Rev.*, **24**, 47–66.

62 Gassbeek, A. and Meinders, A.E. (2005) *Am. J. Med.*, **118**, 1094–1101.

63 Gennari, F.J. (2002) *Crit. Care Clin.*, **18**, 273–288.

64 Fordtran, J.S., Rector, F.C., and Carter, N.W. (1968) *J. Clin. Invest.*, **47**, 884–900.

65 Reynolds, R.M., Padfield, P.L., and Seckl, J.R. (2006) *BMJ*, **332**, 702–705.

66 Ishikawa, S.-E. and Schrier, R.W. (2003) *Clin. Endocrinol. (Oxford)*, **58**, 1–17.

67 Adrogué, H.J. and Madias, N.E. (2000) *New Engl. J. Med.*, **342**, 1581–1589.

68 Gross, P. (2008) *Intern. Med.*, **47**, 885–891.

69 Goh, K.P. (2004) *Am. Fam. Physician*, **69**, 2387–2394.

70 Ten, S., New, M. and Maclaren, N. (2001) *J. Clin. Endocrinol. Metab.*, **86**, 2909–2922.

71 Ellis, S.J. (1995) *QJM*, **88**, 905–909.

72 Adrogué, H.J. and Madias, N.E. (2000) *New Engl. J. Med.*, **342**, 1493–1499.

73 Ball, S.G., Vaidja, B., and Baylis, P.H. (1997) *Clin. Endocrinol. (Oxford)*, **47**, 405–409.

74 Weisinger, J.R. and Bellorín-Font, E. (1998) *Lancet*, **352**, 391–396.

75 Baker, S.B. and Worthley, L.I.G. (2002) *Crit. Care Resusc.*, **4**, 301–306.

76 Vormann, J. (2003) *Mol. Aspects Med.*, **24**, 27–37.

77 Kratz, A., Ferraro, M., Sluss, P.M., and Lewandrowski, K.B. (2004) *New Engl. J. Med.*, **351**, 1548–1563.

78 Vanderhoof, J.A. and Langnas, A.N. (1997) *Gastroenterology*, **113**, 1767–1768.

79 Lainga, C.M., Toye, A.M., Capasso, G., and Unwin, R.J. (2005) *Int. J. Biochem. Cell Biol.*, **37**, 1151–1161.

80 Roth, K.S., Foreman, J.W., and Segal, S. (1981) *Kidney Int.*, **20**, 705–716.

81 Amirlak, I. and Dawson, K.P. (2000) *Q. J. Med.*, **93**, 207–215.

82 Franse, L.V., Pahor, M., Bari, M.D., Grant, W.S., Cushman, W.C., and Applegate, W.B. (2000) *Hypertension*, **35**, 1025–1030.

83 Sevastos, N., Theodossiades, G., and Archimandritis, A.J. (2008) *Clin. Med. Res.*, **6**, 30–32.

84 Wiederkehr, M.R. and Moe, O.W. (2000) *Am. J. Kidney Dis.*, **36**, 1049–1053.

85 Kitamura, K. and Tomita, K. (2005) *Intern. Med.*, **44**, 781–782.

86 Iolascon, A., Stewart, G.W., Ajetunmobi, J.F., Perrotta, S., Delaunay, J., Carella, M., Zelante, L., and Gasparini, P. (1999) *Blood*, **93**, 3120–3123.

87 Allon, M. (1993) *Kidney Int.*, **43**, 1197–1209.

88 Bugg, N.C. and Jones, J.A. (1998) *Anaesthesia*, **53**, 895–902.

89 Stoff, J.S. (1982) *Am. J. Med.*, **72**, 489–495.

90 Garg, R.K. and Tandon, N. (1999) *Indian J. Pediatr.*, **66**, 849–857.

91 Mornet, E. (2000) *Hum. Mutat.*, **15**, 309–315.

92 Whyte, M.P., Murphy, W.A., and Fallon, M.D. (1982) *Am. J. Med.*, **72**, 631–641.

93 Lam, A., Lam, C., Tang, M., Chu, J., and Lam, S. (2006) *H.K. J. Paediatr.*, **11**, 341–346.

94 Davidson, M.B., Thakkar, S., Hix, J.K., Bhandarkar, N.D., Wong, A., and Schreiber, M.J. (2004) *Am. J. Med.*, **116**, 546–554.

95 Bosch, X., Poch, E., and Grau, J.M. (2009) *New Engl. J. Med.*, **361**, 62–72.

96 Vanholder, R., Sever, M.S., Erek, E., and Lameire, N. (2000) *J. Am. Soc. Nephrol.*, **11**, 1553–1561.

97 Kozack, J.K. and MacIntyre, D.L. (2001) *Phys. Ther.*, **81**, 945–951.

98 Locatelli, F., Cannata-Andía, J.B., Drüeke, T.B., Hörl, W.H., Fouque, D., Heimburger, O., and Ritz, E. (2002) *Nephrol. Dial. Transplant.*, **17**, 723–731.

99 Kirk, T.S. and Simon, M.A. (1981) *J. Bone Joint Surg. Am.*, **63**, 1167–1169.

100 Watts, N.B. (1998) *Endocrinol. Metab. Clin. North Am.*, **27**, 419–439.

101 Mohammed, I.A. and Hutchison, A.J. (2008) *Ther. Clin. Risk Manag.*, **4**, 887–893.

102 Schucker, J.J. and Ward, K.E. (2005) *Am. J. Health. Syst. Pharm.*, **62**, 2355–2361.

103 Walsh, J.S. and Fairley, J.A. (1995) *J. Am. Acad. Dermatol.*, **33**, 693–706.

104 Lips, P. (2006) *Prog. Biophys. Mol. Biol.*, **92**, 4–8.

105 Silver, J., Yalcindag, C., Sela-Brown, A., Kilav, R., and Naveh-Many, T. (1999) *Kidney Int.*, **56**, S2–S7.

106 Poole, K.E. and Reeve, J. (2005) *Curr. Opin. Pharmacol.*, **5**, 612–617.

107 Maeda, S.S., Fortes, E.M., Oliveira, U. M., Borba, V.C.Z., and Lazaretti-Castro, M. (2006) *Arq. Bras. Endocrinol. Metabol.*, **50**, 664–673.

108 Jacobs, T.P. and Bilezikian, J.P. (2005) *J. Clin. Endocrinol. Metab.*, **90**, 6316–6322.

109 Wilson, L.C. and Trembath, R.C. (1994) *J. Med. Genet.*, **31**, 779–784.

110 Weinstein, L.S. (1998) in *G Proteins, Receptors, and Disease* (eds A.M. Spiegel), Humana Press, New Jersey, pp. 23–56.

111 Ahmad, R. and Hammond, J.M. (2004) *Otolaryngol. Clin. North Am.*, **37**, 701–713.

112 Khan, A. and Bilezikian, J. (2000) *Can. Med. Assoc. J.*, **163**, 184–187.

113 Francisco, A.L.M.d. (2004) *Clin. Ther.*, **26**, 1976–1993.

114 Eisman, J.A. (1988) *Baillieres Clin. Endocrinol. Metab.*, **2**, 125–155.

115 Basha, B., Rao, D.S., Han, Z.-H., and Parfitt, A.M. (2000) *Am. J. Med.*, **108**, 296–300.

116 Beall, D.P., Henslee, H.B., Webb, H.R., and Scofield, R.H. (2006) *Am. J. Med. Sci.*, **331**, 233–242.

117 Caruso, J.B., Patel, R.M., Julka, K., and Parish, D.C. (2007) *J. Gen. Intern. Med.*, **22**, 1053–1055.

118 Hussmann, J., Russell, R.C., Kucan, J. O., Khardori, R., and Steinau, H.U. (1995) *Ann. Plast. Surg.*, **34**, 138–147.

119 Hussein, M.-R.A., Ali, H.O., Abulwahed, S.R., Argoby, Y., and Tobeigei, F.H. (2009) *Exp. Mol. Pathol.*, **89**, 134–135.

120 Denkhausa, E. and Salnikow, K. (2002) *Crit. Rev. Oncol. Hematol.*, **42**, 35–56.

121 Nielsen, F.H. (1984) *Ann. Rev. Nutr.*, **4**, 21–41.

122 Muyssen, B.T.A., Brix, K.V., DeForest, D.K., and Janssen, C.R. (2004) *Environ. Rev.*, **12**, 113–131.

123 Rajagoplan, K.V. (1988) *Annu. Rev. Nutr.*, **8**, 401–427.

124 Schwarz, G. (2005) *Cell Mol. Life Sci.*, **62**, 2792–2810.

125 Sardesai, V.M. (1993) *Nutr. Clin. Pract.*, **8**, 277–281.

126 Rouault, T.A. and Tong, W.-H. (2005) *Nat. Rev. Mol. Cell Bio.*, **6**, 345–351.

127 Nielsen, F.H. (1991) *FASEB J.*, **5**, 2661 –2667.

128 Anke, M., Illing-Gunther, H., and Schafer, U. (2005) *Biomed. Res. Trace Elem.*, **16**, 208–214.

129 Aurelianoa, M. and Cransb, D.C. (2009) *J. Inorg. Biochem.*, **103**, 536–546.

130 Cohen, M.D., Kargacin, B., Klein, C.B., and Costa, M. (1993) *Crit. Rev. Toxicol.*, **23**, 255–281.

131 Stearns, D.M. (2000) *Biofactors*, **11**, 149 –162.

132 Pechova, A. and Pavlata, L. (2007) *Vet. Med. (Praha)*, **52**, 1–18.

133 Mertz, W. (1993) *J. Nutr.*, **123**, 626–633.

134 Schwarz, K. and Mertz, W. (1959) *Arch. Biochem. Biophys.*, **85**, 292–295.

135 Jeejeebhoy, K.N., Chu, R.C., Marliss, E. B., Greenberg, G.R., and Bruce-Robertson, A. (1997) *Am. J. Clin. Nutr.*, **30**, 531–538.

136 Kobayashi, M. and Shimizu, S. (1999) *Eur. J. Biochem.*, **261**, 1–9.

137 Herbert, V. (1988) *Am. J. Clin. Nutr.*, **48**, 852–858.

138 Allen, R.H., Stabler, S.P., Savage, D.G., and Lindenbaum, J. (1993) *FASEB J.*, **7**, 1344–1353.

139 Fowler, B. (1998) *Eur. J. Pediatr.*, **157** (Suppl. 2), S60–S66.

140 Toh, B.-H., Driel, I.R.v., and Gleeson, P.A. (1997) *New Engl. J. Med.*, **337**, 1441–1448.

141 Takeda, A. (2002) *Brain Res. Brain Res. Rev.*, **41**, 79–87.

142 Barceloux, D.G. and Barceloux, D. (1999) *Clin. Toxicol.*, **37**, 293–307.

143 Crossgrove, J. and Zheng, W. (2004) *NMR Biomed.*, **17**, 544–553.

144 Zeng, H. (2009) *Molecules*, **14**, 1263–1278.

145 Daniels, L.A. (1996) *Biol. Trace Elem. Res.*, **54**, 185–199.

146 Basel, B. (2000) *Cell Mol. Life Sci.*, **57**, 1864–1873.

147 Spallholz, J.E. (1994) *Free Radic. Biol. Med.*, **17**, 45–64.

12
Metal Compounds as Enzyme Inhibitors

Gilles Gasser and Nils Metzler-Nolte

12.1
Introduction

In this chapter we review the use of metal complexes as enzyme inhibitors in medicinal inorganic chemistry. Following the success of cisplatin and its close derivatives like carboplatin and oxaliplatin in the clinics, and the subsequent elucidation of DNA as the cellular target of these compounds (Chapters 3 and 4), it was almost automatically assumed that *any* metal complex would have DNA as its cellular target. Recent research, however, provides solid evidence that such an assumption is premature. Indeed, many metal complexes are specifically designed to act as enzyme inhibitors. It is those complexes, as well as the concurrent biochemical and structural studies, that this chapter covers.

To keep the chapter to a reasonable length, we have deliberately chosen *not* to cover the following topics. Gallium compounds are omitted as they are reviewed in more detail in Chapter 5. Likewise, we are not reviewing Au complexes, which are discussed in detail in Chapter 7 by Berners-Price. There is a growing recognition in the recent literature that Ru(arene) compounds might have protein targets, in addition to the long-suspected DNA interaction of those compounds. This topic is covered elsewhere in this book, and also not mentioned further here (Chapters 1 and 5). We do, however, cover the newly discovered Ru-based kinase inhibitors as highly successful examples of a structure-based design of inorganic enzyme inhibitors. We do not deal with metal complexes that mimic enzymes (such as nuclease or protease mimics), because such compounds are not usually developed for medicinal chemistry applications. There are, however, good reviews available on such compounds [1–4]. This chapter is really concerned with the role of metal ions and complexes in binding to enzymes and their role in enzyme inhibition. Consequently, we do not cover metal complexes (which could have a therapeutic purpose such as radio-metals) that have been covalently linked to a biomolecule (say a peptide) [5, 6], in which the biomolecule, and not the metal moiety, is primarily responsible for target binding. Also not included are approaches where the metal complex has primarily a role to induce a structural change, for example, by working as a peptide turn mimic [7].

Bioinorganic Medicinal Chemistry. Edited by Enzo Alessio
Copyright © 2011 WILEY-VCH Verlag GmbH & Co. KGaA, Weinheim
ISBN: 978-3-527-32631-0

Lastly, a very interesting approach is inhibition of metallo-enzymes by design of a suitable ligand that will specifically bind to the metal ion in the active site, thereby shutting down enzyme activity. This approach has been applied successfully to several classes of metallo-enzymes with medicinal relevance, most recently by Cohen and coworkers to the inhibition of Zn enzymes. The field has been reviewed and is beyond the scope of this chapter [8].

While we have tried to summarize current knowledge in the field within the chosen limitations described above, this chapter certainly also represents the preferences of the authors. Several additional, and partly more specialized, reviews exist on closely related topics, for example by Meade [9], Meggers [10], Fricker [11], and Sadler [12]. Moreover, organometallic compounds constitute a very interesting and promising class of metallodrugs by themselves [13, 14]. For many of them, enzyme inhibition is a valid assumption for the mode of action. Organometallic drug candidates, in particular with antiproliferative activity, are covered in several recent reviews [15–19].

12.2
Kinase Inhibitors

Protein kinases regulate most aspects of cellular life [20]. As their mutations and deregulation play causal roles in many human diseases, kinases are an important therapeutic target [10]. Currently, eight kinase inhibitors are clinically approved, about 40 are in clinical trials, and many more are in earlier stages of development [21]. Most of these inhibitors bind competitively into the ATP binding site, with hydrogen bonding and hydrophobic interactions contributing to the high binding constant and specificity [22]. In 2002, Meggers *et al.* embarked on an original program to rationally design Ru(II) metal complexes that mimic the shape of staurosporine, a lead structure for a natural kinase inhibitor (see Scheme 12.1 for a comparison of the structure of the metal complexes with staurosporine) [10, 24, 25]. Ru metal complexes were found to be ideal candidates for the purpose of this research [10]. They have a great structural variety, far more diverse stereochemistry than organic compounds (for an octahedral complex with six different ligands, 30 stereoisomers exist!). They are also air and water stable – even in aqueous buffers containing millimolar concentrations of thiols. Synthetic routes for their preparation are well-established, and

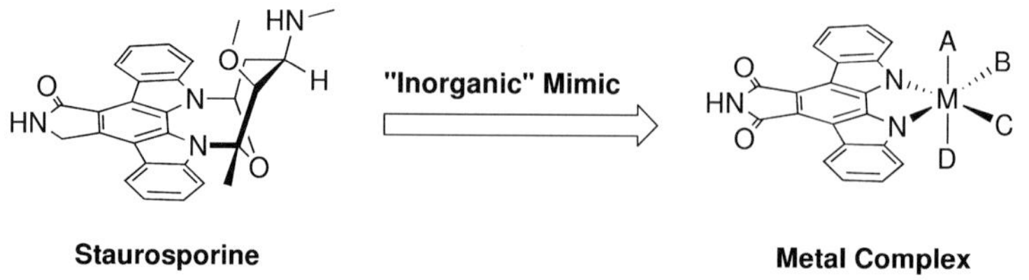

Staurosporine   "Inorganic" Mimic   Metal Complex

Scheme 12.1 Schematic view of how the metal complex mimics the overall shape of staurosporine.
Adapted from References [10, 23].

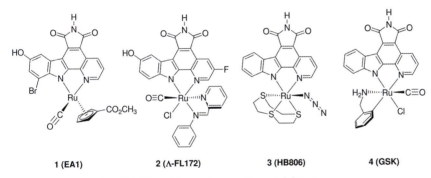

1 (EA1)   **2 (Λ-FL172)**   **3 (HB806)**   **4 (GSK)**

Figure 12.1   Examples of Ru(II) metal complexes as kinase inhibitors.

their general toxicity is low [10]. Importantly, in these compounds, the metal is not playing any direct role in the inhibition; it only allows the spatial organization of the substituents around the metal center as an inert *scaffold*. This original concept increases substantially the opportunity to build complicated three-dimensional enzyme inhibitor structures. This point is crucial for the design of *selective* kinase inhibitors, as kinases form one of the largest families of enzymes with very conserved ATP binding sites, rendering such design very challenging [10].

Meggers *et al.* have prepared numerous metal-containing kinase inhibitors, the majority of them being Ru(II) complex derivatives [20, 26–39], and more recently Pt [23] and Os [27] derivatives. Figure 12.1 presents a selection of metal complexes from this group. Hence, they successfully designed nanomolar and even picomolar ATP-competitive ruthenium-based inhibitors. This concept has been confirmed by, so far, six different X-ray structures of Ru complexes that co-crystallized with protein kinases [see Figure 12.2 for an example of the binding of kinase PAK1 with the

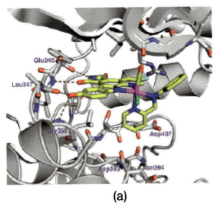

(a)

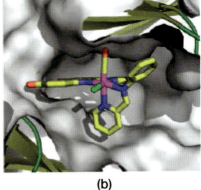

(b)

Figure 12.2   X-ray structure of protein kinase PAK1 that was co-crystallized with the Ru(II) complex Λ-FL172 (**2**). Displayed are the most important H-bonding interactions (a) and a view of the surfaces, displaying the shape match between the inhibitor and the kinase enzyme (b).
Reproduced from Reference [10] with permission of the publisher.

Ru(II) complex Λ-FL172 (**2**)] [28, 33, 34, 38–40]. As expected, the metal ion played a solely structural role. However, the organic ligands can be optimized to occupy the available space in the active site, as well as to provide additional hydrogen bonding interactions, thus making the individual inhibitors highly specific. Moreover, physiological functions as a consequence of kinase inhibition were demonstrated within mammalian cells [26], *Xenopus* embryos [31], and zebrafish embryos [36].

12.3
Proteasome Inhibitors

Most damaged and misfolded proteins in eukaryotic cells are degraded through an ATP- and ubiquitin-dependent pathway. The proteins to be "destroyed" are first conjugated to multiple molecules of the polypeptide ubiquitin – this process is also referred to as tagging. These tagged proteins are then translocated to the 26S proteasome and subsequently degraded by proteolysis [41, 42]. The enzyme activity of the 26S proteasome is mediated by the 20S proteasome core, which contains three pairs of catalytic sites responsible for its chymotrypsin-, trypsin, and caspase-like activities [43–45]. Interestingly, inhibition of 26S proteasomase is currently a field of active research since it has been shown that human cancer cells are more sensitive to such inhibition than normal cells [45]. Indeed, Bortezomib was the first proteasome inhibitor to be approved by the FDA for the treatment of multiple myeloma [46, 47].

Verani, Ping Dou, Peng *et al.* recently embarked on a project to employ Ga(III), Cu(II), Ni(II), and Zn(II) complexes to inhibit purified 20S proteasome and/or 26S proteasome (see Figure 12.3 for a few examples) [45, 48–50]. They discovered that all the complexes presented in Figure 12.3 were indeed good inhibitors with the exception of **8**, which did not show any inhibition effect. As for their mode of action, it has been suggested that they dissociate, at least partially. The metal ion with its free

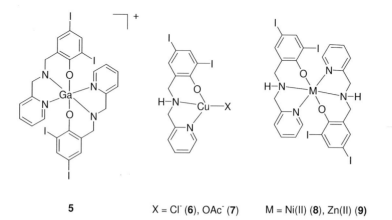

5   X = Cl⁻ (**6**), OAc⁻ (**7**)   M = Ni(II) (**8**), Zn(II) (**9**)

Figure 12.3 Examples of metal complexes as proteasome inhibitors.

coordination site would then bind the N-terminal threonine amino acids present in the active sites, thus inhibiting the protein. In this assumed mode of action, compounds **6** and **7** already possess a free coordination site at the metal, while compounds **5**, **8**, and **9** are actually prodrugs, and the labile ligands are necessary to prevent undesired non-specific interactions during cell uptake [49, 50].

Gallium coordination complexes have been investigated for quite some time. They are covered in more detail in Chapter 5 and in two leading references [51, 52].

12.4
Carbonic Anhydrase Inhibitors (CAIs)

Carbonic anhydrases (CAs) form a family of ubiquitous metalloenzymes, which catalyze the rapid conversion of carbon dioxide into bicarbonate and protons [53, 54]. In mammals, 16 different CA isozymes or related proteins are known with different subcellular localization and tissue distribution [54, 55]. All CAs possess a metal ion, usually Zn(II), in their active site, which is coordinated by three proteinogenic imidazole rings from histidines and a fourth non-proteinogenic ligand (chloride or water). The marine CA in *Thalassiosira weissflogii* is a rare example of a naturally occurring Cd enzyme [56]. CAs are involved in a multitude of crucial physiologic and pathologic processes [57]. Their inhibition is currently clinically exploited, or is thought to have potential, for the treatment of glaucoma agents, obesity, osteoporosis, bacterial and fungal infections, in the treatment of epilepsy or diverse neuromuscular disorders, for the treatment of Alzheimer's disease, as well as for the management of a multitude of tumors [55, 57, 58]. Two main classes of carbonic anhydrase inhibitors (CAIs) are known: the metal complexing anions and the unsubstituted sulfonamides and their bioisosteres (sulfamates and sulfamites). These inhibitors bind to the Zn(II) ion of the enzyme either by substituting the non-proteinogenic zinc ligand to generate a tetrahedral adduct or by addition to the metal coordination sphere, generating trigonal-bipyramidal species (Figure 12.4) [53, 55].

Sulfonamide compounds possessing CA inhibitory properties such as acetazolamide (AAZ), methazolamide (MZA), ethoxzolamide (EZA), dichlorophenamide

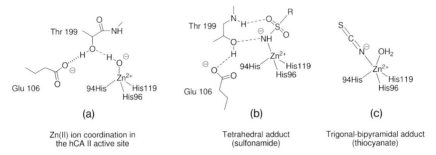

(a)   (b)   (c)

Zn(II) ion coordination in   Tetrahedral adduct   Trigonal-bipyramidal adduct
the hCA II active site   (sulfonamide)   (thiocyanate)

Figure 12.4 Zn(II) ion coordination in the hCAII active site (a); CA inhibition mechanism by sulfonamide (b); and anionic (c) inhibitors. Figure adapted from references [53, 55].

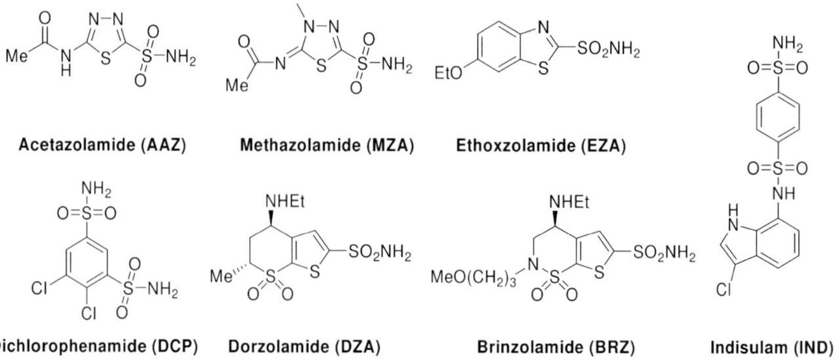

Figure 12.5 Clinically used sulfonamide derivatives as CAIs.

(DCP), and Indisulam (IND) (Figure 12.5) have been used, some of them for more than 40 years, as systemic eye-pressure-lowering drugs in the treatment of open-angle glaucoma as well as other diseases associated with acid/base secretory disequilibria [59]. However, only more recently, the Supuran and Borras groups undertook complexation studies of different main group and transition metal ions with such CAIs [57, 60–66]. Interestingly, they found that these complexes strongly inhibited two CA isoforms, being generally 10–100 times more effective than the parent sulfonamides from which they were prepared [67]! This observation was later explained as a dual enzyme inhibition due to the relative instability of the metal complexes, which are easily dissociated to metal ions and sulfonamidate ions. The sulfonamide anion binds to the usual catalytic Zn(II) ion within the CA active site while the metal cation binds to critical histidine residues of the catalytic cycle, enhancing therefore the inhibitory power of the metal complex compared to the non-metallic derivatives [65, 67].

Nonetheless, the clinically used sulfonamide agents cited above have numerous side-effects, which include augmented diuresis, fatigue, parethesias, anorexia, and others. These drawbacks are the consequence of the inhibition of CAs in other tissues/organs than those targeted. However, these side-effects could be overcome by topical application. Hence, dorzolamide (DZA) and brinzolamide (BRZ) (Figure 12.5) were clinically launched in 1995 and 1999 respectively [59]. In contrast to the other sulfonamide compounds, these two agents contain secondary amine moieties, which can be easily transformed into hydrochloride salts. This protonation increases the water solubility of the compounds, which is needed for topical applications. However, due to the acidity of the salt solutions of DZA (pH 5.5), new undesired adverse effects such as local burning, stinging and reddening of the eyes, blurred vision, and a bitter taste were observed after topical application of DZA [59]. More serious side-effects such as contact allergy, nephrolithiasis, anorexia, depression, dementia as well as corneal decompensation in patients already presenting corneal problems were also reported with DZA [59]. To tackle these major side-effects, Supuran *et al.* envisioned a general approach to obtain water-soluble, high-affinity sulfonamide CAIs, whose solubility is not engendered by formation of hydrochloride salts. They attached diethylenetriaminopentaacetic acid (dtpa) [68] and other polyamino-polycarboxylic acid moieties [59, 67] to aromatic/heterocyclic sulfonamides. These new compounds were shown to be highly water soluble at neutral pH and to display strong CA inhibition. Even more interestingly, Zn(II), Cu(II), and Al(III) complexes of a selection of these derivatives (Figure 12.6) were also investigated and some of them were found to behave as nanomolar CAIs against different CA isozymes [59, 67, 68]. Some of these sulfonamides and their respective Zn(II) and Cu(II) complexes strongly lowered intraocular pressure (IOP) when applied topically, directly into the normotensive/glaucomatous rabbit eye [59, 68].

However, due to the high number of isoforms of CAs and their rather diffuse localization in many tissues/organs, the current CAIs lack selectivity [58]. Recently, an extracellular, transmembrane isoform of CA, Human CA IX, was shown to be a novel and interesting target for anticancer therapy [58]. With this in mind, Winum, Supuran *et al.* prepared a novel class of copper-containing sulfonamides that show excellent CA IX inhibitory properties and, importantly, selectivity over the cytosolic, housekeeping isozymes hCA I and II [58]. The copper complexes inhibited more efficiently CA IX than the corresponding ligand sulfonamides and, importantly, showed membrane impermeability, having therefore the possibility to specifically target the transmembrane CA IX, which has an extracellular active site [58]. Furthermore, the potential incorporation of radioactive copper isotopes in this type of CAI could lead to interesting diagnostic/therapeutic applications [58]. A critical point might be the stability of these complexes *in vivo*, on which, at present, no data are available.

Mallik *et al.* have also demonstrated the use of copper complexes as selective inhibitors of CA over albumin and lysozyme – the copper complexes recognize the unique pattern of histidine residues on the protein surface [69, 70]. Even more

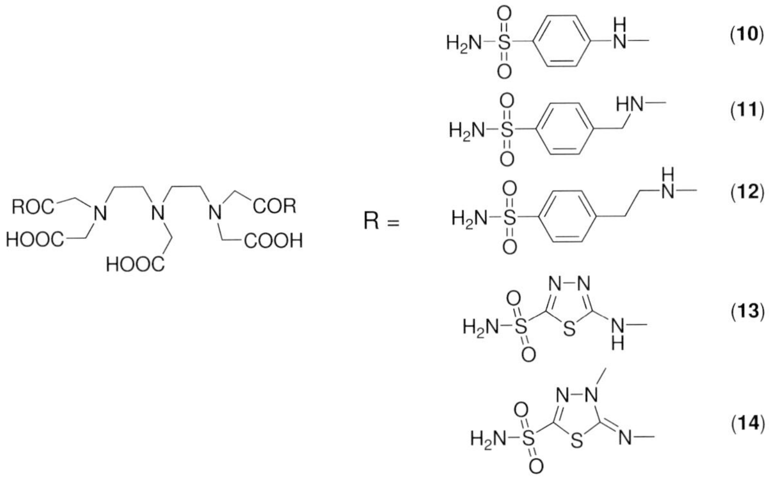

Figure 12.6 Water-soluble sulfonamides, built around the diethylene-triamine-pentacarboxylate (dtpa) core, as CAIs.

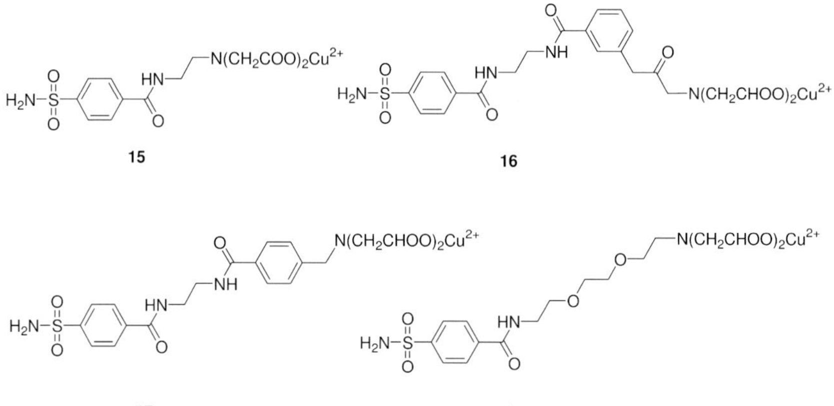

Figure 12.7 Examples of "two-prong" CAs.

interestingly, Srivastava, Mallik *et al.* envisaged a two-prong approach to increase the potency of known CAIs. They attached to a weak CAI (benzenesulfonamide) a Cu^{2+}-chelating iminodiacetic acid (IDA) derivative (Figure 12.7). The metal complex extends beyond the active site pocket of the enzyme and interacts with the surface-exposed histidine residues [71–73]. Hence, they succeeded in converting a poor inhibitor into a good inhibitor. With this concept, the binding energy of the inhibitors is derived not only from the interactions of the sulfonamide to the Zn(II) center of the CA but also from the interactions of the copper complex to the surface-exposed histidine residues [72]. Unsurprisingly, the length of the spacer between the two moieties played a crucial role in the binding affinity [73]. Last but not least, X-ray crystallographic studies of human CA I and II complexed with these "two-prong" inhibitors were also reported by these research groups [74]. As presented in Figure 12.8a for **15**, two inhibitors co-crystallized with the protein: one in the rim of the active site cavity and another one near the N-terminus of the enzyme. As shown in Figure 12.8b, binding near the active site indeed occurred through two different coordination sites: the ionized NH^- group of the benzenesulfonamide coordinates to the active site Zn(II) ion of the CA while the IDA-Cu^{2+} prong binds to a specific histidine residue on the surface of the protein [74]. Recently, these researchers have extended their strategy for designing "multi-prong" enzyme inhibitors by incorporating selective ligands to the liposomal surface [75]. In another study, they have blocked the active site accessibility of metalloproteinases-9 by "multi-prong" surface binding groups [76].

12.5
Cyclooxygenase Inhibitors

Cyclooxygenases (COXs) are enzymes that are responsible for the formation of important biological mediators called prostanoids, including prostaglandins,

(a)

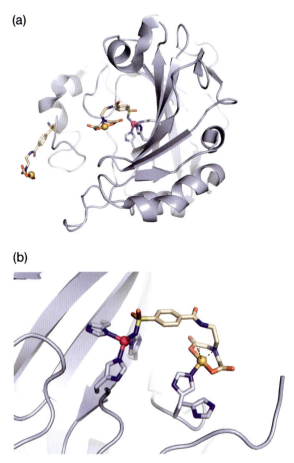

(b)

Figure 12.8 Crystal structure of human carbonic anhydrase II with **15**. (a) Two metal complexes co-crystallized with the enzyme: one in the rim of the active site cavity and another one near the N-terminus of the enzyme; (b) enlargement on the active site. The zinc atom (magenta) binds to three histidines of CAII and to the ionized NH^- group of the benzenesulfonamide of **15**. The copper atom (orange) binds to a specific histidine residue on the surface of the protein. (PDB code 2FOV).

prostacyclin, and thromboxane. COXs convert arachidonic acid (AA) into prostaglandin H2 (PGH2). COXs contain two active sites: a heme with peroxidase activity, responsible for the reduction of the hydroperoxy-endoperoxide prostaglandin G2 (PGG2) to PGH2, and a cyclooxygenase site, where AA is converted into PGG2. Interestingly, two of the most famous common drugs on the market for pain and inflammation relief, namely aspirin and ibuprofen, exert their activity through the inhibition of COX. Both belong to the class of nonsteroidal anti-inflammatory drugs (NSAIDs), which have been recently found to be beneficial for cancer chemoprevention and combination chemotherapy [77, 78]. Clinical studies on aspirin and other non-steroidal anti-inflammatory drugs indicated a correlation between the long-term intake of NSAIDs and positive effects for cancer patients, mainly a substantial decrease of recidive risks [79]. However, the exact mode of chemopreventive action of these NSAIDs is still the subject of debate [80]. Ott, Gust, and coworkers have demonstrated that an alkyne hexacarbonyldicobalt, $Co_2(CO)_6$, derivative of an aspirin-derived ligand [Co-ASS (**19**), Figure 12.9a] is a very effective antiproliferative agent [81, 82]. Co-ASS was found to strongly inhibit the NSAID's main target enzymes COX-1 and COX-2 [82]. Owing to the high stability of Co-ASS [83], it is anticipated that the active species is indeed the intact organometallic complex. Importantly, it was recently confirmed that Co-ASS exhibits several biochemical properties related to the reported antitumoral effects of NSAIDs, including apoptosis induction, inhibition of PGE2, and triggering of anti-angiogenic effects [80]. To evaluate the molecular interaction of Co-ASS with COX-2, peptide fragments generated by trypsin digestion after incubation of the enzyme with either aspirin or Co-ASS were examined by LC-ESI tandem mass spectrometry [80]. The enzyme was also analyzed alone for comparison [80]. It was

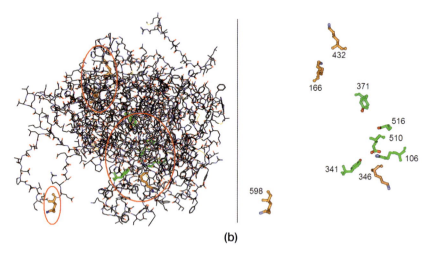

Co-ASS (**19**)

(a)

(b)

Figure 12.9 (a) Structure of Co-ASS (**19**). (b) Theoretical model of COX-2 and the relevant amino acids (catalytic activity: green carbons; Co-ASS acetylation: orange carbons); left-hand side: full view, right-hand side: close-up. Based on the PDB entry 1v0x and reproduced from Reference [80] with permission of the publisher.

demonstrated that Co-ASS acetylated several lysine residues (Lys 166, 346, 432, and 598) of its putative main target COX-2. This is in contrast to aspirin, which mainly acetylates a serine residue (Ser516) close to the active site of the enzyme (Figure 12.9b) [80].

Notably, various other hexacarbonyl-dicobalt species with interesting biological properties have been reported, including several other NSAIDs [82], nucleosides [84], carbohydrate derived complexes [85], peptide derivatives [86], or complexes with hormonally active ligands [87–89]. However, no COX-2 related mechanism, as for Co-ASS, have been observed. Given the great structural diversity of all $Co_2(CO)_6$ derivatives, and the fact that IC_{50} values between 5 and 30 μM were reported for all those compounds, it could also be possible that the biomolecule part serves as a drug carrier for cellular uptake of the Co complex. The metal complex with Co in zero oxidation state is then slowly decomposed by (enzymatic?) oxidation inside the cell. Finally, the cytotoxic action is mediated by the general toxicity of the hydrated Co(II) or Co(III) ions.

12.6
Acetylcholinesterase Inhibitors

Acetylcholinesterase (AChE) is an enzyme that hydrolyzes the ester bond of the neurotransmitter acetylcholine. It is mainly found in cholinergic synapses in the central nervous system where its activity terminates the synaptic signal transmission. In addition to several approved drugs, famous inhibitors of AChE include nerve gases such as the organophosphate Sarin. These molecules inhibit AChE irreversibly by alkylating a serine residue that is essential for catalysis (see also section on serine proteases below). It was recognized very early that cationic metal polypyridyl complexes are able to inhibit AChE at μM concentrations [90, 91]. Since then, the 3D structures of several AChEs were solved and SARs were derived [92]. It is assumed that the metal complexes bind at the so-called peripheral anionic site through a combination of electrostatic and hydrophobic interactions, thereby inhibiting AChE reversibly.

Very recently, Meggers and coworkers developed a solid-phase synthetic strategy for tris-heteroleptic Ru(II) complexes. By using a combination of differently substituted bipyridyl and phenanthroline ligands they synthesized a library of 28 complexes of the type $[Ru(pp)(pp')(pp'')]^{2+}$, which were subsequently screened for AChE inhibition (pp = general polypyridyl ligand) [37]. The complex **20** (Figure 12.10) proved to be the best inhibitor of AChE from *Electrophorus electricus* in this series (IC_{50} = 200 nM), with an IC_{50} value 50 times lower than the parent [Ru (phen)$_3$]$^{2+}$ (IC_{50} = 10 μM). Interestingly, the Δ-enantiomer of **20** had a slightly higher IC_{50} of 400 nM. It seems reasonable to assume that even better, and possibly very specific, AChE inhibitors can be designed by using this versatile synthetic approach in combination with the structural information on AChEs that is already available. Notably, in this context, the same group has recently reported a strategy for the *stereospecific synthesis* of such tris-heteroleptic Ru(II) polypyridyl complexes [93].

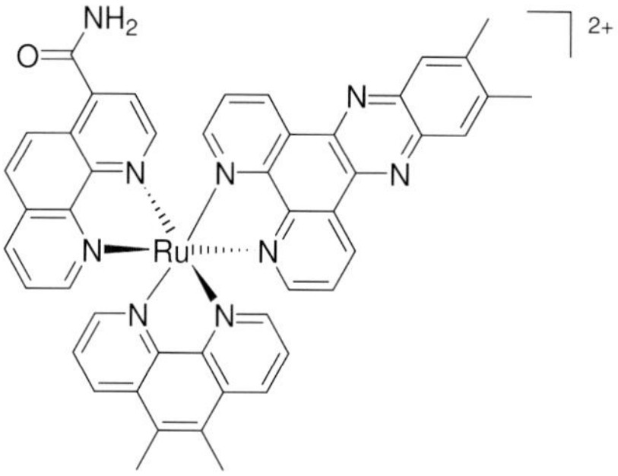

20   **21**

Figure 12.10 A chiral, octahedral Ru(II) AChE inhibitor (**20**) and a vanadyl-bis (hydroxypicolinate) complex as a potent phosphatase inhibitor (**21**).

12.7
Protein Phosphatase Inhibitors

Protein phosphatases (PP) are enzymes that remove or transfer phosphate groups by hydrolysis of phosphate ester bonds. As such, they are the direct opposite of kinases, which were discussed above. They are classified by their substrate specificity (e.g., tyrosine-specific PP, serine/threonine-specific PP), or by their active site such as cysteine PP. Another class of PP is dependent on metal ions as cofactors, such as, *inter alia*, Mg^{2+} or Zn^{2+}. In a more general sense, phosphatases remove phosphate groups from any type of substrate, for example, nucleotides, carbohydrates, or even alkaloids. Names such as the well-known alkaline phosphatase suggest that the optimum pH value for this enzyme is above neutral. Evidently, there are numerous inhibitors for specific phosphatases. Depending on the active site, very different metal complexes can be among phosphatase inhibitors. Here, we restrict our discussion to just a few instructive examples. The general mechanism of phosphate hydrolysis, in particular in comparison with amide or ester bond hydrolysis that is relevant to some other sections in this chapter, has been summarized [94].

Cysteine PPs can be effectively blocked by stable pentacoordinate vanadate ions, which serve as analogs for the pentacoordinate transition state [95]. A crystal structure of the inhibited bovine low molecular weight phosphotyrosyl phosphatase has been obtained and clearly shows the trigonal bipyramidal vanadate ion binding to the catalytic Cys12 residue (PDB code 1Z12) [96]. Considering the available structures of phosphatases and the well-established chemistry of vanadium [97], highly specific vanadate-based phosphatase inhibitors can be synthesized. One recent example is the complex **21** with two 3-hydroxypicolinate ligands (Figure 12.10). Complex **21** selectively inhibits the phosphinositide-3-phosphatase PTEN [98], which has a rather wide catalytic pocket [99], but leaves other phosphatases with smaller active sites untouched. This type of inhibition can be further modified to make the inhibition irreversible, that is, by using pervanadate instead of simple vanadate. Here, the peroxo group serves as a strong oxidant to transform the cysteine sulfhydryl group into a sulfoxide, which is no longer catalytically active, as shown for the protein tyrosine phosphatase PTP1B by mass spectrometry [100]. Further proof for this irreversible mode of action comes from the fact that the inhibition can be prevented if reductants are added to the medium. Following this mode of action [98, 101], inhibition of the above-mentioned phosphatases by vanadate could be related to the well-established insulin-enhancing activity of vanadium complexes [102–104].

12.8
Trypsin and Thrombin Inhibitors

Trypsin and thrombin are both serine proteases, meaning that they are enzymes that cleave peptide bonds in proteins and that serine is an essential amino acid in their active site. The design of inhibitors of these two proteins is of high relevance, notably as new antithrombotic agents in the case of thrombin. Amidine and guanidine derivatives have been shown to be potent inhibitors of trypsin-like enzymes. X-Ray crystallographic studies have demonstrated that the strong inhibition arises from the interaction between the cationic amidinium or guanidinium group of the inhibitor and the anionic carboxyl group of Asp189 in the S1 specificity pocket of trypsin and other similar enzymes [105]. With this in mind, Tanizawa and coworkers investigated amidine-containing Schiff base Cu(II) and Fe(III) chelates as thrombin inhibitors and guanidine-containing Schiff base Cu(II), Zn(II), and Fe(II) chelates as trypsin inhibitors [105–109]. They notably reported a copper chelate (**22**, Figure 12.11a), which inhibits thrombin in the low nanomolar range (27 nM) – this value is comparable to that of argatroban (MD-805), which is a clinically used compound [109]. Furthermore, X-ray crystallographic analyses of bovine β-trypsin with four different amidino-containing Cu(II) and Fe(III) complexes revealed a novel mode of interaction between metal-chelate inhibitors and serine proteases. Similar to the "non-metallic" inhibitors discussed above, the cationic amidino group of chelates forms a salt bridge with the carboxylate of Asp189 in the S1 pocket of trypsin. As shown in Figure 12.11b, the metal ion of the Fe(III) complex **23** did not participate in the binding, thus playing a purely structural role. However, for one copper complex, the data revealed that the metal ion and the phenolic oxygen of the Schiff base also interacted with His57 and Ser195 of trypsin. Upon coordination, the copper center changes its coordination sphere from tetrahedral to distorted octahedral or square pyramidal.

Mead, Gray, and coworkers used a Co(III)-peptide bioconjugate (**24**, Figure 12.12) as a human α-thrombin inhibitor. The peptide employed was known to have a high affinity for the enzyme active site and was used as a reversible inhibitor to direct the metal complex towards the active site [110]. The metal complex was then assumed to coordinate to histidines or other amino acid side-chains in, or near, the

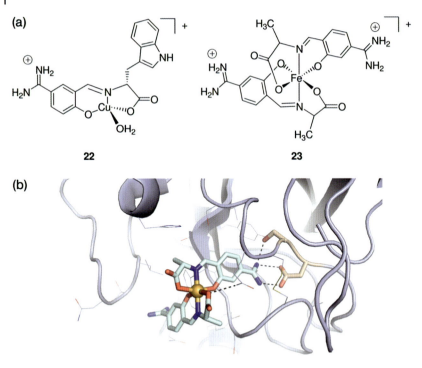

Figure 12.11 (a) Most potent Cu(II) inhibitor of trypsin (**22**) and example of a Fe(III) inhibitor of trypsin (**23**); (b) crystal structure of the iron complex **23** bound to bovine β-trypsin (PDB code 1G3C), in which the iron atom plays only a structural role.

active site through substitution at the axial positions of the Co(III) center. Thus, a selective, irreversible inhibition was obtained.

Notably, Katz, Janc, and coworkers have designed a new class of serine protease inhibitors whose potencies and selectivities were dramatically enhanced (1000-fold) by co-administration of Zn(II) salts [111, 112].

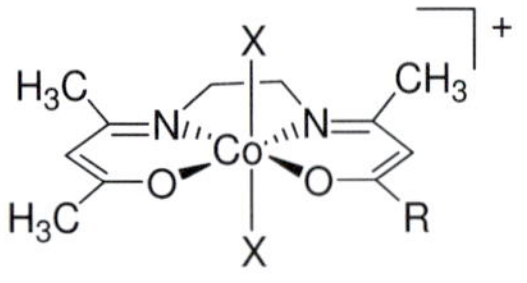

X = NH₃, 2-MeIm
R = peptide (CH₂CH₂CONH-GGG-d-FPR-CO-NH₂ or
CH₂CH₂CONH-GGG-d-FPA-CO-NH₂)

24

Figure 12.12 Co(III) human α-thrombin inhibitor [110].

12.9
Cysteine Protease Inhibitors and Glutathione Transferase Inhibitors

Mammalian cysteine proteases such as cathepsins, caspases, and calpains are involved in many important cellular metabolism processes. As the name implies, and in contrast to the family of serine proteases, they have a cysteine amino acid in their active site that is essential for the catalytic proteolysis of amide bonds. Their malfunctioning is responsible for numerous diseases, making them consequently interesting targets for new drug developments. Notably, the search for selective cathepsin B (cat B) inhibitors is an active research area as cathepsins were implicated in tumor progression, angiogenesis, and arthritis [113]. Several different types of metal complexes were shown to inhibit cat B: ruthenium complexes such as RAPTA-T (**25**) [15, 114], palladium complexes such as the dimeric palladacycles **26** [115, 116], **27** [117], and **29** [113], gold complexes such as Auranofin (**30**) [118] (see also Chapters 1 and 7) and rhenium complexes such as the oxorhenium(V) complex **31** [11, 119] (Figure 12.13).

The ferrocene-containing palladium complex **26** was found to inhibit cat B activity in a reversible fashion. It binds to free cat B (E) as well as to the enzyme–substrate complex (ES) with a dissociation constant of $K_H = 12\ \mu M$ and $\alpha K_H = 2.4\ \mu M$, respectively. Furthermore, **26** has no cytotoxic activity. Even more striking, *in vivo* experiments showed that subcutaneous inoculations of 10^6 tumoral cells into Walker tumor-bearing rats produced solid tumors with a mass of 4.0 g in 12

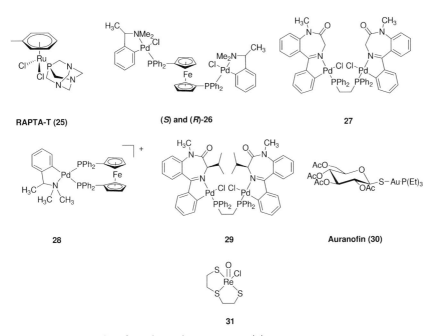

Figure 12.13 Examples of metal complexes as cat B inhibitors.

days. When the rats were treated with **26** (2.0 mg kg^{-1}), the tumoral mass was reduced to 0.3 g! Furthermore, toxicology studies did not show any alteration in red and white blood cell morphology as well as hepatic, kidney, and spleen tissues 14 days after the drug administration, even at high doses [115, 116]. These results have been associated with the cat B inhibition by **26** as several studies have demonstrated that cat B is involved in metastatic tumor development [116]. The same research group also reported the ionic mononuclear palladacycle complex **28** (Figure 12.13), which triggers lysosomal membrane permeabilization with release of lysosomal enzymes, especially of cat B to the cytosol of K562 cells. This cell line lacks the oncoprotein p53 [116, 120]. This finding is of high interest for the development of new anticancer agents with different modes of action, notably to overcome resistance to established drugs. Importantly, toxicology studies have shown that **28** has low toxicity for normal tissues, suggesting a good selectivity for tumor cells. Encouraged by these stunning results, Spencer *et al.* recently investigated other palladacycles such as **27** and **29** (Figure 12.13) and found that these organometallics were cytotoxic, contrary to **26**, but, as for **26**, inhibited cat B with IC$_{50}$ values in the low µM range [113, 117].

The rhenium complex **31** shows a remarkably low nanomolar binding affinity for cat B (IC$_{50}$ = 9 nM) and, interestingly, also a 44-fold selectivity for cat B over the closely related cat K [119]. Mechanistic studies indicated the coordination of **31** to the active site cysteine by substitution of the labile chloride ligand of **31**, as for most organic cysteine protease inhibitors, which form a reversible or irreversible covalent bond with the reactive cysteine residue in the active site. The selectivity for cat B over cat K is most likely due to the specific size and shape of the active site of cat B.

The plant enzyme papain from *Carica papaya* is also a cysteine protease and shares many similarities with cathepsins and calpains. Several studies, notably by Salmain *et al.*, have demonstrated the possibility to use organometallics as papain inhibitors through different approaches. Irreversible inactivation of papain was successfully obtained with organometallic maleimide derivatives **32–39** as a result of Michael addition or nucleophilic substitution reactions (**40** and **41**) to the sulfhydryl group of Cys25, which constitutes the active site of papain together with His159 (Figure 12.14) [121–124]. Interestingly, Rudolf *et al.* studied the inhibition of papain with the η^1-*N*-succinimidato analogs **42–44** of the maleimides **35–37** [125]. As expected, **42–44** did not irreversibly inactivate the enzyme as they lack the ethylenic bond responsible for alkylation of Cys25. However, **42–44** behave as reversible inhibitors with IC$_{50}$ values depending on the metal complex. Docking studies confirmed the binding of the complexes to the enzyme's catalytic pocket [125].

Jaouen and coworkers have also reported that compound **46** (Scheme 12.2) is a mixed competitive and non-competitive inhibitor of glutathione *S*-transferase (GST) – meaning that **46** can bind to either the free enzyme or the enzyme–substrate complex [126]. Interestingly, upon photochemical removal of the organometallic moiety of **46** to give **47**, the inhibition properties of **46** were lost. These data suggest that **46** can enter the electrophilic substrate binding site of GST (H-site) and also, at least partially, the glutathione binding site (G-site). This

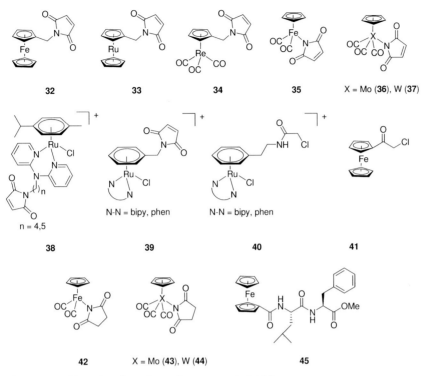

Figure 12.14 Examples of metal complexes as papain inhibitors.

assumption was verified by studying the biochemical behavior of compound **42**, which closely resembles **46**, except that the glutathione part is missing (Scheme 12.2). Compound **42** was found to act as a competitive inhibitor of the conjugation of 1-chloro-2,4-dinitrobenzene (CdNB) to GSH with an inhibition constant K_i of 66 μM. In comparison, **46** was found to be a competitive inhibitor of CdNB ($K_i = 35$ μM), while **47** was found to be an uncompetitive inhibitor ($K_i = 56$ μM). The high inhibition activity of **46** was attributed to the presence of the organometallic $CpFe(CO)_2$ moiety [126].

It should also be stated that Kraatz *et al.* used another approach to inhibit papain, namely, ferrocene derivatives of known peptide inhibitors of papain such as derivative **45** in Figure 12.14 [127–129]. While the metal complex did not

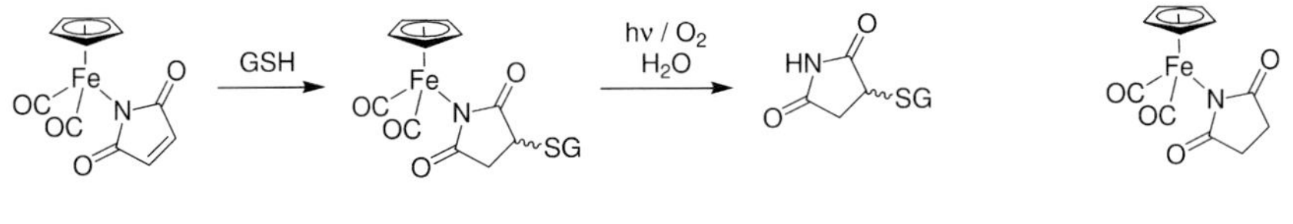

Scheme 12.2 Organometallic inhibitors of GST.

influence the inhibition itself, it served as an electrochemical tag for the detection of the binding between the peptide and papain. This elegant assay system opens possibilities for an alternative screening method for papain and cysteine protease inhibitors in general.

12.10
HIV-1 Reverse Transcriptase and Protease Inhibitors

Human type 1 immunodeficiency virus (HIV-1) has become one of the most studied of all viruses due to its massive threat to health on a global scale. HIV-1 protease (PR) and HIV-1 reverse transcriptase (RT) are two essential enzymes for the life-cycle of HIV-1, the retrovirus that causes AIDS. They are therefore considered as major targets in the treatment of AIDS. Currently, treatment regimes against AIDS infections usually contain a combination of RT and PR inhibitors.

Two main sites have been identified as potential targets for the inhibition of HIV-1 PR: the active site itself and the interface, which is largely responsible for the stabilization of the enzyme's dimeric structure [130]. In an effort to propose original non-peptide HIV-1 PR inhibitors by rational drug design, Lebon, Reboud-Ravaux *et al.* hypothesized that a metal ion could bind HIV-1 PR in place of a water molecule and then further block the activity of the enzyme, presumably by binding additional water molecules. With this in mind, they used the distinct Cu(II) coordination geometry to design compounds that would bind selectively into the HIV-1 PR binding pockets (Figure 12.15a). The most potent copper(II) complex **48** (Figure 12.15b) displayed an IC_{50} value of ~ 1 μM for HIV-1 PR [130]. Molecular modeling suggested that the pyridyl groups of **48** occupy the S_2/S'_2 pockets, while the

Figure 12.15 (a) Pharmacophore for the inhibition of HIV-1 PR by Cu(II) complexes. The enzyme residues are represented in bold style. The octahedral geometry around the metal ion forces the interaction elements into the shape of the active site of the PR. Y represents an H-bond acceptor. S_1/S'_1 and S_2/S'_2 represent the enzyme sub-sites following the Schechter–Berger nomenclature [131]. Figure adapted from reference [130]. (b) Copper complex **48** as competitive inhibitor of HIV-1 PR (X corresponds to CH_3OH) [130].

trimethoxybenzyl groups are located in the bigger S_1/S'_1 sub-sites [132]. However, Cu^{II} complexes such as **48** are kinetically labile and cupric chloride ($CuCl_2$) itself acts as an inhibitor of HIV-1 PR by targeting the thiol group of cysteines [133, 134]. $CuCl_2$ displays the same IC_{50} value as **48**. Therefore, it might well be that the inhibitory effect of **48** was solely due to free cupric ions. Interestingly, however, Lebon, Reboud-Ravaux *et al.* showed that **48** was still able to partially inhibit a recombinant protease devoid of the cysteine residues C67A and C95A, while cupric chloride was not. This suggests a dual effect: part of the inhibition is attributed to free Cu^{2+} and the other part to **48** itself. Other Cu(I) [135] and Cu(II) [136] complexes were also reported as inhibitors of HIV-1 PR, but these compounds did not emerge from a rational design aimed at using the unique structural properties of metal complexes.

Polyoxometalates (POMs) are early transition metal oxygen anion clusters or, more specifically, oligomeric aggregates of metal cations bridged by oxide anions that form by a self-assembly process [137]. Interestingly, these compounds have been shown to exhibit promising *in vitro* and *in vivo* biological activity, ranging from anticancer, antibiotic, antiviral to antidiabetic effects [137–139]. The antiviral activity of POMs has been associated with multiple modes of action, but inhibition of viral enzymes such as reverse transcriptase (RT) [140] and/or protease [141, 142], or inhibition of surface viral proteins such as gp120 for HIV [143, 144], is the most likely cause [137]. The negative charge of POMs is thought to be one of the key features for their interactions with proteins, with POMs binding primarily to positively charged protein regions. However, as most POMs are unstable under physiological conditions, the identification of the active species is problematic and consequently their mechanism of action remains elusive [145]. Nonetheless, Che *et al.* reported recently a ruthenium-oxo oxalate cluster, $Na_7[Ru_4(\mu_3\text{-}O)_4(C_2O_4)_6]$, that is stable under physiologically relevant conditions for 7 days [146]. But, most importantly, they found that this POM was a very strong HIV-1 RT inhibitor (1.9 nM), much more effective than the commonly used drug AZT-triphosphate ($IC_{50} = 68$ nM) [146].

POMs were also found to inhibit or interact with several other enzymes such as ecto-nucleoside triphosphate diphosphohydrolases (E-NTPDases) [147], NADPH oxidase and dehydrogenase, succinoxidase [148], or collagen [149]. However, in no case has the interaction been explored on a structural or atomic level. On the other hand, a recent biochemical paper identifies several POMs as potent ($IC_{50} < 5$ nM) and highly selective inhibitors of protein kinase CK2. This paper also studies in detail possible modes of binding and the cellular response to CK2 inhibition [150].

12.11
Telomerase Inhibitors

Telomerase is a ribonucleoprotein with DNA polymerase activity that maintains the length of telomeric DNA by adding hexameric units to the 3' single strand terminus. The enzyme is crucial for cancer progression, making cancer cells essentially immortal [151]. Knowing that quinoline derivatives show interesting biological properties, especially as enzyme inhibitors [152], Rosenberg, Osella, and

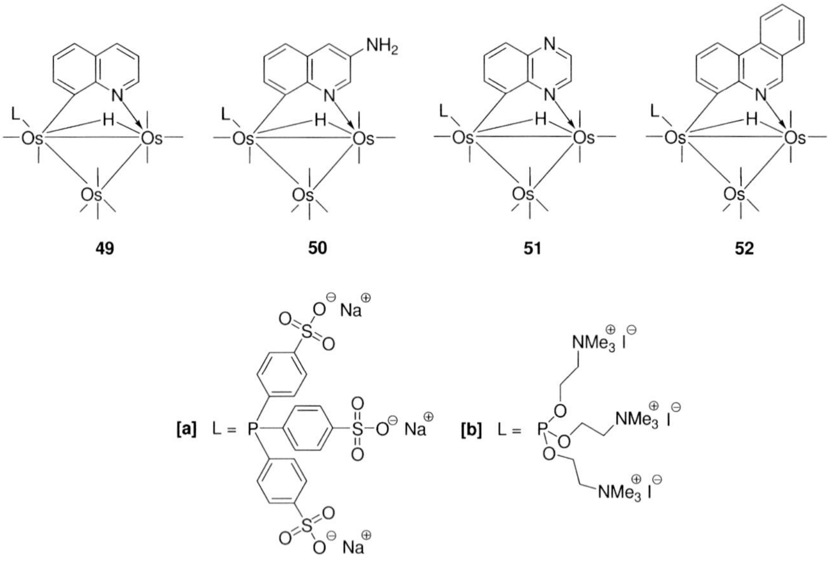

Figure 12.16 Triosmium clusters as potential inhibitors of telomerase enzyme. The symbol (−) indicates the position of additional CO ligands.

coworkers investigated a series of water-soluble benzoheterocycle triosmium clusters as telomerase inhibitors (Figure 12.16) [153]. They discovered that only the negatively charged clusters (by virtue of the sulfonated phosphines) exhibit good anti-telomerase activity when tested on semi-purified enzymes in a cell-free assay. However, unfortunately, these compounds were found to be ineffective *in vitro* on *Taq*, a different DNA polymerase. Furthermore, none of the organometallics tested decreased the telomerase activity in the MCF-7 breast cancer cell line, probably because of the low aptitude of these compounds to cross the cell membrane. But, interestingly, all the clusters were found to be acutely cytotoxic, possibly due to their accumulation on cell membranes.

These workers also examined the interaction of related positively and negatively charged triosmium carbonyl clusters with albumin, using the transverse and longitudinal relaxation times of the hydride resonances as ^{1}H NMR probes of binding to the protein [154]. Evidence of binding was observed for both the positively and negatively charged clusters with, however, distinctly different rotational correlation times [154]. It was assumed that the negatively charged clusters bind more tightly than their positive analogs as albumin is rich in positively charged amino acids [154].

Several other metal complexes were found to be potent telomerase inhibitors, including, for example, Cu(II) [155, 156], Ni(II) [157], Fe(II) [158], Ru(II) [156, 159], Zn(II) [156], Mn(III) [160], and Pt(II) [156, 161–165] complexes (see Figure 12.17 for a few examples).

The cationic porphyrin TMPyP4 (**56**, Figure 12.18) is the most extensively studied G-quadruplex inhibitor to date; **56** inhibits both telomerase (IC$_{50}$ ≈ 0.7–10 μM)

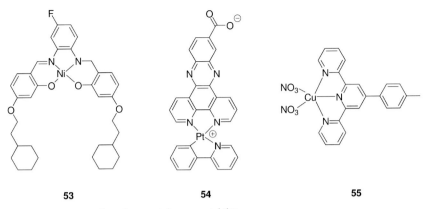

Figure 12.17 Examples of potent telomerase inhibitors.

and *Taq* DNA polymerase (IC$_{50}$ ≈ 2 μM) [151]. However, **56** binds to duplex, triplex, G-quadruplex, single-stranded, and bulk genomic DNA with similar affinities and cannot therefore be considered a structure-selective ligand [151]. Interestingly, Hurley *et al.* analyzed in-depth the influence of metal complexation of the porphyrin on the telomerase inhibition of **56** [166]. They found that the complexes where the porphyrin offered an unhindered face for stacking were the better inhibitors: the square-planar Cu(II) complex and the pyramidal Zn(II) complex [166]. In contrast, octahedral complexes [e.g., Mn(III) and Mg(II)], in which the metal ion is coordinated to two axial ligands, are less active as these ligands block the stacking interactions [166]. A very nice example of the use of metal complexes for enzyme inhibition purposes was recently demonstrated by Luedtke *et al.* They reported a Zn-phthalocyanine (**57**, Figure 12.18), which has the highest affinity reported to

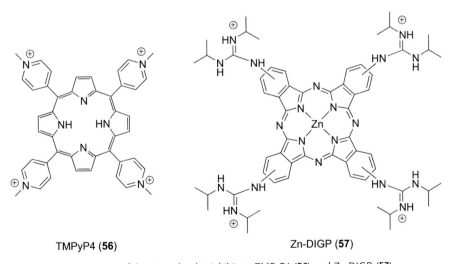

TMPyP4 (**56**)   Zn-DIGP (**57**)

Figure 12.18 Structures of the G-quadruplex inhibitors TMPyP4 (**56**) and Zn-DIGP (**57**).

date to the c-Myc oncogenic promoter G quadruplex DNA ($K_d \leq 2$ nM) [167]. Furthermore, **57** exhibits luminescence upon binding nucleic acids ("turn-on" luminescence) [167].

12.12
Zinc Finger Protein Inhibitors

Zinc finger proteins are involved in various DNA transcriptional and repair functions. Their zinc finger domains, which mediate sequence recognition for a large number of DNA-binding proteins, consist of sequences of amino acids containing cysteine and histidine residues coordinated to a tetrahedral Zn^{2+} ion. Zinc finger proteins have received interest as potential targets for anticancer drugs as well as for HIV inactivation [168, 169]. To inhibit protein function, chelators have been designed to remove the Zn^{2+} ion from its protein coordination environment. For example, Farrell *et al.* have shown that different platinum complexes such as **58** [170, 171] and **59** [168] were indeed able to coordinate to the zinc finger by replacing zinc from the C-terminal knuckle (residues Lys34-Glu53) of the HIV nucleocapsid NCp7 protein (Figure 12.19). A loss of tertiary structure accompanies the loss of the zinc ion [171]. Meade *et al.* have also succeeded in selectively inhibiting zing finger proteins by preparing a Co(III)salen complex (**60**, Figure 12.19), which contained an oligonucleotide as a human transcription factor Sp1 recognition unit. They showed that **60** could selectively inhibit the function of Sp1 in a mixture of six other transcription factors, with no nonspecific inhibition of other DNA-binding proteins [169]. Notably, the gold compound aurothiomalate, which is an antiarthritic drug, inhibits the *in vitro* DNA binding of the model zinc finger transcription factor progesterone to its DNA response element (PRE) at concentrations low enough to be therapeutically relevant [172].

12.13
CXCR4 Inhibitors

The CXCR4 chemokine receptor is a seven-helix transmembrane G-protein coupled receptor with multiple critical functions in both normal and pathological physiology [173, 174]. It is expressed on a wide variety of not only leukocytes, but

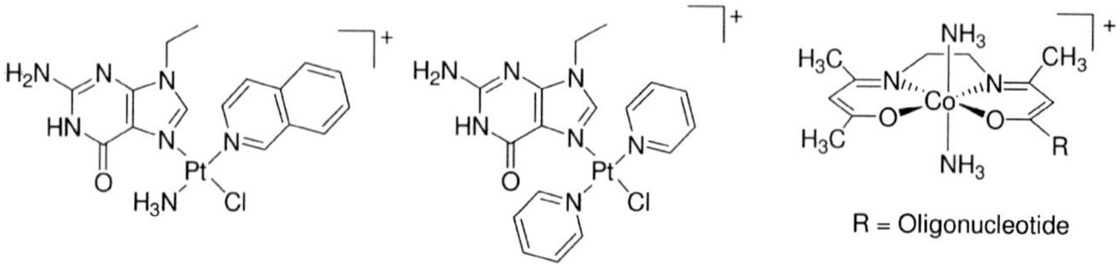

58   **59**   **60**

Figure 12.19 Metal complexes as zinc finger protein inhibitors.

also cells outside the immune system [175]. This cell surface protein also assists in the entry of HIV into cells and anchors stem cells in the bone marrow [10, 176]. Currently, the most potent antagonist available is the nonpeptide bicyclam derivative AMD3100 (**61** in Figure 12.20; also known as Mozobil or Plerixafor) [177, 178]. AMD3100 was approved by the US FDA in December 2008 for use in clinical trials for stem cell mobilization to allow harvesting and transplantation of hematopoietic stem cells in patients with lymphoma and multiple myeloma [179]. It is also a potent anti-HIV agent and has been in clinical trials as an anti-HIV cell entry inhibitor [177, 180, 181]. The affinity of AMD3100 to the CXCR4 receptor was shown to be increased by a factor of 7, 36, and 50 upon complexation with Cu(II), Zn(II), and Ni(II), respectively [175]. This increased binding affinity is obtained through enhanced interaction of one of the cyclam ring systems with the carboxylate group of Asp^{262}, on top of other hydrophobic and H-bonding interactions [175].

Upon metal coordination of AMD3100, the cyclam ring can adopt six possible configurations [182]. Depending on the configurations of the metal complex, its interaction with the protein is strongly influenced. This theme was investigated in-depth by Sadler and coworkers with the Zn(II), Ni(II), and Cu(II) complexes of AMD3100 [180, 183–185]. As a model to study the recognition of proteins by metallomacrocycles, the same researchers also examined the binding of the Cu(II) and Ni(II) complexes of AMD3100 to lysozyme [184, 185]. X-Ray diffraction studies showed that crystals of lysozyme (hen egg white lysozyme – HEWL – from egg white) soaked in Ni(II)-cyclam or Ni$_2$AMD3100 contain two major binding sites. One involves Ni(II) coordination to Asp101 and hydrophobic interactions between the cyclam ring and Trp62 and Trp63. The second is related to hydrophobic interactions with Trp123 [185].

To optimize coordination of the metal drug to the protein, Archibald, Hubin, and coworkers have recently investigated a "trick" to obtain only one chelator "shape" for binding. Hence, they have used configurationally restricted bis-macrocycles such as compound **62** (Figure 12.20) [173, 179, 186]. The bis-copper complex of **62** ([Cu$_2$(**62**)]$^{4+}$) showed improved anti-HIV properties compared to *both* AMD3100 and [Cu$_2$AMD3100]$^{4+}$. This confirms that it is not only the metal coordination that improves the drug interaction – no difference would have been observed between [Cu$_2$(**62**)]$^{4+}$ and [Cu$_2$AMD3100]$^{4+}$ – but also the coordination environment of the copper(II) [179]. A highly rigid chelator such as **62**, with an open face of exchangeable ligands, favors the metal coordination of carboxylate

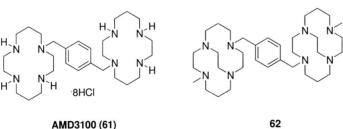

AMD3100 (61)   **62**

Figure 12.20 Structures of the CXCR4 inhibitors AMD3100 (**61**) and **62**.

oxygen atoms from the aspartate or glutamate side chains of CXCR4 [179]. Notably, the same group reported a Cu(II) complex enabling fluorescent labeling of the CXCR4 receptor [187].

12.14
Xanthine Oxidase Inhibitors

Xanthine oxidase (XO) is an enzyme that catalyzes the hydroxylation of hypoxanthine and xanthine to yield uric acid and superoxide anions. The enzyme is responsible for the medical condition known as gout, which is caused by the deposition of uric acid in the joints, leading to painful inflammation [188]. Furthermore, the superoxide anions formed by the enzyme have also been linked to postischemic tissue injury and edema as well as to vascular permeability [188]. The most common potent inhibitor of XO is allopurinol (**63**, Figure 12.21). However, this drug suffers from important disadvantages: its dosing is complex, some patients are hypersensitive to it and it does not prevent the formation of free radicals by the enzyme. To provide alternatives to this drug, Zhu, You, and coworkers used metal complexes to inhibit XO. They reported the preparation and biological evaluation of a series of transition metal complexes derived from Schiff bases, which were obtained from cheap, eco-friendly, and commercially affordable reagents [188, 189]. Among the many compounds reported, the Zn(II) complex **64** and the polymer Cd(II) complex **65** were found to be promising agents with IC_{50} values of 7.23 and 2.16 μM, respectively (Figure 12.21). In comparison, allopurinol displays a value of 10.3 μM. The XO inhibition of the Schiff base metal complexes was assumed to result from the direct interaction of the metal complexes in its "whole complex form" with the active center of the enzyme. Nonetheless, the mechanism of the inhibitory activity requires further investigations.

12.15
Miscellaneous Protein Inhibitors and Conclusions

Interestingly, more and more protein targets are being identified as biochemical experiments are carried out to identify the mode of action of metallodrugs. This

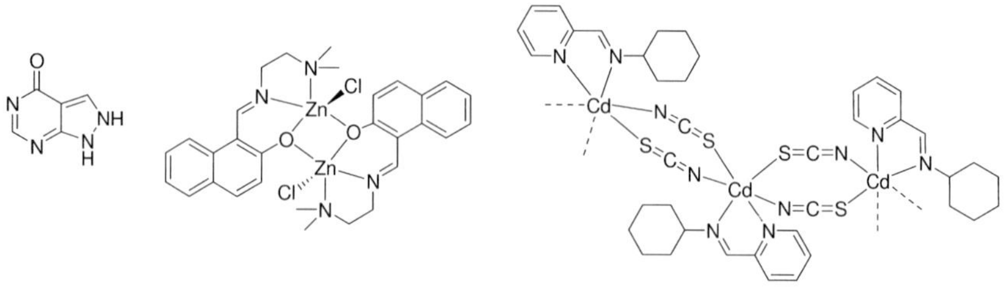

Allopurinol (63)   64   65

Figure 12.21 Structures of xanthine oxidase inhibitors.

amounts almost to a change in paradigm, considering that 20 years ago DNA was automatically assumed as the target for *any* metallodrug candidate, following the work on cisplatin. On the one hand, this change in target identification is a consequence of increasing interdisciplinary work, bringing inorganic chemists and biochemists closer together to work on metallodrugs. It also testifies to new and improved methods for the study of interactions between metal compounds and biomolecules.

In closing, we highlight just three recent examples that underscore this point, two of them related to the cytotoxic Ru(arene) compounds. For this relatively long known class of compounds with anti-proliferative activity, DNA interaction was also originally assumed as the primary target. Recently, Sheldrick and coworkers have used an automated multidimensional protein identification technology (MudPIT), which combines biphasic liquid chromatography with electrospray ionization tandem mass spectrometry (MS/MS), to analyze tryptic peptides from *Escherichia coli*, which were first treated with $[(\eta^6\text{-}p\text{-cymene})RuCl_2(dmso)]$ [190]. They showed that five proteins, namely, the cold-shock protein CspC, the three stress-response proteins ppiD, osmY, and SucC, as well as the DNA damage-inducible helicase dinG were the targets of this Ru-arene compound [190]. Using electrophoretic mobility shift assays, Brabec, Sadler, and coworkers have examined the binding properties of the mismatch repair (MMR) protein MutS in *E. coli* with various DNA duplexes (homoduplexes or mismatched duplexes), all containing a single centrally located adduct of Ru^{II} arene compounds [191]. They showed that the presence of the Ru^{II} arene adducts decreased the affinity of MutS for DNA duplexes that either had a regular sequence or contained a mismatch and that intercalation of the arene contributed considerably to this inhibitory effect [191].

Both examples underscore progress in methodology applied to metal-containing drug candidates, but also an increasing openness of researchers for interdisciplinary work and uncommon ways of looking at seemingly well-known compounds. We are confident that such work will continue to open new vistas on old compounds, and also inspire inorganic chemists to devise wholly new classes of active agents by a biologically informed design. As a final example, a chromium derivative, which mimics a fatty acid biosynthesis enzyme inhibitor in bacteria, was reported recently [192]. This compound draws its inspiration from organic natural product synthesis [193, 194], molecular docking experiments, and biochemical screening. The idea was put into place by a twelve-step organometallic synthesis, hence underscoring the high level of sophistication that medicinal inorganic chemistry has achieved [14, 15].

Acknowledgments

The authors thank the Alexander von Humboldt Foundation (fellowship to G. G.) and the Deutsche Forschungsgemeinschaft (DFG) through the research unit "Biological Function of Organometallic Compounds" (FOR 630, http://www.rub.de/for630/) for funding. Dr. Christian Grütter from the Chemical Genomics

Centre in Dortmund is acknowledged for his kind and expert help with the preparation of the protein structure figures. N. M. N. has benefited greatly from interactions with colleagues through the funding of the European COST D39 action "Metallodrug Design and Action" and wishes to express his gratitude for the existence of this efficient and cost-effective instrument of scientific exchange.

References

1 Trawick, B.N., Daniher, A.T., and Bashkin, J.K. (1998) *Chem. Rev.*, **98**, 939.

2 Liu, C., Wang, M., Zhang, T., and Sun, H. (2004) *Coord. Chem. Rev.*, **248**, 147.

3 Suh, J. and Chei, W.S. (2008) *Curr. Opin. Chem. Biol.*, **12**, 207.

4 Jeon, J.W., Son, S.J., Yoo, C.E., Hong, I.S., and Suh, J. (2003) *Bioorg. Med. Chem.*, **11**, 2901.

5 Metzler-Nolte, N. (2006) in *Bioorganometallics* (ed. G. Jaouen), Wiley-VCH, Verlag GmbH, Weinheim, p. 125.

6 Metzler-Nolte, N. (2007) *Chimia*, **61**, 736.

7 Kirin, S.I., Kraatz, H.-B., and Metzler-Nolte, N. (2006) *Chem. Soc. Rev.*, **35**, 348.

8 Jacobsen, F.E., Lewis, J.A., and Cohen, S.M. (2007) *ChemMedChem.*, **2**, 152.

9 Louie, A.Y. and Meade, T.J. (1999) *Chem. Rev.*, **99**, 2711.

10 Meggers, E. (2009) *Chem. Commun.*, 1001.

11 Fricker, S.P. (2007) *Dalton Trans.*, 903.

12 Berners-Price, S.J. and Sadler, P.J. (1996) *Coord. Chem. Rev.*, **151**, 1.

13 Salmain, M. and Metzler-Nolte, N. (2008) in *Ferrocenes* (ed. P. Stepnicka), John Wiley & Sons, Ltd., Chichester, p. 499.

14 Schatzschneider, U. and Metzler-Nolte, N. (2006) *Angew. Chem. Int. Ed.*, **45**, 1504.

15 Gasser, G., Ott, I., and Metzler-Nolte, N. (2009) *J. Med. Chem.*, accepted.

16 Hartinger, C. and Dyson, P.J. (2009) *Chem. Soc. Rev.*, **38**, 391.

17 Dougan, S.J. and Sadler, P.J. (2007) *Chimia*, **61**, 704.

18 Peacock, A.F.A. and Sadler, P.J. (2008) *Chem. Asian J.*, **3**, 1890.

19 Süss-Fink, G. (2010) *Dalton Trans.*, 1673.

20 Zhang, L., Carroll, P., and Meggers, E. (2004) *Org. Lett.*, **6**, 521.

21 Jänne, P.A., Gray, N., and Settleman, J. (2009) *Nat. Rev. Drug Discovery*, **8**, 709.

22 Bikker, J.A., Brooijmans, N., Wissner, A., and Mansour, T.S. (2009) *J. Med. Chem.*, **52**, 1493.

23 Williams, D.S., Carroll, P.J., and Meggers, E. (2007) *Inorg. Chem.*, **46**, 2944.

24 Meggers, E. (2007) *Curr. Opin. Chem. Biol.*, **11**, 287.

25 Meggers, E., Atilla-Gokcumen, G.E., Bregman, H., Maksimoska, J., Mulcahy, S.P., Pagano, N., and Williams, D.S. (2007) *Synlett*, 1177.

26 Smalley, K.S.M., Contractor, R., Haass, N.K., Kulp, A.N., Atilla-Gokcumen, G. E., Williams, D.S., Bregman, H., Flaherty, K.T., Soengas, M.S., Meggers, E., and Herlyn, M. (2007) *Cancer Res.*, **67**, 209.

27 Maksomiska, J., Williams, D.S., Atilla-Gokcumen, G.E., Smalley, K.S.M., Carroll, P.J., Webster, R.D., Filippakopoulos, P., Knapp, S., Herlyn, M., and Meggers, E. (2008) *Chem. Eur. J.*, **14**, 4816.

28 Maksimoska, J., Feng, L., Harms, K., Yi, C., Kissil, J., Marmorstein, R., and Meggers, E. (2008) *J. Am. Chem. Soc.*, **130**, 15764.

29 Pagano, N., Maksimoska, J., Bregman, H., Williams, D.S., Webster, R.D., Xue, F., and Meggers, E. (2007) *Org. Biomol. Chem.*, **5**, 1218.

30 Bregman, H., Williams, D.S., Atilla, G. E., Carroll, P.J., and Meggers, E. (2004) *J. Am. Chem. Soc.*, **126**, 13594.

31 Atilla-Gokcumen, G.E., Williams, D.S., Bregman, H., Pagano, N., and

Meggers, E. (2006) *ChemBioChem.*, **7**, 1443.

32 Bregman, H., Carroll, P.J., and Meggers, E. (2006) *J. Am. Chem. Soc.*, **128**, 877.

33 Debreczeni, J.E., Bullock, A.N., Atilla, G.E., Williams, D.S., Bregman, H., Knapp, S., and Meggers, E. (2006) *Angew. Chem. Int. Ed.*, **45**, 1580.

34 Xie, P., Williams, D.S., Atilla-Gokcumen, G.E., Milk, L., Xiao, M., Smalley, K.S.M., Herlyn, M., Meggers, E., and Marmorstein, R. (2008) *ACS Chem. Biol.*, **3**, 305.

35 Bregman, H. and Meggers, E. (2006) *Org. Lett.*, **8**, 5465.

36 Williams, D.S., Atilla, G.E., Bregman, H., Arzoumanian, A., Klein, P.S., and Meggers, E. (2005) *Angew. Chem. Int. Ed.*, **44**, 1984.

37 Mulcahy, S.P., Li, S., Korn, R., Xie, X., and Meggers, E. (2008) *Inorg. Chem.*, **47**, 5030.

38 Xie, P., Streu, C., Qin, J., Bregman, H., Pagano, N., Meggers, E., and Marmorstein, R. (2009) *Biochemistry*, **48**, 5187.

39 Anand, R., Maksimoska, J., Pagano, N., Wong, E.Y., Gimotty, P.A., Phyllis, A., Diamond, S.L., Meggers, E., and Marmorstein, R. (2009) *J. Med. Chem.*, **52**, 1602.

40 Atilla-Gokcumen, G.E., Pagano, N., Streu, C., Maksimoska, J., Filippakopoulos, P., Knapp, S., and Meggers, E. (2008) *ChemBioChem.*, **9**, 2933.

41 Goldberg, A.L. (1995) *Science*, **268**, 522.

42 Peters, J.M., Franke, W.W., and Kleinschmidt, J.A. (1994) *J. Biol. Chem.*, **269**, 7709.

43 Seemuller, E., Lupas, A., Stock, D., Lowe, J., Huber, R., and Baumeister, W. (1995) *Science*, **268**, 579.

44 Voges, D., Zwickl, P., and Baumeister, W. (1999) *Annu. Rev. Biochem.*, **68**, 1015.

45 Frezza, M., Hindo, S.S., Tomco, D., Allard, M.M., Cui, Q.C., Heeg, M.J., Chen, D., Ping Dou, Q., and Verani, C. N. (2009) *Inorg. Chem.*, **48**, 5928.

46 Ping Dou, Q. and Goldfarb, R.H. (2002) *IDrugs*, **5**, 828.

47 Kane, R.C., Farrell, A.T., Sridhara, R., and Pazdur, R. (2006) *Clin. Cancer Res.*, **12**, 2955.

48 Chen, D., Frezza, M., Shakya, R., Cui, Q.C., Milacic, V., Verani, C.N., and Ping Dou, Q. (2007) *Cancer Res.*, **67**, 9258.

49 Hindo, S.S., Frezza, M., Tomco, D., Heeg, M.J., Hryhorczuk, L., McGarvey, B.R., Ping Dou, Q., and Verani, C.N. (2009) *Eur. J. Med. Chem.*, **44**, 4353.

50 Daniel, K.G., Gupta, P., Hope Harbach, R., Guida, W.C., and Ping Dou, Q. (2004) *Biochem. Pharmacol.*, **67**, 1139.

51 Rudnev, A.V., Foteeva, L.S., Kowol, C., Berger, R., Jakupec, M.A., Arion, V.B., Timerbaev, A.R., and Keppler, B.K. (2006) *J. Inorg. Biochem.*, **100**, 1819.

52 Jakupec, M.A. and Keppler, B.K. (2004) *Curr. Top. Med. Chem.*, **4**, 575.

53 Supuran, C.T., Scozzafava, A., and Conway, J. (2004) *Carbonic Anhydrase – Its Inhibitors and Activators*, CRC Press, Boca Raton.

54 Supuran, C.T. and Scozzafava, A. (2007) *Bioorg. Med. Chem.*, **15**, 4336.

55 Supuran, C.T. (2008) *Nat. Rev. Drug. Discovery*, **7**, 168.

56 Lane, T.W. and Morel, F.M.M. (2000) *Proc. Natl. Acad. Sci. USA*, **97**, 4627.

57 Puccetti, L., Fasolis, G., Vullo, D., Chohan, Z.H., Scozzafava, A., and Supuran, C.T. (2005) *Bioorg. Med. Chem. Lett.*, **15**, 3096.

58 Rami, M., Cecchi, A., Montero, J.-L., Innocenti, A., Vullo, D., Scozzafava, A., Winum, J.-Y., and Supuran, C.T. (2008) *Chem. Med. Chem.*, **3**, 1780.

59 Scozzafava, A., Menabuoni, L., Mincione, F., and Supuran, C.T. (2002) *J. Med. Chem.*, **45**, 1466.

60 Borras, J., Alzuet, G., Ferrer, S., and Supuran, C.T. (2004) in *Carbonic Anhydrase* (eds C.T. Supuran, A. Scozzafava, and J. Conway), CRC Press LLC, Boca Raton, p. 183.

61 Ferrer, S., Alzuet, G., and Borras, J. (1989) *J. Inorg. Biochem.*, **39**, 297.

62 Alzuet, G., Casella, L., Perotti, A., and Borras, J. (1994) *J. Chem. Soc., Dalton Trans.*, 2347.

63 Supuran, C.T., Mincione, F., Scozzafava, A., Briganti, F., Mincione, G., and Ilies, M.A. (1998) *Eur. J. Med. Chem.*, **33**, 247.

64 Supuran, C.T., Scozzafava, A., Mincione, F., Menabuoni, L., Briganti, F., Mincione, G., and Jitianu, M. (1999) *Eur. J. Med. Chem.*, **34**, 585.

65 Supuran, C.T., Manole, F., and Andruh, M. (1993) *J. Inorg. Biochem.*, **49**, 97.

66 Sumalan, S.L., Casanova, J., Alzuet, G., Borras, J., Castineiras, A., and Supuran, C.T. (1996) *J. Inorg. Biochem.*, **62**, 31.

67 Rami, M., Winum, J.-Y., Innocenti, A., Montero, J.-L., Scozzafava, A., De Simone, G., and Supuran, C.T. (2008) *Bioorg. Med. Chem. Lett.*, **18**, 836.

68 Scozzafava, A., Menabuoni, L., Mincione, F., Mincione, G., and Supuran, C.T. (2001) *Bioorg. Med. Chem. Lett.*, **11**, 575.

69 Roy, B.C., Fazal, M.A., Sun, S., and Mallik, S. (2000) *Chem. Commun.*, 547.

70 Fazal, M.A., Roy, B.C., Sun, S., Mallik, S., and Rogers, K.R. (2001) *J. Am. Chem. Soc.*, **123**, 6283.

71 Roy, B.C., Rodendahl, T., Hegge, R., Peterson, R., Mallik, S., and Srivastava, D.K. (2003) *J. Chem. Soc., Chem. Commun.*, 2328.

72 Roy, B.C., Banerjee, A.L., Swanson, M., Jia, X.G., Haldar, M.K., Mallik, S., and Srivastava, D.K. (2004) *J. Am. Chem. Soc.*, **126**, 13206.

73 Banerjee, A.L., Eiler, D., Roy, B.C., Jia, X., Haldar, M.K., Mallik, S., and Srivastava, D.K. (2005) *Biochemistry*, **44**, 3211.

74 Jude, K.M., Banerjee, A.L., Haldar, M. K., Manokaran, S., Roy, B., Mallik, S., Srivastava, D.K., and Christianson, D. W. (2006) *J. Am. Chem. Soc.*, **128**, 3011.

75 Elegbede, A.I., Haldar, M.K., Manokaran, S., Kooren, J., Roy, B.C., Mallik, S., and Srivastava, D.K. (2007) *Chem. Commun.*, 3377.

76 Banerjee, A.L., Tobwala, S., Haldar, M. K., Swanson, M., Roy, B.C., Mallik, S., and Srivastava, D.K. (2005) *Chem. Commun.*, 2549.

77 Giardello, F.M., Offerhaus, G.J.A., and DuBois, R.N. (1995) *Eur. J. Cancer*, **31A**, 1071.

78 Coussens, L.M. and Werb, Z. (2002) *Nature*, **420**, 860.

79 Ulrich, C.M., Bigler, J., and Potter, J.D. (2006) *Nat. Rev. Cancer*, **6**, 130.

80 Ott, I., Kircher, B., Bagowski, C.P., Vlecken, D.H.W., Ott, E.B., Will, J., Bensdorf, K., Sheldrick, W.S., and Gust, R. (2009) *Angew. Chem. Int. Ed.*, **48**, 1160.

81 Schmidt, K., Jung, M., Keilitz, R., Schnurr, B., and Gust, R. (2000) *Inorg. Chim. Acta*, **306**, 6.

82 Ott, I., Schmidt, K., Kircher, B., Schumacher, P., Wiglenda, T., and Gust, R. (2005) *J. Med. Chem.*, **48**, 622.

83 Ott, I. and Gust, R. (2005) *Biometals*, **18**, 171.

84 Sergeant, C.D., Ott, I., Sniady, A., Meneni, S., Gust, R., Rheingold, A.L., and Dembinski, R. (2008) *Org. Biomol. Chem.*, **6**, 73.

85 Ott, I., Koch, T., Shorafa, H., Bai, Z., Poeckel, D., Steinhilber, D., and Gust, R. (2005) *Org. Biomol. Chem.*, **3**, 2282.

86 Neukamm, M.A., Pinto, A., and Metzler-Nolte, N. (2008) *Chem. Commun.*, 232.

87 Schlenk, M., Ott, I., and Gust, R. (2008) *J. Med. Chem.*, **51**, 7318.

88 Osella, D., Galeotti, F., Cavigiolio, G., Nervi, C., Hardcastle, K.I., Vessières, A., and Jaouen, G. (2002) *Helv. Chim. Acta*, **85**, 2918.

89 Top, S., El Hafa, H., Vessières, A., HuchÅ, M., Vaissermann, J., and Jaouen, G. (2002) *Chem. Eur. J.*, **8**, 5241.

90 Dwyer, F.P., Gyarfas, E.C., Rogers, W. P., and Koch, J.H. (1952) *Nature*, **170**, 190.

91 Dwyer, F.P., Gyarfas, E.C., Wright, R. D., and Shulman, A. (1957) *Nature*, **179**, 435.

92 Bourne, Y., Taylor, P., Radic, Z., and Marchot, P. (2003) *EMBO J.*, **22**, 1.

93 Gong, L., Mulcahy, S.P., Harms, K., and Meggers, E. (2009) *J. Am. Chem. Soc.*, **131**, 9602.

94 Kirin, S., Krämer, R., and Metzler-Nolte, N. (2006) in *Concepts and Models in Bioinorganic Chemistry* (eds H.-B. Kraatz and N. Metzler-Nolte), Wiley-VCH Verlag, GmbH, Weinheim, p. 159.

95 Davies, D.R. and Hol, W.G.J. (2004) *FEBS Lett.*, **577**, 315.

96 Zhang, M., Zhou, M., VanEtten, R.L., and Stauffacher, C.V. (1997) *Biochemistry*, **36**, 15.

97 Rehder, D. (2008) *J. Inorg. Biochem.*, **102**, 1152.

98 Rosivatz, E., Matthews, J.G., McDonald, N.Q., Mulet, X., Ho, K.K., Lossi, N., Schmid, A.C., Mirabelli, M., Pomeranz, K.M., Erneux, C., Lam, E. W.F., Vilar, R., and Woscholski, R. (2006) *ACS Chem. Biol.*, **1**, 780.

99 Lee, J.O., Yang, H.J., Georgescu, M.M., Di Cristofano, A., Maehama, T., Shi, Y. G., Dixon, J.E., Pandolfi, P., and Pavletich, N.P. (1999) *Cell*, **99**, 323.

100 Huyer, G., Liu, S., Kelly, J., Moffat, J., Payette, P., Kennedy, B., Tsaprailis, G., Gresser, M.J., and Ramachandran, C. (1997) *J. Biol. Chem.*, **272**, 843.

101 Nakai, M., Obata, M., Sekiguchi, F., Kato, M., Shiro, M., Ichimura, A., Kinoshita, I., Mikuriya, M., Inohara, T., Kawabe, K., Sakurai, H., Orvig, C., and Yano, S. (2004) *J. Inorg. Biochem.*, **98**, 105.

102 Thompson, K.H. and Orvig, C. (2006) *J. Inorg. Biochem.*, **100**, 1925.

103 Thompson, K.H. and Orvig, C. (2004), in *Metal Ions and Their Complexes in Medication* (eds A. Sigel and H. Sigel) Metal Ions in Biological Systems, Vol. 41, Marcel Dekker, New York, p. 221.

104 Thompson, K.H., McNeill, J.H., and Orvig, C. (1999) *Chem. Rev.*, **99**, 2561.

105 Toyota, E., Ng, K.K.S., Sekizaki, H., Itoh, K., Tanizawa, K., and James, M. N.G. (2001) *J. Mol. Biol.*, **305**, 471.

106 Toyota, E., Chinen, C., Sekizaki, H., Itoh, K., and Tanizawa, K. (1996) *Chem. Pharm. Bull.*, **44**, 1104.

107 Toyota, E., Miyazaki, H., Itoh, K., Sekizaki, H., and Tanizawa, K. (1999) *Chem. Pharm. Bull.*, **47**, 116.

108 Toyota, E., Sekizaki, H., Itoh, K., and Tanizawa, K. (2003) *Chem. Pharm. Bull.*, **51**, 625.

109 Toyota, E., Sekizaki, H., Takahashi, Y., Itoh, K., and Tanizawa, K. (2005) *Chem. Pharm. Bull.*, **53**, 22.

110 Takeuchi, T., Böttcher, A., Quezada, C. M., Simon, M.I., Meade, T.J., and Gray, H.B. (1998) *J. Am. Chem. Soc.*, **120**, 8555.

111 Katz, B.A., Clark, J.M., Finer-Moore, J. S., Jenkins, T.E., Johnson, C.R., Luong, C., Moore, W.R., and Stroud, R.M. (1998) *Nature*, **391**, 608.

112 Janc, J.W., Clark, J.M., Warne, R.L., Elrod, K.C., Katz, B.A., and Moore, W. R. (2000) *Biochemistry*, **39**, 4792.

113 Spencer, J., Casini, A., Zava, O., Rathnam, R.P., Velhanda, S.K., Hursthouse, M.B., and Dyson, P.J. (2009) *Dalton Trans.*, 10731.

114 Casini, A., Gabbiani, C., Sorrentino, F., Rigobello, M.P., Bindoli, A., Geldbach, T.J., Marrone, A., Re, N., Hartinger, C. G., Dyson, P.J., and Messori, L. (2008) *J. Med. Chem.*, **51**, 6773.

115 Bincoletto, C., Tersariol, I.L.S., Oliveira, C.R., Dreher, S., Fausto, D.M., Soufen, M.A., Nascimento, F.D., and Caires, A. C.F. (2005) *Bioorg. Med. Chem. Lett.*, **13**, 3042.

116 Caires, A.C. (2007) *Anti-Cancer Agents Med. Chem.*, **7**, 484.

117 Spencer, J., Rathnam, R.P., Motukuri, M., Kotha, A.K., Richardson, S.C.W., Hazrati, A., Hartley, J.A., Male, L., and Hursthouse, M.B. (2009) *Dalton Trans.*, 4299.

118 Chircorian, A. and Barrios, A.M. (2004) *Bioorg. Med. Chem. Lett.*, **14**, 5113.

119 Mosi, R., Baird, I.R., Cox, J., Anastassov, V., Cameron, B., Skerlj, R. T., and Fricker, S.P. (2006) *J. Med. Chem.*, **49**, 5262.

120 Barbosa, C.M.V., Oliveira, C.R., Nascimento, F.D., Smith, M.C.M., Soufen, M.A., Sena, E., Araújo, R.C., Tersariol, I.L.S., Bincoletto, C., and Caires, A.C. (2006) *Eur. J. Pharmacol.*, **542**, 37.

121 Douglas, K.T., Ejim, O.S., and Taylor, K. (1992) *J. Enzyme Inhib. Med. Chem.*, **6**, 233.

122 Haquette, P., Salmain, M., Svedlung, K., Martel, A., Rudolf, B., Zakrzewski, J., Cordier, S., Roisnel, T., Fosse, C., and Jaouen, G. (2007) *ChemBioChem.*, **8**, 224.

123 Haquette, P., Talbi, B., Canaguier, S., Dagorne, S., Fosse, C., Martel, A.,

Jaouen, G., and Salmain, M. (2008) *Tetrahedron Lett.*, **49**, 4670.

124 Haquette, P., Dumat, B., Talbi, B., Arbabi, S., Renaud, J.-L., Jaouen, G., and Salmain, M. (2009) *J. Organomet. Chem.*, **694**, 937.

125 Rudolf, B., Salmain, M., Martel, A., Palusiak, M., and Zakrzewski, J. (2009) *J. Inorg. Biochem.*, **103**, 1162.

126 Salmain, M., Jaouen, G., Rudolf, B., and Zakrzewski, J. (1999) *J. Organomet. Chem.*, **589**, 98.

127 Plumb, K. and Kraatz, H.-B. (2003) *Bioconjugate Chem.*, **14**, 601.

128 Mahmoud, K.A. and Kraatz, H.-B. (2007) *Chem. Eur. J.*, **13**, 5885.

129 Kerman, K. and Kraatz, H.-B. (2009) *Analyst*, **134**, 2400.

130 Lebon, F., Boggetto, N., Ledecq, M., Durant, F., Benatallah, Z., Sicsic, S., Lapouyade, R., Kahn, O., Mouithys-Mickalad, A., Deby-Dupont, G., and Reboud-Ravaux, M. (2002) *Biochem. Pharmacol.*, **63**, 1863.

131 Schechter, I. and Berger, A. (1967) *Biochem. Biophys. Res. Commun.*, **27**, 157.

132 Ledecq, M., Lebon, F., Durant, F., Giessner-Prettre, C., Marquez, A., and Gresh, N. (2003) *J. Phys. Chem. B*, **107**, 10640.

133 Karlström, A.R. and Levine, R.L. (1991) *Proc. Natl. Acad. Sci. USA*, **88**, 5552.

134 Karlström, A.R., Shames, B.D., and Levine, R.L. (1993) *Arch. Biochem. Biophys.*, **304**, 163.

135 Davis, D.A., Branca, A.A., Pallenberg, A.J., Marschner, T.M., Patt, L.M., Chatlynne, L.G., Humphrey, R.W., Yarchoan, R., and Levine, R.L. (1995) *Arch. Biochem. Biophys.*, **322**, 127.

136 DeCamp, D.L., LBabe, L.M., Salto, R., Lucich, J.L., Koo, M.-S., Kahl, S.B., and Craik, C.S. (1992) *J. Med. Chem.*, **35**, 3426.

137 Rhule, J.T., Hill, C.L., Judd, D.A., and Schinazi, R.F. (1998) *Chem. Rev.*, **98**, 327.

138 Hasenknopf, B. (2005) *Front. Biosci.*, **10**, 275.

139 Pope, M.T. and Müller, A. (2001) *Polyoxometalate Chemistry: From Topology via Self-Assembly to Applications*, Kluwer, Dordrecht-Boston-London.

140 Sarafianos, S.G., Kortz, U., Pope, M.T., and Modak, M.J. (1996) *Biochem. J.*, **319**, 619.

141 Hill, C.L., Judd, D.A., Tang, J., Nettles, J., and Schinazi, R.F. (1997) *Antiviral Res.*, **34**, A43.

142 Judd, D.A., Nettles, J.H., Nevins, N., Snyder, J.P., Liotta, D.C., Tang, J., Ermolieff, J., Schinazi, R.F., and Hill, C.L. (2001) *J. Am. Chem. Soc.*, **123**, 886.

143 Judd, D.A., Schinazi, R.F., and Hill, C. L. (1994) *Antiviral Chem. Chemother.*, **5**, 410.

144 Take, Y., Tokutake, Y., Inouye, Y., Yoshida, T., Yamamoto, A., Yamase, T., and Nakamura, S. (1991) *Antiviral Res.*, **15**, 113.

145 Hill, C.L., Weeks, M.S., and Schinazi, R.F. (1990) *J. Med. Chem.*, **33**, 2767.

146 Wong, E.L.-M., Sun, R.W.-Y., Chung, N.P.-Y., Lin, C.-L.S., Zhu, N., and Che, C.-M. (2006) *J. Am. Chem. Soc.*, **128**, 4938.

147 Müller, C.E., Iqbal, J., Baqi, Y., Zimmermann, H., Röllich, A., and Stephan, H. (2006) *Bioorg. Med. Chem. Lett.*, **16**, 5943.

148 Fessenden-Raden, J.M. (1971) *J. Biol. Chem.*, **246**, 6745.

149 Nemetschek, T., Riedl, H., and Jonak, R. (1979) *J. Mol. Biol.*, **133**, 67.

150 Prudent, R., Moucadel, V., Laudet, B., Barette, C., Lafanechere, L., Hasenknopf, B., Li, J., Bareyt, S., Lacote, E., Thorimbert, S., Malacria, M., Gouzerh, P., and Cochet, C. (2008) *Chem. Biol.*, **15**, 683.

151 Luedtke, N.W. (2009) *Chimia*, **63**, 134.

152 Kim, J.H., Lee, G.E., Kim, S.W., and Chung, I.K. (2003) *Biochem. J.*, **373**, 523.

153 Colangelo, D., Ghiglia, A.L., Ghezzi, A. R., Ravera, M., Rosenberg, E., Spada, F., and Osella, D. (2005) *J. Inorg. Biochem.*, **99**, 505.

154 Nervi, C., Gobetto, R., Milone, L., Viale, A., Rosenberg, E., Spada, F., Rokhsana, D., and Fiedler, J. (2004) *J. Organomet. Chem.*, **689**, 1796.

155 Evans, S.E., Mendez, M.A., Turner, K. B., Keating, L.R., Grimes, R.T.,

Melchoir, S., and Szalai, V.A. (2007) *J. Biol. Inorg. Chem.*, **12**, 1235.

156 Bertrand, H., Monchaud, D., De Cian, A., Guillot, R., Mergny, J.L., and Teulade-Fichou, M.P. (2007) *Org. Biomol. Chem.*, **5**, 2555.

157 Reed, J.E., Arnal, A.A., Neidle, S., and Vilar, R. (2006) *J. Am. Chem. Soc.*, **128**, 5992.

158 Tuntiwechapikul, W. and Salazar, M. (2001) *Biochemistry*, **40**, 13652.

159 Shi, S., Liu, J., Yao, T., Geng, X., Jiang, L., Yang, Q., Cheng, L., and Ji, L. (2008) *Inorg. Chem.*, **47**, 2910.

160 Dixon, I.M., Lopez, F., Tejera, A.M., Esteve, J.-P., Blasco, M.A., Pratviel, G., and Meunier, B. (2007) *J. Am. Chem. Soc.*, **129**, 1502.

161 D.-L. Ma, C.-M. Che, and Yan, S.-C. (2009) *J. Am. Chem. Soc.*, **131**, 1835.

162 Bertrand, H., Bombard, S., Monchaud, D., and Teulade-Fichou, M.P. (2007) *J. Biol. Inorg. Chem.*, **12**, 1003.

163 Kieltyka, R., Fakhoury, J., Moitessier, N., and Sleiman, H.F. (2008) *Chem. Eur. J.*, **14**, 1145.

164 Kieltyka, R., Englebienne, P., Fakhoury, J., Autexier, C., Moitessier, N., and Sleiman, H.F. (2008) *J. Am. Chem. Soc.*, **130**, 10040.

165 Wu, P., Ma, D.-W., Leung, C.-H., Yan, S.-C., Zhu, N., Abagyan, R., and Che, C.-M. (2009) *Chem. Eur. J.*, **15**, 13008.

166 Shi, D.-F., Wheelhouse, R.T., Sun, D., and Hurley, L.H. (2001) *J. Med. Chem.*, **44**, 4509.

167 Alzeer, J., Vummidi, B.R., Roth, P.J.C., and Luedtke, N.W. (2009) *Angew. Chem. Int. Ed.*, **48**, 9362.

168 Liu, Q., Golden, M., Darensbourg, M. Y., and Farrell, N. (2005) *Chem. Commun.*, 4360.

169 Louie, A.Y. and Meade, T.J. (1998) *Proc. Natl. Acad. Sci. USA*, **95**, 6663.

170 Sartori, D.A., Miller, B., Bierbach, U., and Farrell, N. (2000) *J. Biol. Inorg. Chem.*, **5**, 575.

171 Anzellotti, A.I., Liu, Q., Bloemink, M. J., Scarsdale, J.N., and Farrell, N. (2006) *Chem. Biol.*, **13**, 539.

172 Handel, M.L., deFazio, A., Watts, C.K., Day, R.O., and Sutherland, R.L. (1991) *Mol. Pharmacol.*, **40**, 613.

173 McRobbie, G., Valks, G.C., Empson, C. J., Khan, A., Silversides, J.D., Pannecouque, C., De Clercq, E., Fiddy, S.G., Bridgeman, A.J., Young, N.A., and Archibald, S.J. (2007) *Dalton Trans.*, 5008.

174 Khan, A., Greenman, J., and Archibald, S.J. (2007) *Curr. Med. Chem.*, **14**, 2257.

175 Gerlach, L.O., Jakobsen, J.S., Jensen, K. P., Rosenkilde, M.R., Skerlj, R.T., Ryde, U., Bridger, G.J., and Schwartz, T.W. (2003) *Biochemistry*, **42**, 710.

176 Ronconi, L. and Sadler, P.J. (2007) *Coord. Chem. Rev.*, **251**, 1633.

177 De Clercq, E., Yamamoto, N., Pauwels, R., Baba, M., Schols, D., Nakashima, H., Balzarini, J., Debyser, Z., Murrer, B., Schwartz, D., Thornton, D., Bridger, G., Fricker, S., Henson, G., Abrams, M., and Picker, D. (1992) *Proc. Natl. Acad. Sci. USA*, **89**, 5286.

178 De Clercq, E. (2003) *Nat. Rev. Drug Discovery*, **2**, 581.

179 Khan, A., Nicholson, G., Greenman, J., Madden, L., McRobbie, G., Pannecouque, C., De Clercq, E., Ullom, R., Maples, D.L., Maples, R.D., Silversides, J.D., Hubin, T.J., and Archibald, S.J. (2009) *J. Am. Chem. Soc.*, **131**, 3416.

180 Liang, X., Parkinson, J.A., Weishaeupl, M., Gould, R.O., Paisey, S.J., Park, H.-S., Hunter, T.M., Blindauer, C.A., Parsons, S., and Sadler, P.J. (2002) *J. Am. Chem. Soc.*, **124**, 9105.

181 Bridger, G.J., Skerlj, R.T., Thornton, D., Padmanabhan, S., Martellucci, S.A., Henson, G.W., Abrams, M.J., Yamamoto, N., De Vreese, K., Pauwels, R., and De Clercq, E. (1995) *J. Med. Chem.*, **38**, 366.

182 Bosnich, B., Poon, C.K., and Tobe, M. L. (1965) *Inorg. Chem.*, **4**, 1102.

183 Liang, X., Weishaeupl, M., Parkinson, J.A., Parsons, S., McGregor, P.A., and Sadler, P.J. (2003) *Chem. Eur. J.*, **9**, 4709.

184 Hunter, T.M., McNae, I.W., Liang, X., Bella, J., Parsons, S., Walkinshaw, M. D., and Sadler, P.J. (2005) *Proc. Natl. Acad. Sci. USA*, **102**, 2288.

185 Hunter, T.M., McNae, I.W., Simpson, D.P., Smith, A.M., Moggach, S., White,

F., Walkinshaw, M.D., Parsons, S., and Sadler, P.J. (2007) *Chem. Eur. J.*, **13**, 40.

186 Valks, G.C., McRobbie, G., Lewis, E.A., Hubin, T.J., Hunter, T.M., Sadler, P.J., Pannecouque, C., De Clercq, E., and Archibald, S.J. (2006) *J. Med. Chem.*, **49**, 6162.

187 Khan, A., Silversides, J.D., Madden, L., Greenman, J., and Archibald, S.J. (2007) *Chem. Commun.*, 416.

188 You, Z.-L., Shi, D.-H., Xu, C., Zhang, Q., and Zhu, H.-L. (2008) *Eur. J. Med. Chem.*, **43**, 862.

189 You, Z.-L., Shi, D.-H., and Zhu, H.-L. (2006) *Inorg. Chem. Commun.*, **9**, 642.

190 Will, J., Kyas, A., Sheldrick, W.S., and Wolters, D. (2007) *J. Biol. Inorg. Chem.*, **12**, 883.

191 Castellano-Castillo, M., Kostrhunova, H., Marini, V., Kasparkova, J., Sadler, P.J., Malinge, J.-M., and Brabec, V. (2008) *J. Biol. Inorg. Chem.*, **13**, 993.

192 Patra, M., Gasser, G., Pinto, A., Merz, K., Ott, I., Bandow, J.E., and Metzler-Nolte, N. (2009) *ChemMedChem.*, **4**, 1930.

193 Wang, J., Soisson, S.M., Young, K., Shoop, W., Kodali, S., Galgoci, A., Painter, R., Parthasarathy, G., Tang, Y. S., Cummings, R., Ha, S., Dorso, K., Motyl, M., Jayasuriya, H., Ondeyka, J., C. Herath. L. Zhang. J. Hernandez. A. Basilio Allocco, Tormo, J.R., Genilloud, O.O., Vicente, F., Pelaez, F., Colwell, L., Lee, S.H., Michael, B., Felcetto, T., Gill, C., Silver, L.L., Hermes, J.D., Bartizal, K., Barrett, J., Schmatz, D., Becker, J.W., Cully, D., and Singh, S.B. (2006) *Nature*, **441**, 358.

194 Nicolaou, K.C., Stepan, A.F., Lister, T., Li, A., Montero, A., Tria, G.S., Turner, C.I., Tang, Y., Wang, J., Denton, R.M., and Edmonds, D.J. (2008) *J. Am. Chem. Soc.*, **130**, 13110.

13
Biomedical Applications of Metal-Containing Luminophores

Albert Ruggi, David N. Reinhoudt, and Aldrik H. Velders

13.1
Introduction: Luminescence in Diagnostics and Imaging

The development of medical imaging techniques can be considered a major milestone in medicine, ranking with crucial discoveries like antibiotics and vaccines. Indeed, during the past 20 years the advancements in techniques and processes used to create images of the human body for clinical purposes (revealing, diagnosing, and examining diseases) have substantially improved the quality of medical care [1]. The possibility of visualizing the inside of the human body, without invasive techniques like explorative surgery and with a high level of detail, permits nowadays an early and less traumatic detection of major diseases (e.g., cancer) at its more curable stages. The most widely used techniques for medical imaging [2, 3] (Table 13.1) are currently based on X-rays (like radiography and computed axial tomography, CAT), radiofrequency interacting with nuclei (like magnetic resonance imaging, MRI), ultrasounds (e.g., obstetric sonography), and nuclear emission from radionuclides (like positron emission tomography, PET, and single photon emission computed tomography, SPECT) [4].

Optical imaging (i.e., imaging techniques based on the emission of light from biological or chemical moieties) is a rapidly developing field and a central research line is the development of specific imaging contrast agents suitable for biological and diagnostic applications [5–8]. Despite the wide range of technologies available for cell and small animals applications (e.g., fluorescent proteins [9] and bioluminescence [10]) fluorophore-labeled agents (i.e., labeled with a fluorescent moiety) are the most promising for human diagnostic. Owing to their high sensitivity, optical imaging techniques are already widely used for *ex vivo* analysis of tissues (e.g., after biopsy), for *in vitro* screening (e.g., in drug discovery), and are a very promising tool for intra-operative imaging [11].

The emission of light that follows the absorption of photons by a species (luminophore) is called photoluminescence [12]. Generally speaking, it is possible to distinguish between two forms of photoluminescence, namely fluorescence (spin-allowed emission of light from an electronic excited state) and phosphorescence (spin-forbidden emission of light from an electronic excited state). These two

Bioinorganic Medicinal Chemistry. Edited by Enzo Alessio
Copyright © 2011 WILEY-VCH Verlag GmbH & Co. KGaA, Weinheim
ISBN: 978-3-527-32631-0

Table 13.1 Examples of imaging techniques in clinical use.

Imaging technique	Detection	Common contrast agents[a]	Clinical applications
CAT	X-rays	Lopamidol™, Ioversol™	Intracranial hemorrhage
MRI	Magnetic field	Gadoteridol™, Gadodiamide™	Cerebral and coronary angiography
PET	γ-Rays	^{18}FDG, ^{15}OH$_2$	Cerebral blood flow, degenerative diseases
SPECT	γ-Rays	^{99m}Tc-HMPAO, ^{99m}Tc-ECD	Ischemia, cardiac imaging
Ultrasonography	Ultrasonic waves	Microbubbles	Echocardiography

[a]DG = deoxyglucose, HMPAO = hexamethylpropylene amine oxime, ECD = ethyl cysteinate dimer.
Adapted from Reference [3].

emission mechanisms can be schematically illustrated with a Jablonski diagram (Figure 13.1) in which the excitation and the relaxations pathways are shown.

Quantum yield and fluorescence lifetime are the most important characteristics of a luminophore. Fluorescence anisotropy (the extent to which a luminophore rotates during the excited-state lifetime) is another interesting property shown by some molecules, which can be used for the determination of molecular volume of labeled species [12]. Some examples of these anisotropy probes will be discussed in the following sections. The quantum yield (ϕ) can be defined as in Eq. (13.1):

$$\phi = \frac{k_r}{k_r + \sum k_{nr}} \tag{13.1}$$

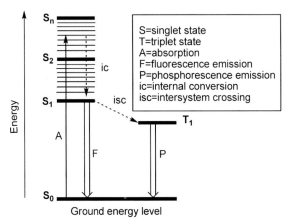

Figure 13.1 Simplified Jablonski diagram for a generic luminophore, showing the possible pathways of excitation and relaxation.

where k_r is the radiative kinetic constant and k_{nr} are the non-radiative kinetic constants. The quantum yield can be close to unity if the radiationless decay rate is much smaller than the emissive one. The lifetime (τ) of the excited state is the average time the molecule spends in the excited state prior to return to the ground state, and is defined as in Eq. (13.2):

$$\tau = \frac{1}{k_r + \sum k_{nr}} \qquad (13.2)$$

Luminophores can be divided into different families, according to their chemical nature [13]. Organic and coordination compound-based dyes are among the most popular classes of luminescent compounds. More recently semiconductor nano-particles (quantum dots) and lanthanide-doped nanoparticles have received great attention due to their remarkable luminescent properties. Organic dyes [14] (Figure 13.2) are the most used fluorophores, due to their generally high quantum yield (often close to unit) and due to the large number of commercially available com-pounds, which virtually cover the entire visible spectrum, from blue to near-infrared (NIR). Among the most used organic labels we can cite fluorescein and Rhodamine derivatives, cyanines, BODIPY®, and Alexa® dyes. All these compounds are widely used in biological applications, for instance in protein labeling [15–17]. Despite the good luminescence and the availability of a large number of fluorescent derivatives, organic dyes show several serious drawbacks, like the small Stokes shift (the difference between the excitation and emission maxima), which often causes

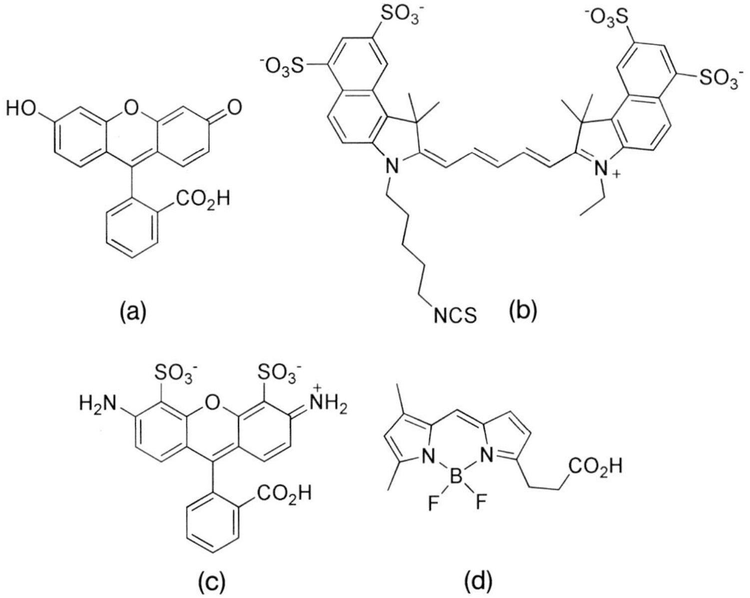

Figure 13.2 Examples of organic dyes: (a) fluorescein, (b) Cy5.5®, (c) Alexa®, and (d) BODIPY®.

self-absorption with a consequent decrease of the fluorescence emission upon multiple labeling, the generally poor photochemical stability, the decomposition under repeated excitation (photobleaching), and the very short lifetime (which makes these compound unsuitable for time-resolved detection techniques).

Owing to their interesting properties and to their versatility (which allows coverage of the entire spectrum from blue to NIR), coordination compound-based dyes are currently among the most promising agents for the future development of luminescent imaging dyes. Most of the above-mentioned drawbacks of organic dyes can be overcome by using inorganic dyes [18, 19], which actually show very interesting properties like a large Stokes shift (which permits us to label a biological molecule with multiple luminophores without reduced fluorescent intensity due to the self quenching), a long-lived emission (which permits us to remove the background interference of the biological molecules – autofluorescence – by using time-resolved detection) [20], interesting anisotropic properties (which can be used for the hydrodynamic study of proteins) [21], and a sensitive response to the local environment (which can serve as luminescent reporter of the biological surroundings) [22]. Despite these interesting properties, current diagnostic applications that take advantage of inorganic luminophores are relatively limited: among the most important we can cite fluoroimmunoassay [23] and electrochemiluminescence (i.e., immunoassay techniques based on the luminescence generated by electrochemical reactions) [24].

In this chapter we summarize the utilization of luminescent inorganic dyes for biodiagnostics, showing the advantages and the open challenges and giving a short perspective on the new frontiers in metallic-luminophore based imaging. In Section 13.2 we give an overview of the most representative examples of transition metal-luminescent complexes (mainly based on ruthenium, iridium, rhenium, and platinum) developed for biological imaging. In Section 13.3 selected examples of luminescent lanthanide complexes used as labels for bioimaging are summarized. Section 13.4 is a short overview of inorganic nanoparticles (quantum dots and lanthanide-doped nanoparticles) that could find application in biological imaging. Lastly, in Section 13.5 the most promising perspectives in the field of metal-containing luminophores for biomedical applications are discussed. Non-imaging related applications, like photodynamic therapy, are omitted, as are the metal-containing systems in which luminescence does not derive from the metallic moiety (e.g., sensors with an organic dye and a non-fluorescent metal complex).

13.2
Transition-Metal Containing Luminescent Agents

13.2.1
Photophysical Properties of Transition Metal Complexes

Transition metal complexes have unique properties due to their unique electronic structure, compared to an organic molecule [25]. The electronic structure of transition metal complexes can be described with the molecular orbital (MO) theory, in which the molecular orbitals of a complex are constructed as a linear combination of metal and ligand orbitals. A typical MO diagram of a d^6, low-spin metal ion complex with octahedral geometry is shown in Figure 13.3, together with the possible electronic transitions and their nomenclature. A metal complex is emissive only if there is a large energy gap between the lowest excited state and the ground state. The magnitude of the splitting between these two states is denoted as Δ (ligand-field splitting parameter) and depends on the metal ion (for instance Δ is proportional to the d orbitals quantum number) and on the strength of the ligands (i.e., their position in the spectrochemical series). The presence of a heavy atom induces a certain degree of spin–orbit coupling. Because of this effect, the spin-forbidden electronic relaxation pathways may become partly allowed and, for this reason, most of the luminescent transition metal complexes show an emission in which spin-allowed (fluorescence) and spin-forbidden (phosphorescence) transitions are combined in varying degrees. The first consequence of this effect is the usually longer lifetimes of the excited state in transition metal complexes compared to organic dyes. In this section some examples of transition metal-based luminophores based on Ru(II), Ir(III), Re(I), and Pt(II) are reported, together with the photophysical properties and diagnostic applications.

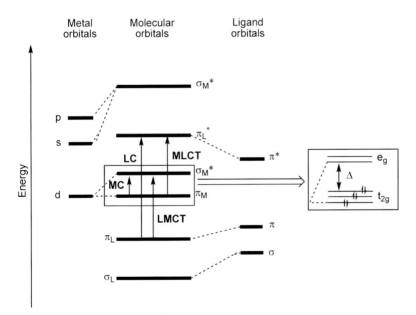

Figure 13.3 MO diagram of a d^6, low-spin metal ion complex with octahedral geometry, showing possible electronic transitions: metal centered (MC), ligand centered (LC), ligand to metal charge transfer (LMCT), and metal to ligand charge transfer (MLCT) electronic transitions are shown. The splitting of d orbitals e_g (LUMO) and t_{2g} (HOMO) and the ligand-field splitting parameter (Δ) are also shown.

13.2.2
Ruthenium(II) Complexes

Ruthenium(II) complexes have a low-spin d^6 electronic structure and generally form octahedral complexes. Complexes with bidentate chelating ligands (chel) are the most common; these complexes have the general formula $[Ru(chel)_3]^{2+}$ and are chiral (Δ, Λ isomerism, Figure 13.4). Since the first observation of [Ru(bpy)$_3$]$^{2+}$ (**1a**) luminescence (Figure 13.4, bpy = 2,2′-bipyridine) [26], ruthenium complexes have been the most widely studied among the transition metal-based luminophores [27], especially for their applications in dye-sensitized solar cells (Grätzel cells) [28]. The usually straightforward synthesis and the remarkable photophysical properties make these $[Ru(chel)_3]^{2+}$ complexes very interesting for biological imaging, despite their relatively low quantum yield. Ruthenium complexes generally show absorption bands both in the UV region (200–300 nm) and in the visible region (400–450 nm) with a typical emission centered around 600 nm. In this section we report some selected applications of Ru(II) luminophores for DNA binding, protein labeling, and cell imaging.

13.2.2.1 **DNA Binding**
The first example of interaction of Ru(II) luminophores with DNA was reported in 1984 by Barton *et al.* [29], who described the stereoselective non-covalent binding of $[Ru(chel)_3]^{2+}$ complexes to DNA. Since then research on Ru(II) luminophores as DNA labels focused on two different approaches: the interaction with ds-DNA oligonucleotides by non-covalent interactions (e.g., intercalation, groove binding,

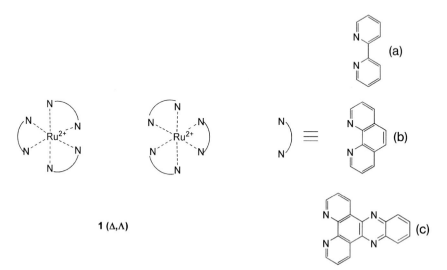

1 (Δ,Λ)

Figure 13.4 $[Ru(chel)_3]^{2+}$ complexes with chel = bipyridine (bpy) (a), phenanthroline (phen) (b), and phenazine (c). The latter has been used as molecular probe for DNA. Δ and Λ enantiomers of the complexes are shown on the left.

electrostatic interaction, etc.), and the synthesis of nucleosides labeled with luminescent $[Ru(chel)_3]^{2+}$ [30].

The first approach concerns the selective non-covalent interaction of luminescent Ru(II) probes with oligonucleotides. $[Ru(phen)_3]^{2+}$ (**1b**, phen = 1,10-phenanthroline) (Figure 13.4) can intercalate into double stranded DNA [29], but only with low changes in the absorption/emission properties of the luminophore and with relatively weak binding constant ($K = 10^3$ M^{-1}). In 1990 Friedman *et al.* [31] described an improved molecular probe for DNA based on the Ru(II)-phenazine complex **1c** (Figure 13.4). This new probe has two main advantages over the $[Ru(phen)_3]^{2+}$ dye: the intercalation in between DNA bases is stronger ($K = 10^6$ M^{-1}) and in aqueous buffer the compound is luminescent only upon DNA intercalation. Furthermore, the probe shows a sensitive response to the changes of the helix structure: the emission maximum varies from 628 to 640 to 650 nm in the presence of B, Z, and A-form helices, respectively. A very recent example of an RNA probe has been reported by O'Connor *et al.*: by functionalizing a $[Ru(bpy)_3]^{2+}$ luminophore with an intercalating ethidium moiety the authors were able to achieve a ninefold increase of luminescence upon RNA binding, both in solution and in living cells [32].

The second approach of DNA labeling involves the synthesis of nucleosides modified with luminescent Ru(II) derivatives that are then incorporated into oligonucleotides by solid-phase synthesis. For instance, the modified uridine compound **2** (Figure 13.5), reported in 1999 by Kahn *et al.*, was incorporated into an oligonucleotide by using an automated DNA synthesizer with a standard coupling protocol [33]. More recently, Hurley *et al.* have described the synthesis of Ru(II)- and Os(II)-modified nucleosides **3** (Figure 13.5) and their incorporation into DNA oligonucleotides [34]. The authors studied also the donor–acceptor interactions of the dyes on different ds-DNA [35]. Another strategy for post-modification of nucleotides is the introduction of a reactive unit (e.g., an activated ester) suitable for the reaction with a modified nucleoside previously introduced in the DNA sequence [36].

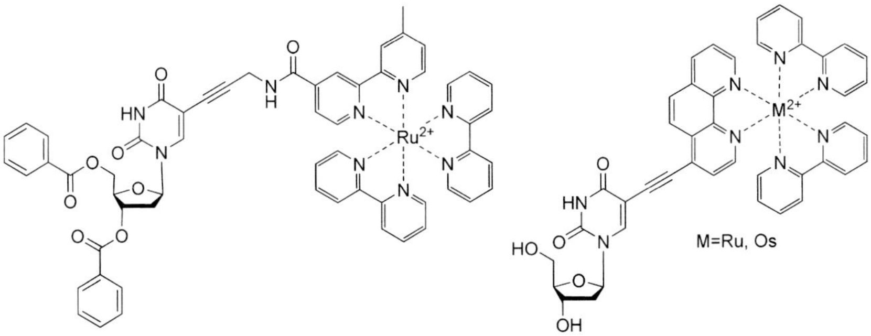

Figure 13.5 Nucleosides modified with $[Ru(bpy)_3]^{2+}$ moieties (**2**) and with Ru(II)/Os(II) complexes (**3**).

13.2.2.2 Protein Binding

The functionalization of a luminophore with a chemical reactive moiety is the easiest way to bind the luminescent moiety to a target molecule. Despite the apparent simplicity of this strategy, the first reactive luminophore based on transition metals was described in 1992 by Ryan *et al.* [37]: by introducing on phenanthroline an isothiocyanate (NCS) group that reacts easily with primary amines the authors were able to decorate different albumins, immunoglobulins, and polylysines with the $[Ru(bpy)_2(phen)]^{2+}$ unit 4 (Figure 13.6). More recently, luminescent Ru(II) derivatives of 5 decorated with biotin (suitable for avidin binding) [38] and estradiol moieties [39] were described by Lo *et al.* These probes (Figure 13.6) can be used to selectively bind proteins and receptors for *in vitro* imaging or bioassays.

13.2.2.3 Cell Staining

The mechanism of the interaction of Ru(II)-based dyes with living cells and small animals has been thoroughly studied [40] to elucidate the cellular uptake and toxicity mechanisms. $[Ru(phen)_3]^{2+}$ has been used by Paxian *et al.* [41] for *in vivo* visualization of oxygen distribution in liver by exploiting the oxygen quenching of the Ru(II) fluorescence. These compounds show potential for future *in vivo* applications.

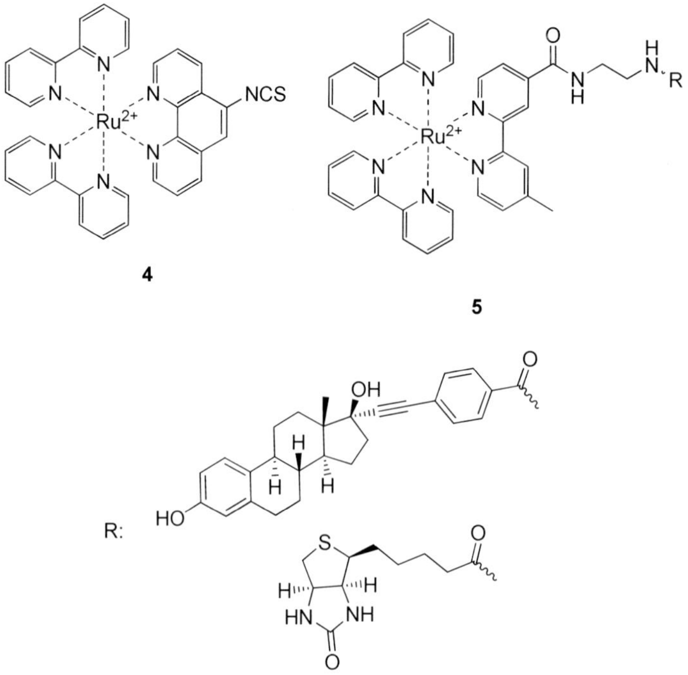

Figure 13.6 Ru(II) complexes suitable for protein labeling. From References [37–39].

13.2.3
Iridium(III) Complexes

Iridium(III) complexes have a low-spin d^6 electronic structure and form octahedral complexes, often with covalent carbon–iridium bonds, beside nitrogen–iridium coordinative bonds. Although the first example of fluorescence emission from the iridium cyclometallate compound [Ir(ppy)$_3$] (ppy = 2-phenylpyridine) was already reported in 1985 [42], iridium(III)-fluorescent derivatives [43] have received major interest only in the last ten years, especially for their application in organic light emitting diode (OLED) technology [44]. Compared to the other transition metal complexes, iridium(III) complexes possess unique photochemical and photophysical properties such as the possibility of tuning the emission along the whole visible spectrum by changing the electronic density of the substituents on the aromatic ring. The excitation is usually in the UV (200–300 nm) or in the visible (400 nm) region [45]. Furthermore, the iridium derivatives normally posses higher emission quantum yield (up to 60%) and longer emission lifetimes (on the order of milliseconds) due to the higher spin–orbit coupling; the excited state of the complex is usually a combination of singlet and triplet states. On the other hand, in general, Ir(III) complexes are synthetically more demanding than other coordination complexes with chelate ligands (like bpy) because of the high activation energies required for the formation of the Ir–carbon bond. Surprisingly, there are only relatively few examples of application of Ir(III) derivatives as biological labels. The group of Lo, in particular, has worked extensively on this topic and has reported most of the examples available in literature.

The first use of the cyclometallated Ir(III)-based fluorophore **6** (Figure 13.7) as biological label was reported by Lo *et al.* in 2001 [46]. By exploiting the reactivity of the peripheral isothiocyanate and iodoacetate towards primary amines and thiols,

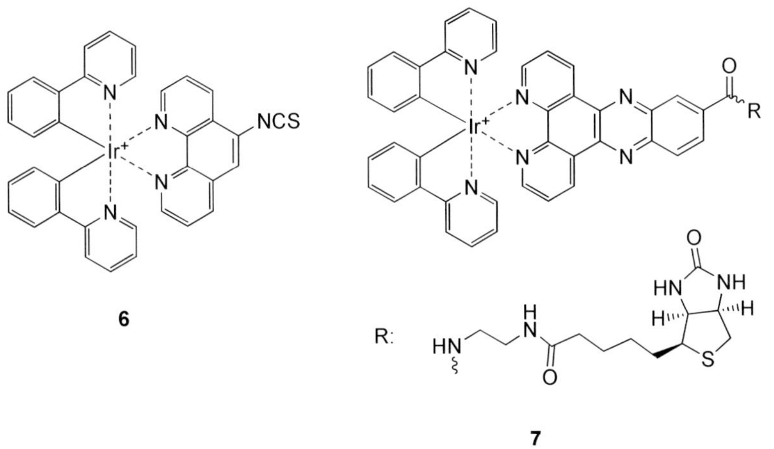

Figure 13.7 Ir(III) complexes suitable for protein labeling. From References [46, 47].

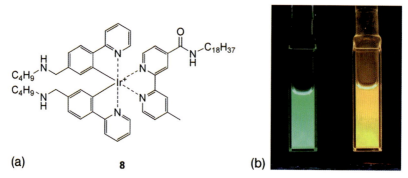

(a) **8** **(b)**

Figure 13.8 (a) Dual emissive Ir(III) complex (**8**); (b) luminescence in CH₃CN and CH₂Cl₂. Reproduced from reference [49] with permission from Wiley VCH.

respectively, the authors prepared some oligonucleotides and human serum albumin (HSA) conjugates that show intense fluorescence in deareated buffers. In particular, the HSA conjugate shows only a small decrease in emission intensity and lifetime in the presence of molecular oxygen, due to the shielding effect given by the protein matrix. The same group also reported different examples of luminescent Ir (III)-cyclometallated complexes functionalized with biotin (**7**) [48] (Figure 13.7).

In 2008, Lo *et al.* reported a dual emissive cyclometallated Ir(III) complex **8** (Figure 13.8), which can be used as an environment-responsive biological probe for polarity. In fact, the compound shows a shift of emission from 500 to 600 nm in apolar (e.g., CH₂Cl₂) and polar (e.g., CH₃CN) solvents, respectively [49]. In the same year, Lai *et al.* reported the synthesis of Fe_3O_4/SiO_2 nanoparticles decorated with a luminescent $[Ir(piq)_2(pp)]^+$ (piq = 1-phenylisoquinoline and pp = pyridyl pyrazole) complex [50]. This luminescent complex is used both as luminophore for optical imaging and as $^1O_2$ photosensitizer for photodynamic therapy.

13.2.4
Rhenium(I) Complexes

Rhenium(I) complexes have a low-spin d^6 electronic structure and form octahedral complexes. Since the first description of their photophysical properties [51], Re(I)-based complexes have gained interest for their high stability and versatility. However, the first examples of application in biomolecular imaging are relatively recent [52, 53]. Re(I)-luminophores can be excited in the UV (300 nm) and in the visible (400 nm) region and usually show an emission band at around 500 nm. In 2005 Wei *et al.* reported the synthesis of thymidine and uridine functionalized with a Re(I) tricarbonyl luminophore (**9**, Figure 13.9) [55]. In 2001 Lo *et al.* reported the Re(I) tricarbonyl functionalized with an isothiocyanate unit **10** (Figure 13.9) suitable for biomolecule labeling [54]. The compound was conjugated with nucleosides and albumin. In 2002 the same group reported the functionalization of the luminophore with a maleimide moiety [56]. The reactivity of the maleimide towards thiols was exploited to functionalize a thiolated oligonucleotide and

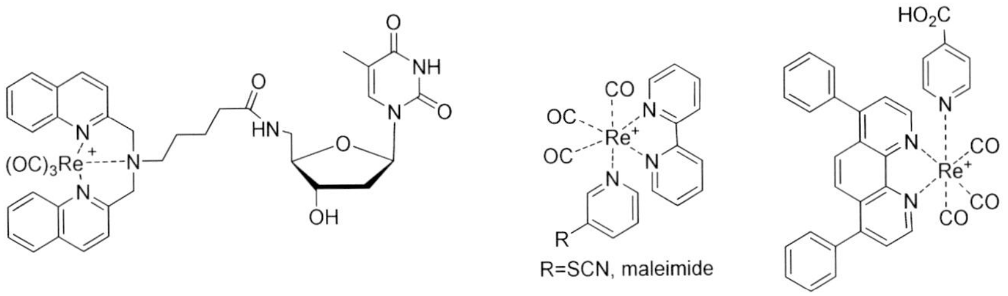

Figure 13.9 Re(I) complexes for nucleosides (**9**) and protein (**10, 11**) labeling. From References [54, 55, 47].

cysteine containing peptides and proteins. In 1997 Guo *et al.* showed Re(I)-tricarbonyl derivative **11** (Figure 13.9) as luminescent and anisotropy probe [47], which was conjugated with biomolecules (immunoglobulin, albumin, and lipid), and the anisotropy properties were studied to evaluate the hydrodynamic radii.

13.2.5
Platinum(II) Complexes

Platinum(II) complexes have a d^8 electronic structure and generally form low-spin square-planar complexes. Platinum(II) complexes with amine and carboxylic acid ligands are well known for the possibility of strong DNA interaction and have been extensively investigated as antitumor drugs [57]. These compounds, however, are usually not luminescent and therefore not interesting for labeling purposes. In contrast, several luminescent Pt(II) complexes with aromatic ligands have been developed and used as probes for DNA. Platinum(II) luminophores can be excited around 400 nm and usually show luminescence around 600 nm. Polypyridine complexes, for instance, can intercalate DNA [58] or form covalent [59] or hydrogen bonds (**12**, Figure 13.10) [60]. Liu *et al.* reported the cyclometallated complex [Pt(II)(dpp)(CH$_3$CN)]$^+$ (dpp = 2,9-diphenyl-1,10-phenanthroline), which can be utilized as a molecular probe for DNA [61]. The complex [Pt(ppy)(dppz-CO$_2$H)]$^+$ (dppz-CO$_2$H = 11-carboxydipyrido[3,2-*a*:2′,3′-*c*]phenazine) has been recently reported as luminescent probe for the G-quadruplex structural motif [62]. Luminescent platinum(II) complexes have also been used for protein labeling, like the Pt(II)-terpyridyl complex **13** (Figure 13.10) used by Wong *et al.* for albumin labeling [63]. Che *et al.* have reported the water-soluble complex [Pt(hbpp)Cl], hbpp = 4-(4-hydroxyphenyl)-6-phenyl-2,2′-bipyridine, supported on a poly(ethylene glycol) polymer that can be used as a protein detector [64].

13.2.6
Other Metal Complexes

As shown before, Ru(II), Ir(III), Re(I), and Pt(II) complexes have been the most popular choice for imaging luminophores. Although many more examples of

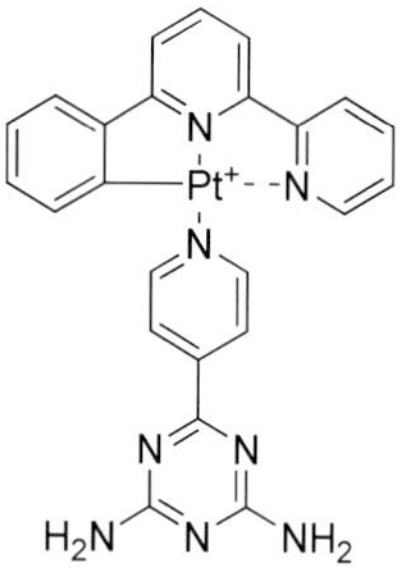

Figure 13.10 Pt(II) complexes for DNA (**12**) and protein (**13**) labeling.

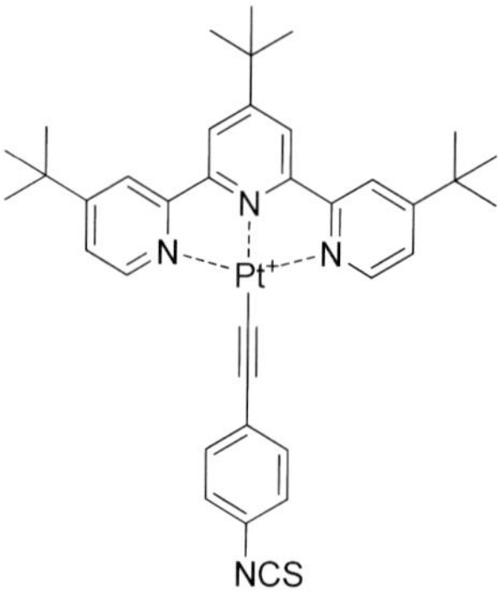

luminescent complexes of transition metals are described in the literature [44, 65, 66], only a few examples of luminophores for biological applications based on other transition metals are known. Osmium(II) compounds can be considered as a logical extension of ruthenium(II) systems, due to the chemical similarity of these two metals. In fact, Os(II)-polypyridine complexes have been exploited as DNA probes for energy and electron transfer studies [67]. Hurley *et al.* reported the synthesis of a nucleoside decorated with an $[Os(bpy)_3]^{2+}$ luminophore [34, 35] and its application in energy transfer between Ru(II) and Os(II) (dyads), which can be used for the evaluation of the distance between the two luminophores. [Os (phen)$_3]^{2+}$ and $[Os(bpy)_3]^{2+}$ labels have also been employed as anisotropic probes [68, 69]. Copper(II)-based luminophores have also been studied as DNA probes and some examples of these luminophores are based on phenanthroline [70] and porphyrin [71] derivatives. In both cases, a luminescence increase upon DNA intercalation has been observed, due to the shielding effect of the nucleic acid that prevents solvent quenching. Few examples of palladium(II)-based luminophores for biological imaging have been reported, mainly based on coproporphyrin ligands [72, 73].

13.3
Lanthanide-Based Luminophores

13.3.1
Chemical and Photophysical Properties of Lanthanide Complexes

Lanthanide complexes are perhaps the most investigated class of inorganic luminophores since the 1980s [74]. Lanthanide compounds have been extensively used for many different applications [75], especially in the field of phosphors preparation (e.g., the phosphors of TV screens and lighting tubes are made of inorganic oxides doped with Eu^{2+} and Eu^{3+}) [76]. Complexes of Ln(III) ions posses unique photophysical properties [77]. First of all, due to the shielding of the 4f orbitals by the filled 5p and 5s orbitals, the selection rules forbidding f–f transitions are weakly relaxed. As a consequence, the transitions keep their atomic nature: the emission appears sharp and the transition energy is almost not influenced by the metal surroundings. On the other hand, the same shielding results in a very low molecular absorption coefficient (very often $< 1 \ M^{-1} \ cm^{-1}$). As a consequence, the direct excitation of Ln^{3+} ions is not a suitable strategy for achieving efficient luminophores, and a sensitization process ("antenna excitation," *vide infra*) is necessary. Most of the Ln(III) ions form luminescent complexes (Figure 13.11), but some are more emissive than others: the emission properties are directly related to the availability of non-radiative processes. Therefore, the ions with a small energy gap between the lowest emissive state and the highest sublevel of its ground multiplet (e.g., Nd^{3+}, Er^{3+}, and Yb^{+3}, Figure 13.11) usually show low quantum yields in aqueous solutions. With respect to the energy gap requirement, Gd^{3+}, Eu^{3+}, and Tb^{3+} ions are the best emitters; however, Gd^{3+} shows UV emission and therefore is not very useful for luminescence applications in biological systems. To overcome the low molecular absorption coefficient of Ln(III) ions it is possible to exploit energy transfer from the surroundings of the metal ion with a three-step mechanism: absorption of the light by a suitable ligand ("antenna"), transfer onto the lanthanide ion, and consequent light emission. This sensitization strategy has also the advantage of remarkably increasing the Stokes shift, which is very useful in biological applications.

Lanthanide(III) luminophores are ideal luminescent probes for biological systems [78]: the sharp emission bands are usually well separated from the fluorescence emission of the organic moieties in the matrix and the long lifetimes of the excited states (often in the order of ms) permit the use of time-resolved techniques. For preparation of effective lanthanide probes for biological applications there are three main problems to overcome: the choice of a system with effective sensitization, prevention of non-radiative deactivation processes, and the design of kinetically and thermodynamically stable systems [79] (to avoid the toxicity of free Ln^{3+} ions [80]). Among the most commonly used ligands are polydentate [81] (in particular DTPA and DOTA derivatives, β-diketonates, and acyclic Schiff bases) and macrocyclic ligands [82]. As in the case of organic and transition metal-based

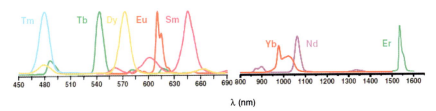

Figure 13.11 Typical luminescence of selected lanthanide(III) ions.
Reproduced from reference [78] with permission from the Chemical Society of Japan.

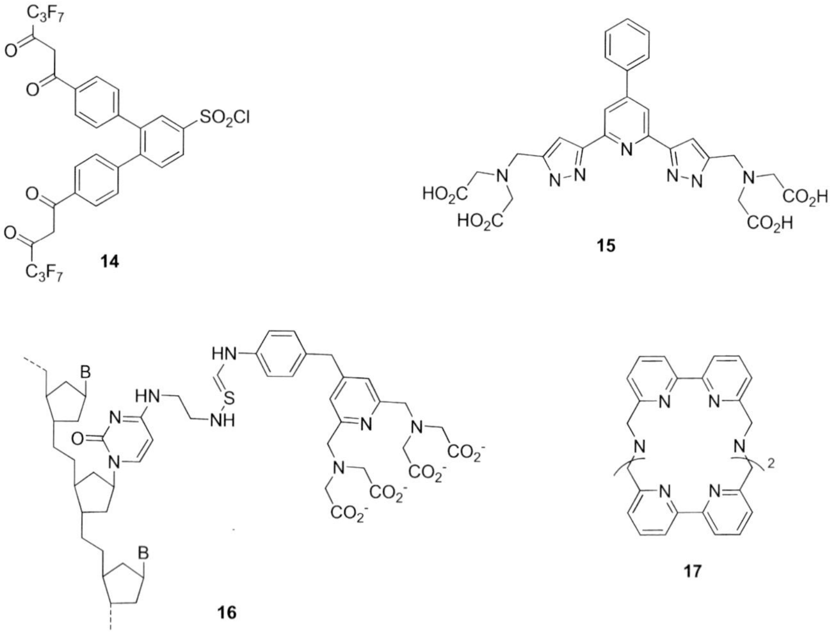

Figure 13.12 Some common ligands for luminescent lanthanide(III) complexes.

dyes, the easiest way to conjugate lanthanide luminophores with a biological substrate is the introduction of a reactive group on the dye [83]. Many examples have been reported in literature and some of them are shown in Figure 13.12. Owing to the wide application of these luminophores in biological sensing [84], below we focus our attention on the different fields of application, that is, immunoassay, nucleic acid probes, and molecular imaging. A selection of the most relevant examples of clinical applications of those luminophores is also reported.

13.3.2
Europium(III) and Terbium(III) Complexes

13.3.2.1 Immunoassays

Europium(III) and terbium(III) have f^6 and f^8 electron configurations, respectively, and form eight- or nine-coordinated complexes with chelating ligands. If ligands with less coordinating atoms are used, the coordination sphere can be filled by water molecules, which causes a fluorescence quenching. Most applications in biomedical diagnosis are based on immunoassays, often with detection by time-resolved luminescence. The most common approach makes use of antibodies labeled with luminescent Ln(III) (usually Eu^{3+}) complexes. In the first developed process (heterogeneous immunoassay, Figure 13.13a) [85], the labeled antibody interacts with the immobilized analyte and, after washing, the pH is lowered and the Eu^{3+} is released and quantified by spectrofluorimetry after complexation with a suitable ligand. A more advanced technique, homogeneous immunoassay,

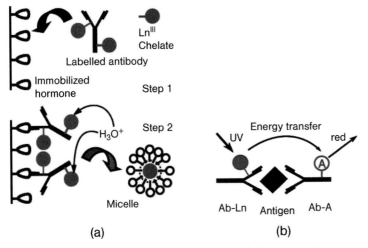

Figure 13.13 Heterogeneous (a) and homogeneous (b) luminescent immunoassay. Reproduced from reference [75] with permission from RCS Publishing.

Figure 13.13b) [86] is based on the energy transfer between an Eu^{3+} luminophore and a red-emitting protein (allophycocyanin). Each luminescent dye is conjugated to a different, specific monoclonal antibody. Upon UV irradiation, the excitation energy is transferred in a non-radiative way (Förster resonance energy transfer, FRET) from the lanthanide to the fluorescent protein. Since the latter is excited by the de-excitation of the long-lived state of the Eu^{3+}, the long lifetime typical of the lanthanides is preserved. The strict distance requirements (7–10 nm between Eu^{3+} and the acceptor) for an effective energy transfer make the assay quite specific (Figure 13.14). Many examples of clinical applications of these techniques have been reported [87]: assays for diabetes-related antibodies, kallikrein for prostate screening, prion protein on animal tissues, and detection of different infectious microbes as potential terror and warfare agents [88].

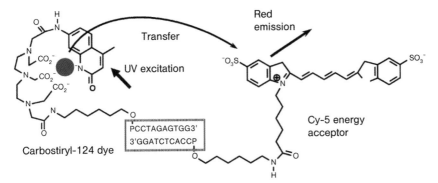

Figure 13.14 A FRET based nucleic acids probe. See text for details. Reproduced from reference [75] with permission from RCS Publishing.

13.3.2.2 Nucleic Acid Probes

Nucleic acid probes are widely used in medicine, for instance for the recognition of genetic predisposition to diseases and for virus detection [89]. Radioisotopes have been the traditional probe for such applications, but lanthanide luminophores have proved to be a safer alternative [90]. An energy-transfer-based DNA assay was reported in 1990 by Oser and Valet [91], who exploited the energy transfer between two complementary DNA sequences decorated with an Eu^{3+} chelate and a Cy-5® dye, respectively. Upon hybridization of the two DNA strands, the donor and the acceptor are brought within a suitable distance for an effective FRET and the emission of the sensitized Cy-5® can be detected without interference from the remaining Eu^{3+} luminescence (Figure 13.14). Chemical labeling of DNA with lanthanide luminophores has also been reported with several different strategies, for instance by enzymatic transamination of deoxycytidine residues of DNA with an Eu^{3+} chelate [92]. Luminescent lanthanide complexes have also been used to detect PCR products [93].

13.3.2.3 Molecular Imaging

The first example of cellular staining with a luminescent lanthanide complex was reported in 1969 by Scaff et al., who described the internalization of luminescent [Eu (tta)] (tta = 4,4,4-trifluoro-1-(2-thienyl)-1,3-butanedione) into *Escherichia coli* [94]. Terbium(III) complexes have been used for time-resolved imaging [95], exploiting the long luminescence lifetime of the complex. Lanthanide complexes have been used for the selective labeling of cancer cells *in vivo*. For instance, Bornhop *et al.* reported the [Tb(**20**)] complex (Figure 13.15) suitable for the early-stage detection of oral cancer lesions [96]. Manning *et al.* reported in 2004 the [Tb(**21**)] complex

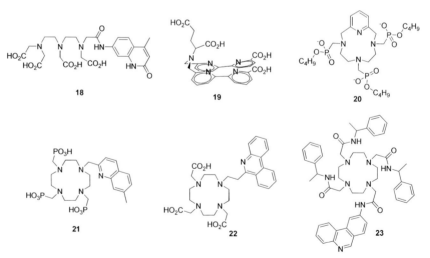

Figure 13.15 Selected polydentate ligands for lanthanide ions for *in vitro* sensing applications.

(Figure 13.15) which shows interesting properties for use as abnormal tissue marker [97]. An extensive series of luminescent lanthanide chelates has been reported by the group of Parker, who developed many examples of Eu^{3+} and Tb^{3+} luminophores both for cellular imaging and for *in vitro* sensing of pH, pO_2, and anions (Figure 13.15) [98]. A luminescent Tb^{3+} complex has recently been used by Akiba *et al.* to selectively detect phosphotyrosine in the presence of other phosphate-containing biomolecules [99].

13.3.3
Neodymium(III) and Ytterbium(III) Complexes

Above we focused on lanthanide complexes that emit in the visible region of the electromagnetic spectrum (i.e., Eu^{3+} and Tb^{3+}). On the other hand, Nd^{3+} and Yb^{3+} complexes have also been of interest because of their emission in the NIR region. NIR emitters have recently found application in diagnostics, especially for non-invasive *in vivo* imaging [100], exploiting the low absorption of biological tissues in the NIR region [6]. The first example of NIR luminophores based on Yb^{3+} was described in 1990 with porphyrin-based complexes [101]. Despite the efficient accumulation of these complexes in cancer cells, a high phototoxicity and low contrast were serious drawbacks. In more recent years, Zhang *et al.* have reported that the tropolone ligand is suitable for NIR sensitization of several lanthanide ions (Yb^{3+}, Nd^{3+}, Tm^{3+}, Er^{3+}, Ho^{3+}) in solution [102].

13.4
Nanoparticle-Based Luminophores

13.4.1
Chemical and Photophysical Properties of Luminescent Nanoparticles

The study of nanoparticle properties is an extremely popular field, related to the "nanotechnology revolution" observed in the last ten years [3]. Nanoparticle (e.g., semiconductor quantum dots (QDs), transition metal and lanthanide-doped silica nanoparticles, nanobeads) applications as contrast or imaging agents have been reported by many groups, but here we summarize only the achievements in the field of luminescent nanoparticles for biological applications. The emission of transition metal and lanthanide-doped silica nanoparticles follows the same mechanism given in the corresponding previous sections, so here only the emission mechanism of semiconductor QDs is discussed.

QDs [103, 104] are semiconductor nanocrystals with dimensions of $1-10$ nm, made mainly of II–V group elements (CdS, CdSe, InP, etc.). The light emission mechanism is based on the phenomenon of quantum confinement. Rapidly, upon excitation, an electron is promoted from the filled valence band to the conduction band, creating a positive vacancy (hole) in the valence band (Figure 13.16). The spatial separation (Bohr radius) of this electron–hole pair (exciton) is typically of

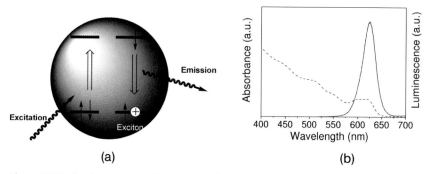

Figure 13.16 Luminescence mechanism (a) and typical QD absorption—emission spectrum (b). Reproduced from reference [105] with permission from ACS Publications.

the order of 1–10 nm and the quantum confinement effect is due to the fact that the size of the semiconductor crystal is of the same order as that of the Bohr radius. Fluorescence occurs upon recombination of an electron–hole pair usually with lifetimes in the order of milliseconds. The emission bands are usually very narrow (30–40 nm full width at half maximum) and the emission wavelength strictly depends on the crystal size: the larger the crystal, the lower the emission energy. QDs have very interesting properties, which are highly valuable for optical imaging: the emission spectrum is narrow whereas the excitation spectrum is extremely broad (Figure 13.16) [103].

13.4.2
Semiconductor Quantum Dots

It is, virtually, possible to excite QDs at every wavelength shorter than the emission peak and this feature can be used for multiple labeling. Furthermore, QDs do not show photobleaching (which is a typical problem of organic dyes) even under strong irradiation [106]. QDs usually show also very large molar extinction coefficients and high quantum yield (up to 80%), which make them bright probes. Toxicity of QDs is still a main controversial point and many studies have been carried out to evaluate this crucial parameter [107]. "Naked" QDs (without any capping agent) are known to be toxic [108] because of a UV-induced photolysis process that oxidizes the crystal with consequent release of toxic ions (e.g. Cd^{2+}). On the other hand, studies on capped QDs (capping is actually necessary also to obtain water solubility, which is convenient for biological applications) proved that the capping is useful in preventing crystal photooxidation, resulting in non-observable toxic effects in a mouse model even over a period of months after injection [109].

The first examples of biological application of QDs were reported in 1998 [104, 110]. In the last ten years, QDs conjugated with different moieties (e.g., streptavidin, thiolated peptides, proteins, antibodies) have extensively been used for *in vitro* and *in vivo* imaging [111]. For instance, multiple color imaging of living cells was reported by Jaiswal *et al.* in 2003 (Figure 13.17) [106]. Antibody-labeled QDs were used for *in vivo* cancer targeting and imaging by Gao *et al.* in 2004 [112] and in the same year NIR-QDs emitters were used for intraoperative imaging of sentinel lymph nodes by Kim *et al.* (Figure 13.17b) [100]. These three examples show the impressive results

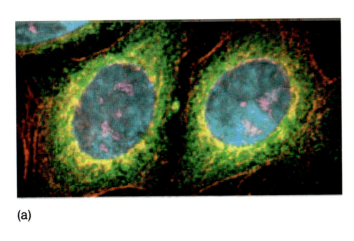

(a)

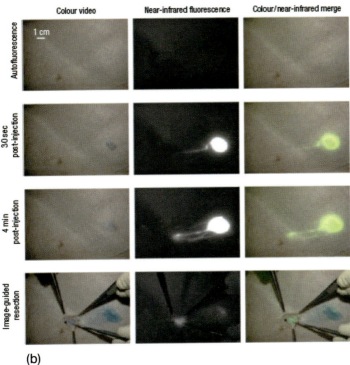

(b)

Figure 13.17 Multicolor cell imaging with QDs (a) and intraoperative imaging of a sentinel lymph node labeled with a NIR-QD (b).
Reproduced from references [100], [111] with permission from Nature Publishing Group.

achieved using QDs as luminophores, to date. Many more examples are available in the literature and are summarized in the cited reviews.

13.4.3
Metal-Doped Nanoparticles

Lanthanide-doped nanoparticles (LNs) generally consist of an inorganic matrix (e.g., SiO_2) doped with Ln^{3+} ions [113]. The inorganic matrix acts as lanthanide sensitizer and the emission retains the same characteristics of Ln^{3+} complexes (i.e., narrow emission band and long lifetimes) and can be tuned by adding different dopant ions to obtain different emission colors. In addition, some LNs show the capability of emission in the visible spectrum upon irradiation at IR wavelengths (up-conversion) [114]. This unique feature can be usefully exploited for biological applications due to the high penetration of IR radiation in tissue and the low optical background (owing to the absence of biomolecules' autofluorescence upon NIR irradiation). LNs have been used as nucleic acid reporters [115] and for the labeling of antibodies, albumins [116], and avidin [117].

Lastly, SiO_2 nanoparticles doped with luminescent $[Ru(bpy)_3]^{2+}$ [118] have also been reported. The latter shows an improvement of Ru(II) emission in terms of quantum yield, due to the shielding effect of the SiO_2 shell against solvent quenching. Ruthenium(II)-doped SiO_2 nanoparticles have been used for the synthesis of multimodality agents: for instance, silica nanoparticles doped with $[Ru(bpy)_3]^{2+}$ and functionalized with [Gd-Si-DTTA] (Si-DTTA = (trimethoxysilylpropyl)diethylene-triaminetetraacetate) have been investigated [119]. The silica matrix prevents the oxygen quenching and the photobleaching of $[Ru(bpy)_3]^{2+}$ and provides a reactive surface (SiO_2) that can be easily functionalized with the Gd^{3+} complex to act as MRI contrast agent.

13.5
Conclusions and Perspectives

Metal-based luminophores present many interesting chemical and photophysical properties and can be usefully applied in biomedical diagnostics and imaging. Despite the numerous examples reported in the literature, the *in vivo* application in humans is still an unachieved goal. In fact, we are still far from a "perfect" label for diagnostic optical imaging and the research in this field requires a focus on a compromise between luminophore performances and the drawbacks that are incidental to each luminescent species. Transition metal-based luminophores show, generally speaking, a relatively modest quantum yield, which is a serious limitation for their *in vivo* application because of the high concentrations needed. On the other hand, these complexes can be usually excited with visible light, which constitutes a strong point for their biological application, since UV radiation is harmful. Lanthanide-based luminophores usually show relatively high quantum yields, but the wavelengths used for the antenna's excitation are mostly in the UV region. Both transition metal and lanthanide complexes show quite long lifetimes (in the range of microseconds to milliseconds), which makes them eligible for time-resolved techniques; moreover, these complexes are in general very stable, which avoids release of free heavy metal ions in the body. Finally, coordination complexes can readily be modified with an exact and known stoichiometry, which is a great advantage for the assembly of molecular architectures suitable for *in vivo* biological applications like targeting. Quantum dots show extremely high quantum yields, but the possibility of releasing the heavy metals of which QDs are constituted is the most serious obstacle to their application *in vivo*. Furthermore, functionalization of QDs cannot be completely controlled with an exact and known stoichiometry, which constitutes a second obstacle to their application in diagnostic imaging.

The combination of different imaging agents for the development of multimodal agents has recently received great attention [120]. The possibility of conjugating different agents in one system allows us to overcome the intrinsic drawbacks of the separate imaging techniques and allows the design of a "perfect" imaging agent. The most popular multimodality agents developed till now are based on the conjugation of moieties suitable for MRI imaging and optical imaging. Other techniques (e.g., based on radionuclides) have also been implemented in multimodality agents and the availability of many different kinds of imaging agents and scaffolds make this field extremely wide and interesting. In conclusion, metal-containing luminophores play a crucial role in the development of luminescent labels for *in vitro* and *in vivo* biomedical diagnostic and imaging applications.

References

1 Licha, K. and Olbrich, C. (2005) *Adv. Drug Delivery Rev.*, **57**, 1087–1108.

2 Gemmel, F., Dumarey, N., and Welling, M. (2009) *Semin. Nucl. Med.*, **39**, 11–26.

3 Sharrna, P., Brown, S., Walter, G., Santra, S., and Moudgil, B. (2006) *Adv. Colloid Interface Sci.*, **123**, 471–485.

4 Massoud, T.F. and Gambhir, S.S. (2003) *Gene. Dev.*, **17**, 545–580.

5 Kovar, J.L., Simpson, M.A., Schutz-Geschwender, A., and Olive, D.M. (2007) *Anal. Biochem.*, **367**, 1–12.

6 Weissleder, R. and Ntziachristos, V. (2003) *Nat. Med.*, **9**, 123–128.

7 Andersson Engels, S., af Klinteberg, C., Svanberg, K., and Svanberg, S. (1997) *Phys. Med. Biol.*, **42**, 815–824.

8 Licha, K. (2002) *Top. Curr. Chem.*, **222**, 1–29.

9 Shaner, N.C., Steinbach, P.A., and Tsien, R.Y. (2005) *Nat. Methods*, **2**, 905–909.

10 Chaudhari, A.J., Darvas, F., Bading, J.R., Moats, R.A., Conti, P.S., Smith, D.J., Cherry, S.R., and Leahy, R.M. (2005) *Phys. Med. Biol.*, **50**, 5421–5441.

11 Veiseh, M., Gabikian, P., Bahrami, S.B., Veiseh, O., Zhang, M., Hackman, R.C., Ravanpay, A.C., Stroud, M.R., Kusuma, Y., Hansen, S.J., Kwok, D., Munoz, N.M., Sze, R.W., Grady, W.M., Greenberg, N.M., Ellenbogen, R.G., and Olson, J.M. (2007) *Cancer Res.*, **67**, 6882–6888.

12 Lakowicz, J.R. (ed.) (2006) *Principles of Fluorescence Spectroscopy*, Springer, Singapore.

13 Wang, F., Tan, W.B., Zhang, Y., Fan, X.P., and Wang, M.Q. (2006) *Nanotechnology*, **17**, R1–R13.

14 Haugland, R.P. (ed.) (2005) *The Handbook – A Guide to Fluorescent Probes and Labeling Technologies*, Invitrogen Corp., Eugene, OR.

15 Banks, P.R. and Paquette, D.M. (1995) *Bioconjugate Chem.*, **6**, 447–458.

16 Mishra, A., Behera, R.K., Behera, P.K., Mishra, B.K., and Behera, G.B. (2000) *Chem. Rev.*, **100**, 1973–2011.

17 Panchuk-Voloshina, N., Haugland, R. P., Bishop-Stewart, J., Bhalgat, M.K., Millard, P.J., Mao, F., Leung, W.Y., and Haugland, R.P. (1999) *J. Histochem. Cytochem.*, **47**, 1179–1188.

18 Lees, A.J. (1987) *Chem. Rev.*, **87**, 711–743.

19 Selvin, P.R. (2002) *Annu. Rev. Biophys. Biomol. Struct.*, **31**, 275–302.

20 Dickson, E.F.G., Pollak, A., and Diamandis, E.P. (1995) *J. Photochem. Photobiol. B*, **27**, 3–19.

21 Terpetschnig, E., Szmacinski, H., and Lakowicz, J.R. (1997) in *Methods in Enzymology*, Vol. 278, (eds J.N. Abelson, M.I. Simon, L. Brand, and M. L. Johnson), Academic Press, New York, pp. 295–321.

22 Lees, A.J. (1998) *Coord. Chem. Rev.*, **177**, 3–35.

23 Youn, H.J., Terpetschnig, E., Szmacinski, H., and Lakowicz, J.R. (1995) *Anal. Biochem.*, **232**, 24–30.

24 Miao, W.J. (2008) *Chem. Rev.*, **108**, 2506–2553.

25 Balzani, V., Bergamini, G., Campagna, S., and Puntoriero, F. (2007) *Top. Curr. Chem.*, **280**, 1–36.

26 Paris, J.P. and Brandt, W.W. (1959) *J. Am. Chem. Soc.*, **81**, 5001–5002.

27 Campagna, S., Puntoriero, F., Nastasi, F., Bergamini, G., and Balzani, V. (2007) *Top. Curr. Chem.*, **280**, 117–214.

28 Oregan, B. and Gratzel, M. (1991) *Nature*, **353**, 737–740.

29 Barton, J.K., Danishefsky, A.T., and Goldberg, J.M. (1984) *J. Am. Chem. Soc.*, **106**, 2172–2176.

30 Lo, K.K.W. (2007) *Struct. Bond.*, **123**, 205–245.

31 Friedman, A.E., Chambron, J.C., Sauvage, J.-P., Turro, N.J., and Barton, J.K. (1990) *J. Am. Chem. Soc.*, **112**, 4960–4962.

32 O'Connor, N.A., Stevens, N., Samaroo, D., Solomon, M.R., Marti, A.A., Dyer, J., Vishwasrao, H., Akins, D.L., Kandel, E.R., and Turro, N.J. (2009) *Chem. Commun.*, 2640–2642.

33 Khan, S.I., Beilstein, A.E., Tierney, M. T., Sykora, M., and Grinstaff, M.W. (1999) *Inorg. Chem.*, **38**, 5999–6002.

34 Hurley, D.J. and Tor, Y. (2002) *J. Am. Chem. Soc.*, **124**, 3749–3762.

35 Hurley, D.J. and Tor, Y. (2002) *J. Am. Chem. Soc.*, **124**, 13231–13241.

36 Ortmans, I., Content, S., Boutonnet, N., Kirsch-De Mesmaeker, A., Bannwarth, W., Constant, J.F., Defrancq, E., and Lhomme, J. (1999) *Chem. Eur. J.*, **5**, 2712–2721.

37 Ryan, E.M., Okennedy, R., Feeney, M. M., Kelly, J.M., and Vos, J.G. (1992) *Bioconjugate Chem.*, **3**, 285–290.

38 Lo, K.K.W. and Lee, T.K.M. (2004) *Inorg. Chem.*, **43**, 5275–5282.

39 Lo, K.K.W., Lee, T.K.M., Lau, J.S.Y., Poon, W.L., and Cheng, S.H. (2008) *Inorg. Chem.*, **47**, 200–208.

40 Puckett, C.A. and Barton, J.K. (2008) *Biochemistry*, **47**, 11711–11716.

41 Paxian, M., Keller, S.A., Cross, B., Huynh, T.T., and Clemens, M.G. (2004) *Am. J. Physiol.-Gastr. L.*, **286**, G37–G44.

42 King, K.A., Spellane, P.J., and Watts, R.J. (1985) *J. Am. Chem. Soc.*, **107**, 1431–1432.

43 Flamigni, L., Barbieri, A., Sabatini, C., Ventura, B., and Barigelletti, F. (2007) *Top. Curr. Chem.*, **281**, 143–203.

44 Evans, R.C., Douglas, P., and Winscom, C.J. (2006) *Coord. Chem. Rev.*, **250**, 2093–2126.

45 Lowry, M.S. and Bernhard, S. (2006) *Chem. Eur. J.*, **12**, 7970–7977.

46 Lo, K.K.W., Ng, D.C.M., and Chung, C. K. (2001) *Organometallics*, **20**, 4999–5001.

47 Guo, X.Q., Castellano, F.N., Li, L., Szmacinski, H., Lakowicz, J.R., and Sipior, J. (1997) *Anal. Biochem.*, **254**, 179–186.

48 Lo, K.K.W., Li, C.K., and Lau, J.S.Y. (2005) *Organometallics*, **24**, 4594–4601.

49 Lo, K.K.W., Zhang, K.Y., Leung, S.K., and Tang, M.C. (2008) *Angew. Chem. Int. Ed.*, **47**, 2213–2216.

50 Lai, C.W., Wang, Y.H., Lai, C.H., Yang, M.J., Chen, C.Y., Chou, P.T., Chan, C. S., Chi, Y., Chen, Y.C., and Hsiao, J.K. (2008) *Small*, **4**, 218–224.

51 Wrighton, M. and Morse, D.L. (1974) *J. Am. Chem. Soc.*, **96**, 998–1003.

52 Kirgan, R.A., Sullivan, B.P., and Rillema, D.P. (2007) *Top. Curr. Chem.*, **281**, 45–100.

53 Coleman, A., Brennan, C., Vos, J.G., and Pryce, M.T. (2008) *Coord. Chem. Rev.*, **252**, 2585–2595.

54 Lo, K.K.W., Ng, D.C.M., Hui, W.K., and Cheung, K.K. (2001) *J. Chem. Soc., Dalton Trans.*, 2634–2640.

55 Wei, L.H., Babich, J., Eckelman, W.C., and Zubieta, J. (2005) *Inorg. Chem.*, **44**, 2198–2209.

56 Lo, K.K.W., Hui, W.K., Ng, D.C.M., and Cheung, K.K. (2002) *Inorg. Chem.*, **41** (1), 40–46.

57 Jamieson, E.R. and Lippard, S.J. (1999) *Chem. Rev.*, **99**, 2467–2498.

58 Arena, G., Scolaro, L.M., Pasternack, R. F., and Romeo, R. (1995) *Inorg. Chem.*, **34**, 2994–3002.

59 Peyratout, C.S., Aldridge, T.K., Crites, D.K., and McMillin, D.R. (1995) *Inorg. Chem.*, **34**, 4484–4489.

60 Ma, D.L. and Che, C.M. (2003) *Chem. Eur. J.*, **9**, 6133–6144.

61 Liu, H.Q., Cheung, T.C., and Che, C.M. (1996) *Chem. Commun.*, 1039–1040.

62 Ma, D.L., Che, C.M., and Yan, S.C. (2009) *J. Am. Chem. Soc.*, **131**, 1835–1846.

63 Wong, K.M.C., Tang, W.S., Chu, B.W. K., Zhu, N.Y., and Yam, V.W.W. (2004) *Organometallics*, **23**, 3459–3465.

64 Che, C.M., Zhang, J.L., and Lin, L.R. (2002) *Chem. Commun.*, 2556–2557.

65 Balzani, V. and Campagna, S. (eds) (2007) *Photochemistry and Photophysics of Coordination Compounds I*, Springler, Berlin/Heidelberg.

66 Balzani, V. and Campagna, S. (eds) (2007) *Photochemistry and Photophysics of Coordination Compounds II*, Springer, Berlin/Heidelberg.

67 Holmlin, R.E., Stemp, E.D.A., and Barton, J.K. (1996) *J. Am. Chem. Soc.*, **118**, 5236–5244.

68 Murtaza, Z., Herman, P., and Lakowicz, J.R. (1999) *Biophys. Chem.*, **80**, 143–151.

69 Terpetschnig, E., Szmacinski, H., and Lakowicz, J.R. (1996) *Anal. Biochem.*, **240**, 54–59.

70 Tamilarasan, R., Ropartz, S., and McMillin, D.R. (1988) *Inorg. Chem.*, **27**, 4082–4084.

71 Hudson, B.P., Sou, J., Berger, D.J., and McMillin, D.R. (1992) *J. Am. Chem. Soc.*, **114**, 8997–9002.

72 de Haas, R.R., van Gijlswijk, R.P.M., van der Tol, E.B., Veuskens, J., van Gijssel, H.E., Tijdens, R.B., Bonnet, J., Verwoerd, N.P., and Tanke, H.J. (1999) *J. Histochem. Cytochem.*, **47**, 183–196.

73 Soini, A.E., Yashunsky, D.V., Meltola, N.J., and Ponomarev, G.V. (2003) *Luminescence*, **18**, 182–192.

74 Bünzli, J.-C.G. and Choppin, G.R. (eds) (1989) *Lanthanide Probes in Life, Chemical and Earth Sciences: Theory and Practice*, Elsevier Science, Amsterdam.

75 Bünzli, J.-C.G. and Piguet, C. (2005) *Chem. Soc. Rev.*, **34**, 1048–1077.

76 Levine, A.K. and Palilla, F.C. (1964) *Appl. Phys. Lett.*, **5**, 118–120.

77 Leonard, J.P., Nolan, C.B., Stomeo, F., and Gunnlaugsson, T. (2007) *Top. Curr. Chem.*, **281**, 1–43.

78 Bünzli, J.-C.G. (2009) *Chem. Lett.*, **38**, 104–109.

79 Martell, A.E., Hancock, R.D., and Motekaitis, R.J. (1994) *Coord. Chem. Rev.*, **133**, 39–65.

80 Bulman, R.A. (2003) in *The Lanthanides and Their Interrelations with Biosystems, The Lanthanides and Their Interrelations with Biosystems*, Metal Ions in Biological Systems, Vol. **40**, (eds A. Sigel and H. Sigel), Marcel Dekker Inc., New York, pp. 323–353.

81 Choppin, G.R. (1995) *J. Alloys Compd.*, **225**, 242–245.

82 Izatt, R.M., Bradshaw, J.S., Nielsen, S.A., Lamb, J.D., and Christensen, J.J. (1985) *Chem. Rev.*, **85**, 271–339.

83 Mayer, A. and Neuenhofer, S. (1994) *Angew. Chem. Int. Ed.*, **33**, 1044–1072.

84 Bünzli, J.-C.G., Comby, S., Chauvin, A.-S., and Vandevyver, C.D.B. (2007) *J. Rare Earths*, **25**, 257–274.

85 Hemmila, I., Dakubu, S., Mukkala, V. M., Siitari, H., and Lovgren, T. (1984) *Anal. Biochem.*, **137**, 335–343.

86 Mathis, G. (1995) *Clin. Chem.*, **41**, 1391–1397.

87 Hemmila, I. and Laitala, V. (2005) *J. Fluoresc.*, **15**, 529–542.

88 Beeby, A., Botchway, S.W., Clarkson, I. M., Faulkner, S., Parker, A.W., Parker, D., and Williams, J.A.G. (2000) *J. Photochem. Photobiol. B*, **57**, 83–89.

89 Bünzli, J.-C.G. (2004) in *Metal Complexes in Tumor Diagnosis and as Anticancer Agents* (ed. A. Sigel and H. Sigel) Metal Ions in Biological Systems, Vol. **42**, Marcel Dekker, New York, pp. 39–75.

90 Syvanen, A.C., Tchen, P., Ranki, M., and Soderlund, H. (1986) *Nucleic Acids Res.*, **14**, 1017–1028.

91 Oser, A. and Valet, G. (1990) *Angew. Chem. Int. Ed.*, **29**, 1167–1169.

92 Hurskainen, P. (1995) *J. Alloys Compd.*, **225**, 489–491.

93 Lopez, E., Chypre, C., Alpha, B., and Mathis, G. (1993) *Clin. Chem.*, **39**, 196–201.

94 Scaff, W.L., Dyer, D.L., and Mori, K. (1969) *J. Bacteriol.*, **98**, 246–248.

95 Harsha, E.R., Reddy, D.R., Shabnam, M., Nathaniel, G.B., and Lawrence, W. M. (2009) *Angew. Chem. Int. Ed.*, **48**, 4990–4992.

96 Bornhop, D.J., Griffin, J.M.M., Goebel, T.S., Sudduth, M.R., Bell, B., and Motamedi, M. (2003) *Appl. Spectrosc.*, **57**, 1216–1222.

97 Manning, H.C., Goebel, T., Thompson, R.C., Price, R.R., Lee, H., and Bornhop, D.J. (2004) *Bioconjugate Chem.*, **15**, 1488–1495.

98 Pandya, S., Yu, J.H., and Parker, D. (2006) *Dalton Trans.*, 2757–2766.

99 Akiba, H., Sumaoka, J., and Komiyama, M. (2009) *ChemBioChem.*, **10**, 1773–1776.

100 Kim, S., Lim, Y.T., Soltesz, E.G., De Grand, A.M., Lee, J., Nakayama, A., Parker, J.A., Mihaljevic, T., Laurence, R.G., Dor, D.M., Cohn, L.H., Bawendi, M.G., and Frangioni, J.V. (2004) *Nat. Biotechnol.*, **22**, 93–97.

101 Gaiduk, M.I., Grigoryants, V.V., Mironov, A.F., Rumyantseva, V.D., Chissov, V.I., and Sukhin, G.M. (1990) *J. Photochem. Photobiol. B*, **7**, 15–20.

102 Zhang, J., Badger, P.D., Geib, S.J., and Petoud, S. (2005) *Angew. Chem. Int. Ed.*, **44**, 2508–2512.

103 Drbohlavova, J., Adam, V., Kizek, R., and Hubalek, J. (2009) *Int. J. Mol. Sci.*, **10**, 656–673.

104 Bruchez, M., Moronne, M., Gin, P., Weiss, S., and Alivisatos, A.P. (1998) *Science*, **281**, 2013–2016.

105 Dorokhin, D., Tomczak, N., Han, M.Y., Reinhoudt, D.N., Velders, A.H., and Vancso, G.J. (2009) *ACS Nano*, **3**, 661–667.

106 Jaiswal, J.K., Mattoussi, H., Mauro, J. M., and Simon, S.M. (2003) *Nat. Biotechnol.*, **21**, 47–51.

107 Cai, W.B., Hsu, A.R., Li, Z.B., and Chen, X.Y. (2007) *Nanoscale Res. Lett.*, **2**, 265–281.

108 Derfus, A.M., Chan, W.C.W., and Bhatia, S.N. (2004) *Nano Lett.*, **4**, 11–18.

109 Chen, H.Y., Wang, Y.Q., Xu, J., Ji, J.Z., Zhang, J., Hu, Y.Z., and Gu, Y.Q. (2008) *J. Fluoresc.*, **18**, 801–811.

110 Chan, W.C.W. and Nie, S.M. (1998) *Science*, **281**, 2016–2018.

111 Medintz, I.L., Uyeda, H.T., Goldman, E.R., and Mattoussi, H. (2005) *Nat. Mater.*, **4**, 435–446.

112 Gao, X.H., Cui, Y.Y., Levenson, R.M., Chung, L.W.K., and Nie, S.M. (2004) *Nat. Biotechnol.*, **22**, 969–976.

113 Riwotzki, K. and Haase, M. (1998) *J. Photochem. Photobiol. B*, **102**, 10129–10135.

114 Auzel, F. (2004) *Chem. Rev.*, **104**, 139–173.

115 van de Rijke, F., Zijlmans, H., Li, S., Vail, T., Raap, A.K., Niedbala, R.S., and Tanke, H.J. (2001) *Nat. Biotechnol.*, **19**, 273–276.

116 Yi, G.S., Lu, H.C., Zhao, S.Y., Yue, G., Yang, W.J., Chen, D.P., and Guo, L.H. (2004) *Nano Lett.*, **4**, 2191–2196.

117 Meiser, F., Cortez, C. and Caruso, F. (2004) *Angew. Chem. Int. Ed.*, **43**, 5954–5957.

118 Ogawa, M., Nakamura, T., Mori, J.I., and Kuroda, K. (2001) *Microporous Mesoporous Mat.*, **18**, 159–164.

119 Rieter, W.J., Kim, J.S., Taylor, K.M.L., An, H.Y., Lin, W.L., Tarrant, T., and Lin, W.B. (2007) *Angew. Chem. Int. Ed.*, **46**, 3680–3682.

120 Kim, J., Piao, Y., and Hyeon, T. (2009) *Chem. Soc. Rev.*, **38**, 372–390.

Index